Daniela Dragoman Mircea Dragoman

Optical Characterization of Solids

Springer
Berlin
Heidelberg
New York
Barcelona
Hong Kong
London
Milan
Paris
Tokyo

D. Dragoman M. Dragoman

Optical Characterization of Solids

With 184 Figures

Springer

Professor Daniela Dragoman
University of Bucharest, Physics Department
P.O. Box MG-11
76900 Bucharest, Romania
e-mail: danieladragoman@yahoo.com

Professor Mircea Dragoman
National Institute for Research and Development in Microtechnologies (IMT)
P.O. Box 38-160
72225 Bucharest, Romania
e-mails: mircead@imt.ro *and* mdragoman@yahoo.com

ISBN 3-540-41803-2 Springer-Verlag Berlin Heidelberg New York

Library of Congress Cataloging-in-Publication Data applied for

Die Deutsche Bibliothek - CIP-Einheitsaufnahme
Dragoman, Daniela:
Optical characterization of solids / D. Dragoman ; M. Dragoman. - Berlin ; Heidelberg ; New York ; Barcelona ; Hong Kong ; London ; Milan ; Paris ; Tokyo : Springer, 2002
(Physics and astronomy online library)
ISBN 3-540-41803-2

Springer-Verlag Berlin Heidelberg New York
a member of BertelsmannSpringer Science+Business Media GmbH
http://www.springer.de

Printed in Germany

Typesetting: Camera-ready copy from authors
Cover concept: *design & production* GmbH, Heidelberg

Printed on acid-free paper SPIN: 10760539 57/3020 CU – 5 4 3 2 1 0 –

Preface

The interaction between the electromagnetic field and matter is an essential issue in modern physics. The physical concepts describing the field–matter interaction are of paramount importance, being at the origin of quantum mechanics and quantum field theories. On the other hand, the field–matter interaction is the main tool used nowadays to investigate the structure and properties of matter at the microscopic level.

The subject of this book is the characterization of the solid-state through the interaction with the electromagnetic field. The main properties of different types of solids, such as dielectrics, semiconductors, disordered materials, impurified solids, low-dimensional structures, nanostructures, metals or superconductors, are revealed through their interaction with the electromagnetic field whose wavelength lies within the optical spectrum. Therefore, the book is focused on the investigation of the main properties of different types of solids that are obtained through light emission, absorption or scattering.

There is already an immense literature dedicated to light–solid-state interactions or, in other words, the optical spectroscopy of solids, due to its major role in understanding the physics and properties of the solid-state. This literature is continuously growing and flourishing, due also to the paramount importance of solid-state devices in the development of any conceivable area of technology, including the booming information technology. However, this literature is mainly written emphasizing a certain type of materials (semiconductors, high-temperature superconductors, etc.) or a certain type of optical spectroscopy method (Raman scattering, luminescence, etc.). We have tried a new, hybrid and at the same time unifying approach.

The theory of light–solid interaction based on light absorption, emission and scattering is very complex, being mainly based on quantum mechanical concepts, although sometimes semi-classical approaches are still satisfactory. A simple question comes immediately to mind: how can we relate the results of the theory of light–solid interactions with the properties of solids? In other words, how can one determine basic solid characteristics such as electronic band structure or important quantitative parameters such as the effective mass of an electron or a hole, the magnitude of electron-phonon coupling, the exciton lifetime, etc., from the absorption, emission or the scattering of light in a solid? There is no simple or direct answer to this straightforward question. We hope that the reader will be able to find an answer to this question after reading this book, which is a first attempt to unify disparate theories and experimental facts in a single and coherent way, able to connect light–matter interaction theories with solid-state characteristics and parameters. The practical significance of this view is of great importance for all readers who work in the area of solid-state physics, electronics, optoelectronics or nanoscience.

The book can be read by anyone with some elementary knowledge of solid-state physics. Chapters 1 and 2 provide the reader with all basic information needed to understand later chapters. The book is inspired by a course presented by one of us (D.D) at the University of Bucharest, Physics Faculty, Solid State Department, at the Master in Science level. Therefore, although rigorous, the mathematical treatment is presented in quite a simple and straightforward manner. The book style is somewhere between a pedagogical and a review-like style, our focus being on the physical explanations.

We had no desire to write a book on solid-state theory, which would have been a hopeless project due to the immensity of theories and concepts applicable to various types of materials and interactions, so that a balance had to be found between theory and the practical methods for characterizing the parameters of different materials. Throughout the book many exotic types of materials have been introduced without a detailed definition and associated mathematical model. The mention of the fact that a material belongs to a specific category was made only to help interested readers to go deeper in the study of that material. We could not have insisted on rigorous models for all materials; however, we have always specified the parameter under scrutiny and the physical basis of its determination through optical means. Frequently, in the area of light–matter interaction the mathematical treatments and models are so complex that many times the physical insights become obscure. Therefore, we have not insisted on mathematical details, the theoretical treatment serving mainly for the introduction and explanation of the fundamental physical effects involved in the light–matter interaction. This approach is partially justified because many experimental results are not compared with rigorous and sophisticated mathematical models but with simplified phenomenological models or simple mathematical fitting formulas. We have provided in the entire book, whenever possible, simple formulas helping to elucidate the main subject of this book, i.e. the connection between light–solid interactions and solid-state characteristics and properties.

The same criterion was applied in the design of the figures. Almost always the figures do not reproduce exactly the experimental details, but are only sketches of main trends and typical situations encountered in practice. We have chosen this manner of illustrating the book since we considered the figures as graphical supports of physical explanations rather than raw scientific data. Although in most cases we refer to specific experiments, on specific materials, these examples are chosen to evidence a concept or a trend typical for a larger category of materials, the figures underlining this character of the examples. On the other hand, even for the same material, the experimental results can differ because of the fabrication method of the sample, because of the method used for optical characterization, environmental conditions, etc.; no two experimental results are alike. However, the phenomenon under investigation is still present, except for some details; its characteristics have to be extracted from experimental data, so the trends rather than details are important in the description of the way in which the parameters of the samples are obtained through light–matter interactions.

Chapter 1 reviews the main elementary excitations encountered in the solid state such as phonons, excitons, coupled and collective excitations. It is important to read this chapter because these excitations are referred to throughout the remainder of the book, their behavior on interaction with the electromagnetic field being in direct relation with many solid-state properties.

Chapter 2 deals with the main physical effects of light–solid interaction and constitutes the main theoretical foundation of the book. The calculations are presented in detail and explained in such a way that the reader should be able to follow them easily. Also, the main experimental methods are reviewed here in depth, together with the physical effect that has originated them. We have not provided very detailed set-ups, many of them are only block schemes or simplified versions. The reason is that an actual set-up is done with what is available in a certain laboratory. However, the block scheme is the same in any experimental arrangement. This chapter is about one quarter of the entire book and must be read carefully by graduated students and young researchers before tackling the subjects of the following chapters.

The next six chapters of the book are dedicated to different types of solid-state materials that interact with optical fields and so reveal their properties. The connection between different solid-state properties and the light–solid interaction theories and experimental facts are discussed in detail.

Chapter 3 is dedicated to doped solids, exemplified through different types of ions introduced in crystalline hosts. The optical properties of rare-earth ions, transition-metal ions, diluted magnetic semiconductors, manganites, color-centers, filled-shell ions and donors and acceptors in semiconductors are presented.

Chapter 4 is focused on bulk solids, such as semiconductors or metals, and their optical properties. A special emphasis is put here on new materials, for example GaN or metal nanoparticles, and on special physical phenomena such as anharmonic, isotope or many-body effects. The relaxation phenomena in bulk solids are presented extensively, considering the excitons, carriers and phonons relaxation behavior at excitation with ultrashort pulses.

Chapter 5 reviews briefly the optical properties of interfaces and thin films, whereas *Chapter 6* is dedicated to the optical characterization and parameter determination of low-dimensional structures. These micro- and nanostructures are the subject of numerous recent studies due to their unique electronic and optical properties, not encountered in any other type of solid, and with important applications in electronics and optoelectronics. Therefore, this chapter constitutes a substantial part of the book. Optical properties of low-dimensional semiconductors, in particular optical absorption, luminescence, and scattering are explained in detail, followed by optical characterization methods. Emphasis is put on confinement effects, electronic structure and internal fields, excitons, biexcitons, coupled excitations and microcavities, relaxation phenomena and many-body effects. This long chapter ends with optical properties of coupled quantum wells, superlattices and nanoparticles. The newly developed carbon nanotubes, with huge and still not fully exploited applications in electronics, are also mentioned briefly.

Chapter 7 is dedicated to disordered solids, in particular to the optical properties of disordered structures, such as disordered alloys and rough surfaces, and disordered materials, such as polycrystalline and different amorphous states.

The last chapter, *Chapter 8*, deals with optical methods for the investigation of high-temperature superconductors (HTS). Since there is no satisfactory theory to explain the HTS behavior, the optical investigation is used to test the new theories on HTS, giving extremely valuable information about the symmetry of the superconducting gap, the BCS or non-BCS behavior of HTS, and the pseudogap. The book keeps pace with the most recent developments in the field, including the unexpected discovery in March 2001 of the superconducting MgB_2 intermetallic compound.

We have written this book with the goal of giving a comprehensive view of the determination of solid parameters through the interaction with optical fields. The reader is helped in his efforts to understand the main issues of the book by some hundreds of references listed at the end of each chapter. The references were selected from the *most recent publications* on the basis of their direct relevance to the main issues of the book; of course, not all studies could be included and there are most probably other papers worth mentioning, which are not referred to here. On the other hand, some important subjects could have been missed because they have not made a significant progress during 1996–2001, which is the period from which we have gathered the material of the book. This is unavoidable. We have always respected the point of view of the authors with respect to the models and the interpretation of various experiments. We are conscious that at least some of these models and interpretations are still under debate, and that other versions have appeared regarding the same subject. Therefore, it is possible that some members of the scientific community are not entirely happy with our presentation; others will be unhappy with a different interpretation – this is how science progress.

We hope that this book will be a helpful instrument for graduated students, Ph.D. students and researchers in the area of solid-state physics, materials, nanoscience, electronics, optics and optoelectronics.

We would like to thank to all the librarians of the National Atomic Physics Institute Library-Bucharest, especially Mrs. Stela Emilia Mihalcea and Mrs. Natalita Feroiu for their continuous efforts to provide us with hundreds of references in a very short time. For the same reasons we are indebted to the librarians of Univ. of Bucharest, Physics Faculty, especially to Mrs. Laura Vlaescu.

We would like to thank to Dr. Claus Ascheron and Dr. Angela Lahee from Springer Verlag for their continuous trust and support regarding our scientific work. Their sincere encouragement and confidence has helped us to overcome the difficulties naturally encountered in writing books with thorny subjects like this one.

Bucharest,
May 2001

Daniela Dragoman,
Mircea Dragoman

Contents

1. Elementary Excitations in Solids ... 1
1.1 Energy Band Structure in Crystalline Materials ... 2
1.2 $\boldsymbol{k} \cdot \boldsymbol{p}$ Method ... 11
1.3 Numerical Methods of Electron Energy Band Calculation ... 14
1.4 Phonons ... 16
1.4.1 Bose–Einstein Statistics ... 19
1.4.2 Fermi–Dirac Statistics ... 19
1.5 Plasmons ... 20
1.6 Magnetic Excitations ... 22
1.6.1 Magnons ... 24
1.7 Excitons ... 26
1.7.1 Frenkel Excitons ... 26
1.7.2 Wannier–Mott Excitons ... 27
1.7.3 Interactions Between Excitons ... 30
1.7.4 Bound Excitons and Excitonic Complexes ... 31
1.8 Polarons ... 31
1.9 Polaritons ... 35
References ... 36

2. Optical Transitions ... 37
2.1 The Quantization of the Electromagnetic Radiation ... 37
2.2 Transition Probabilities ... 39
2.2.1 Optical Properties of Crystalline Materials ... 42
2.2.2 Mechanisms for Spectral Line Broadening ... 44
2.2.2.1 Lorentzian Broadening of Spectral Lines ... 44
2.2.2.2 Gaussian Broadening of Spectral Line ... 45
2.2.2.3 Voigt Profile ... 46
2.2.3 Multipolar Contributions to the Interaction Hamiltonian ... 46
2.3 Optical Constants of Solids ... 47
2.3.1 Kramers–Krönig Relations ... 50
2.3.2 Drude–Lorentz Theory of the Electrically Charged Oscillators ... 51
2.3.3 Sum Rule ... 54

2.3.4 Microscopic Origin of Absorption.................................. 55
2.3.5 Absorption Mechanisms.. 56
2.3.5.1 Band-to-Band Light Absorption: Direct and Allowed Absorption.................................. 56
2.3.5.2 Band-to-Band Direct, Forbidden Transitions.......... 57
2.3.5.3 Band-to-Band Indirect, Allowed Transitions.......... 57
2.3.5.4 Band-to-Band Indirect, Forbidden Transitions........ 58
2.3.5.5 Effects That Can Appear at Band-to-Band Absorption.................................. 59
2.3.5.6 Intraband Absorption.. 60
2.3.5.7 Absorption on Free Carriers.............................. 61
2.3.5.8 Absorption on Wannier–Mott Excitons................. 62
2.3.5.9 Indirect Excitonic Absorption............................ 63
2.3.5.10 Biexciton Production Through Photon Absorption.... 64
2.3.5.11 Light Absorption on Phonons............................ 65
2.3.6 Optical Properties of Solids in External Magnetic Fields....... 66
2.3.6.1 Energy Levels in Magnetic Field........................ 66
2.3.6.2 Interaction of Electromagnetic Radiation with a System in a Magnetic Field.......................... 67
2.3.6.3 Light Interaction with Metals in Magnetic Field...... 71
2.3.6.4 Phenomena at Low Frequencies......................... 73
2.3.6.5 The Shubnikov–de Haas Effect.......................... 75
2.4 Photoluminescence... 75
2.5 Inelastic Light Scattering... 78
2.5.1 Raman Scattering.. 81
2.5.2 Brillouin Scattering.. 88
2.6 Nonlinear Optical Response of Solids................................... 92
2.6.1 Nonlinear Raman Effect... 95
2.6.1.1 Stimulated Raman Effect.................................. 95
2.6.1.2 Inverse Raman Scattering................................. 97
2.6.1.3 Hyper-Raman Effect....................................... 97
2.6.1.4 Coherent Anti-Stokes Raman Scattering (CARS)..... 98
2.7 Optical Methods for the Characterization of Light-Matter Interaction.. 99
2.7.1 Transmission/Reflection Spectroscopy............................ 101
2.7.2 Modulated Reflectance and Transmittance....................... 103
2.7.3 Pump-Probe Spectroscopy... 104
2.7.4 Amplitude and Phase Spectroscopy................................ 105
2.7.5 Photoluminescence.. 108
2.7.6 Light Scattering.. 110
2.7.7 Four-Wave Mixing... 112
2.7.7.1 Optical Bloch Equations.................................. 114
2.7.7.2 Quantum Beats and Polarization Beats................. 117
2.7.8 Spectral Hole Burning... 118

2.7.9 Fluorescence Line Narrowing.. 120
2.7.10 Magneto-Optical Kerr Effect.. 121
2.8 Line Identification.. 122
References.. 124

3. Optical Properties of Impurities in Solids.............................. 127
3.1 Electronic Structure of Isolated Atoms..................................... 127
3.2 Ions in Crystals... 129
3.3 Frank–Condon Principle for Defects with Large Orbits....................... 137
3.4 Jahn–Teller Effect... 138
3.5 Interactions Between Ions.. 138
3.6 Spatial Migration.. 142
3.7 Optical Properties of the Rare-Earth Ions.................................. 143
3.7.1 Judd–Ofelt Theory of One and Two Photon Transitions........................ 144
3.7.2 Optical Spectra of Rare-Earth Ions.. 146
3.7.3 Interaction Processes for Rare-Earth Ions................................. 153
3.8 Transition Metal Ions.. 154
3.8.1 Optical Properties of Transition Metal Ions............................... 155
3.8.2 Iron-Group Ion Pairs.. 158
3.9 Diluted Magnetic Semiconductors... 158
3.10 Manganites.. 165
3.11 Color Centers... 167
3.12 Filled-Shell Ions... 169
3.13 Optical Properties of Donors and Acceptors in Semiconductors.............. 170
References.. 177

4. Bulk Materials... 181
4.1 Electronic Structure... 181
4.2 Excitons in Bulk Semiconductors.. 186
4.3 Anharmonic Effects... 190
4.4 Isotope Effects.. 192
4.5 Effects Due to Alloying.. 193
4.6 Deformation Potentials... 194
4.7 Optical Determination of Magnetic Properties............................... 196
4.8 Rotational Spectra of Solids... 199
4.9 Many-Body Effects.. 200
4.10 Coupled Excitations... 202
4.11 Relaxation Phenomena in Bulk Semiconductors............................... 205
4.11.1 Exciton Relaxation... 205
4.11.2 Carrier Relaxation... 208
4.11.3 Phonon Relaxation.. 213
4.12 Optical Properties of Metals.. 214
4.13 Phase Transitions... 220
References.. 231

5. Optical Properties of Interfaces and Thin Films....................235
5.1 Surface Effects.. 236
5.2 Thin Films... 243
5.2.1 Magnetic Thin Films.. 246
References.. 250

6. Low-Dimensional Structures.......................................253
6.1 Low-Dimensional Semiconductor Heterostructures.................... 254
6.1.1 Quantum Well Excitons.. 261
6.2 Optical Properties of Low-Dimensional
Semiconductor Heterostructures.. 263
6.2.1 Intraband Absorption... 263
6.2.2 Excitonic Absorption... 266
6.2.3 Luminescence... 267
6.2.4 Scattering in Low-Dimensional Heterostructures................ 268
6.2.5 Intersubband Optical Transitions............................... 269
6.2.6 Depolarization Shift and Coulomb Renormalization.............. 270
6.2.7 Radiative Intraband Transitions in Heterostructures............ 270
6.2.8 Collective Excitations.. 271
6.2.9 Nonlinear Optics... 272
6.3 Optical Characterization of Low-Dimensional
Semiconductor Heterostructures.. 272
6.3.1 Confinement Effects.. 272
6.3.2 Optical Determination Electronic Structure Parameters 278
6.3.3 Determination of Internal Electric Fields...................... 286
6.3.4 Diffusion.. 288
6.3.5 Optical Properties of Excitons................................. 289
6.3.6 Biexcitons... 296
6.3.7 Magnetic Properties.. 298
6.3.8 Collective Excitations.. 306
6.3.9 Coupled Excitations.. 307
6.3.10 Quantum Microcavities... 308
6.3.11 Relaxation Phenomena in Low Dimensional Structures........ 311
6.3.12 Many-Body Effects... 318
6.3.13 Coupled Quantum Wells... 320
6.3.14 Carbon Nanotubes.. 324
6.4 Optical Properties of Superlattices.................................. 325
6.4.1 Band Structure... 328
6.4.2 Excitons in Superlattices....................................... 336
6.4.3 Other Parameters... 340
6.5 Nanoparticles... 342
References.. 347

7. Optical Properties of Disordered Materials 353
7.1 Disordered Alloys 354
7.2 Disordered Superlattices 359
7.3 Rough Surfaces 361
7.4 Polycrystalline and Nanocrystalline Materials 362
7.5 Continuous-Network Structures 371
7.6 Optical Properties of Amorphous Materials 371
7.6.1 Theory of Electronic States in Amorphous Semiconductors ... 372
7.6.2 General Optical Properties of Amorphous Semiconductors ... 374
7.6.3 Tetrahedrally Coordinated Amorphous Semiconductors 380
7.6.4 Chalcogenide Glasses 391
7.6.5 Other Amorphous Materials 401
References 404

8. Optical Properties of High-Temperature Superconductors 407
8.1 High-Temperature Superconductors 410
8.2 Optical Methods for HTS Investigation 417
8.2.1 Anisotropy of Optical Response 417
8.2.2 Observation of Superconducting Transition 419
8.2.3 Symmetry of Superconducting Gap 422
8.2.4 Non-BCS Behavior 426
8.2.5 BCS-Like Behavior 430
8.2.6 HTS Dynamics 430
8.2.7 Pseudogap 431
References 438

Index 441

Index of Materials 447

1. Elementary Excitations in Crystalline Solids

Most solid-state materials described throughout this book, with the notable exception of low-dimensional heterostructures and amorphous materials, have a crystalline structure, i.e. they are distinguished by a long-range order characterized by a periodic arrangement of atoms in a 3D lattice. In this chapter we present the model describing crystalline materials, and the most often encountered elementary excitations that can build in them.

Even in the simplified case of a periodic lattice, the properties of a collection of atoms, each of which is composed of several electrons surrounding a nucleus, is quite difficult. One approximation that considerably simplifies the problem is the separation of the nucleus and electron motions, justified by their huge mass difference; this separation is known as the adiabatic approximation. The approximate separation of variables in the Schrödinger equation serves also as the basis for a systematized analysis of experimental data.

Expanding the energy operator in powers of the small parameter m/M where m is the mass of the electron and M that of the nucleus, the total energy of an atom can be approximated by

$$E \cong E_0 + (m/M)^{1/2} E_1 + (m/M) E_2 + \dots . \tag{1.1}$$

The various terms in (1.1) correspond to various types of energy levels, i.e. electronic, vibrational, rotational, translational, and so on, which reflect different processes occurring in the atomic system. More specifically, the zero-order term E_0 gives the electronic energy for a fixed core, the first-order term $(m/M)^{1/2} E_1$ describes the vibrational energy and the second-order term includes the rotational energy and a proportion of the vibrational energy.

Neglecting the rotation of atoms, the total Hamiltonian of the system containing n electrons and N nuclei may be written as

$$H = \sum_{i=1}^{n} \boldsymbol{p}_i^2 / 2m + \sum_{s=1}^{N} \boldsymbol{P}_s^2 / 2M_s + V(\boldsymbol{r}_i, \boldsymbol{q}_s), \tag{1.2}$$

where the first two terms are the kinetic energy operators for the motion of electrons and the vibrations of nuclei, respectively, with corresponding coordinates $\boldsymbol{r}_i$ and $\boldsymbol{q}_s$, and V is the potential energy associated with the Coulomb interaction between electrons and nuclei. A suitable zero-order solution of (1.2) in the adiabatic approximation, which assumes that the motion of electrons is much faster (by one or two orders of magnitude) than the vibrations of nuclei, can be written as a product of eigenfunctions of the electronic and vibrational Hamiltonians as $\Psi_{nj}(\boldsymbol{r},\boldsymbol{q}) = \varphi_n^{\text{el}}(\boldsymbol{r},\boldsymbol{q})\varphi_{nj}^{\text{vib}}(\boldsymbol{q})$, where

$$[-\Sigma_i(\hbar^2/2m)\nabla_i^2 + V(\boldsymbol{r},\boldsymbol{q})]\varphi_n^{\text{el}}(\boldsymbol{r},\boldsymbol{q}) = E_n^{\text{el}}(\boldsymbol{q})\varphi_n^{\text{el}}(\boldsymbol{r},\boldsymbol{q}), \tag{1.3.a}$$

$$[-\Sigma_s(\hbar^2/2M_s)\nabla_s^2 + E_n^{\text{el}}(\boldsymbol{q})]\varphi_{nj}^{\text{vib}}(\boldsymbol{q}) = E\varphi_{nj}^{\text{vib}}(\boldsymbol{q}). \tag{1.3.b}$$

$\varphi_n^{\text{el}}(\boldsymbol{r},\boldsymbol{q})$ is a rapidly varying function of $\boldsymbol{r}$, with the nuclear coordinate $\boldsymbol{q}$ as parameter and $E_n^{\text{el}}(\boldsymbol{q})$ is the energy of electrons corresponding to a certain configuration of the nuclei. As $\boldsymbol{q}$ changes in time, both $E_n^{\text{el}}(\boldsymbol{q})$ and $\varphi_n^{\text{el}}(\boldsymbol{r},\boldsymbol{q})$ follow the change adiabatically, the system remaining in the state represented by the vibrational quantum number n, i.e. the electrons in a certain state n move in a potential $V(\boldsymbol{r},\boldsymbol{q})$ and follow the lattice motion without making any transitions to any other electronic state $m \neq n$.

During optical transitions between vibrational sublevels of two electronic levels of an atom the system obeys the so-called Franck–Condon principle which expresses the fact that optical transitions occur in a time much too short for the lattice to relax. Mathematically, it is expressed as follows: the sum of transition probabilities from a vibrational sublevel j of the original electron-vibrational state characterized by the quantum numbers n, j, to all the vibrational sublevels of the final state is independent of j. Furthermore, the intensity of individual vibrational bands depends only on the properties of the vibrational eigenfunctions of the initial and final states, the system undergoing transitions with a large probability only to that vibrational level of the final electronic state for which the most probable values of the coordinates are the same as for the initial vibrational level. Physically, this corresponds to electronic-vibrational transitions in which the distance between nuclei remains constant.

1.1 Energy Band Structure in Crystalline Materials

The ideal crystalline material is formed from an infinite succession in all spatial directions of an elementary block – the unit cell. Mathematically, this periodicity is described by an appropriately chosen set of three noncoplanar vectors $\boldsymbol{a}_1$, $\boldsymbol{a}_2$, $\boldsymbol{a}_3$, such that the position $\boldsymbol{R}_i$ of any unit cell i can be expressed as

$$\boldsymbol{R}_i = n_{i1}\boldsymbol{a}_1 + n_{i2}\boldsymbol{a}_2 + n_{i3}\boldsymbol{a}_3, \tag{1.4}$$

with n_{ij} · integers. The atoms that form the solid state are usually placed in positions of high symmetry in this unit cell: in the corners, centers of sides, center of faces, or in the center of the unit cell. There can be one or more atoms in the unit cell, in the latter case the position of the atom being specified by a set of basis vectors $\boldsymbol{d}_j$, with j ranging from 1 to the number of atoms in the unit cell. If the unit cell has one atom it is called a primitive cell, the noncoplanar vectors in this case being called primitive translation vectors of the lattice.

The properties of the lattice are invariant with respect to translation operators along any vector $\boldsymbol{R}_i$. Besides invariance to translation operators, the lattice can also be invariant to rotations through angles $2\pi/n$ around certain symmetry axes, described by the operators C_n, and/or invariant to reflection or inversions around certain symmetry planes. The symmetry operations that do not involve any translations form the point group of the lattice. There are 32 possible crystal point groups, identified through the symbols (in the Schönflies notation):

- C_n if the lattice is invariant to only rotation operators C_n with possible values $n = 1, 2, 3, 4, 6$,
- C_{nv} if, besides the rotation operators C_n, the point group contains also a reflection operator with respect to a plane containing the axis of highest symmetry, defined as the vertical plane. For this point group the allowed values for n are 2, 3, 4, 6,
- C_{nh} if besides C_n the point group contains a horizontal reflection plane, perpendicular to the axis of highest symmetry. The allowed values of n are in this case n = 1, 2, 3, 4, 6, the inversion operation being also contained in the point group for even n values,
- S_n for point groups containing an n-fold rotation operator, combined with reflection in a plane perpendicular to the rotation axis (improper rotations). The distinct n values are 2, 4, 6,
- D_n, when there are n twofold axes (three for D_2) normal to the highest symmetry axis C_n. n can take the values 2, 3, 4, 6,
- D_{nd}, when the point group contains, in addition to the operations in D_n, reflections in a diagonal plane which contains the symmetry axis and bisects the angle between the twofold axis. In this case n = 2, 3,
- D_{nh} for groups containing the horizontal reflection plane in addition to the symmetry operations contained in D_n,
- T for point groups containing the twelve rotations that carry a regular tetrahedron into itself,
- T_d, if the group includes also the improper rotations which carry a regular tetrahedron into itself,
- T_h for groups including the inversion operators in addition to the operations of T,
- O, for groups containing the rotations that carry the cube into itself,
- O_h if the point group also includes the improper rotations that carry the cube into itself.

The latter five point groups are of higher symmetry and characterize the cubic crystal system. In particular the ordinary cubic, body-centered cubic and face-centered cubic lattices, as well as the diamond lattice, are all of O_h symmetry. The symmetry operations are important since they define the symmetry of the

Hamiltonian, and thus of the wavefunction of electrons. To completely account for the symmetry of the electron wavefunction, for example, the spin operator must also be included, the point groups being replaced in this case by double groups.

The classification of crystals in various point groups is essential in establishing the form of the wavefunction for the quantum system. The wavefunction must form bases for the irreducible representations of the group of operators that commute with the Hamiltonian. In particular, the degeneracy of energy levels and the transformation properties of eigenfunctions may be derived once the symmetry of the Hamiltonian is known. The irreducible representations also determine the selection rules for various transitions in the material system. There are several monographs that treat the optical interactions in solids based on group theory, for example the excellent monographs by di Bartolo (1968) and Ivchenko and Pikus (1997). Therefore, we will not insist throughout this book on the group theoretical description of light interaction with solids, although reference will be made whenever appropriate to the point group of several materials and of their relevance to optical spectra.

The optical properties of solids are mainly determined by the free electrons, which can move throughout the crystal. Besides these there are core electrons, localized inside the atom. In the following we understand through electrons the free electrons only, which feel the core electrons and the nuclei through an effective potential $V(\boldsymbol{r})$. Neglecting the interaction between electrons, each of them can be described by its own wavefunction, and by its own set of quantum numbers. Due to lattice periodicity, the Hamiltonian of each electron is

$$H = -(\hbar^2/2m)\nabla^2 + V(\boldsymbol{r}), \tag{1.5}$$

where m is the free electron mass and the potential energy satisfies the relation

$$V(\boldsymbol{r}+\boldsymbol{R}_i) = V(\boldsymbol{r}), \tag{1.6}$$

for any $\boldsymbol{R}_i$. This means that the Hamiltonian commutes with the set of translation operators $T(\boldsymbol{R}_i)$ defined as

$$T(\boldsymbol{R}_i)\varphi(\boldsymbol{r}) = \varphi(\boldsymbol{r}+\boldsymbol{R}_i), \tag{1.7}$$

so that the eigenfunctions of the Hamiltonian are also eigenfunctions of the translation operator with complex eigenvalues of unit modulus (since the electron charge density $\rho(\boldsymbol{r}) = |\varphi(\boldsymbol{r})|^2$ must be conserved by a translation with $\boldsymbol{R}_i$). This requirement (Bloch's theorem) leads to the separation of the wavefunction in

$$\varphi(\boldsymbol{k},\boldsymbol{r}) = \exp(\mathrm{i}\boldsymbol{k}\cdot\boldsymbol{r})u(\boldsymbol{k},\boldsymbol{r}), \tag{1.8}$$

where the quantum number $\boldsymbol{k}$ characterizes a particular eigenfunction of the translation operator and determines the corresponding eigenvalue:

$$\varphi(\boldsymbol{k},\boldsymbol{r}+\boldsymbol{R}_i)=\exp(\mathrm{i}\boldsymbol{k}\cdot\boldsymbol{R}_i)\varphi(\boldsymbol{k},\boldsymbol{r}). \tag{1.9}$$

The function $u(\boldsymbol{k},\boldsymbol{r})$ is called the cell periodic function.

The quantum number $\boldsymbol{k}$ is specified by imposing periodic boundary conditions on the lattice (equivalent to avoiding the inclusion of the surface in the treatment), which implies that the wavefunction repeats itself after translation with vectors $(2N_1+1)\boldsymbol{a}_1$, $(2N_2+1)\boldsymbol{a}_2$, $(2N_3+1)\boldsymbol{a}_3$ along the respective primitive translation vectors $\boldsymbol{a}_i$. N_i are for now arbitrary integer numbers that specify the allowed values of $\boldsymbol{k}$ as those for which $\exp[\mathrm{i}(2N_i+1)\boldsymbol{k}\cdot\boldsymbol{a}_i]=1$, for each $i=1$, 2, 3. The discrete $\boldsymbol{k}$ values defined in this way can be thought of as belonging to a reciprocal lattice whose primitive vectors are

$$\boldsymbol{b}_1=2\pi(\boldsymbol{a}_2\times\boldsymbol{a}_3)/\Omega,\quad \boldsymbol{b}_2=2\pi(\boldsymbol{a}_3\times\boldsymbol{a}_1)/\Omega,\quad \boldsymbol{b}_3=2\pi(\boldsymbol{a}_1\times\boldsymbol{a}_2)/\Omega, \tag{1.10}$$

such that $\boldsymbol{b}_j\cdot\boldsymbol{a}_i=2\pi\delta_{ij}$. Here $\Omega=\boldsymbol{a}_1\cdot(\boldsymbol{a}_2\times\boldsymbol{a}_3)$ is the volume of the primitive cell. If the integers N_i are sufficiently large, the $\boldsymbol{k}$ values can be approximated as belonging to a continuous spectrum; this assumption is used throughout the book.

The set of vectors $\boldsymbol{K}_j$ defined by

$$\boldsymbol{K}_j=g_{j1}\boldsymbol{b}_1+g_{j2}\boldsymbol{b}_2+g_{j3}\boldsymbol{b}_3, \tag{1.11}$$

and which satisfy $\boldsymbol{K}_j\cdot\boldsymbol{R}_i=2\pi m_{ij}$, where g_{ji} and m_{ij} are integers, define the reciprocal lattice. For an arbitrary reciprocal lattice vector $\boldsymbol{K}_s$, any two wavevectors $\boldsymbol{k}$ and $\boldsymbol{k}'$ such that $\boldsymbol{k}'=\boldsymbol{k}+\boldsymbol{K}_s$ are equivalent, in the sense that for any $\boldsymbol{R}_i$ the wavefunctions $\varphi(\boldsymbol{k},\boldsymbol{r})$ and $\varphi(\boldsymbol{k}',\boldsymbol{r})$ have the same eigenvalue $\exp(\mathrm{i}\boldsymbol{k}\cdot\boldsymbol{R}_i)=\exp(\mathrm{i}\boldsymbol{k}'\cdot\boldsymbol{R}_i)$ for all lattice translations. $\boldsymbol{k}$ is thus restricted to points that lie within or on a geometric figure – the Brillouin zone – inside which no two points are equivalent. The Brillouin zone is constructed by choosing first a reciprocal lattice point as origin. Then, a set of planes is constructed which are perpendicular bisectors of the vectors that connect the chosen reciprocal lattice point with all other reciprocal lattice points. The Brillouin zone is the smallest geometric figure that results from this construction and that contains the origin.

The different solution $\varphi_n(\boldsymbol{k},\boldsymbol{r})$, labeled by n, of the equation

$$H\varphi_n(\boldsymbol{k},\boldsymbol{r})=E_n(\boldsymbol{k})\varphi_n(\boldsymbol{k},\boldsymbol{r}) \tag{1.12}$$

are called Bloch functions, and $E_n(\boldsymbol{k})=E_n(\boldsymbol{k}+\boldsymbol{K}_s)$ is the energy band function. The Bloch functions belonging to different bands are orthogonal, i.e. for $\boldsymbol{q}$, $\boldsymbol{k}$ confined in the Brillouin zone,

$$\int\varphi_n{}^*(\boldsymbol{k},\boldsymbol{r})\varphi_l(\boldsymbol{q},\boldsymbol{r})\mathrm{d}\boldsymbol{r}=\delta_{nl}\delta(\boldsymbol{q}-\boldsymbol{k})\,. \tag{1.13}$$

To understand the energy band structure in crystalline solids, we write $V(\boldsymbol{r})$ as a sum of N identical terms, each corresponding to a unit cell (Callaway, 1991):

$$V(\boldsymbol{r}) = \Sigma_i V_c(\boldsymbol{r} - \boldsymbol{R}_i). \tag{1.14}$$

Considering a plane wave expansion of the Bloch functions

$$\varphi_n(\boldsymbol{k},\boldsymbol{r}) = (1/\sqrt{N\Omega})\Sigma_s b_n(\boldsymbol{k}+\boldsymbol{K}_s)\exp[\mathrm{i}(\boldsymbol{k}\cdot\boldsymbol{K}_s)\cdot\boldsymbol{r}] \tag{1.15}$$

the coefficients b_n are solutions of the set of equations

$$\Sigma_s\{[\hbar^2(\boldsymbol{k}+\boldsymbol{K}_s)^2/(2m) - E_n(\boldsymbol{k})]\delta_{st} + V(\boldsymbol{K}_t - \boldsymbol{K}_s)\}b_n(\boldsymbol{k}+\boldsymbol{K}_s) = 0, \tag{1.16}$$

where $V(\boldsymbol{K}) = (1/\Omega)\int\exp(-\mathrm{i}\boldsymbol{K}\cdot\boldsymbol{r})V_c(\boldsymbol{r})\mathrm{d}\boldsymbol{r}$ is the Fourier coefficient of the crystal potential. Assuming that $V(\boldsymbol{K}) = V^*(-\boldsymbol{K})$ is small enough, the second-order perturbation theory can be applied, the lowest-energy eigenvalue being

$$E(\boldsymbol{k}) = V(0) + \hbar^2\boldsymbol{k}^2/2m + (2m/\hbar^2)\Sigma_s|V(\boldsymbol{K}_s)|^2/[\boldsymbol{k}^2 - (\boldsymbol{k}+\boldsymbol{K}_s)^2]. \tag{1.17}$$

For small $k = |\boldsymbol{k}|$ (parabolic approximation), i.e. close to the center of the Brillouin zone, the energy depends quadratically on k as

$$E(k) = V(0) - \Sigma_s|V(\boldsymbol{K}_s)|^2/\boldsymbol{K}_s^2 + k^2[1 - 4\Sigma_s|V(\boldsymbol{K}_s)|^2\cos^2\theta_s/\boldsymbol{K}_s^4], \tag{1.18}$$

where θ_s is the angle between $\boldsymbol{k}$ and $\boldsymbol{K}_s$.

Expression (1.18) for the energy band cannot be applied when $\boldsymbol{k}^2 = (\boldsymbol{k}+\boldsymbol{K}_s)^2$, a condition satisfied on the faces of the Brillouin zone. Near such points the degenerate perturbation theory gives

$$E(\boldsymbol{k}) = V(0) + \frac{\hbar^2[\boldsymbol{k}^2 + (\boldsymbol{k}+\boldsymbol{K}_s)^2]}{4m} \pm \sqrt{\left(\frac{\hbar^2[\boldsymbol{k}^2 - (\boldsymbol{k}+\boldsymbol{K}_s)^2]}{4m}\right)^2 + |V(\boldsymbol{K}_s)|^2}. \tag{1.19}$$

So, on the faces of the Brillouin zone the energy band has discontinuities; gaps of width $2|V(\boldsymbol{K}_s)|^2$ appear in the energy spectrum. These are forbidden values for the electron energy, which appear due to the periodicity of the potential.

The energy bands, separated by energy gaps, are occupied by electrons in the increasing order of their energy, according to the Pauli principle, which states that there are no two electrons with the same set of quantum numbers. Depending on the number of electrons that are free to move throughout the lattice at low temperatures, three situations can be encountered:

1) The electrons partially occupy the highest-energy band. This situation characterizes the metals.
2) The electrons completely occupy the highest-energy band. This is the case for dielectric materials, in which the last occupied band is called the valence band and the first empty band is termed the conduction band. When the energy gap between the conduction and valence bands is not very large (a qualitative estimation gives for the upper limit a value of 3 eV) the materials are called semiconductors. In semiconductors it is relatively easy to induce transitions of electrons from the valence to the conduction band by applying external fields (in particular, electromagnetic fields) or by heating the material.
3) The valence and conduction bands overlap each other over a certain range of $\boldsymbol{k}$ vectors. These materials are called semimetals.

Many optical properties of crystalline materials depend on the density of states $\rho(E)$ defined such that $\rho(E)\mathrm{d}E$ is the number of electron states per unit volume with energies between E and $E+\mathrm{d}E$. Accounting for the two possible electron spin directions, the density of states per unit volume is defined as

$$\rho(E)=[2/(2\pi)^3]\Sigma_n\int \mathrm{d}\boldsymbol{k}\delta[E-E_n(\boldsymbol{k})]=[2/(2\pi)^3]\Sigma_n\int \mathrm{d}S(E)/|\nabla_{\boldsymbol{k}}E_n(\boldsymbol{k})|, \quad (1.20)$$

where the sum is performed over different energy bands. The density of states is singular when $|\nabla_{\boldsymbol{k}}E(\boldsymbol{k})|=0$; the points in the Brillouin zone where this happens are called critical points or van Hove singularities. Some of these critical points can be identified from symmetry considerations. Others can only be predicted after the energy surfaces are calculated with an adequate model for the wavefunction. If there is a critical point for a given $\boldsymbol{k}$, then there are critical points for all bands and for all symmetrical points equivalent in the $\boldsymbol{k}$-space.

Points with high symmetry in the Brillouin zone can be critical points. They are found from the condition that $E(\boldsymbol{k})$ is a continuous function of $\boldsymbol{k}$, with continuous derivatives and that $E(\boldsymbol{k})$ remains unchanged at symmetry operations. For example, a critical point is always the Γ point ($\boldsymbol{k}=0$, the center of the Brillouin zone) because the time-invariance condition implies that $E(\boldsymbol{k})=E(-\boldsymbol{k})$, so for $\boldsymbol{k}\to 0$, $\partial E(k_i)/\partial k_i=0$ for all i = 1,2,3. Other points result from the symmetry conditions for each crystallographic symmetry class. The symmetry elements of the crystal's point-group transform every $\boldsymbol{k}$ vector in the Brillouin zone into a certain number of wavevectors obtained by applying upon $\boldsymbol{k}$ all the symmetry operations (rotations, reflections, inversions). The ensemble of all these wavevectors forms the star of $\boldsymbol{k}$. For some $\boldsymbol{k}$ values the operations of the point group of the crystal can carry $\boldsymbol{k}$ into itself or into an equivalent $\boldsymbol{k}$ vector. This usually happens for $\boldsymbol{k}$ vectors with extremities corresponding to a singular point of the Brillouin zone (symmetry axis, symmetry plane, zone boundary). In this case the star of $\boldsymbol{k}$ is called degenerate; otherwise it is nondegenerate. Critical points, including the center of the Brillouin zone, correspond to $\boldsymbol{k}$ values with degenerate stars. Examples of points and lines of symmetry are given below for two frequently encountered cases: the body-centered and face-centered cubic lattices.

The energy dispersion relations for electrons in such crystals are frequently given in terms of the points and lines of symmetries, so that these are labeled in the following examples in accordance with standard notations.

The body-centered cubic crystal is shown in Fig. 1.1 where $\boldsymbol{i}$, $\boldsymbol{j}$, $\boldsymbol{k}$ are orthogonal unit vectors (versors).

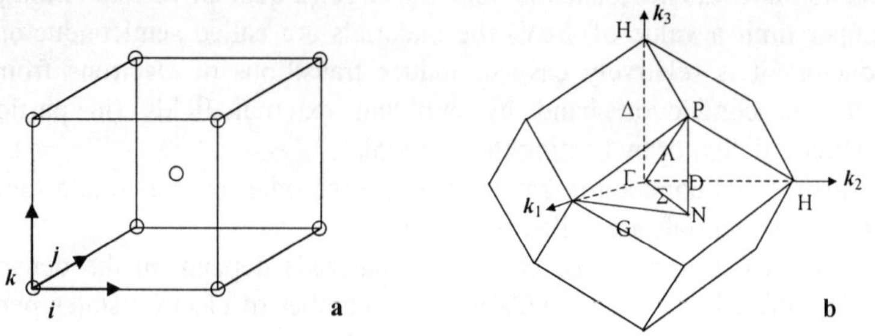

Fig. 1.1. (**a**) The body-centered cubic crystal and (**b**) its Brillouin zone

The primitive cell is spanned by the primitive vectors $\boldsymbol{a}_1 = a(\boldsymbol{i}+\boldsymbol{j}+\boldsymbol{k})/2$, $\boldsymbol{a}_1 = a(\boldsymbol{i}+\boldsymbol{j}-\boldsymbol{k})/2$, $\boldsymbol{a}_3 = a(\boldsymbol{i}-\boldsymbol{j}+\boldsymbol{k})/2$, where a is the lattice constant. They are identical to the reciprocal primitive cell vectors of the face-centered cubic crystal.

The face-centered cubic crystal is presented in Fig. 1.2.

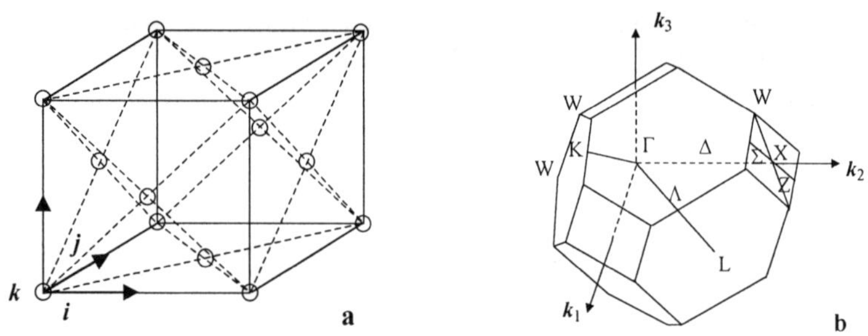

Fig. 1.2. (**a**) The face-centered cubic crystal and (**b**) its Brillouin zone

The primitive cell is formed by three vectors $\boldsymbol{a}_1 = a(\boldsymbol{j}+\boldsymbol{k})/2$, $\boldsymbol{a}_2 = a(\boldsymbol{i}-\boldsymbol{j})/2$, $\boldsymbol{a}_3 = a(\boldsymbol{i}-\boldsymbol{k})/2$, obtained by joining the atom at one corner to the atoms in the center of the adjacent faces of the elementary cell. The primitive vectors of the reciprocal lattice are identical to the primitive vectors of the body-centered cubic crystal. As can be seen from Figs. 1.1 and 1.2, the form of the Brillouin zone can be quite different from the elementary cell. An exception to this is, for example, the cubic crystal, whose Brillouin zone is also cubic.

Based on the face-centered cubic crystal configuration, two more complicated structures can be obtained: the NaCl structure, formed by two face-centered cubic crystal structures translated one with respect to the other by $a/2$ along one axis, and the diamond, or zincblende structure formed from two face-centered cubic structures translated one with respect to the other by $a/4$ along the volume diagonal. Another frequently encountered crystalline structure is that of wurtzite, formed from two compact hexagonal structures translated one with respect to the other along the c axis (the C_3 rotation axis).

As we have already specified, the Γ point ($\boldsymbol{k}$ = 0) is a critical point in all cases because of the symmetry at time inversion, which requires that $E(\boldsymbol{k}) = E(-\boldsymbol{k})$. For other points with high symmetry on the faces of the Brillouin zone this condition can be also satisfied – it must be checked for every particular case. For example, at the Brillouin zone of a face-centered cubic crystal, critical points due to symmetry appear at Γ, X, L, W. Among the critical points there are analytical critical points for which the energy is a quadratic form when expanded in a power series around the critical point $\boldsymbol{k}_0$ in a properly chosen coordinate system:

$$\begin{aligned} E(\boldsymbol{k}) &= (E_c - E_v)(\boldsymbol{k}) = E_0(\boldsymbol{k}_0) + \textstyle\sum_{i=1}^{3} (\mathrm{d}^2 E/\mathrm{d}k_i^2)_{\boldsymbol{k}=\boldsymbol{k}_0} (k_i - k_{0i})^2 \\ &= E_0(\boldsymbol{k}_0) + \textstyle\sum_{i=1}^{3} m_i (k_i - k_{0i})^2 . \end{aligned} \tag{1.21}$$

The analytical behavior of the density of states at critical points depends on the signs of m_i. There are four types of critical points (Callaway, 1991): M_0, M_1, M_2, and M_3.

Table 1.1 Critical points in the three dimensional case

Type of critical points	Notation	m_1	m_2	m_3	$\rho(E)$	
					$E < E_0$	$E > E_0$
Minimum	M_0	+	+	+	0	$C_0\sqrt{E - E_0}$
Saddle point	M_1	+	+	–	$C_1 - C_1'\sqrt{E_0 - E}$	C_1
Saddle point	M_2	+	–	–	C_2	$C_2 - C_2'\sqrt{E - E_0}$
Maximum	M_3	–	–	–	$C_3\sqrt{E_0 - E}$	0

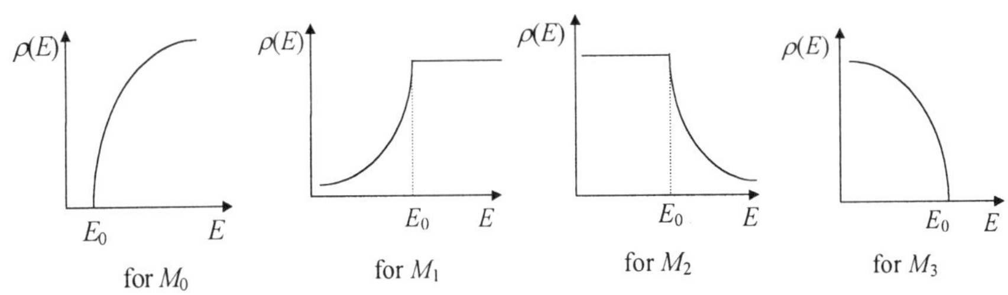

Fig. 1.3. The density of states for the critical points M_0, M_1, M_2, and M_3

M_0 exists always at the $\boldsymbol{k}$ value for which the bandgap energy in semiconductors is minimum; this value, denoted by E_g, is the threshold for interband transitions. The expressions for the density of states in Table 1.1 are obtained in the isotropic mass approximation; their behavior at critical points is shown in Fig. 1.3.

For a two-dimensional case there are only three types of critical points (see Table 1.2), the density of states at these points being shown in Fig. 1.4.

Table 1.2 Critical points in the two-dimensional case

Type of critical points	Notation	m_1	m_2	$\rho(E)$
Minimum	M_0	+	+	Step function
Saddle point	M_1	+	−	$C'\left[\ln\left(C+\sqrt{\lvert\hbar\omega-E_0\rvert+C^2}\right)-\frac{1}{2}\ln\lvert\hbar\omega-E_0\rvert\right]$
Maximum	M_2	−	−	Step function

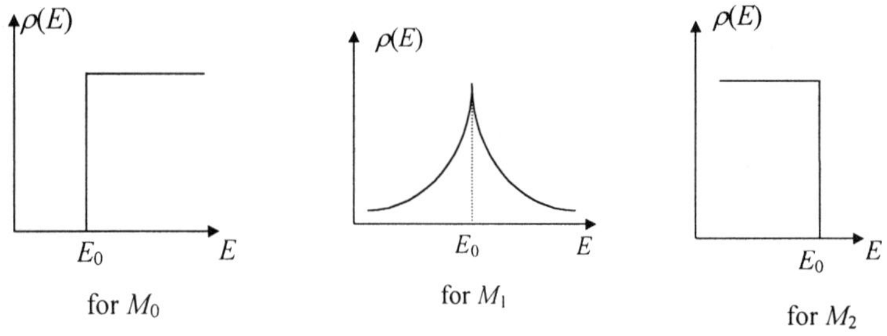

Fig. 1.4. The density of states for the critical points in the two-dimensional case

The number of critical points of various types are not independent if $E(\boldsymbol{k})$ is multiply periodic in $\boldsymbol{k}$. The following relations exist for three-dimensional lattices:

$$N_0 \geq 1,\ \ N_1 - N_0 \geq 2,\ \ N_2 - N_1 + N_0 \geq 1,\ \ N_3 - N_2 + N_1 - N_0 = 0, \tag{1.22}$$

where N_i denotes the number of critical points of the M_i type. These relations are obtained from topological considerations. For example, the minimum set of critical points in a band is one point of type M_0 (in which the energy is minimum), one point of type M_3 (in which the energy is maximum) and three critical points of each saddle type (M_1 and M_2).

1.2 $\boldsymbol{k}\cdot\boldsymbol{p}$ Method

The $\boldsymbol{k}\cdot\boldsymbol{p}$ perturbation theory is used to relate the energies and wavefunctions at nearby points in the Brillouin zone, i.e. it can be used to calculate the wave-function $\varphi_n(\boldsymbol{k},\boldsymbol{r})$ in the nth band at position $\boldsymbol{k}$ if $\varphi_n(\boldsymbol{k}_0,\boldsymbol{r})$, at a nearby point $\boldsymbol{k}_0$ is known. To this end, the unknown $\varphi_n(\boldsymbol{k},\boldsymbol{r})$ is expanded as

$$\varphi_n(\boldsymbol{k},\boldsymbol{r}) = \sum_j A_{nj}(\boldsymbol{k})\exp(\mathrm{i}\boldsymbol{k}\cdot\boldsymbol{r})u_j(\boldsymbol{k}_0,\boldsymbol{r}), \tag{1.23}$$

the equation satisfied by A_{nj} being determined by substituting this expansion in the Schrödinger equation. The result is

$$\begin{aligned}\sum_j A_{nj}(\boldsymbol{k})[E_j(\boldsymbol{k}_0) + (\hbar/m)(\boldsymbol{k}-\boldsymbol{k}_0)\cdot\boldsymbol{p} + (\hbar^2/2m)(\boldsymbol{k}^2-\boldsymbol{k}_0^2)]u_j(\boldsymbol{k}_0,\boldsymbol{r})\\ = E_n(\boldsymbol{k})\sum_j A_{nj}(\boldsymbol{k})u_j(\boldsymbol{k}_0,\boldsymbol{r}).\end{aligned} \tag{1.24}$$

By multiplying the above equation by $u_l{}^*(\boldsymbol{k}_0,\boldsymbol{r})$, integrating over a unit crystal cell with volume Ω, and using the orthogonality relation

$$\int \exp[\mathrm{i}(\boldsymbol{q}-\boldsymbol{k})\cdot\boldsymbol{r}]u_n{}^*(\boldsymbol{k},\boldsymbol{r})u_l(\boldsymbol{q},\boldsymbol{r})\mathrm{d}\boldsymbol{r} = \delta_{nl}\delta(\boldsymbol{k}-\boldsymbol{q}) \tag{1.25}$$

we finally obtain

$$\sum_j\{[E_j(\boldsymbol{k}_0) - E_n(\boldsymbol{k}) + (\hbar^2/2m)(\boldsymbol{k}^2-\boldsymbol{k}_0^2)]\delta_{jl} + (\hbar/m)(\boldsymbol{k}-\boldsymbol{k}_0)\cdot\boldsymbol{p}_{lj}\}A_{nj}(\boldsymbol{k}) = 0, \tag{1.26}$$

where

$$\boldsymbol{p}_{lj} = [(2\pi)^3/\Omega]\int_\Omega \mathrm{d}\boldsymbol{r}u_l{}^*(\boldsymbol{k}_0,\boldsymbol{r})\boldsymbol{p}u_j(\boldsymbol{k}_0,\boldsymbol{r}). \tag{1.27}$$

$\boldsymbol{p}_{lj}/m$ is sometimes called the interband velocity matrix. There is one equation for each band index l. This infinite set of linear homogeneous equations has a solution only if the determinant of the coefficients vanishes, the general element of the determinant having the form $H_{jl} - E(\boldsymbol{k})\delta_{jl}$ with

$$H_{jl} = [E_j(\boldsymbol{k}_0) + (\hbar^2/2m)(\boldsymbol{k}^2-\boldsymbol{k}_0^2)]\delta_{jl} + (\hbar/m)(\boldsymbol{k}-\boldsymbol{k}_0)\cdot\boldsymbol{p}_{lj}. \tag{1.28}$$

If enough states are included in the Hamiltonian and if the eigenvalues are obtained accurately, the bands can be obtained throughout the whole Brillouin zone, not only at nearby points.

One of the simplest applications of the $\boldsymbol{k}\cdot\boldsymbol{p}$ method is to calculate the effective mass using perturbation theory. In the case when the off-diagonal terms of the Hamiltonian are small (the coupling to other bands is negligible), and there are no degenerate levels at $\boldsymbol{k}_0$, to second-order

$$\begin{aligned} E_n(\boldsymbol{k}) = E_n(\boldsymbol{k}_0) + (\hbar/m)(\boldsymbol{k}-\boldsymbol{k}_0)\cdot \boldsymbol{p}_{nn} + (\hbar^2/2m)(\boldsymbol{k}^2-\boldsymbol{k}_0^2) \\ +(\hbar^2/m^2)\Sigma_{j\neq n}\{[(\boldsymbol{k}-\boldsymbol{k}_0)\cdot \boldsymbol{p}_{nj}][(\boldsymbol{k}-\boldsymbol{k}_0)\cdot \boldsymbol{p}_{jn}]/[E_n(\boldsymbol{k}_0)-E_j(\boldsymbol{k}_0)]\}. \end{aligned} \tag{1.29}$$

If the band n has an extremum at $\boldsymbol{k}_0$ so that the term linear in $\boldsymbol{k}-\boldsymbol{k}_0$ vanishes (this implies $\boldsymbol{p}_{nn}+\hbar\boldsymbol{k}_0 = 0$) then second-order terms dominate and a reciprocal effective mass tensor can be introduced as

$$\begin{aligned} (m/m_n)_{\alpha\beta} &= (m/\hbar^2)\partial^2 E_n/\partial k_\alpha \partial k_\beta \\ &= \delta_{\alpha\beta} + (1/m)\Sigma_{j\neq n}\{(p_{nj}^\alpha p_{jn}^\beta + p_{nj}^\beta p_{jn}^\alpha)/[E_n(\boldsymbol{k}_0)-E_j(\boldsymbol{k}_0)]\}, \end{aligned} \tag{1.30}$$

where α, β label rectangular components with respect to some fixed axes. By an appropriate choice of these axes the reciprocal effective mass tensor can be diagonalized. Since $p_{jn}^\alpha = (p_{nj}^\alpha)^*$, from the above relation it follows that interaction with a lower energy level $E_j < E_n$ decreases the effective mass with respect to the principal axis, whereas interaction with higher-energy states increases the effective mass. The effective mass can even take negative values for some ranges of the wavevector.

The $\boldsymbol{k}\cdot\boldsymbol{p}$ method can be extended to include spin-orbit coupling by adding to the Hamiltonian the term (Callaway (1991))

$$H_{\mathrm{so}} = (\hbar/4m^2c^2)\boldsymbol{\sigma}\cdot(\nabla V\times \boldsymbol{p}), \tag{1.31}$$

where $\boldsymbol{\sigma}$ is the Pauli spin operator. If the set of basis states at $\boldsymbol{k}_0$ include the spin-orbit coupling, the treatment is identical, except that $\boldsymbol{p}_{lj}$ in (1.28) is replaced by

$$\boldsymbol{p}_{lj} + [(2\pi)^3/\Omega](\hbar/4m^2c^2)\int_\Omega u_l{}^*(\boldsymbol{k}_0,\boldsymbol{r})(\boldsymbol{\sigma}\times\nabla V)u_j(\boldsymbol{k}_0,\boldsymbol{r})\mathrm{d}\boldsymbol{r}. \tag{1.32}$$

Alternatively, the spin-orbit coupling can be treated as a perturbation that splits the otherwise degenerate states at $\boldsymbol{k}_0$. The calculation of the set of basis states is quite a difficult problem; it is not our aim to describe it here so that we refer the interested reader to Callaway (1991) for a detailed description of the various methods used to this end.

Especially in the theory of semiconductors an approximate form of the $\boldsymbol{k}\cdot\boldsymbol{p}$ method is used, which includes the spin-orbit interaction, and considers the interaction between four bands only: the conduction, heavy-hole, light-hole and spin-orbit split-off bands, all of them being doubly degenerate with respect to the spin. In this so-called Kane model for the band structure (see Chuang (1995)), by taking $\boldsymbol{k}_0 = 0$, the Schrödinger equation for the cell periodic function is

$$\left\{-\frac{\hbar^2}{2m}\nabla^2+V(\boldsymbol{r})+\frac{\hbar}{m}\boldsymbol{k}\cdot\boldsymbol{p}+\frac{\hbar}{4m^2c^2}[\nabla V\times\boldsymbol{p}]\cdot\boldsymbol{\sigma}+\frac{\hbar^2}{4m^2c^2}\nabla V\times\boldsymbol{k}\cdot\boldsymbol{\sigma}\right\}u_n(\boldsymbol{k},\boldsymbol{r})$$
$$=E'u_n(\boldsymbol{k},\boldsymbol{r})=[E_n(\boldsymbol{k})-\hbar^2\boldsymbol{k}^2/(2m)]u_n(\boldsymbol{k},\boldsymbol{r}),\tag{1.33}$$

where the last term on the left-hand side is neglected because the crystal momentum $\hbar\boldsymbol{k}$ is small compared to the atomic momentum $\boldsymbol{p}$ in the interior of the atom where most of the spin-orbit interaction occurs. As in the $\boldsymbol{k}\cdot\boldsymbol{p}$ theory, we look for solutions for the eigenvalue E' with corresponding eigenfunctions

$$u_n(\boldsymbol{k},\boldsymbol{r})=\textstyle\sum_{n'}A_{n'}u_{n'}(0,\boldsymbol{r}).\tag{1.34}$$

Since the electron wavefunctions are p-like near the top of the valence band and s-like near the bottom of the conduction band, the basis functions are $|\mathrm{i}S\downarrow\rangle$, $2^{-1/2}|X-\mathrm{i}Y,\uparrow\rangle$, $|Z\downarrow\rangle$, $-2^{-1/2}|X+\mathrm{i}Y,\uparrow\rangle$ and $|\mathrm{i}S\uparrow\rangle$, $-2^{-1/2}|X+\mathrm{i}Y,\downarrow\rangle$, $|Z\uparrow\rangle$, $2^{-1/2}|X-\mathrm{i}Y,\downarrow\rangle$, where the first four functions are, respectively, degenerate with respect to the last four. The band-edge functions $|S\downarrow\rangle$ and $|S\uparrow\rangle$ which characterize the conduction band have eigenenergies E_s with respect to the Hamiltonian $H_0=-\hbar^2\nabla^2/(2m)+V(\boldsymbol{r})$, whereas the valence band-edge functions $|X\downarrow\rangle$, $|Y\downarrow\rangle$, $|Z\downarrow\rangle$, $|X\uparrow\rangle$, $|Y\uparrow\rangle$, $|Z\uparrow\rangle$ have eigenenergies E_p with respect to H_0.

Defining the reference energy as $E_\mathrm{p}=-\Delta/3$, $E_\mathrm{s}=E_\mathrm{g}$, and taking $\boldsymbol{k}=k\hat{z}$, the Hamiltonian becomes an 8×8 interaction matrix

$$\begin{bmatrix}\boldsymbol{H} & 0\\ 0 & \boldsymbol{H}\end{bmatrix},\tag{1.35}$$

where

$$\boldsymbol{H}=\begin{bmatrix}E_\mathrm{g} & 0 & kP & 0\\ 0 & -2\Delta/3 & \sqrt{2}\Delta/3 & 0\\ kP & \sqrt{2}\Delta/3 & -\Delta/3 & 0\\ 0 & 0 & 0 & 0\end{bmatrix}.\tag{1.36}$$

In (1.36) P is the Kane parameter and Δ is the spin-orbit split-off energy, defined respectively, as

$$P=-\mathrm{i}\hbar\langle S|p_z|Z\rangle/m,\quad \Delta=3\mathrm{i}\hbar\langle X|p_y(\partial V/\partial x)-p_x(\partial V/\partial y)|Y\rangle/(4m^2c^2).\tag{1.37}$$

The four eigenvalues for E' determined from the equation $\det(\boldsymbol{H} - E'\boldsymbol{I}) = 0$ in the quadratic approximation in k, are, for the four bands:

$$E_c(k) = E_g + \hbar^2 k^2/2m + k^2 P^2 (E_g + 2\Delta/3)/[E_g(E_g + \Delta)], \quad E_{hh}(k) = \hbar^2 k^2/2m,$$
$$E_{lh}(k) = \hbar^2 k^2/2m - 2k^2 P^2/3E_g, \quad E_{so}(k) = -\Delta + \hbar^2 k^2/2m - k^2 P^2/[3(E_g + \Delta)]. \tag{1.38}$$

As can be seen from the above relations, the effective mass in the heavy-hole band is identical to the free-electron mass. This is obviously an incorrect result, due mainly to the fact that higher bands have not been included. They are taken into consideration in the Luttinger–Kohn model, which includes the doubly degenerate heavy-hole, light-hole and spin-orbit split-off valence bands, into one category – class A, and the other bands in another – class B. The effect of the bands in class B is treated as a perturbation on those in class A, so that the equation to be solved in order to obtain the eigenvalues is $\det(\boldsymbol{U} - E\boldsymbol{I}) = 0$ where

$$U_{jj'} = E_j(0)\delta_{jj'} + (\hbar^2/2m)\Sigma_{\alpha,\beta} k_\alpha k_\beta (m/m_{\text{eff}})^{\alpha\beta}_{jj'}, \tag{1.39}$$

with

$$(m/m_{\text{eff}})^{\alpha\beta}_{jj'} = \delta_{jj'}\delta_{\alpha\beta} + \Sigma_{\gamma\in B}(p^{\alpha}_{j\gamma} p^{\beta}_{\gamma j'} + p^{\beta}_{j\gamma} p^{\alpha}_{\gamma j'})/[m(E_0 - E_\gamma)], \tag{1.40}$$

where the j indices belong to class A. For $j = j'$, $(m/m_{\text{eff}})^{\alpha\beta}_{jj'}$ can be viewed as an analog to the effective mass tensor for a single band. The Luttinger–Kohn band structure parameters are defined as

$$-\gamma_1 = [(m/m_{\text{eff}})^{xx}_{xx} + 2(m/m_{\text{eff}})^{yy}_{xx}]/3, \quad -\gamma_2 = [(m/m_{\text{eff}})^{xx}_{xx} - (m/m_{\text{eff}})^{yy}_{xx}]/6,$$
$$-\gamma_3 = (m/m_{\text{eff}})^{xy}_{xy}/6. \tag{1.41}$$

1.3 Numerical Methods of Electron Energy Band Calculation

Although we have extensively used the plane wave basis for the illustration of Bloch's theorem, this set of functions is not suitable in numerical calculations due to their poor convergence. The electron wavefunction varies rapidly with the distance when close to the nucleus, and is slowly varying with the distance in the outer part of the atom, where the nuclear potential is screened by inner electrons; this variation requires a large number of plane waves to be modeled. Therefore, in practical computations, one looks for either a change of the plane wave basis, a modification of the potential, or a modification of the geometry of the problem from a periodic lattice to a single cell. All these methods are well reviewed in Callaway (1991). Here we will only point out the physical principles on which these methods are based.

The pseudopotential is introduced starting from the consideration that the electron energy is mostly determined by the behavior of the wavefunction in outer portions of the cell, where bonding occurs. So, the real potential can be replaced with a, generally not unique, nonlocal and non-Hermitian pseudopotential that gives the same energy bands. A suitable form is $V(r) = A\exp(-\beta r)/r$, although other forms can also be used. This method was especially useful for band-structure calculations in metals in which d states do not contribute to occupied bands.

Orthogonalized plane waves behave as plane waves at large distance from the atom and are rapidly varying, as atomic wavefunctions, near any nucleus. They are constructed by forming first Bloch functions for atomic core states, and then by defining plane waves of wavevector $\boldsymbol{k}$ orthogonal to all core states of the same $\boldsymbol{k}$. Although these functions differ from the plane waves, still a large number of them are required in calculus. A smaller basis set may be employed in the tight-binding method, the numerical calculations being, however, more complex. More exactly, in the tight-binding method a set of functions localized on the atoms in the crystal is defined, the ith function at site $\boldsymbol{R}_\mu$ being denoted by $u_i(\boldsymbol{r} - \boldsymbol{R}_\mu)$. Each such function is separable into a radial part and spherical harmonics; a linear combination of these functions that obey the Bloch theorem is defined as $\phi_i(\boldsymbol{k},\boldsymbol{r}) = N^{-1/2}\sum_\mu \exp(\mathrm{i}\boldsymbol{k}\cdot\boldsymbol{R}_\mu)u_i(\boldsymbol{r} - \boldsymbol{R}_\mu)$. The exact Bloch wavefunction for the band n and wavevector $\boldsymbol{k}$ is then $\psi_n(\boldsymbol{k},\boldsymbol{r}) = \sum_i c_{in}\phi_i(\boldsymbol{k},\boldsymbol{r})$, the coefficients c_{in} and the energy $E_n(\boldsymbol{k})$ being obtained by a simultaneous diagonalization of the Hamiltonian and the overlap matrices on the basis of functions ϕ_i.

Another method of simplifying energy band calculations is by solving the Schrödinger equation in a single atomic cell, rather than in the periodic lattice, subject to boundary conditions on the surface of the cell implied by the Bloch theorem. This can be achieved by expanding the wavefunction in a sum over spherical harmonics defined on a sphere. The sphere can have the same volume as the polyhedral cell, in which case it is called the Wigner–Seitz sphere, or it can be the largest sphere that can be inscribed within the atomic cell. In the latter case it is called a muffin-tin sphere. Then, the problem reduces to matching the wavefunction in the muffin-tin sphere to the wavefunction in the region between the muffin-tin and the cell boundaries, the region in which the wavefunction can be taken as a plane wave. In a related, augmented plane wave (APW) method, the wavefunction is expanded in a set of plane waves outside the atomic cell (the muffin-tin sphere) and spherical waves inside. The spherical waves are generally energy dependent. Attempts using energy-independent basis of functions have resulted in the linear augmented plane wave (LAPW) method and linear muffin-tin orbitals (LMTO).

Still another way of overcoming the numerical difficulties is by replacing the differential Schrödinger equation and the associated boundary conditions with an equivalent integral equation. This widely used method is known as the Green's function method.

All the methods we referred to up to now are based on the single-particle Schrödinger equation with a periodic potential. The associated wavefunction of a system of N electrons can be found by either the Hartree–Fock method or the density-functional theory. In the Hartree–Fock method one starts by associating a wavefunction to each electron, the so-called spin orbitals, and then the total antisymmetric wavefunction of the system is constructed from them such that relevant symmetry operations are preserved. The single-electron wavefunctions must be chosen such that the total energy of the system is minimum, variational calculus being employed to this end. In improved versions of the method the number of single-electron wavefunctions is larger than the number of electrons, more parameters becoming available for variational calculations. In the density-functional theory the external potential that contributes to the total Hamiltonian of the system of N electrons is taken as a sum of identical contributions for each electron i, depending only on the coordinates of the ith electron. This external potential is determined uniquely by the electron density in the ground state (or even in excited states in some generalized versions). The correct potential, and thus wavefunction, is that for which the ground-state energy is minimum.

1.4 Phonons

In the last section we have presented some basic properties of the electronic band structure of the solid, which are consequences of the periodicity of the crystal. The lattice itself can, however, be characterized by elementary excitations, called phonons. Phonons are quanta of the collective lattice movement, as photons are quanta of the electromagnetic radiation. Their introduction is possible due to the symmetry of the crystal at translations, which allows the classification of stationary states of the crystal in terms of the quasi-momentum $\boldsymbol{p} = \hbar\boldsymbol{k}$.

Let us consider a crystal in which the position of each cell is given by $\boldsymbol{R}_i$, and the position of the kth atom in the cell (k = 1…r) is $\boldsymbol{x}_{ik} = \boldsymbol{R}_i + \boldsymbol{d}_k$. In a classical treatment, by denoting with $\boldsymbol{u}_{ik}$ the displacement of the atom from the equilibrium position, the equation of motion along the direction α is

$$M_k \ddot{u}_{\alpha,ik} = -\partial\Phi / \partial u_{\alpha,ik} = -\sum_{\beta j\nu} \Phi_{\alpha\beta,ik,j\nu} u_{\beta,j\nu}, \tag{1.42}$$

where Φ is the potential energy of the crystal in nonequilibrium, and $\Phi_{\alpha\beta,ik,j\nu} = (\partial^2\Phi / \partial u_{\alpha,ik}\partial u_{\beta,j\nu})_0$, the subscript 0 indicating that these parameters are calculated at the equilibrium position. The periodic solutions of (1.42) are

$$u_{\alpha,ik} = [u_{\alpha,k}(\boldsymbol{k}) / M_k^{1/2}]\exp(-\mathrm{i}\omega t + \mathrm{i}\boldsymbol{k}\cdot\boldsymbol{R}_i), \tag{1.43}$$

where $\boldsymbol{k}$ is the wavevector of lattice vibrations. Introducing these solutions in (1.42), it becomes

$$\omega^2 u_{\alpha,k}(\boldsymbol{k}) = \sum_{\nu\beta} D_{\alpha\beta,k\nu}(\boldsymbol{k}) u_{\beta,\nu}(\boldsymbol{k}), \tag{1.44}$$

where (Callaway (1991))

$$D_{\alpha\beta,k\nu}(\boldsymbol{k}) = (M_k M_\nu)^{-1/2} \sum_{\boldsymbol{R}_i - \boldsymbol{R}_j} \Phi_{\alpha\beta,ik,j\nu} \exp[-\mathrm{i}\boldsymbol{k} \cdot (\boldsymbol{R}_i - \boldsymbol{R}_j)]. \tag{1.45}$$

From the condition

$$\det[\omega^2 \delta_{\alpha\beta} \delta_{k\nu} - D_{\alpha\beta,k\nu}(\boldsymbol{k})] = 0 \tag{1.46}$$

one obtains $3r$ eigenvalues $\omega_j^2(\boldsymbol{k})$, where $j = 1 \ldots 3r$ labels the vibration branches (see Fig. 1.5). These are the dispersion relations in the crystal. For each eigenvalue $\omega_j^2(\boldsymbol{k})$ there is an eigenvector of D with elements

$$\omega_j^2(\boldsymbol{k}) e_{\alpha\nu}^{(j)}(\boldsymbol{k}) = \sum_{\beta k} D_{\alpha\beta,\nu k}(\boldsymbol{k}) e_{\beta k}^{(j)}(\boldsymbol{k}). \tag{1.47}$$

Since D is a Hermitic matrix the eigenvalues are real and $\omega_j^2(\boldsymbol{k}) = \omega_j^2(-\boldsymbol{k})$.

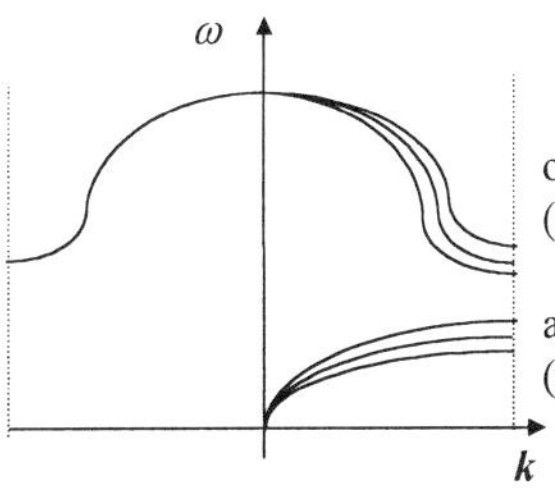

Fig. 1.5. Dispersion relations for phonons for a crystal with two atoms per unit cell

If $\boldsymbol{k} \to 0$ there are three branches of the spectrum for which $\omega \to 0$. These are the acoustic modes. For them

$$\omega_j^2(0) e_{\alpha k}^{(j)}(0) = \sum_{\beta j \nu} \Phi_{\alpha\beta,ik,j\nu} (M_k M_\nu)^{-1/2} e_{\beta\nu}^{(j)}(0). \tag{1.48}$$

In a rigid lattice for which $u_{\beta,j\nu}$ are independent of j and ν, and $e_{\beta\nu}(0)/M_\nu^{1/2}$ is independent of ν, for acoustic modes $\boldsymbol{u}_{ik} = \boldsymbol{e}_k(0)/M_k^{1/2} = \text{const.}$, so that all r particles from each elementary cell move in parallel with equal amplitudes. The acoustic modes correspond to the movement of the cell as a whole. For $q \neq 0$ different primitive cells suffer different displacements, analogously to the deformations produced when acoustic waves pass through the medium. The small-$\boldsymbol{k}$ behavior of acoustic branches is characterized by the proportionality between ω and $\boldsymbol{k}$, the proportionality constant being the sound velocity. For example, in a lattice with one atom per unit cell, for small $\boldsymbol{k}$,

$$\omega^2(\boldsymbol{k})u_\alpha(\boldsymbol{k}) = \sum_{\nu\delta\beta} c_{\alpha\beta,\nu\delta} k_\nu k_\delta u_\beta(\boldsymbol{k}), \tag{1.49}$$

where $c_{\alpha\beta,\gamma\delta} = (-1/2M)\sum_i \Phi_{\alpha\beta}(\boldsymbol{R}_i)\cdot(\boldsymbol{R}_i)_\gamma \cdot (\boldsymbol{R}_i)_\delta$. If the dependence of u_α on $|\boldsymbol{k}|$ can be neglected, this equation is similar to that determining the vibration frequencies of an elastic continuum.

The remaining $3r-3$ modes are optical modes, for which the ions in the elementary cell vibrate in antiphase. If there are, for example, two ions of opposite charges in the unit cell, their movement is such that the center-of-mass of the cell remains fixed. Because the two ions have charges with different sign, there is a net dipolar moment that oscillates in the crystal and can interact with external fields.

To quantify the oscillations of the lattice one must introduce *normal coordinates* $Q_j(\boldsymbol{k},t)$ for which the total Hamiltonian is identical to that of a harmonic oscillator. These coordinates, which satisfy the relation $Q_j{}^*(\boldsymbol{k},t) = Q_j(-\boldsymbol{k},t)$ are defined by

$$u_{\alpha,lk} = (\Omega/8\pi^3)(N/M_k)^{1/2}\sum_j \int \mathrm{d}\boldsymbol{k} e^{(j)}_{\alpha,k}(\boldsymbol{k})Q_j(\boldsymbol{k},t)\exp(\mathrm{i}\boldsymbol{k}\cdot\boldsymbol{R}_l), \tag{1.50}$$

the total Hamiltonian $H = (1/2)\sum_{\alpha ik} M_k \dot{u}^2_{\alpha,ik} + (1/2)\sum_{\alpha ik,\beta j\nu} \Phi_{\alpha\beta,ik,j\nu} u_{\alpha,ik} u_{\beta,j\nu}$ being written in terms of them as

$$H = [V/2(2\pi)^3]\sum_j \int \mathrm{d}\boldsymbol{k}[P_j{}^*(\boldsymbol{k})P_j(\boldsymbol{k}) + \omega_j^2(\boldsymbol{k})Q_j{}^*(\boldsymbol{k})Q_j(\boldsymbol{k})]. \tag{1.51}$$

Here $V = N\Omega$ and $P_j(\boldsymbol{k}) = \dot{Q}_j(\boldsymbol{k})$ is the momentum conjugate to $Q_j(\boldsymbol{k})$. The total Hamiltonian is now a sum of Hamiltonians of uncoupled harmonic oscillators, each of them obeying the equation of motion $\ddot{Q}_j(\boldsymbol{k}) + \omega_j^2(\boldsymbol{k})Q_j(\boldsymbol{k}) = 0$.

To quantify the lattice motion $Q_j(\boldsymbol{k})$ and $P_j(\boldsymbol{k})$ are first replaced by operators that satisfy the commutation relation

$$[Q_j^+(\boldsymbol{k}), P_{j'}(\boldsymbol{q})] = Q_j^+(\boldsymbol{k})P_{j'}(\boldsymbol{q}) - P_{j'}(\boldsymbol{q})Q_j^+(\boldsymbol{k}) = \mathrm{i}\hbar\delta(\boldsymbol{k}-\boldsymbol{q})\delta_{jj'}, \tag{1.52}$$

if $\boldsymbol{k}$, $\boldsymbol{q}$ are in the Brillouin zone; otherwise $\delta(\boldsymbol{k}-\boldsymbol{q})$ should be replaced by $\sum_s \delta(\boldsymbol{k}-\boldsymbol{q}-\boldsymbol{K}_s)$. The annihilation and creation operators are then introduced, respectively, as

$$\begin{aligned} b_j(\boldsymbol{k}) &= \sqrt{\omega_j(\boldsymbol{k})/2\hbar}Q_j(\boldsymbol{k}) + \mathrm{i}P_j(\boldsymbol{k})/\sqrt{2\hbar\omega_j(\boldsymbol{k})}, \\ b_j^+(\boldsymbol{k}) &= \sqrt{\omega_j(\boldsymbol{k})/2\hbar}Q_j^+(\boldsymbol{k}) - \mathrm{i}P_j^+(\boldsymbol{k})/\sqrt{2\hbar\omega_j(\boldsymbol{k})} \end{aligned} \tag{1.53}$$

so that they satisfy the following relations

$$[b_j(\boldsymbol{k}), b_{j'}^+(\boldsymbol{q})] = \delta_{jj'}\delta(\boldsymbol{k}-\boldsymbol{q}), \quad [b_j(\boldsymbol{k}), b_{j'}(\boldsymbol{q})] = [b_j^+(\boldsymbol{k}), b_{j'}^+(\boldsymbol{q})] = 0. \tag{1.54}$$

The total Hamiltonian (1.55) can be expressed in terms of the new operators as

$$H = [V/(2\pi)^3]\Sigma_j \int \mathrm{d}\boldsymbol{k}[b_j^+(\boldsymbol{k})b_j(\boldsymbol{k}) + 1/2]\hbar\omega_j(\boldsymbol{k}), \tag{1.55}$$

i.e. as a sum of independent quantum harmonic oscillators, called phonons. If $\boldsymbol{k}$ takes discrete values, (1.55) can be written in the more common form

$$H = \Sigma_{jk}(b_{jk}^+ b_{jk} + 1/2)\hbar\omega_{jk}, \tag{1.56}$$

or even $H = \Sigma_{jk} b_{jk}^+ b_{jk} \hbar\omega_{jk}$ when the zero-point energy $(1/2)\Sigma_{jk}\hbar\omega_{jk}$ is neglected.

1.4.1 Bose–Einstein Statistics

The number of phonons in a branch j with wavevector $\boldsymbol{k}$, denoted by n_{jk}, is the eigenvalue of the number operator $\hat{n}_{jk} = b_{jk}^+ b_{jk}$. The total energy corresponding to (1.56) is then

$$E = \Sigma_{jk}(n_{jk} + 1/2)\hbar\omega_{jk}, \tag{1.57}$$

the annihilation and creation operators acting upon the eigenstates $|n_{jk}\rangle$ of $\hat{n}_{jk}$, respectively, as

$$b_{jk}|n_{jk}\rangle = \sqrt{n_{jk}}\,|n_{jk} - 1\rangle, \quad b_{jk}^+|n_{jk}\rangle = \sqrt{n_{jk} + 1}\,|n_{jk} + 1\rangle. \tag{1.58}$$

These operators decrease, respectively increase, the phonon number in a given mode by 1.

n_{jk} can have any integer value greater than or equal to 0. This is a characteristic of an ideal boson system of particles, i.e. particles or quasiparticles with integer spin values. The expectation value of the number operator of an ideal Bose gas at the equilibrium temperature T is called the Bose–Einstein distribution function, and is given by

$$\langle\hat{n}_{jk}\rangle = \overline{n}(\omega_{jk}) = 1/[\exp(\hbar\omega_{jk}/k_{\mathrm{B}}T) - 1], \tag{1.59}$$

where k_{B} is the Boltzmann constant.

1.4.2 Fermi–Dirac Statistics

Electrons have noninteger spin values, so if they do not interact with one another, an ideal fermion gas is the best model to describe them in the second quantization. Creation and annihilation operators can be introduced even in this case; they satisfy the commutation relations for fermions

$$\{c_{jk}, c^{+}_{j'q}\} = c_{jk}c^{+}_{j'q} + c_{jk}, c^{+}_{j'q} = \delta_{jj'}\delta_{kq}, \quad \{c_{jk}, c_{j'q}\} = \{c^{+}_{jk}, c^{+}_{j'q}\} = 0. \tag{1.60}$$

As for phonons, the single-particle Hamiltonian (neglecting the zero-point energy) can be written as

$$H = \sum_{jk} E_{jk} c^{+}_{jk} c_{jk}, \tag{1.61}$$

where E_{jk} is the energy of an electron in band j with wavevector $\boldsymbol{k}$. The spin index σ is sometimes explicitly displayed. The main difference from the boson case is that the number of electrons in band j with wavevector $\boldsymbol{k}$ and spin σ, defined as the eigenvalue of the operator $\hat{n}_{j\sigma k} = c^{+}_{j\sigma k} c_{j\sigma k}$, can take, at low temperatures, only values 1 or 0, in agreement with the Pauli principle. $n_{j\sigma k} = 1$ if $E_{j\sigma k} < \mu$ where μ is the chemical potential of the fermion gas, and $n_{j\sigma k} = 0$ if $E_{j\sigma k} > \mu$. The chemical potential depends on both temperature and the number of fermions in the system; its value for $T = 0$ is called the Fermi energy and is denoted by E_{F}. The expectation value of the number operator of an ideal fermion gas in equilibrium at temperature T is called the Fermi–Dirac distribution function, and is given by

$$\langle \hat{n}_{j\sigma k} \rangle = f(E_{j\sigma k}) = 1/\{\exp[(E_{j\sigma k} - \mu)/k_{\mathrm{B}}T] + 1\}. \tag{1.62}$$

For small electron energies or high temperatures (1.62) goes into the Maxwell–Boltzmann distribution function

$$f_{MB}(E_{j\sigma k}) \cong \exp(\mu / k_{\mathrm{B}}T)\exp(-E_{j\sigma k} / k_{\mathrm{B}}T). \tag{1.63}$$

1.5 Plasmons

Plasmons are quanta of the collective motion of a gas of carriers, usually electrons, with respect to a rigid background of oppositely charged ions. Plasmons are usually observed in metals, and/or in strongly correlated electron systems. Displacing the gas of carriers with density n, mass m and charge e by an amount $\boldsymbol{u}(t)$ results in an electric field $\boldsymbol{E} = ne\boldsymbol{u}(t) / \varepsilon_{\mathrm{b}}\varepsilon_0$, where ε_{b} is the relative dielectric constant of the background for frequencies above the plasma frequency, which we define below. The motion of the ensemble of free carriers is described classically by (Klingshirn, 1997)

$$m\partial^2 \boldsymbol{u}(t) / \partial t^2 = ne^2 \boldsymbol{u}(t) / \varepsilon_0\varepsilon_{\mathrm{b}}, \tag{1.64}$$

an equation similar to that of an harmonic oscillator with a frequency

$$\omega_{\mathrm{p}} = \sqrt{e^2 n/(m\varepsilon_0\varepsilon_{\mathrm{b}})}, \tag{1.65}$$

called the plasma frequency. This is a longitudinal motion at $\boldsymbol{q} = 0$, since a gas of carriers has no static shear stiffness and so does not support transverse vibrations. As will be seen in Sect. 2.3.2, the frequency of this longitudinal motion corresponds to the zeros of the dielectric constant. The quantization of these oscillations leads to new quasiparticles, called plasmons, which obey Bose–Einstein statistics. The quantization procedure is, however, different from the case of phonons since the interactions between electrons and also, generally, the oppositely charged ions, must be accounted for. More precisely, in an electron gas in which Coulomb interactions are present, the total Hamiltonian can be described by

$$H = \sum_k E_k c_k^+ c_k + (1/2)\sum_{q\neq 0,k,k'} V_q c_{k-q}^+ c_{k'+q}^+ c_{k'} c_k. \tag{1.66}$$

The two-operator expectation values $\langle c_{k-q}^+ c_k \rangle$ related to the expectation value of the electron charge density operator as $\langle \rho_q \rangle = -(e/V)\sum_k \langle c_{k-q}^+ c_k \rangle$ with V the volume, satisfy the equation of motion

$$\hbar \frac{\mathrm{d}}{\mathrm{d}t}\langle c_{k-q}^+ c_k \rangle \cong i(E_{k-q} - E_k)\langle c_{k-q}^+ c_k \rangle + iV_q (f_k - f_{k-q})\sum_n \langle c_{n-q}^+ c_n \rangle. \tag{1.67}$$

Here f_k is the Fermi–Dirac distribution function, V_q is the interaction potential, and the calculations are done in the random phase approximation which assumes that the kinetic energy of electrons is much higher than the potential energy. For damped harmonic solutions of (1.67), for which $\langle c_{k-q}^+ c_k \rangle(t) = \exp[(\mathrm{i}\omega - \delta)t]\langle c_{k-q}^+ c_k \rangle(0)$, the eigenfrequencies ω_q of the collective equation of motion are to be determined from the condition

$$V_q \sum_k (f_{k-q} - f_k)/(\hbar\omega_q + E_{k-q} - E_k) = 1, \tag{1.68}$$

in the limit of small damping. The long-wavelength $q \to 0$ solution of (1.68) is found to be identical to the classical expression of plasma frequency in (1.65), as can be seen from the fact that in this limit for parabolic dispersion relations (1.68) reduces to $1 = V_q (q^2/m\omega_q^2)\sum_k f_k$. The demonstration is completed by noting that $V_q = 4\pi e^2/(q^2 \varepsilon V)$ is the Fourier transform of the Coulomb interaction potential in real space $V(r) = e^2/\varepsilon r$ with ε the electric permeability of the medium. The Coulomb interaction is dominant in the long-wavelength limit in which q is small. For more details on the quantization of plasma oscillations see Haug and Kock (1990). At shorter wavelengths, the dispersion relation for plasmons is found to depend quadratically on $\boldsymbol{q}$. Longitudinal plasma waves can decay into electron-pair excitations when the plasma frequency is in the range of these excitations, in the mechanism known as Landau damping. Electron-pair excitations appear in a certain range of $\boldsymbol{q}$ values and represent the transfer (scattering) of an electron between a pair of states: from a state $\boldsymbol{k}$ to another state $\boldsymbol{k} \pm \boldsymbol{q}$. The plasma frequency is situated in the infrared (IR) or far-IR in typical semiconductors, and in the visible or ultraviolet range in metals.

1.6 Magnetic Excitations

Phonons are not the only quanta of collective excitations that can be present in a solid. Other quasiparticles that appear in solids with magnetic properties are the magnons. These are spin waves in magnetically ordered solids.

From the point of view of magnetic properties, the magnetism of crystalline solids can be disordered (diamagnetism and paramagnetism) or ordered (ferromagnetism, antiferromagnetism, ferrimagnetism). Disordered magnetism appears in those materials where the induced magnetization $\boldsymbol{M} = \hat{\chi}\boldsymbol{H}$ is proportional to the applied field $\boldsymbol{H}$, i.e. the magnetic susceptibility tensor $\hat{\chi}$ does not depend on $\boldsymbol{H}$. In diamagnetic materials the susceptibility is small, negative, and independent of temperature, whereas in paramagnetism the susceptibility is small, positive and proportional to the inverse of the temperature. In ordered magnetic solids, all individual spins are parallel in ferromagnetic materials, or can be divided in different sublattices (usually two), in each sublattice the spins being parallel. If the magnetic moments of different sublattices compensate each other, the material is called antiferromagnetic, otherwise we have a ferrimagnetic material. Ferromagnetic materials are composed of domains of spontaneous magnetization, separated by walls, each domain being magnetized at saturation, with all spins aligned in a certain direction that varies randomly from one domain to another. If an external magnetic field is applied, the walls between domains can be shifted and/or the spins in each domain can rotate. When a very small field is applied on a ferromagnetic material with an initially zero magnetization, the walls between domains are shifted (spatially) irreversibly from an energy minimum to another, so that when the field is removed the material attains a nonvanishing remanent magnetization. This magnetization value can be maintained indefinitely at low applied magnetic fields, and depends on the history of the probe. To bring again the magnetization at zero, a magnetic field – the coercitive magnetic field H_{C} – must be applied; this field is a measure of the necessary field to displace a wall over an energy barrier.

To study the properties of magnetically ordered solids, let us denote by $\boldsymbol{S}_j$ the spin on site $\boldsymbol{R}_j$ of a crystal in which the only relevant degrees of freedom are the individual atomic spins; the rectangular components of the quantum mechanical operator $\boldsymbol{S}_j$ satisfy the commutation rule

$$[S_{j\alpha}, S_{l\beta}] = \mathrm{i}\hbar\delta_{jl} \sum\nolimits_{\gamma} \varepsilon_{\alpha\beta\gamma} S_{j\gamma}, \tag{1.69}$$

where $\varepsilon_{\alpha\beta\gamma}$ is the completely antisymmetric Levi–Civita symbol. The operator $\boldsymbol{S}_j^2$ commutes with the so-called Heisenberg Hamiltonian that describes magnetically ordered solids (Callaway, 1991)

$$H_{\mathrm{H}} = -2\sum_{i>j} J_{ij}\boldsymbol{S}_i \cdot \boldsymbol{S}_j, \tag{1.70}$$

its eigenvalue $S(S+1)$ – independent of j – determining the magnitude of the spin on a certain site. The exchange parameters J_{ij} depend only on the distance $|\boldsymbol{R}_i - \boldsymbol{R}_j|$ between sites i and j. The calculation of exchange parameters depends on the mechanism responsible for the magnetic ordering: for example, there can be direct coupling of spins of well-separated atoms at large distance, or a superexchange in insulating crystals containing a single species of transition metal atoms. Different models describe these mechanisms, and subsequently the exchange parameters, as described in Callaway (1991).

The properties of a system of spins coupled by the Heisenberg Hamiltonian can be most easily studied in the molecular field theory. For ferromagnetic crystals, the molecular theory assumes that coupling occurs only between atoms and their Z nearest-neighbors. Then the interaction is described by a single-atom Hamiltonian (Callaway, 1991)

$$H = -2J\boldsymbol{S}_i \cdot \sum_{j=1}^{Z} \boldsymbol{S}_j, \tag{1.71}$$

i.e. we can consider that the atom is embedded in an effective magnetic field $\boldsymbol{H}_{\mathrm{eff}}$ such that $H = -g\mu_{\mathrm{B}}\boldsymbol{S}_i \cdot \boldsymbol{H}_{\mathrm{eff}}$ where g is the gyromagnetic ratio (Landé factor) and μ_{B} denotes the Bohr magneton.

In the molecular field theory $\boldsymbol{H}_{\mathrm{eff}} = (2ZJ / g\mu_{\mathrm{B}})\langle \boldsymbol{S}_j \rangle = (2ZJ / Ng^2\mu_{\mathrm{B}}^2)\boldsymbol{M}$, proportional to the magnetization of the crystal $\boldsymbol{M} = Ng\mu_{\mathrm{B}}\langle \boldsymbol{S}_j \rangle$, where N is the number of atoms per unit volume. When an external field $\boldsymbol{H}_0$ is added to $\boldsymbol{H}_{\mathrm{eff}}$, the total magnetization is $M = Ng\mu_{\mathrm{B}}SB_S(x)$, with

$$B_S(x) = [(2S+1)/2S]\coth[x(2S+1)/2S] - (1/2S)\coth(x/2S) \tag{1.72}$$

the Brillouin function, and $x = g\mu_{\mathrm{B}}SH / k_{\mathrm{B}}T$, where H is the magnitude of the total field. The total field defines the z axis, along which the spins tend to line up. If $\boldsymbol{H}_{\mathrm{eff}} = 0$, the same theory describes an isolated magnetic ion in a crystal.

When the effective magnetic field is parallel to the external field, so that $H = H_0 + H_{\mathrm{eff}}$ the argument of the Brillouin function can be written as

$$x = g\mu_{\mathrm{B}}SH_0 / k_{\mathrm{B}}T + 2ZSJM / (Ng\mu_{\mathrm{B}}k_{\mathrm{B}}T). \tag{1.73}$$

Equations (1.72) and (1.73) can be simultaneously satisfied, i.e. there is spontaneous magnetization, only when the temperature is lower than a critical value $T_{\mathrm{c}} = (2/3)[JZS(S+1)/k_{\mathrm{B}}]$. The magnetic susceptibility $\chi = M / H_0$ should have the form $\chi = C/(T - T_{\mathrm{c}})$, with Curie constant $\mathrm{C} = Ng^2\mu_{\mathrm{B}}^2 S(S+1)/3k_{\mathrm{B}}$; however, in most materials, the magnetic susceptibility obeys the empirical rule $\chi = C/(T - \Theta_c)$ with $\Theta_c = T_c + T_0$ the paramagnetic Curie temperature. It differs in general from T_c through a phenomenological parameter T_0. For temperatures higher than T_c the material is paramagnetic – it has no spontaneous magnetization, its spins tending to align along the applied external field.

Antiferromagnetic ordering of spins occurs when the negative values of the exchange parameter predominate. The simplest model of an antiferromagnet assumes that the crystal is composed of two sublattices, in each of them the spins being parallel (ferromagnetically ordered), but antiparallel compared to the spins in the other sublattice. We must therefore introduce two exchange parameters: J_{11}, which describes the interaction between nearest-neighbor spins on the same sublattice, and J_{12}, characterizing the interaction between neighboring spins on different lattices (Callaway, 1991). The total magnetic field that acts on the ith sublattice is now $\boldsymbol{H}_i = \boldsymbol{H}_0 + \sum_j \gamma_{ij} \boldsymbol{M}_j$, where $\gamma_{ij} = 2Z_{ij}J_{ij}/(N/n)g^2\mu_B^2$ with n (=2) the number of sublattices, and Z_{ij} the number of neighbors of an atom on sublattice i, situated on sublattice j and which act on it through J_{ij}. If the two lattices are identical, and if the nearest-neighbors of an atom on one lattice are all on the other, from symmetry considerations it follows that $Z_{12} = Z_1$ – the number of nearest-neighbors, $Z_{11} = Z_{22} = Z_2$ – the number of second-nearest-neighbors, $J_{12} = J_1$, $J_{11} = J_{22} = J_2$, $\gamma_{12} = \gamma_{21} = 4Z_1J_1/Ng^2\mu_B^2$ and $\gamma_{11} = \gamma_{22} = 4Z_2J_2/Ng^2\mu_B^2$. As for the ferromagnetic case, $M_i = (N/n)g\mu_B SB_S(x_i)$ with $x_i = g\mu_B SH_i/k_BT$. At high temperatures the material is paramagnetic, the transition to the antiferromagnetic regime occurring for the Néel temperature $T_N = [2S(S+1)/3k_B](Z_2J_2 - Z_1J_1)$. T_N must be positive, so that if $J_1 < 0$, J_2 must be either positive or negative, but with $|J_2| < (Z_1/Z_2)|J_1|$. Otherwise, this simple model cannot be used.

1.6.1 Magnons

Magnons are introduced in a quantum treatment of the Heisenberg Hamiltonian (see Callaway (1991)). For each spin operator $\boldsymbol{S}_i$, raising and lowering operators S_i^+, S_i^- are introduced as $S_i^{\pm} = S_i^x \pm iS_i^y$; they satisfy the commutation relations

$$[S_i^z, S_j^{\pm}] = \pm\hbar\delta_{ij}S_i^{\pm}, \quad [S_i^+, S_j^-] = 2\hbar\delta_{ij}S_i^z, \tag{1.74}$$

operators for different sites commuting. In terms of these operators, the Heisenberg Hamiltonian can be written as

$$H_H = -2\sum_{i>j} J_{ij}\boldsymbol{S}_i \cdot \boldsymbol{S}_j = -\sum_{i,j} J_{ij}(S_i^z S_j^z + S_i^- S_j^+). \tag{1.75}$$

In a ferromagnetic crystal, with positive exchange parameters, the ground state $|0\rangle$ of the coupled spins is that in which each S_i^z attains its maximum value S. The energy of this highly degenerate state is $E_0 = -S^2\sum_{i,j} J_{ij}$. In the absence of an external field, which would also specify the z direction, the degeneracy of the ground state is $(2NS+1)$, equal to the number of possible orientations of the total spin with the same energy. The lowest excited state of the system corresponds to the situation in which a single spin deviates from the complete alignment. A set of orthonormal states can be defined for the situation when the spin at the lth site has been lowered: $|l\rangle = (1/\sqrt{2S})S_l^- |0\rangle$. The matrix elements calculated between two such states are

$$\langle n| H_{\mathrm{H}} |l\rangle = -\delta_{nl}(S^2 \Sigma_{ij} J_{ij} - 2S\Sigma_j J_{nj}) - 2SJ_{nl}. \tag{1.76}$$

An eigenfunction $|E\rangle$ of the Hamitonian, with corresponding energy E can be expanded over these localized states as $|E\rangle = \Sigma_l \phi(\boldsymbol{R}_l)|l\rangle$, where the coefficients $\phi(\boldsymbol{R}_l)$ satisfy the set of equations

$$2S\Sigma_l(\delta_{nl}\Sigma_j J_{lj} - J_{nl})\phi(\boldsymbol{R}_l) = (E - E_0)\phi(\boldsymbol{R}_n). \tag{1.77}$$

The wavelike solutions $\phi(\boldsymbol{R}_l) \approx \exp(\mathrm{i}\boldsymbol{k}\cdot\boldsymbol{R}_l)$ of the above set of equations are called spin waves or magnons. They are not spin deviations localized on a certain site, but spin deviations spread throughout the system. Their dispersion relation, in the hypothesis that $J_{nl} = J(|\boldsymbol{R}_n - \boldsymbol{R}_l|)$ is

$$E(k) = E_0 + 2S\Sigma_l J(\boldsymbol{R}_l)[1 - \exp(\mathrm{i}\boldsymbol{k}\cdot\boldsymbol{R}_l)]. \tag{1.78}$$

As for phonons, the wavevector $\boldsymbol{k}$ is restricted to the first Brillouin zone.

A spin wave state with a given $\boldsymbol{k}$ has thus the form

$$|\boldsymbol{k}\rangle = (1/\sqrt{N})\Sigma_l \exp(\mathrm{i}\boldsymbol{k}\cdot\boldsymbol{R}_l)|l\rangle \tag{1.79}$$

and is an eigenstate of the squared total spin operator $\boldsymbol{S}^2 = (\Sigma_i \boldsymbol{S}_i)^2 = \Sigma_{i,j}\boldsymbol{S}_i\cdot\boldsymbol{S}_j$ with eigenvalue $[NS(NS-1) + 2NS\delta_{\boldsymbol{k},0}]$. In the presence of an external magnetic field $\boldsymbol{H}_0 = H_0\hat{z}$ an additional term $H_{\mathrm{m}} = -g\mu_{\mathrm{B}}H_0\Sigma S_{lz}$ must be added to the Hamiltonian, which changes the dispersion relation to

$$E(k) = E_0 + 2S\Sigma_l J(\boldsymbol{R}_l)[1 - \exp(\mathrm{i}\boldsymbol{k}\cdot\boldsymbol{R}_l)] + g\mu_{\mathrm{B}}H_0. \tag{1.80}$$

Still another term $H_{\mathrm{d}} = (g^2\mu_{\mathrm{B}}^2/2)\Sigma_{i,j}(1/\boldsymbol{R}_{ij}^3)[\boldsymbol{S}_i\cdot\boldsymbol{S}_j - (\boldsymbol{S}_i\cdot\boldsymbol{R}_{ij})(\boldsymbol{S}_j\cdot\boldsymbol{R}_{ij})/\boldsymbol{R}_{ij}^3]$ accounts for the dipole-dipole interaction of spins. The dispersion relation in this case may be put in the form

$$E(\boldsymbol{k}) = [E(\boldsymbol{k}) - E_0 + g\mu_{\mathrm{B}}H_0]\sqrt{1 + \vartheta(\boldsymbol{k})\sin^2\theta_k}, \tag{1.81}$$

where $\vartheta(\boldsymbol{k}) = 4\pi NSg^2\mu_{\mathrm{B}}^2/[E(\boldsymbol{k}) - E_0 + g\mu_{\mathrm{B}}H_0]$ and θ_k is the angle between the wavevector of the magnon, $\boldsymbol{k}$, and the magnetization.

The quantization of magnons is more difficult than for phonons since the commutation rules for the spin operators are identical to neither the Bose nor Fermi field. However, at low temperatures, when a small number of spins are excited, they can be approximated as independent bosons. A correspondence between the localized spin operators $\boldsymbol{S}_l$ and boson operators can be introduced as $S_l^+ \to \sqrt{2S}a_l$, $S_l^- \to \sqrt{2S}a_l^+$, $S_l^z \to S - a_l^+ a_l$, where $[a_j, a_l^+] = \delta_{jl}$, $[a_j^+, a_l^+] = [a_j, a_l] = 0$, such that the creation and annihilation operators for magnons are, respectively,

$$a^+(\boldsymbol{k}) = (1/\sqrt{N})\Sigma_l \exp(-\mathrm{i}\boldsymbol{k}\cdot\boldsymbol{R}_l)a_l^+, \quad a(\boldsymbol{k}) = (1/\sqrt{N})\Sigma_l \exp(\mathrm{i}\boldsymbol{k}\cdot\boldsymbol{R}_l)a_l. \tag{1.82}$$

They satisfy the commutation relations $[a(\boldsymbol{k}), a^+(\boldsymbol{q})] = \delta_{\boldsymbol{kq}}$, $[a(\boldsymbol{k}), a(\boldsymbol{q})] = [a^+(\boldsymbol{k}), a^+(\boldsymbol{q})] = 0$, and the state $|\boldsymbol{k}\rangle = a^+(\boldsymbol{k})|0\rangle$ is indeed an eigenstate of the Hamiltonian. However, states in which more than one spin are excited, for example the two spin waves, are not eigenstates of the Hamiltonian; the interaction between spins must be accounted for. The approximation with a system of bosons, which works at low temperatures, allows us to identify the average number of magnons with wavevector $\boldsymbol{k}$ at a temperature T with the Bose–Einstein distribution function

$$\langle a^+(\boldsymbol{k})a(\boldsymbol{k})\rangle_T = 1/\{\exp[(E(\boldsymbol{k}) - E_0)/k_{\mathrm{B}}T] - 1\}. \tag{1.83}$$

Creation and annihilation operators can also be introduced for antiferromagnetic interaction, in terms of the respective operators on each sublattice (see Callaway (1991)). The treatment of this case is much more difficult.

1.7 Excitons

The exciton is a neutral quasiparticle that satisfies the Bose statistics and describes the electron-hole bound state due to mutual Coulomb attraction; it has no contribution to electric conduction. If the electron-hole pair has a small radius, the exciton is termed Frenkel and corresponds to a bound electron-hole pair localized on an atom of the crystal. For bound states with radius much larger than the lattice constant, the exciton is termed Wannier–Mott and is no longer described in terms of states of an isolated atom, but in terms of Bloch states of the electron and hole. The Frenkel and Wannier–Mott excitons are limit models of a real exciton.

1.7.1 Frenkel Excitons

These appear in molecular or ionic crystals in which the forces between atoms are weak in comparison to the forces between atoms and electrons in the molecule. In this case the total Hamiltonian of the crystal with one atom per elementary cell is

$$H = \sum_n H_n + (1/2)\sum_{n \neq m} V_{nm}, \tag{1.84}$$

where the indices n, m denote elementary cells, H_n is the Hamiltonian of the atom and V_{nm} the operator which describes the Coulomb interaction between the n and m atoms. By denoting with φ_n^0 the wavefunction of the unexcited atom of energy W_0, and with φ_n^f the wavefunction of the atom in the excited state (exciton) f of energy W_f, the wavefunctions of the crystal in unexcited and excited states, respectively, are

$$\Psi_0 = \prod_n \varphi_n^0, \quad \Psi_f = \sum_n a_n \phi_n^f, \tag{1.85}$$

where

$$\phi_n^f = \varphi_n^f \prod_{m \neq n} \varphi_m^0. \tag{1.86}$$

φ_n^f are orthonormal. If Ψ_f is an eigenfunction of the translation operator, the coefficients a_n are given by $a_m = N^{-1/2} \exp(\mathrm{i}\boldsymbol{k} \cdot \boldsymbol{R}_m)$ where N is the total number of atoms in the crystal.

The energy of an exciton with wavevector $\boldsymbol{k}$ is then $E(\boldsymbol{k}) = E_f(\boldsymbol{k}) - E_0$ where $E_f(\boldsymbol{k}) = \langle \Psi_f(\boldsymbol{k}) | H | \Psi_f(\boldsymbol{k}) \rangle$ and $E_0 = \langle \Psi_0 | H | \Psi_0 \rangle$ are the energies of the crystal in the excited and unexcited states, respectively. Calculations show that

$$E(\boldsymbol{k}) = W_f - W_0 + D_f + \sum_{m,n} M_{mn}^{f0} \exp[\mathrm{i}\boldsymbol{k} \cdot (\boldsymbol{R}_m - \boldsymbol{R}_n)] \tag{1.87}$$

is a continuous function, with D_f describing the change of the interaction energy of the crystal at the transition from the unexcited to the excited state (it is independent of $\boldsymbol{k}$).

$$M_{mn}^{f0} = \int (\varphi_m^0)^* (\varphi_n^f)^* V_{mn} \varphi_m^f \varphi_n^0 \mathrm{d}V \tag{1.88}$$

are the matrix elements of the excitation transfer operator from atom m to atom n. The interatomic interaction has as a result the transfer of atomic terms in an excitonic band $E(\boldsymbol{k})$. The $\boldsymbol{k}$ dependence of the band and its width are completely determined by the value of M_{mn}^{f0} and by the crystal structure.

If the atomic state has degeneracy r, instead of one excitonic band there are r – a phenomenon known as Bethe splitting. Analogously, if there are σ atoms in an elementary cell, even in the nondegenerate case, there are σ excitonic bands – an effect known as Davydov splitting.

1.7.2 Wannier–Mott Excitons

This type of exciton appears in semiconductors with covalent bonding (group IV materials or ionic III-V compounds). When the Coulomb interaction is strong, the electron-hole pair must be considered as an entity, denoted as an exciton. Because these excitons are spread over many cells, the notion of effective mass has meaning.

If there were no Coulomb interaction, the separate movement of electrons and holes in the exciton would be described by (the zero energy is taken at the upper edge of the valence band) (Basu, 1997):

$$[E_g - (\hbar^2/2m_c)\nabla_c^2]\varphi_c(\boldsymbol{r}_c) = E_c\varphi_c(\boldsymbol{r}_c),$$
$$-(\hbar^2/2m_v)\nabla_v^2\varphi_v(\boldsymbol{r}_v) = E_v\varphi_v(\boldsymbol{r}_v), \tag{1.89}$$

and the wavefunction of the exciton would be $\varphi(\boldsymbol{r}_c,\boldsymbol{r}_v) = \varphi_c(\boldsymbol{r}_c)\varphi_v(\boldsymbol{r}_v)$.

When Coulomb interaction is taken into account, $\varphi(\boldsymbol{r}_c,\boldsymbol{r}_v) \neq \varphi_c(\boldsymbol{r}_c)\varphi_v(\boldsymbol{r}_v)$ and it satisfies the equation

$$[E_g - (\hbar^2/2m_c)\nabla_c^2 - (\hbar^2/2m_v)\nabla_v^2 - (e^2/4\pi\varepsilon|\boldsymbol{r}_c - \boldsymbol{r}_v|)]\varphi(\boldsymbol{r}_c,\boldsymbol{r}_v) = E\varphi(\boldsymbol{r}_c,\boldsymbol{r}_v). \tag{1.90}$$

Note that, whether the simple description of Coulomb interaction in (1.90) is satisfactory for excitons in bulk materials, and even in many cases of low-dimensional structures, sometimes it is needed to consider both long-range and short-range terms in the Coulomb potential. Since these corrections are necessary mainly in low-dimensional structures, we postpone their introduction until then. The solution $\varphi(\boldsymbol{r}_c,\boldsymbol{r}_v)$ of (1.90) can be separated into a term corresponding to the movement of the mass center, described by the coordinate $\boldsymbol{R} = (m_c\boldsymbol{r}_c + m_v\boldsymbol{r}_v)/M$ with $M = m_c + m_v$, and a term corresponding to the relative motion, described by $\boldsymbol{r} = \boldsymbol{r}_c - \boldsymbol{r}_v$. In the new coordinates (1.90) becomes

$$[E_g - (\hbar^2/2M)\nabla_{\boldsymbol{R}}^2 - (\hbar^2/2m_r)\nabla_{\boldsymbol{r}}^2 - (e^2/4\pi\varepsilon r)]\varphi(\boldsymbol{r},\boldsymbol{R}) = E\varphi(\boldsymbol{r},\boldsymbol{R}), \tag{1.91}$$

where m_r is the relative mass given by $m_r^{-1} = m_c^{-1} + m_v^{-1}$. Assuming a separable form for the solution $\varphi(\boldsymbol{r},\boldsymbol{R}) = g(\boldsymbol{R})f(\boldsymbol{r})$, we obtain two distinct equations. One,

$$-(\hbar^2/2M)\nabla_{\boldsymbol{R}}^2 g(\boldsymbol{R}) = E_1 g(\boldsymbol{R}) \tag{1.92}$$

represents the movement of a free particle of mass M and energy $E_1 = (\hbar^2K^2)/2M$, where $\boldsymbol{K} = \boldsymbol{k}_c + \boldsymbol{k}_v$ is the total wavenumber of the exciton. The other equation is

$$[E_g - (\hbar^2/2m_r)\nabla_{\boldsymbol{r}}^2 - (e^2/4\pi\varepsilon r)]f(\boldsymbol{r}) = E_2 f(\boldsymbol{r}), \tag{1.93}$$

where $E_2 - E_g$ is the energy of a particle with mass m_r around a point which attracts it through the Coulomb force. So, as for the hydrogen atom,

$$E_2 - E_g = -(m_r e^4 / 2\varepsilon^2\hbar^2)/n^2 = -E_b / n^2 \tag{1.94}$$

describes the discrete energy spectrum, with energy levels inside the bandgap (see Fig. 1.6). The Coulomb interaction leads to the formation of a series of discrete levels (for excitons) under the continuum two-particle energy spectrum. Unlike other Coulomb bound states, such as impurity states (see Sect. 3.2), only the relative motion of the electron and hole is affected by the Coulomb attraction; the excitons are not localized around a particular position in the crystal, the center-of-mass motion of the electron-hole pair being described by a completely delocalized plane-wave wavefunction. Another difference between excitons and charged impurity states is that the former have a finite lifetime due to the possibility of (radiative or nonradiative) recombination of the electron and hole. Although the energy levels of excitons are similar to those in a hydrogen atom, the interaction between electrons and holes can be screened in the solid by the valence electrons of other atoms, no such screening being present in the case of the interaction between the nucleus and the electron in a hydrogen atom.

The electron-hole binding energy E_b is given by the difference between the energy necessary for the creation of a free electron-hole pair and of an exciton. It is also possible to have $E_2 - E_g > 0$, a situation that corresponds, as in the case of a hydrogen atom, to a continuous spectrum.

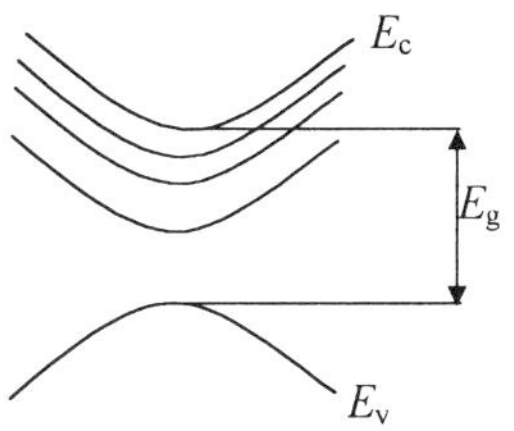

Fig. 1.6. The discrete spectrum of exciton energy

The radius of the exciton has a similar definition as the Bohr radius of the hydrogen atom:

$$a_{ex} = \varepsilon\hbar^2 / e^2 m_r = a_B (m / m_r)\varepsilon_r . \tag{1.95}$$

For example, in GaAs $a_{ex} = 120$ Å and in Ge $a_{ex} = 80$ Å, both much greater than the respective lattice constants, which justifies the use of the macroscopic characteristic of the medium ε. In semiconductors with large dielectric constants E_b is of the order of 10^{-2} eV, so that excitons break easily (for example, by absorbing phonons) even at moderate temperatures and decompose into a free electron and a free hole. In molecular crystals (benzene, naphtalene, anthracene, etc.), not treated in this book, $E_b \cong 1$ eV.

In most semiconductors the valence band is split into two or more subbands due to lower crystal symmetry and spin-orbit coupling.

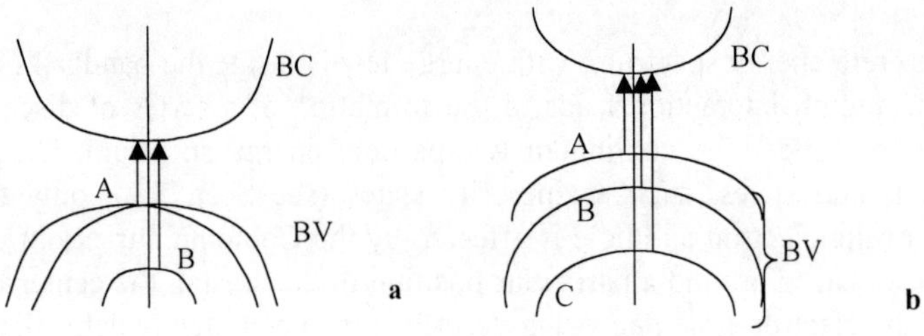

Fig. 1.7. Types of transitions in materials with split valence subbands

In $GaInP_2$ for example, a crystal with rhombohedral point group in the ordered phase, the structure of the valence band looks like that in Fig. 1.7a. The two higher valence bands are degenerate at $\boldsymbol{k} = 0$, the transitions between them and the conduction band being called A transitions. The associated excitons are also called A excitons. From the two higher valence bands the one with a smaller curvature is called the heavy hole subband, the other being the light hole subband. The lowest-lying valence band is separated from the other two at $\boldsymbol{k} = 0$ by the spin-orbit split-off parameter Δ. The transitions between the conduction band and this valence subband, including exciton transitions, are called B transitions.

In CdS the highest valence band is threefold split due to the spin-orbit coupling and the internal hexagonal crystal field, the exciton series between these three subbands of the valence band and the conduction band being denoted by A, B, C (see Fig. 1.7b). The A valence subband has Γ_9 symmetry, whereas the B and C subbands have Γ_7 symmetry. The separations between the three valence subbands are determined by the values of the crystal-field and spin-orbit splitting parameters. The same band structure is also found in GaN.

1.7.3 Interactions Between Excitons

The simplest method to create excitons is by absorption of electromagnetic radiation. For powerful laser sources, the exciton concentration can reach $N_{ex} \cong 10^{24}$ m^{-3}, the distance between excitons ($\approx N_{ex}^{-1/3}$) becoming comparable with a_{ex}. In this situation the excitons can no longer be approximated as independent particles and we have a nonideal exciton gas (in interaction). By a further increase of exciton concentration and at low temperatures (comparable to liquid He temperature) the exciton gas condenses and excitonic droplets appear with linear dimensions of hundreds of micrometers. The light suffers an additional diffraction process on these droplets. The condensation of free excitons is not a Bose–Einstein condensation, but can be seen as a phase transition between the free electron gas and electron-hole drop. The origin of this condensation is the difference between the total number of repulsive interactions between carriers in system with N electron-hole pairs, equal to $N(N-1)$, and the total number of attractive interactions $N^2 > N(N-1)$. Electron-hole drops are usually located in regions of maximum shear stress (Curie, 1979).

1.7.4 Bound Excitons and Excitonic Complexes

Excitons are bound when at least one of its constituents, the electron or hole, is trapped in a potential minimum. This situation is encountered, for example, in doped semiconductors, where excitons can be trapped by ionized impurities (Curie, 1979); especially in II-VI semiconductors the excitons are bound to hydrogenic impurities. Moreover, excitons can be trapped by defects in the material. The fraction of excitons which are trapped depends on the excitation intensity and frequency: at low excitation intensities and for radiation with energy very close to the exciton energy such that the kinetic energy of the exciton is small, almost all excitons created by irradiation become trapped. Biexcitons can form in semiconductors with a large charge density, by bonding two excitons in a so-called excitonic molecule, or biexciton. Biexcitons can also form by two-photon absorption, at large radiation intensities. If the semiconductor is n-doped or p-doped, charged excitons appear, called trions. They are formed, respectively, from a negative or a positive charge bounded to an exciton.

1.8 Polarons

In any crystal, longitudinal optical modes involve displacements in opposite directions of the constituents of the unit cell. In ionic crystals these are oppositely charged ions, so that the excitation of longitudinal optical modes produces long-range dipole fields that perturb the motion of electrons in the system. The coupling between the electron and the longitudinal optical phonon produces a new quasiparticule – the polaron. The electron-phonon interaction is described by the Fröhlich Hamiltonian (Callaway, 1991):

$$H = \boldsymbol{p}^2 / 2m_{\mathrm{eff}} + \hbar\omega\Sigma_q b_q^+ b_q + (\mathrm{i}/\sqrt{N})\Sigma_q V(\boldsymbol{q})[b_q^+ \exp(-\mathrm{i}\boldsymbol{q}\cdot\boldsymbol{r}) - b_q \exp(\mathrm{i}\boldsymbol{q}\cdot\boldsymbol{r})], \tag{1.96}$$

where $V(q) = (4\pi\alpha)^{1/2}(\hbar\omega / q)[\hbar/(2m_{\mathrm{eff}}\Omega^2\omega)]^{1/4}$, N is the number of unit cells, $\alpha = (e^2 / 8\pi\hbar\omega\varepsilon_0)(2m_{\mathrm{eff}}\omega/\hbar)^{1/2}[(1/\varepsilon_{\mathrm{high}}) - (1/\varepsilon_{\mathrm{low}})]$ is the dimensionless coupling constant, and m_{eff} the effective electron mass. $\varepsilon_{\mathrm{high}}$, $\varepsilon_{\mathrm{low}}$ are the high- and low-frequency dielectric constants. The electron-phonon interaction can be treated quite easily in two extreme limits: the weak coupling limit, in which α is small – a situation encountered in many semiconductors, and the strong coupling case, for very large α. The intermediate case is far more difficult.

In the weak coupling limit the self-energy and effective mass of the polaron is calculated using perturbation theory. Since $\alpha \approx \omega^{-1/2}$, the weak coupling limit corresponds to high frequencies, where the polarization follows the motion of the electron. In the noninteracting system the electron energy is $E_0(\boldsymbol{k}) = \hbar^2 \boldsymbol{k}^2 / 2m$; the interaction is modeled as the emission of a phonon of wavevector $\boldsymbol{q}$ followed by its reabsorption. The electron energy changes to

$$E(\boldsymbol{k}) = E_0(\boldsymbol{k}) - (2m_{\text{eff}} / \hbar^2 N)\Sigma_q |V(\boldsymbol{q})|^2 /(-2\boldsymbol{k} \cdot \boldsymbol{q} + \boldsymbol{q}^2 + 2m_{\text{eff}}\omega / \hbar) \quad (1.97)$$
$$= -\alpha\hbar\omega + (\hbar^2 \boldsymbol{k}^2 / 2m_{\text{eff}})(1 - \alpha / 6).$$

The net effect is the lowering of the ground state with $-\alpha\hbar\omega$ and the effective mass, in first order in α, is modified to $m_{\text{pol}} = m_{\text{eff}}(1 + \alpha / 6)$. In the limit of small $\boldsymbol{k}$, it is also found that the average number of phonons accompanying a slow electron is $\alpha / 2$. The polaron effect is stronger the lower the dimensionality of the structure; for example, it is about 3 times stronger in 2D than in bulk.

The strong coupling limit corresponds to low frequencies, where the electron becomes bound in a potential well, and it is treated by a variational method. The wavefunction is assumed to be $\psi(\boldsymbol{r} - \boldsymbol{X}, \boldsymbol{X}) = \Theta(\boldsymbol{r} - \boldsymbol{X})\Phi(\boldsymbol{X})$ where $\boldsymbol{r}$ is the electron coordinate, $\Theta(\boldsymbol{r} - \boldsymbol{X})$ is a bound-state wavefunction for the electron centered at some point $\boldsymbol{X}$, which can be viewed as the center of the potential well in which the electron traps itself. $\Phi(\boldsymbol{X})$ describes the state of the polarization field of the lattice. Other wavefunctions can be chosen to account for the translational invariancy. The polaron binding energy in this model is $E_{\text{pol}} = -(\alpha^2 / 3\pi)\hbar\omega$.

The polaron effect in solids depends on the model used to calculate the electron bands. For electrons described by the tight-binding theory, the most appropriate model for the electron-lattice interaction is the small polaron. In the small-polaron model, valid for strong interaction, an excess electron or hole on a site $\boldsymbol{R}_\mu$ alters the charge of the site, which in turn induces a lattice deformation to accommodate for this charge modification. The final effect is band narrowing; the bandwidth decreases with increasing temperature, leading to hopping transport. For a single particle strongly coupled to a deformable lattice, as in semiconductors or insulators, the total Hamiltonian is $H = H_{\text{el}} + H_{\text{ph}} + H_{\text{el-ph}} + V(\boldsymbol{r})$ where the electron Hamiltonian is $H_{\text{el}} = p^2 / 2m + U(\boldsymbol{r} - \boldsymbol{R}_\mu)$, $U(\boldsymbol{r})$ (in the tight-binding approximation) being the potential of an electron bound to site 0. The Hamiltonian of the system of phonons is $H_{\text{ph}} = \Sigma_q \hbar\omega(\boldsymbol{q})[a^+(\boldsymbol{q})a(\boldsymbol{q}) + 1/2]$ $= (1/2)\Sigma_q \hbar\omega(\boldsymbol{q})[Q^2(\boldsymbol{q}) + P^2(\boldsymbol{q})]$ and the electron-phonon interaction is modeled by $H_{\text{el-ph}} = \Sigma_q [P(\boldsymbol{q})g(\boldsymbol{q}, \boldsymbol{r}) + Q(\boldsymbol{q})f(\boldsymbol{q}, \boldsymbol{r})]$ where *f*, *g* are functions which can be determined but whose form is not essential. The potential $V(\boldsymbol{r}) = \Sigma_{\nu \neq \mu} U(\boldsymbol{r} - \boldsymbol{R}_\nu)$ that acts on the electron from other sites is considered as a perturbation. The presence of this perturbation leads to a dispersion relation of the form

$$E(\boldsymbol{k}) = E_0 + \Sigma_\rho J(\boldsymbol{R}_\rho)\exp(\mathrm{i}\boldsymbol{k} \cdot \boldsymbol{R}_\rho), \quad (1.98)$$

where $J(\boldsymbol{R}_\rho)=\int\Psi(\boldsymbol{r})\sum_{\mu\neq 0}V(\boldsymbol{r}-\boldsymbol{R}_\mu)\Psi(\boldsymbol{r}-\boldsymbol{R}_\rho)\mathrm{d}\boldsymbol{r}$, with $\Psi(\boldsymbol{r},\boldsymbol{Q})=\Phi(\boldsymbol{r})X(\boldsymbol{Q})$ the eigenfunction of the unperturbed Hamiltonian $(H_{\mathrm{el}}+H_{\mathrm{ph}}+H_{\mathrm{el-ph}})\Psi(\boldsymbol{r},\boldsymbol{Q})$ $=E\Psi(\boldsymbol{r},\boldsymbol{Q})$. At low temperatures, when all phonon oscillators are in their lowest state, $J(\boldsymbol{R}_\rho)=M(\boldsymbol{R}_\rho)\exp(-S)$ decreases from its value $M(\boldsymbol{R}_\rho)$ $=\int\Phi(\boldsymbol{r})\sum_{\mu\neq 0}V(\boldsymbol{r}-\boldsymbol{R}_\mu)\Phi(\boldsymbol{r}-\boldsymbol{R}_\rho)\mathrm{d}\boldsymbol{r}$ when no electron-phonon interaction is present, the factor S being $S=\sum_q[\omega(\boldsymbol{q})/(2\pi)][f(\boldsymbol{q},0)-f(\boldsymbol{q},\boldsymbol{R}_\rho)]$. At higher temperatures the reduction in the bandwidth is also a function of the degree of occupation of the modes, decreasing rapidly with temperature from its maximum at $T=0$. Electron hopping is always accompanied by phonon emission or absorption.

The electron-phonon interaction can even lead to a static deformation of the lattice and of the electron charge distribution, called charge density wave (CDW). Although CDW can occur in principle also in materials with two- and three-dimensional band structures, they are characteristic of one-dimensional structures, encountered for example in organic semiconductors, layered transition metal compounds, or any other solids with strongly anisotropic Fermi surface. CDW can even be related to superconductivity. A review of the experimental state of affairs related to CDW can be found in Grüner (1988). The electron density in the presence of the CDW is modified with respect to that in a normal or ordinary metal, $\rho_0(\boldsymbol{r})$, by

$$\rho(\boldsymbol{r})=\rho_0(\boldsymbol{r})+\rho_1\cos(\boldsymbol{Q}\cdot\boldsymbol{r}+\phi), \tag{1.99}$$

the last term characterizing a condensate, a collective mode, whose wavevector modulus is $|\boldsymbol{Q}|=2k_{\mathrm{F}}$, with k_{F} the distance between the zone center and the Fermi surface in the direction of $\boldsymbol{Q}$. This perturbation favors the opening of a gap at k_{F} in the one-dimensional energy band of a perfect crystal. The atomic arrangement with lattice constant a is thus perturbed by the introduction of a new periodicity with wavelength $\lambda=\pi/k_{\mathrm{F}}$. The one-dimensional metal becomes unstable and a gap opens in the energy band, similar to the case of semicondutors. The system is, however, not a usual semiconductor since the CDW can move rigidly and carry current; impurities, interchain coupling or commensurability can pin the rigid motion of the CDW. The commensurability pinning occurs when the CDW wavelength is commensurate with the lattice, i.e. when $2k_{\mathrm{F}}/(2\pi/a)$ $=N/M$ is a rational number. In the commensurate phase the ion-displacement pattern repeats after M unit cells and is phase locked to the electron gas. The energy of the incommensurate CDW is independent of the phase ϕ.

Electrons can couple not only to phonons (delocalized lattice excitations), but also to localized oscillations of the lattice, which appear mainly around impurities. Which of these two types of interaction dominates depends also on the localization degree of the electron. Free electrons, or those localized at shallow impurities couple mainly to extended lattice vibrations, whereas charge carriers localized at deep centers interact mainly with localized lattice vibrations. In the latter case the electron-phonon coupling is no longer described by the Fröhlich

Hamiltonian, but the deformation potential must be accounted for. The physical origin of the deformation potential is that the impurities or defects present in the lattice lead to displacements of the equilibrium position of ions, so that the lattice becomes polarized. The net effect is equivalent to the introduction of a stress, which displace the electron energy of a state of wavevector $\boldsymbol{k}$ in band n with (Callaway, 1991)

$$\delta E_n(k) = C_n \Delta, \tag{1.100}$$

for electrons restricted to a small portion of the Brillouin zone, where Δ is the dilatation. For free electrons near the Fermi surface $C \cong -(2/3)E_{\mathrm{F}}$. The electron-phonon Hamiltonian in the more general case, where the dilatation is related to the displacement of lattice vibrations, is

$$H_{\text{el-ph}} = -\sum_{nk\boldsymbol{K}j}[D_{nj}(\boldsymbol{K})c_n^{+}(\boldsymbol{k}+\boldsymbol{K})c_n(\boldsymbol{k})b_j(\boldsymbol{K}) + D_{nj}(-\boldsymbol{K})c_n^{+}(\boldsymbol{k}-\boldsymbol{K})c_n(\boldsymbol{k})b_j^{+}(\boldsymbol{K})], \tag{1.101}$$

with $D_{nj}(\boldsymbol{K}) = i(\hbar/2\omega_j NM)^{1/2}\boldsymbol{K}\cdot\boldsymbol{e}_j(\boldsymbol{K})C_n$. As for the case of Fröhlich polarons, one important effect of electron-phonon interaction is the electron mass renormalization from m_{eff} to $m_{\text{pol}} = m_{\text{eff}}(1+\overline{\alpha})$, where for electrons near the Fermi surface $\overline{\alpha}$ is the average of

$$\alpha(\boldsymbol{k}) = [2\Omega/(2\pi)^3]\sum_j[\mathrm{d}S_{k'}/|\nabla E(\boldsymbol{k}')|][|D_j(\boldsymbol{k}-\boldsymbol{k}')|^2/\hbar\omega_j(\boldsymbol{k}-\boldsymbol{k}')] \tag{1.102}$$

over the Fermi surface. Typical values for $\overline{\alpha}$ are 10–30%.

Similar to the case of electrons, excitation localization due to strong coupling to phonons occurs also for excitons. In this mechanism the creation of an exciton results in a spontaneous relaxation of the lattice in the vicinity of the excitation, which produces a Stokes shift, taking the excited atom out of resonance with its neighbors. This, in turn, inhibits the motion of the exciton and may even result in complete energy localization, in the form of a self-trapped exciton. Self-trapping occurs if the exciton-phonon coupling constant G is larger than the nearest-neighbor interaction $H_{\boldsymbol{R}_{l'}\boldsymbol{R}_l}$. In terms of lifetime, self-trapping occurs if the time needed for lattice relaxation $\tau_{\mathrm{r}} \cong \hbar/G$ is much smaller than the jump time $\tau_{\mathrm{j}} \cong \hbar/H_{\boldsymbol{R}_{l'}\boldsymbol{R}_l}$. The self-trapped exciton may decay by radiative transitions, multiphonon decay, or energy transfer to another site. In the latter case the excitation moves in a migrating, hopping fashion, the transfer being thermally activated by absorption of phonons. The migration may end by energy transfer to an impurity acting as a sink.

1.9 Polaritons

In direct-gap semiconductors, excitons and photons can become coupled, a new quasiparticle – the polariton – being formed. For example, the center-of-mass energy dispersion curve $E = E_g + E_n + \hbar^2 K^2 / 2M$ of a Wannier exciton can intersect the light dispersion in the material, $\hbar\omega_k = \hbar ck / n(\omega)$, with n the background refractive index. At the intersection point the dispersions are degenerate, degeneracy that can be removed by exciton-photon interaction. The quasiparticles associated with this new dispersion law are called exciton-polaritons. It is also possible that phonon-polaritons form as a result of the interaction between transverse optical phonons and photons – we do not discuss this case here. Since $n^2(\omega) = \varepsilon(\omega)$, in the vicinity of a resonance, the polariton dispersion is given by (Haug and Koch, 1990)

$$c^2 k^2 / \omega^2 = \varepsilon_b [1 - \Delta / (\omega - \omega_0 + i\delta)], \tag{1.103}$$

where ε_b is the background dielectric constant and the right-hand side term represents the exciton dielectric function near resonance. Neglecting for the moment the damping constant ($\delta = 0$) the above dispersion relation for $\omega << \omega_0$ is photon-like: $\omega \cong ck / \sqrt{\varepsilon_b (1 + \Delta / \omega_0)}$, with a velocity smaller than that in the medium $c / \sqrt{\varepsilon_b}$. No interaction is present in this case. Between ω_0 and $\omega_0 + \Delta$ (1.103) has no real solutions, so there is a gap separating the lower and upper polariton branches. For $\omega >> \omega_0$ the dispersion relation is again photon-like, with a velocity $c / \sqrt{\varepsilon_b}$. Since ω_0 is the transverse exciton frequency and $\omega_0 + \Delta$ is the longitudinal exciton frequency, Δ is called the longitudinal-transverse splitting.

In terms of creation and annihilation operators, the generation of an exciton in the state ν with total momentum $\boldsymbol{K}$ is described by the operator

$$B^+_{\nu,\boldsymbol{K}} = \sum_k \psi_\nu (\boldsymbol{k} - \boldsymbol{K}/2) \alpha^+_{\boldsymbol{k}} \beta^+_{\boldsymbol{K}-\boldsymbol{k}}, \tag{1.104}$$

where ψ_ν is the Fourier transform of the exciton wavefunction for the relative motion, and α^+_k, β^+_k are creation operators of an electron in the conduction band with wavevector $\boldsymbol{k}$, and a hole in the valence band with wavevector $\boldsymbol{k}$, respectively. With this notation, the total exciton-photon Hamiltonian which includes the Hamiltonians of free excitons, photons, and the interaction, is

$$H = \hbar \sum_{q,\nu} \omega_{ex,\nu q} B^+_{\nu q} B_{\nu q} + \hbar \sum_q \omega_q a^+_q a_q - i\hbar \sum_{q,\nu} g_{\nu q} (B^+_{\nu q} b_q - B_{\nu q} b^+_q), \tag{1.105}$$

if the nonresonant interaction terms are small. It can be diagonalized to

$$H = \hbar \sum_q \Omega_q p^+_q p_q \tag{1.106}$$

by introducing polariton annihilation and creation operators as

$$p_q = u_q B_q + v_q a_q, \quad p_q^+ = u_q^* B_q^+ + v_q^* a_q^+. \tag{1.107}$$

They obey the Bose–Einstein commutation rules if $|u_q|^2 + |v_q|^2 = 1$.

The polariton frequencies for the upper and lower branches are

$$\Omega_{q,1,2} = (1/2)[(\omega_{\text{ex},q} + \omega_q) \pm \sqrt{(\omega_{\text{ex},q} - \omega_q)^2 + 4g_q^2}, \tag{1.108}$$

the coefficients u, v being for the upper and lower branches, respectively,

$$u_{q,1} = \sqrt{\frac{\Omega_{q,1} - \omega_q}{2\Omega_{q,1} - \omega_{\text{ex},q} - \omega_q}}, \quad v_{q,1} = \mathrm{i}\sqrt{\frac{\Omega_{q,1} - \omega_{\text{ex},q}}{2\Omega_{q,1} - \omega_{\text{ex},q} - \omega_q}},$$

$$u_{q,2} = \sqrt{\frac{\Omega_{q,2} - \omega_q}{2\Omega_{q,2} - \omega_{\text{ex},q} - \omega_q}}, \quad v_{q,2} = -\mathrm{i}\sqrt{\frac{\Omega_{q,2} - \omega_{\text{ex},q}}{2\Omega_{q,2} - \omega_{\text{ex},q} - \omega_q}}. \tag{1.109}$$

The upper branch polariton is exciton-like ($|u_{q,1}|^2 \cong 1$) for small q-values, and changes to a photon-like excitation by passing through the resonance, whereas the lower branch polariton has a reverse behavior.

The case of non-small nonresonant interaction terms is treated similarly, with the difference that the polariton operators depend on both creation and annihilation exciton and photon operators. They are introduced by the so-called Hopfield polariton transformations.

References

Basu, P.K. (1997) *Theory of Optical Processes in Semiconductors. Bulk and Microstructures*, Clarendon Press, Oxford.

Callaway, J. (1991) *Quantum Theory of the Solid State*, 2nd edition, Academic Press, Boston.

Chuang, S.L. (1995) *Physics of Optoelectronic Devices*, John Wiley & Sons, New York.

Curie, D. (1979) in *Luminescence of Inorganic Solids*, ed. B. di Bartolo, Plenum Press, New York, p.337.

Grüner, G. (1988) Rev. Mod. Phys. **60**, 1129.

di Bartolo, B. (1968) *Optical Interactions in Solids*, Wiley, New York.

Haug, H. and S.W. Koch (1990) *Quantum Theory of the Optical and Electronic Properties of Semiconductors*, World Scientific, Singapore.

Ivchenko, E.L. and G. Pikus (1997) *Superlattices and Other Heterostructures. Symmetry and Optical Phenomena*, 2nd edition, Springer, Berlin, Heidelberg.

Klingshirn, C.F. (1997) *Semiconductor Optics*, Springer, Berlin, Heidelberg.

2. Optical Transitions

At a macroscopic level, the result of the interaction between an incident electromagnetic field and a solid-state material depends on the relation between the wavelength λ of the electromagnetic radiation and the scale of the internal inhomogeneity of the solid. If these two parameters are comparable, light scattering occurs; if λ is much larger than the scale of internal inhomogeneity, the electromagnetic field propagates as in a homogeneous medium with well-defined optical constants, while if λ is smaller than this scale, spatially varying optical constants must be introduced. An example of the latter case is light reflection at the surface between the solid-state material and the surrounding medium. In all cases, information about the internal structure and material parameters of the solid state can be gained by analyzing the spectral and/or time dependence of the light after it interacts with the matter. The macroscopic result of the interaction can in each case be related to microscopic transitions between different levels/bands in the solid. In this chapter we present briefly the relation between the microscopic transition probabilities and the macroscopic optical constants, as well as the methods used to measure them.

2.1 The Quantization of Electromagnetic Radiation

Electromagnetic radiation is introduced in classical electromagnetism as a vector field, which satisfies the Maxwell's equations with specific boundary conditions, whereas in quantum electrodynamics it is treated as a collection of harmonic oscillators, each represented by creation and annihilation operators. As the solid state can only be treated quantum mechanically, the result of its interaction with the electromagnetic field is more suitably described by considering also a quantum description of the radiation field.

The quantization of the electromagnetic field can be introduced in a similar manner to the quantization of lattice vibrations. Defining the energy of the electromagnetic field as

$$E = (1/2)\int(\varepsilon_0 \boldsymbol{E}^2 + \boldsymbol{B}^2 / \mu_0)\mathrm{d}\boldsymbol{r}, \tag{2.1}$$

we can express it in terms of the vector potential $\boldsymbol{A}$, related to the electric and magnetic fields through $\boldsymbol{E} = -\partial \boldsymbol{A}/\partial t$ and $\boldsymbol{B} = \nabla \times \boldsymbol{A}$, respectively (the scalar potential of the electromagnetic field is neglected). By defining the vector potential as (Callaway, 1991)

$$\boldsymbol{A}(\boldsymbol{r},t) = c(\mu_0^{1/2}/8\pi^3)\Sigma_\rho \int \boldsymbol{e}_\rho(\boldsymbol{k}) q_\rho(\boldsymbol{k},t) \exp(\mathrm{i}\boldsymbol{k}\cdot\boldsymbol{r}) \mathrm{d}\boldsymbol{k}, \tag{2.2}$$

and by following the same procedure as in the case of phonons, with q_ρ playing the role of normal coordinate and $\boldsymbol{e}_\rho$ the polarization vector, the total Hamiltonian is found to be

$$H = [1/2(2\pi)^3]\Sigma_\rho \int \mathrm{d}\boldsymbol{k}[|p_\rho(\boldsymbol{k},t)|^2 + \omega^2(\boldsymbol{k})|q_\rho(\boldsymbol{k},t)|^2]. \tag{2.3}$$

In (2.3) $\omega^2(\boldsymbol{k}) = c^2\boldsymbol{k}^2$ with c the light velocity in vacuum, and $p_\rho(\boldsymbol{k},t) = \dot{q}_\rho(\boldsymbol{k},t)$ is the canonical momentum associated to $q_\rho(\boldsymbol{k},t)$. The time-averaged Hamiltonian can be written as a collection of quantum uncoupled harmonic oscillators by the introduction of annihilation and creation operators for photons, defined in terms of the operators $p_\rho(\boldsymbol{k})$ and $q_\rho(\boldsymbol{k})$ as, respectively

$$a_\rho(\boldsymbol{k}) = [2\hbar\omega(\boldsymbol{k})]^{-1/2}[\omega(\boldsymbol{k})q_\rho(\boldsymbol{k}) + \mathrm{i}p_\rho(\boldsymbol{k})], \tag{2.4}$$

$$a_\rho^+(\boldsymbol{k}) = [2\hbar\omega(\boldsymbol{k})]^{-1/2}[\omega(\boldsymbol{k})q_\rho^+(\boldsymbol{k}) - \mathrm{i}p_\rho^+(\boldsymbol{k})]. \tag{2.5}$$

The total Hamiltonian can be expressed in terms of these operators as

$$H = (2\pi)^{-3}\Sigma_\rho \int \mathrm{d}\boldsymbol{k}\hbar\omega(\boldsymbol{k})[a_\rho^+(\boldsymbol{k})a_\rho(\boldsymbol{k}) + 1/2], \tag{2.6}$$

or

$$H = \Sigma_{\rho k}\hbar\omega_k(a_{\rho k}^+ a_{\rho k} + 1/2) = \Sigma_{\rho k}\hbar\omega_k(\hat{n}_{\rho k} + 1/2), \tag{2.7}$$

when the wavevector $\boldsymbol{k}$ takes discrete values. Here $\hat{n}_{\rho k} = a_{\rho k}^+ a_{\rho k}$ is the number of photons operator with integer eigenvalues $n_{\rho k}$. $a_{\rho k}$, $a_{\rho k}^+$ are the operators of annihilation and creation, respectively, of waves, which differ with respect to their wavevector $\boldsymbol{k}$ or their polarization $\boldsymbol{e}_{\rho k}$. Each of these waves, characterized by a certain $(\boldsymbol{k}, \boldsymbol{e}_{\rho k})$ is a *mode*. $a_{\rho k}$, $a_{\rho k}^+$ satisfy the commutation relations for bosons, but the creation and annihilation operators commute if they refer to different (independent) modes.

2.2 Transition Probabilities

At the interaction of the electromagnetic radiation with a solid, the total Hamiltonian $H = H_s + H_{em}$ is the sum of the Hamiltonian

$$H_s = (1/2)\Sigma_i (\boldsymbol{p}_i - e_i\boldsymbol{A})^2 / m_i \tag{2.8}$$

of the solid and the Hamiltonian of the electromagnetic field

$$H_{em} = \Sigma_{\rho k} \hbar\omega_k (a_{\rho k}^+ a_{\rho k} + 1/2), \tag{2.9}$$

where $\boldsymbol{p}_i$ is the momentum operator of particle i (usually electron or ion), situated at position $\boldsymbol{r}_i$, and m_i and e_i are its mass and charge, respectively. The total Hamiltonian can be written also as $H = H_0 + H_{int}$ where

$$H_0 = (1/2)\Sigma_i \boldsymbol{p}_i^2 / m_i + \Sigma_{\rho k} \hbar\omega_k (a_{\rho k}^+ a_{\rho k} + 1/2) \tag{2.10}$$

is the Hamiltonian of the total system when *no* interaction between radiation and matter is taken into account and

$$H_{int} = -(1/2)\Sigma_i e_i (\boldsymbol{p}_i \cdot \boldsymbol{A} + \boldsymbol{A} \cdot \boldsymbol{p}_i) / m_i + (1/2)\Sigma_i (e_i / m_i)\boldsymbol{A}^2 = H_A + H_{AA}. \tag{2.11}$$

The operators $\boldsymbol{p}_i$ and $\boldsymbol{A}$ do not generally commute. However, we can simplify the linear part in $\boldsymbol{A}$ of the Hamiltonian $H_A = -(1/2)\Sigma_i e_i (\boldsymbol{p}_i \cdot \boldsymbol{A} + \boldsymbol{A} \cdot \boldsymbol{p}_i) / m_i$ to $H_A = -\Sigma_i e_i \boldsymbol{A} \cdot \boldsymbol{p}_i / m_i$ if we choose the gauge transformation $\nabla \boldsymbol{A} = 0$. The linear part in $\boldsymbol{A}$ of the Hamiltonian is usually dominant; the term quadratic in $\boldsymbol{A}$, $H_{AA} = (1/2)\Sigma_i e_i \boldsymbol{A}^2 / m_i$, needs only to be taken into account in the calculation of multiphoton processes such as, for example, the scattering of light.

If the interaction between the electromagnetic field and the solid is neglected, $H \cong H_0$, and the total wavefunction $\Psi = \varphi_n \prod_{\rho k} \phi_{\rho k}$ is the product between the eigenfunction of the particle, φ_n, with corresponding energy E_n, and the eigenfunction of the electromagnetic field (equal to the product of the eigenfunctions of each radiation mode $\phi_{\rho k}$ with corresponding energies $\hbar\omega_k (n_{\rho k} + 1/2)$). The total energy of the system is $E = E_n + \Sigma_{\rho k} \hbar\omega_k (n_{\rho k} + 1/2)$. Transitions are possible only if we take into account the interaction between the electromagnetic field and matter.

During a transition, the energy of the electromagnetic field varies only with integer multiples of $\hbar\omega_k$. The increase or decrease in $n_{\rho k}$ corresponds to emission and absorption of photons, respectively. In its turn, the particle can make transitions between states characterized by different quantum numbers *n, m* which can be: i) both discrete, ii) one discrete, the other belonging to a continuum of states or iii) between energy bands. All these three types of transitions can be observed in a solid. Due to the periodicity of the crystalline structure, the energy levels of charged carriers degenerate into bands, but at the same time there are discrete levels due to impurities, defects, etc.

The action of the electromagnetic field, seen as a *perturbation*, changes the system from the initial state Ψ_i such that after a time t it is in the final state Ψ_f. We are interested in calculating the transition probability between the initial and final states, supposing that the orthonormal eigenfunctions Ψ_{0m} of the unperturbed system with the Hamiltonian H_0 and the corresponding energies E_{0m} *are known*. In particular, m can represent a system of quantum numbers. In the first order of the perturbation theory only the linear part in $\boldsymbol{A}$ of the Hamiltonian contributes, i.e. $H_{\text{int}} \equiv H_A$.

The wavefunction of the final state in the first-order approximation can be expressed as:

$$\Psi_f(t,q) = \sum_m a_m(t)\Psi_{0m}(q,t), \tag{2.12}$$

where it was assumed that Ψ_i is one of the Ψ_{0m} eigenfunctions, denoted Ψ_{0i}, and q is any variable, except time. When H_0 does not explicitly depend on time

$$\Psi_{0m}(q,t) = \Psi_{0m}(q)\exp(-\mathrm{i}tE_{0m}/\hbar). \tag{2.13}$$

Introducing the last two relations in the Schrödinger equation for the total system

$$(\hbar/\mathrm{i})\partial\Psi/\partial t + H\Psi = 0, \tag{2.14}$$

one obtains, after scalar multiplying (2.14) with Ψ_{0l} and using the orthogonality relation $\langle\Psi_{0l}|\Psi_{0m}\rangle = \delta_{lm}$, the following equation for the coefficients a_l

$$-\mathrm{i}\hbar(\mathrm{d}a_l/\mathrm{d}t) + \sum_m a_m(t)H_{A,lm}\exp[\mathrm{i}(E_{0m} - E_{0l})t/\hbar] = 0. \tag{2.15}$$

If the perturbation is *small*, i.e. if $\langle\Psi_{0l}|H_A|\Psi_{0m}\rangle << E_{0l} - E_{0m}$, $\mathrm{d}a_l/\mathrm{d}t$ is also small and a_l vary slowly in time. We can obtain then an approximate solution for $a_l(t)$ by replacing $a_m(t)$ in the sum after m with their values at the initial moment t_i (this approximation is not valid for long time intervals, when the variations of a_m can no longer be neglected). Assuming that at t_i the quantum system is with certainty in state Ψ_{0i}, i.e. $a_m(t_i) = 0$ for $m \neq i$ and $a_i(t_i) = 1$, and focusing our attention to a final state denoted with f, different from the initial one, we get:

$$a_f(t_f) = -(\mathrm{i}/\hbar)\int_{t_i}^{t_f} H_{A,fi}\exp[\mathrm{i}(E_{0f} - E_{0i})t/\hbar]\mathrm{d}t. \tag{2.16}$$

$a_i(t_f)$ can then be deduced from the normalization condition of the total wavefunction. The expression for $a_f(t_f)$ is also valid when H_A depends explicitly on time. The transition probability between the initial state i and the final state f when the time varies from t_i to t_f is defined as $P_{fi} = |a_f(t_f)|^2$. For a time-independent Hamiltonian

$$P_{fi} = |a_f(t_f)|^2 = -2|H_{A,fi}|^2 \{1 - \cos[(E_{0f} - E_{0i})(t_f - t_i)/\hbar]\}/(E_{0f} - E_{0i})^2, \tag{2.17}$$

the transition probability depending *only on the duration* $t = t_f - t_i$ of the process. In particular, for small t the probability of finding the system in the final state is quadratic in t – this quadratic dependence is rarely found in practice.

The transition probability per unit time (or the rate of transition) is obtained by differentiating the above relation with respect to t

$$\mathrm{d}P_{fi}/\mathrm{d}t = (2/\hbar)|H_{A,fi}|^2 \sin[(E_{0f} - E_{0i})t/\hbar]/E_{0f} - E_{0i}, \tag{2.18}$$

an expression which reduces for $t \to \infty$ to

$$\mathrm{d}P_{fi}/\mathrm{d}t = (2\pi/\hbar)|H_{A,fi}|^2 \delta(E_{0f} - E_{0i}). \tag{2.19}$$

Equation (2.19) is known as the Fermi golden rule.

The above equation shows that the rate of transition is *constant* in time, and is different from zero only when the final and initial states have the same energy, i.e. when the law of energy conservation is satisfied. On the other hand the limit $t \to \infty$ is in contradiction with our hypothesis that the time is short enough such that $a_f(t)$ do not vary considerably. When this assumption is not valid, $\sin(\alpha t)/(\pi\alpha)$ is a good approximation to $\delta(\alpha)$ only when a large number of oscillations occur in the interval t. Since the duration of an oscillation is $\Delta t = 2\pi/\alpha$, this condition implies $t/\Delta t >> 1$ or $(E_{0f} - E_{0i}) >> h = 2\pi\hbar$, i.e. the uncertainty relation between energy and duration.

Equation (2.19) gives the first-order approximation for a_f. If we substitute it into (2.15) we get a system of equations for the second-order approximation, in which the energy of the final state is

$$E_{0f} = E_{0i} + H_{A,ii} + \sum_m H_{A,im} H_{A,mi}/(E_{0i} - E_{0m}) + H_{AA,ii}, \tag{2.20}$$

$H_{A,ii}$ usually being neglected. The sum after m in (2.20) can be interpreted as virtual excitations and de-excitations in intermediate states characterized by wavefunctions Ψ_{0m}. The last term describes transitions between initial and final states in which two photons participate simultaneously. The wavefunction of the final state in the second-order approximation becomes

$$\Psi_f = \Psi_{0i} + \sum_{m \neq i} \left(\frac{H_{A,mi} + H_{AA,mi}}{E_{0i} - E_{0m}} + \sum_{n \neq i} \frac{H_{A,mn} H_{A,ni}}{(E_{0i} - E_{0n})(E_{0i} - E_{0m})} \right) \Psi_{0m}, \tag{2.21}$$

confirming the formal result that in the second-order approximations, $H_{A,mi}$ must be replaced with

$$H_{A,mi} + \sum_{n \neq i} H_{A,mn} H_{A,ni} / (E_{0i} - E_{0n}) + H_{AA,mi}. \tag{2.22}$$

In the following we understand by $H_{\text{int},mi}$ either $H_{A,mi}$ in the first-order approximation, or the above expression in the second-order approximation in $\boldsymbol{A}$.

Up to now we have assumed that both initial and final states have discrete, well-defined energies. It is possible, however, that either the initial or final state between which the transition occurs, or even both, belong to a continuous spectrum of energy. In the case when, for example, the final state belongs to a group of states with density $\rho(E)$, then (2.18) becomes

$$\mathrm{d}P_{fi} / \mathrm{d}t = (2\pi / \hbar) | H_{\text{int},fi} |^2 \rho(E_{0f} = E_{0i}). \tag{2.23}$$

If we further separate the energy of the total system into the energy of the solid, denoted by E_i, E_f in the initial and final states, respectively, and the energy of the electromagnetic radiation, (2.23) can be rewritten as

$$\mathrm{d}P_{fi} / \mathrm{d}t = (2\pi / \hbar) | H_{\text{int},fi} |^2 \rho(E_f = E_i \pm \hbar\omega_{\boldsymbol{k}}). \tag{2.24}$$

If both initial and final states belong to continuous energy spectra, characterized by densities of states, the product of both initial and final density of states should appear in (2.24).

In the above expressions no allowance is made for the occupation probability of energy bands. These can also be incorporated in the formalism, depending on the particular type of transition we are dealing with. For example, when a photon is absorbed, an electron or a phonon from the material system is elevated to a higher-energy level or band. So, the optical transitions in this case take place from an occupied to an empty energy level. In the case of electrons, this means that we must multiply the transition probability with $f(E_i)[1 - f(E_f)]$, a similar expression holding for the case of transitions that involve phonons.

2.2.1 Optical Properties of Crystalline Materials

The translational invariance, characteristic of crystalline materials, has important consequences on their optical properties, since it imposes the conservation of total momentum at optical transitions, besides energy conservation.

To see this, let us calculate $| H_{\text{int},fi} |^2$ in the first-order approximation with $H_{\text{int}} = -\sum_l e_l \boldsymbol{A} \cdot \boldsymbol{p}_l / m_l$. We assume for simplicity that the electromagnetic field has only one mode with polarization s and wavevector $\boldsymbol{k}$, and that it induces the transition of one electron from an initial state in band i, with wavefunction $\psi_i(\boldsymbol{k}_i, \boldsymbol{r})$ to a final state in band f with wavefunction $\psi_f(\boldsymbol{k}_f, \boldsymbol{r})$. The total wavefunctions in the initial and final states are $\Psi_i = \psi_i(\boldsymbol{k}_i, \boldsymbol{r}) | n_{s\boldsymbol{k}}^i \rangle$,

$\Psi_f = \psi_f(\boldsymbol{k}_f, \boldsymbol{r}) | n_{sk}^f \rangle$, where $| n_{sk}^i \rangle$, $| n_{sk}^f \rangle$ are wavefunctions of the electromagnetic field in the basis of the occupation number operator. Writing $\boldsymbol{A}$ as (the expression of A_0 is easily deduced from (2.2))

$$\boldsymbol{A}(\boldsymbol{r}, \boldsymbol{k}) = A_0 (\hat{\boldsymbol{e}}_s / \omega_{sk}^{1/2})[a_{sk} \exp(\mathrm{i}\boldsymbol{k} \cdot \boldsymbol{r}) + a_{sk}^+ \exp(-\mathrm{i}\boldsymbol{k} \cdot \boldsymbol{r})], \tag{2.25}$$

where $\hat{\boldsymbol{e}}_s$ is now the unit polarization vector, the operators $a_{s\mathbf{k}}$ and $a_{s\mathbf{k}}^+$ act only on the electromagnetic field wavefunctions. The only situations when $| H_{\text{int}, fi} |^2 \neq 0$ are those when absorption or emission of a photon takes place. In these cases, $\langle n_{sk}^f | a_{sk} n_{sk}^i \rangle = (n_{sk})^{1/2}$ and $E_f = E_i + \hbar\omega$ (with E_f, E_i the electron energies), or $\langle n_{sk}^f | a_{sk}^+ n_{sk}^i \rangle = (n_{s\mathbf{k}} + 1)^{1/2}$ and $E_f = E_i - \hbar\omega$, respectively. Considering only the absorption process, and taking into account that the electron initial and final states are described by Bloch functions $\psi_i(\boldsymbol{k}_i, \boldsymbol{r}) = u_i(\boldsymbol{k}_i, \boldsymbol{r}) \exp(\mathrm{i}\boldsymbol{k}_i \cdot \boldsymbol{r})$, $\psi_f(\boldsymbol{k}_f, \boldsymbol{r}) = u_f(\boldsymbol{k}_f, \boldsymbol{r}) \exp(\mathrm{i}\boldsymbol{k}_f \cdot \boldsymbol{r})$,

$$\begin{aligned} | H_{\text{int}, fi} |^2 \approx & \left| \int_V u_i^* \exp[-\mathrm{i}(\boldsymbol{k}_i - \boldsymbol{k} - \boldsymbol{k}_f) \cdot \boldsymbol{r}] \hat{\boldsymbol{e}}_s \nabla u_f \mathrm{d}\boldsymbol{r} \right. \\ & \left. + \mathrm{i}\boldsymbol{k}_f \int_V \exp[-\mathrm{i}(\boldsymbol{k}_i - \boldsymbol{k} - \boldsymbol{k}_f) \cdot \boldsymbol{r}] u_i^* u_f \hat{\boldsymbol{e}}_s \mathrm{d}\boldsymbol{r} \right|^2, \end{aligned} \tag{2.26}$$

where the integral must be performed over the crystal volume V. Since u_i, u_f are periodic in the crystal, the volume integral can be separated into a sum on crystalline cells multiplied by the integral on a unit cell with volume Ω. The second term in (2.26) can then be written as (Callaway, 1991)

$$\begin{aligned} \int_V \exp[-\mathrm{i}(\boldsymbol{k}_i - \boldsymbol{k} - \boldsymbol{k}_f) \cdot \boldsymbol{r}] u_i^* u_f \mathrm{d}\boldsymbol{r} &= [(2\pi)^3 / \Omega] \delta(\boldsymbol{k}_i - \boldsymbol{k} - \boldsymbol{k}_f) \\ \times \int_{\text{cell}} \exp[-\mathrm{i}(\boldsymbol{k}_i - \boldsymbol{k} - \boldsymbol{k}_f) \cdot \boldsymbol{r}] u_i^*(\boldsymbol{k}_i, \boldsymbol{r}) u_f(\boldsymbol{k}_f, \boldsymbol{r}) \mathrm{d}\boldsymbol{r} &= \delta(\boldsymbol{k}_i - \boldsymbol{k} - \boldsymbol{k}_f) \delta_{if}, \end{aligned} \tag{2.27}$$

where the factor δ_{if} appears due to the orthogonality of Bloch functions. For $i \neq f$, (2.26) becomes

$$| H_{\text{int}, fi} |^2 \approx | [(2\pi)^3 / \Omega] \hat{\boldsymbol{e}}_s \delta(\boldsymbol{k}_i - \boldsymbol{k} - \boldsymbol{k}_f) \int_{\text{cell}} u_i^* \nabla u_f \mathrm{d}\boldsymbol{r} |^2 . \tag{2.28}$$

This equation expresses the law of momentum conservation at transitions, which in the most general case can be written as $\boldsymbol{k}_i - \boldsymbol{k}_f - \boldsymbol{k} = \boldsymbol{K}$, with $\boldsymbol{K}$ a vector of the reciprocal lattice. A similar selection rule holds also for the transition probabilities between two phonon states (see Callaway (1991)). This selection rule appears whenever the system is periodic. Because $| \boldsymbol{k} | = 2\pi / \lambda$ is small in comparison with the dimension of the Brillouin zone ($k = | \boldsymbol{k} | \approx 1/\mu\text{m}$, $k_i = | \boldsymbol{k}_i | \approx 1/\text{Å}$), $\boldsymbol{k}_i \cong \boldsymbol{k}_f$ and only vertical transitions in the $\boldsymbol{k}$-space (direct transitions) are allowed.

Other selection rules result from the symmetry of the wavefunction; from (2.28) it follows that $H_{\text{int}, fi}$ is zero if transitions occur between states with wavefunctions of the same parity. This is why, in general, $H_{\text{int}, ii}$ is equal to zero. Also, due to the momentum conservation law indirect transitions (oblique in the $\boldsymbol{k}$-

space) can only take place if mediated by phonons. In this case, $H_{\text{int},fi} = 0$ and the transition probability is calculated using the higher-order approximation in which $H_{\text{int},fi}$ is replaced by $\sum_n H_{\text{int},fn} H_{\text{int},ni} /(E_{0i} - E_{0n})$.

2.2.2 Mechanisms for Spectral Line Broadening

Due to the fact that, in general, the interaction time is finite, the frequencies of spectral lines $\omega = \pm(E_f - E_i)/\hbar$ are not exactly determined but satisfy the uncertainty relation $\omega t > 2\pi$; this relation defines the *natural width of the spectral line*. Besides this natural width, a spectral line can be broadened by several mechanisms (see di Bartolo (1968)).

2.2.2.1 Lorentzian Broadening of Spectral Lines

This type of broadening, also called homogeneous line broadening, appears when the interaction between the radiation and the system *is the same for all constituents* (atoms, electrons, etc.) of the system. One cause of homogeneous line broadening is the finite lifetime of the excited state. If a transition takes place between a higher (excited) state of the material system to the ground state (with the lowest energy), by emission of a photon, the transition amplitude is given by

$$\mathrm{i}\hbar(\mathrm{d}a_f / \mathrm{d}t) = H_{\text{int},fi} \exp[\mathrm{i}\Delta\omega t - (\gamma/2)t], \tag{2.29}$$

where $\Delta\omega = (E_{0f} - E_{0i})/\hbar$, and the additional term $\exp(-\gamma t/2)$ with a complex constant $\gamma = \gamma_{\text{re}} + \mathrm{i}\gamma_{\text{im}}$ accounts for the finite lifetime of the initial state. Then, the transition probability for $H_{\text{int},fi}$ independent of time and $t >> 1/\operatorname{Re}\gamma$ is,

$$|a_f|^2 = |H_{\text{int},fi}|^2 / \hbar^2 = \{[(E_{0i}/\hbar + \gamma_{\text{im}}/2) - (E_{0f}/\hbar)]^2 + \gamma_{\text{re}}^2/4\}^{-1}. \tag{2.30}$$

According to (2.30) the transition does not occur exactly at $E_{0f} = E_{0i}$, but has a dependence on energy with a Lorentzian shape.

The real part of γ describes the broadening of the transition caused by the fact that the atom in the initial state decays in a longer time than its lifetime. For times smaller than the lifetime of the initial state $a_i \cong \text{const.}$ and the transition is very narrow in energy. The imaginary part of γ produces a shift in the transition energy, called the Lamb shift.

If the interaction lasts a sufficiently long time, $i \to f \to i$ transitions from the final state back to the initial state of the system can also occur. The transition probability in this case is described by the same relation

$$\mathrm{i}\hbar(\mathrm{d}a_i / \mathrm{d}t) = a_f H_{\text{int},if} \exp(-\mathrm{i}\Delta\omega t), \tag{2.31}$$

from which it follows, since $H_{\text{int},fi} = H_{\text{int},if}^*$, that for $\gamma << \Delta\omega$

$$\gamma_{\text{re}} + \mathrm{i}\gamma_{\text{im}} = (2/\hbar^2)|H_{\text{int},fi}|^2 (1 - \cos\Delta\omega t + \mathrm{i} \times \sin\Delta\omega t)/\mathrm{i}\Delta\omega, \tag{2.32}$$

i.e. $\gamma = \gamma(t)$ is time dependent, and

$$d|a_f|^2/dt \cong (2/\hbar^2)|H_{\text{int},fi}|^2 \sin\Delta\omega t/\Delta\omega. \tag{2.33}$$

Thus, γ_{re} can be identified with the *probability transition per unit time*, which determines also the width (in energy) of the transition line at half the intensity: $\Delta E = \hbar\gamma_{\text{re}}$. If we define the lifetime of the initial state as $\tau = \gamma_{\text{re}}^{-1}$, the uncertainty relation reads $\Delta E = \hbar/\tau$.

More than one mechanism of homogeneous line broadening can act simultaneously (for example: thermal vibrations, radiation damping, etc.). In this case

$$\Delta E = \hbar\textstyle\sum_i \gamma_{\text{re},i}, \tag{2.34}$$

where $\gamma_{\text{re},i}$ is the transition probability associated to the ith mechanism. In particular, if the transition takes place between two excited levels with widths γ_{re1} and γ_{re2}, respectively, the width of the transition is $\gamma_{\text{re1}} + \gamma_{\text{re2}}$.

2.2.2.2 Gaussian Broadening of the Spectral Line

The mechanisms that produce a Gaussian distribution of the spectral line act differently for different constituents of the system. One example is the Doppler effect, caused by the fact that the charged particles in highly doped semiconductors or metals do not have the same velocity. Namely, the Maxwell distribution function of velocities for a gas of charged particles with mass M, at an equilibrium temperature T is

$$F(v_x, v_y, v_z) = (M/2\pi k_{\text{B}}T)^{3/2}\exp[-M(v_x^2 + v_y^2 + v_z^2)/2k_{\text{B}}T], \tag{2.35}$$

where $F(v_x, v_y, v_z)\text{d}v_x\text{d}v_y\text{d}v_z$ is the probability that the velocity vector $\boldsymbol{v}$ of a particle is between $\boldsymbol{v}$ and $\boldsymbol{v} + \text{d}\boldsymbol{v}$.

Since in the nonrelativistic approximation, the light emitted during transition is detected in the x direction with a frequency displaced by $\nu - \nu_0 = \nu_0 v_x/c$, the probability $f(\nu)\text{d}\nu$ that the transition frequency is between ν and $\nu + \text{d}\nu$ is equal to the probability that v_x is between $c(\nu - \nu_0)/\nu_0$ and $c(\nu + d\nu - \nu_0)/\nu_0$, irrespective of the values of v_y and v_z. After performing the integrals,

$$f(\nu) = (c/\nu_0)(M/2\pi k_{\text{B}}T)^{1/2}\exp[-(M/2k_{\text{B}}T)c^2(\nu - \nu_0)^2/\nu_0^2]. \tag{2.36}$$

The Gaussian profile can be considered as a result of the superposition of a large number of independent spectral lines, each of them corresponding to transitions which involve a certain number of particles.

Another mechanism, besides the Doppler effect that can produce a Gaussian broadening of the spectral line, is the influence of the surrounding medium upon the ions in a crystal, reflected in the presence of a crystal field that depends on the position of the ion. The variations of this field are completely random in space. Inhomogeneous broadening can also be caused by fluctuations of different parameters, for example, the static strain. Although a demonstration of the Gaussian form of line broadening would be much more difficult in these cases, this spectral form of broadening is encountered for all inhomogeneous mechanisms of broadening.

2.2.2.3 Voigt Profile

If the homogeneous and inhomogeneous spectral linewidths are comparable, the lineshape has a Voigt profile (di Bartolo, 1968). Denoting by $g_L(\nu)$ the Lorentzian line shape characteristic for homogeneous broadening, by $\Delta\nu_L$ its width, and by $g_G(\nu)$, $\Delta\nu_G$ the corresponding quantities for an inhomogeneous broadened line with a Gaussian shape, the Voigt profile is

$$g_V(\nu) = \int_{-\infty}^{\infty} g_G(\nu') g_L(\nu - \nu') d\nu' = (\sqrt{\ln 2/\pi}\,/\Delta\nu_G) K(x, y), \tag{2.37}$$

where $K(x,y) = (y/\pi)\int_{-\infty}^{\infty} \mathrm{d}z \exp(-z^2)/[y^2 + (x-z)^2]$, $x = (\ln 2)^{1/2}(\nu - \nu_0)/\Delta\nu_G$, $y = (\ln 2)^{1/2}(\Delta\nu_L/\Delta\nu_G)$, and ν_0 is the line center frequency. Equation (2.40) is valid if the inhomogeneous broadening is not strongly correlated with different optical transitions.

2.2.3 Multipolar Contributions to the Interaction Hamiltonian

Since the wavenumber of the electromagnetic radiation is extremely small compared to the length of the reciprocal lattice vectors, the term $\exp(\mathrm{i}\boldsymbol{k}\cdot\boldsymbol{r})$ that appears in the product $\boldsymbol{A}\cdot\boldsymbol{p}$ can be approximated by

$$\exp(\mathrm{i}\boldsymbol{k}\cdot\boldsymbol{r})\boldsymbol{p} \cong \boldsymbol{p} + \mathrm{i}(\boldsymbol{k}\cdot\boldsymbol{r})\boldsymbol{p}. \tag{2.38}$$

The first term in the expansion corresponds to electric dipolar transitions. The second term encompass both magnetic dipolar and electric quadrupole transitions, as can be seen by expressing it as $(\boldsymbol{k}\cdot\boldsymbol{r})\boldsymbol{p} = [(\boldsymbol{k}\cdot\boldsymbol{r})\boldsymbol{p} + (\boldsymbol{k}\cdot\boldsymbol{p})\boldsymbol{r}]/2$ $+[(\boldsymbol{k}\cdot\boldsymbol{r})\boldsymbol{p} - (\boldsymbol{k}\cdot\boldsymbol{p})\boldsymbol{r}]/2$, where

$$[(\boldsymbol{k}\cdot\boldsymbol{r})\boldsymbol{p} - (\boldsymbol{k}\cdot\boldsymbol{p})\boldsymbol{r}]/2 = -\boldsymbol{k}\times(\boldsymbol{r}\times\boldsymbol{p})/2 = -\boldsymbol{k}\times\boldsymbol{L}/2, \tag{2.39}$$

with $\mu_B = -e\boldsymbol{L}/(2mc)$ the orbital magnetic momentum. When the spin is also considered the relevant term is $-[\boldsymbol{k}\times(\boldsymbol{L}\times 2\boldsymbol{S})]/2$ where $-e\boldsymbol{S}/(mc)$ is the spin magnetic moment. On the other hand,

$$\begin{aligned}[(\boldsymbol{k}\cdot\boldsymbol{r})\boldsymbol{p} + (\boldsymbol{k}\cdot\boldsymbol{p})\boldsymbol{r}]/2 &= m[(\boldsymbol{k}\cdot\boldsymbol{r})\dot{\boldsymbol{r}} + (\boldsymbol{k}\cdot\dot{\boldsymbol{r}})\boldsymbol{r}]/2 = (m/2)\mathrm{d}[(\boldsymbol{k}\cdot\boldsymbol{r})\boldsymbol{r}]/\mathrm{d}t \\ &= \mathrm{i}m\omega[(\boldsymbol{k}\cdot\boldsymbol{r})\boldsymbol{r}]/2\end{aligned} \tag{2.40}$$

corresponds to the electric quadrupole interaction (for a particle with mass m, in the Hamiltonian theory, $\boldsymbol{p} = m\mathrm{d}\boldsymbol{r}/\mathrm{d}t = \mathrm{i}m[H,\boldsymbol{r}]/\hbar$).

The term corresponding to electric dipolar transitions can also be written as

$$\begin{aligned}\langle \psi_f \mid \boldsymbol{p} \mid \psi_i \rangle &= (m\mathrm{i}/\hbar)\langle \psi_f \mid H\boldsymbol{x} - \boldsymbol{x}H \mid \psi_i \rangle = (m\mathrm{i}/\hbar)(E_f - E_i)\langle \psi_f \mid \boldsymbol{x} \mid \psi_i \rangle \\ &= \mathrm{i}m\omega_k \langle \psi_f \mid x \mid \psi_i \rangle,\end{aligned} \tag{2.41}$$

where $\psi_{f,i}$ are the final and initial particle wavefunctions, $E_{f,i}$ the final and initial particle energies, and ω_k the frequency of the electromagnetic radiation.

The electric dipole operator is an odd operator and the magnetic dipole and electric quadrupole operators are even. This leads to the Laporte rule: electric dipole transitions can take place only between quantum states of opposite parity, whereas magnetic dipole and electric quadrupole transitions occur only between states with the same parity. A system has states of definite parity when it presents a center of symmetry: this is the case of free atoms and ions, centrosymmetric molecules and impurity ions in centrosymmetric solids. The ratio between the probabilities of the magnetic dipole and the electric dipole transitions is about 10^{-8}, whereas the ratio between the probabilities of the electric quadrupole and electric dipole transitions is about 10^{-7}.

2.3 Optical Constants of Solids

There are two basic processes that occur at interaction of radiation with matter: absorption and emission of photons, accompanied by a change of state of the material system. Absorption of photons of frequency ω inducing transitions of a bosonic system between a lower energy level i and an upper energy level j produces a rate variation of the respective populations of $\mathrm{d}N_i(t)/\mathrm{d}t = -\mathrm{d}N_j(t)/\mathrm{d}t = -B_{ij}N_i\rho(\omega)$, where B_{ij} is the Einstein coefficient for absorption and $\rho(\omega)$ the energy density of photons. Light absorption is a stimulated process since it can only take place in the presence of radiation. The inverse transition, from the upper to the lower energy level of the material system accompanied by photon emission can be either spontaneous or stimulated. In the first case $\mathrm{d}N_i(t)/\mathrm{d}t = -\mathrm{d}N_j(t)/\mathrm{d}t = A_{ji}N_j$, whereas in the second case $\mathrm{d}N_j(t)/\mathrm{d}t = -B_{ji}N_j\rho(\omega)$. The three Einstein coefficients are related through $B_{ji} = B_{ij}$, $A_{ji}/B_{ji} = \omega^3 n^3/\pi^2 c^3$ where n is the refractive index. In the stimulated emission case the emitted photons have the same frequency, polarization and wavevector as the initial photon that stimulated the light emission. The refractive index of the material system not only influences the rate of light emission and absorption, it is also affected by it. Our task in this section is to relate the refractive index n or the dielectric constant ε – the optical constants of the material system – to both the microscopic structure of the matter, expressed by the transition probabilities between different states, and to the characteristics of the electromagnetic radiation, in particular its wavelength.

The measured parameters in optical experiments are not directly ε or n but usually the absorption coefficient α, or the reflectivity (also called reflectance, reflection coefficient) R. Based on the information provided in this section we will see in the following chapters how to extract the microscopic characteristics of the solid from measurements of α or R.

In order to establish the relations between α, R, and n, ε we start from Maxwell's equations (without charges):

$$\begin{aligned} &\nabla \boldsymbol{D} = 0, && \nabla \times \boldsymbol{E} = -\partial \boldsymbol{B} / \partial t, \\ &\nabla \boldsymbol{B} = 0, && \nabla \times \boldsymbol{H} = \boldsymbol{j} + \partial \boldsymbol{D} / \partial t, \end{aligned} \tag{2.42}$$

and the material equations

$$\boldsymbol{D} = \varepsilon_0 \boldsymbol{E} + \boldsymbol{P} = \varepsilon_0 (1 + \hat{\chi}) \boldsymbol{E}, \quad \boldsymbol{j} = \hat{\sigma} \boldsymbol{E}, \tag{2.43}$$

where $\hat{\chi}$, $\hat{\sigma}$ are the tensors of susceptibility and electrical conductivity, respectively. $\hat{\alpha}_{\mathrm{p}} = \varepsilon_0 \hat{\chi}$ is called the polarizability tensor. We consider non-magnetic materials in which $\boldsymbol{B} = \mu_0 \boldsymbol{H}$ and isotropic media where the susceptibility and conductivity tensors become scalars.

In the coordinate system aligned along the principal axes of the medium, the polarizability tensor, for example, can be diagonalized and has three nonvanishing components, $\alpha_{\mathrm{p}x}$, $\alpha_{\mathrm{p}y}$, and $\alpha_{\mathrm{p}z}$. If $\alpha_{\mathrm{p}x} \neq \alpha_{\mathrm{p}y} \neq \alpha_{\mathrm{p}z}$ the medium polarizes differently for different orientations of the electric field and therefore, for any direction which does not coincide with one of the principal polarization axes, the polarization vector is not collinear with the electric field. If $\alpha_{\mathrm{p}x} \neq \alpha_{\mathrm{p}y} \neq \alpha_{\mathrm{p}z}$ the polarizability tensor is invariant at rotations around the z axis. For isotropic media, with a high degree of symmetry, $\alpha_{\mathrm{p}x} = \alpha_{\mathrm{p}y} = \alpha_{\mathrm{p}z}$ and the electric dipolar moment is collinear with the electric field. In the general case, one can introduce two parameters to characterize the isotropy of the polarizability tensor: one, b, characterizes the spherical (or isotropic) part

$$b = (\alpha_{\mathrm{p}xx} + \alpha_{\mathrm{p}yy} + \alpha_{\mathrm{p}zz}) / 3. \tag{2.44}$$

The other part, g, characterizes the anisotropy:

$$g^2 = [(\alpha_{\mathrm{p}xx} - \alpha_{\mathrm{p}yy})^2 + (\alpha_{\mathrm{p}yy} - \alpha_{\mathrm{p}zz})^2 + (\alpha_{\mathrm{p}zz} - \alpha_{\mathrm{p}xx})^2 + 6(\alpha_{\mathrm{p}xy}^2 + \alpha_{\mathrm{p}yz}^2 + \alpha_{\mathrm{p}zx}^2)] / 2. \tag{2.45}$$

In terms of principal polarizability values these parameters can be written as $b = (\alpha_{\mathrm{p}x} + \alpha_{\mathrm{p}y} + \alpha_{\mathrm{p}z}) / 3$, $g^2 = [(\alpha_{\mathrm{p}x} - \alpha_{\mathrm{p}y})^2 + (\alpha_{\mathrm{p}y} - \alpha_{\mathrm{p}z})^2 + (\alpha_{\mathrm{p}z} - \alpha_{\mathrm{p}x})^2] / 2$.

The wavenumber $\boldsymbol{k}$ of the electromagnetic radiation and the refractive index are involved in the monochromatic and plane-wave-type solutions of Maxwell's equations of frequency ω,

$$\boldsymbol{E} = \boldsymbol{E}_0 \exp(\mathrm{i}\omega t) \exp(\mathrm{i}\boldsymbol{k} \cdot \boldsymbol{r}) = \boldsymbol{E}_0 \exp(\mathrm{i}\omega t) \exp(\mathrm{i}n\boldsymbol{k}_0 \cdot \boldsymbol{r}), \tag{2.46}$$

where $k_0 = |\boldsymbol{k}_0| = \omega / c$. The absorption coefficient α is defined as

$$\mathrm{d}I / \mathrm{d}z = -\alpha I, \tag{2.47}$$

where $I \approx |\boldsymbol{E}|^2$ is the intensity of the electromagnetic radiation and z is the direction of propagation. Since ω is real, a variation of the intensity takes place only if n is complex, i.e. if $n = n_{\mathrm{re}} + \mathrm{i}n_{\mathrm{im}}$. Under this assumption it follows that:

$$I = |\boldsymbol{E}|^2 \exp(-2k_0 n_{\mathrm{im}} z), \text{ or } 2k_0 n_{\mathrm{im}} = \alpha. \tag{2.48}$$

The reflection coefficient r of the electromagnetic field at normal incidence is defined as

$$r = (n-1)/(n+1), \tag{2.49}$$

and the reflectivity is given by

$$R = |r|^2 = [(n_{\mathrm{re}} - 1)^2 + n_{\mathrm{im}}^2]/[(n_{\mathrm{re}} + 1)^2 + n_{\mathrm{im}}^2]. \tag{2.50}$$

So, α and R can be written as *functions of the complex refractive index n.*

If n is complex, then ε, defined as $n^2 = \varepsilon / \varepsilon_0$, is also complex, $\varepsilon = \varepsilon_{\mathrm{re}} + \mathrm{i}\varepsilon_{\mathrm{im}}$, with

$$n_{\mathrm{re}}^2 - n_{\mathrm{im}}^2 = \varepsilon_{\mathrm{re}} / \varepsilon_0, \quad 2n_{\mathrm{re}}n_{\mathrm{im}} = \varepsilon_{\mathrm{im}} / \varepsilon_0. \tag{2.51}$$

Moreover, $\varepsilon = \varepsilon_0 + \alpha_{\mathrm{p}} = \varepsilon_{\mathrm{re}} + \mathrm{i}\varepsilon_{\mathrm{im}}$, so α_{p} (the scalar polarizability) is also *complex*.

In the plane-wave, monochromatic approximation, two currents appear in the equation $\nabla \times \boldsymbol{H} = \boldsymbol{j} + \partial \boldsymbol{D} / \partial t$: $\boldsymbol{j}_{\mathrm{c}} = \sigma \boldsymbol{E}$ (the *conduction* current) and $\boldsymbol{j}_{\mathrm{d}} = \partial \boldsymbol{D} / \partial t = \varepsilon \partial \boldsymbol{E} / \partial t = \mathrm{i}\omega\varepsilon \boldsymbol{E}$ (the *displacement* current), dephased with respect to each other with 90° (a fact indicated by the i factor) and having *different* effects:
(i) the power associated with the displacement current $P_{\mathrm{d}} = \boldsymbol{j}_{\mathrm{d}} \cdot \boldsymbol{E} = \omega\varepsilon(\mathrm{i}\boldsymbol{E}) \cdot \boldsymbol{E} = 0$, so that it cannot be related to absorption
(ii) the power associated to the conduction current $P_{\mathrm{c}} = \boldsymbol{j}_{\mathrm{c}} \cdot \boldsymbol{E} = \sigma \boldsymbol{E}^2 \neq 0$, so that it *produces* light absorption.

Now, we can write

$$\varepsilon = \varepsilon_0 + \alpha_{\mathrm{p,re}} + \mathrm{i}\alpha_{\mathrm{p,im}} = \varepsilon_0 + \alpha_{\mathrm{p,re}} + \mathrm{i}\sigma / \omega = \varepsilon_{\mathrm{re}} + \mathrm{i}\varepsilon_{\mathrm{im}}, \tag{2.52}$$

(only σ is associated to the absorption and has dimensions of $\omega\varepsilon$, or $\omega\alpha_{\mathrm{p}}$). So, $\alpha = 2n_{\mathrm{im}}\omega / c = \varepsilon_{\mathrm{im}}\omega / (\varepsilon_0 n_{\mathrm{re}} c) = \sigma / (\varepsilon_0 n_{\mathrm{re}} c)$.

In conclusion, the interaction between the electromagnetic radiation and the matter is characterized by a *single complex physical quantity – n or* ε. In what follows, we demonstrate that the real and imaginary part of n or ε are also connected between them, the interaction of the electromagnetic radiation with matter being, in fact, characterized by a *single real number, a single independent quantity from the solutions of the Maxwell's equation.*

2.3.1 Kramers–Krönig Relations

These relations connect the real and imaginary part of any parameter that relates two fields in a linear and causal way. For example, ε relates in a linear and causal manner $\boldsymbol{D}$ and $\boldsymbol{E}$, α_{p} relates in the same way $\boldsymbol{P}$ and $\boldsymbol{E}$, r – the incident and reflected field, etc. In what follows we will focus on the connection between the real and imaginary part of ε.

Supposing that the excitation $\boldsymbol{E}$ is a plane wave $\boldsymbol{E}(t) = \boldsymbol{E}_0 \exp(\mathrm{i}\omega t)$, the causal response $\boldsymbol{D}$ is given by

$$\boldsymbol{D}(t) = \int_{-\infty}^{t} f(t-\tau)\boldsymbol{E}(\tau)\mathrm{d}\tau + \varepsilon_0 \boldsymbol{E} = \boldsymbol{E}(t)\int_0^{\infty} f(t')\exp(-\mathrm{i}\omega t')\mathrm{d}t' + \varepsilon_0 \boldsymbol{E}(t), \tag{2.53}$$

where the substitution $t' = t - \tau$ is used. Since by definition $\boldsymbol{D}(t) = \boldsymbol{E}(t)\varepsilon(\omega)$, it follows that

$$\varepsilon(\omega) = \int_0^{\infty} f(t')\exp(-\mathrm{i}\omega t')\mathrm{d}t' + \varepsilon_0 . \tag{2.54}$$

The last relation allows us to identify $f(t)$ with the Fourier transform of $\varepsilon(\omega) - \varepsilon_0$. Therefore, $f(t') = (1/2\pi)\int_{-\infty}^{\infty}[\varepsilon(\omega) - \varepsilon_0]\exp(\mathrm{i}\omega t')\mathrm{d}\omega$ and

$$\varepsilon(\omega) = (1/2\pi)\int_0^{\infty} \mathrm{d}t' \int_{-\infty}^{\infty} \mathrm{d}\omega' [\varepsilon(\omega') - \varepsilon_0]\exp[\mathrm{i}t'(\omega' - \omega)] + \varepsilon_0 . \tag{2.55}$$

After performing the integral over t', identical to the Fourier transform of the step function, and then over ω', and after separating the real and imaginary parts of the dielectric constant, we arrive finally at the Kramers–Krönig relations for ε

$$\varepsilon_{\mathrm{re}}(\omega) = \varepsilon_0 + \frac{1}{\pi} P \int_{-\infty}^{\infty} \frac{\varepsilon_{\mathrm{im}}(\omega')}{\omega' - \omega} \mathrm{d}\omega', \quad \varepsilon_{\mathrm{im}}(\omega) = -\frac{1}{\pi} P \int_{-\infty}^{\infty} \frac{\varepsilon_{\mathrm{re}}(\omega') - \varepsilon_0}{\omega' - \omega} \mathrm{d}\omega', \tag{2.56}$$

where P is the principal part of the integral. Since $\omega > 0$ it is desirable to transform (2.60) in integrals over the domain $(0, \infty)$. To this end we use the relation $\varepsilon(-\omega) = \varepsilon_{\mathrm{re}}(-\omega) + \mathrm{i}\sigma(-\omega)/(-\omega) = \varepsilon_{\mathrm{re}}(\omega) - \mathrm{i}\varepsilon_{\mathrm{im}}(\omega) = \varepsilon^*(\omega)$ ($\varepsilon_{\mathrm{re}}$, σ are dependent on $|\omega|$ from physical reasons). Thus, it follows that $\varepsilon_{\mathrm{re}}(\omega)$ is an even function of ω, and $\varepsilon_{\mathrm{im}}(\omega)$ is an odd function of ω. By multiplying now both the numerator and denominator of the Kramers–Krönig relations by $\omega + \omega'$ and using the parity relations of $\varepsilon_{\mathrm{re}}$, $\varepsilon_{\mathrm{im}}$, we finally obtain

$$\varepsilon_{\mathrm{re}}(\omega) = \varepsilon_0 + \frac{2}{\pi} P \int_0^{\infty} \frac{\omega' \varepsilon_{\mathrm{im}}(\omega')}{\omega'^2 - \omega^2} \mathrm{d}\omega', \quad \varepsilon_{\mathrm{im}}(\omega) = -\frac{2\omega}{\pi} P \int_0^{\infty} \frac{\varepsilon_{\mathrm{re}}(\omega') - \varepsilon_0}{\omega'^2 - \omega^2} \mathrm{d}\omega'. \tag{2.57}$$

So, $\varepsilon_{\rm re}$ can be computed if $\varepsilon_{\rm im}$ is known at all frequencies and vice-versa.

At very low frequencies, in the static case for which $\omega \to 0$, (2.57) becomes $\varepsilon_{\rm re}(0) = \varepsilon_0 + (2/\pi)P\int_0^\infty {\rm d}\omega' \varepsilon_{\rm im}(\omega')/\omega'$. If $\varepsilon_{\rm re}(0) \neq \varepsilon_0$, $\varepsilon_{\rm im}$ (the absorption) must be different from zero at least at some frequencies. Moreover, if the solid absorbs at low frequencies (ω' small), $\varepsilon_{\rm re}$ is large. On the contrary, at high frequencies $\omega >> \omega'$, $\varepsilon_{\rm re}(\omega) = \varepsilon_0 - (2/\pi\omega^2)P\int_0^\infty \omega' \varepsilon_{\rm im}(\omega'){\rm d}\omega'$. Introducing the plasma frequency $\omega_{\rm p}$ through $P\int_0^\infty \omega' \varepsilon_{\rm im}(\omega'){\rm d}\omega' = \pi\omega_{\rm p}^2\varepsilon_0/2$, for this frequency range

$$\varepsilon_{\rm re}(\omega) = \varepsilon_0(1 - \omega_{\rm p}^2/\omega^2), \tag{2.58}$$

the latter formula being known as the Drude expression. It describes the free-carrier contribution to ε at high frequencies in doped nonpolar semiconductors or metals. On the other hand, the dielectric constant in ionic insulators, with cubic structure and a single IR active transverse optical (TO) phonon of frequency $\omega_{\rm TO}$ has the form $\varepsilon(\omega) = \varepsilon_\infty + \Omega_{\rm p}^2/(\omega_{\rm TO}^2 - \omega^2)$, where ε_∞ is the background dielectric constant due to interband electronic transitions and $\Omega_{\rm p}^2$ measures the oscillator strength for TO phonons. In doped ionic semiconductors such as n-type GaAs, the contributions from TO phonons and free carriers are present simultaneously.

Another example of a Kramers–Krönig relation for the reflection coefficient $r = |r|\exp({\rm i}\theta)$, or $\ln r = \ln|r| + {\rm i}\theta$, reads as

$$\theta(\omega) = -(\omega/\pi)\int_0^\infty {\rm d}\omega' \ln[R(\omega')]/(\omega'^2 - \omega^2). \tag{2.59}$$

The Kramers–Krönig relations do not imply any knowledge of the interactions *inside* the solid, and therefore their validity is practically unlimited.

2.3.2 Drude–Lorentz Theory of the Electrically Charged Oscillator

A simple model which describes the general behavior of $\varepsilon(\omega)$ is the Drude–Lorentz theory which supposes that each electron in the dielectric medium is subjected to a forced movement when the electric field $\boldsymbol{E} = \hat{y}E$ of an incident electromagnetic radiation is applied along the y direction (whose versor we denote by $\hat{y}$), and that in the same time a restoring force towards equilibrium (generated by the ensemble of all other charges) is acting on the electron. Under these conditions the equation of the electron movement along the y direction is:

$$m({\rm d}^2u/{\rm d}t^2) - m\delta({\rm d}u/{\rm d}t) + Ku = -eE_0\exp({\rm i}\omega t), \tag{2.60}$$

where u is the displacement from the equilibrium position, δ is the amortization constant of the oscillations, and K the Hook constant. Looking for solutions of the type $u \approx \exp({\rm i}\omega t)$ we arrive at:

$$u = -(eE_0/m)\exp({\rm i}\omega t)/(\omega_0^2 - \omega^2 - {\rm i}\omega\delta), \tag{2.61}$$

where $\omega_0 = (K/m)^{1/2}$ is the characteristic frequency. Since the electron is moving from the equilibrium position, a total dipolar moment per unit volume $P = -N_0 e u \hat{=} (\alpha_{\text{p,re}} + \mathrm{i}\alpha_{\text{p,im}})E$ is induced along the y direction, with N_0 the number of oscillators per unit volume. By identification we obtain that

$$\varepsilon_{\text{re}}(\omega) = \varepsilon_0 + \alpha_{\text{re}}(\omega) = \varepsilon_0 + (N_0 e^2/m)(\omega_0^2 - \omega^2)/[(\omega_0^2 - \omega^2)^2 + \omega^2\delta^2],$$
$$\varepsilon_{\text{im}}(\omega) = \alpha_{\text{im}}(\omega) = (N_0 e^2/m)\omega\delta/[(\omega_0^2 - \omega^2)^2 + \omega^2\delta^2]. \quad (2.62)$$

In the case of multiple oscillators with characteristic frequencies ω_j and amortization factors δ_j,

$$\varepsilon = \varepsilon_0 + \sum_j (e^2/m)N_j/(\omega_j^2 - \omega^2 - \mathrm{i}\omega\delta_j), \quad (2.63)$$

with $\sum_j N_j = N_0$.

The dependence of ε_{re}, ε_{im} on ω is shown in Fig. 2.1.

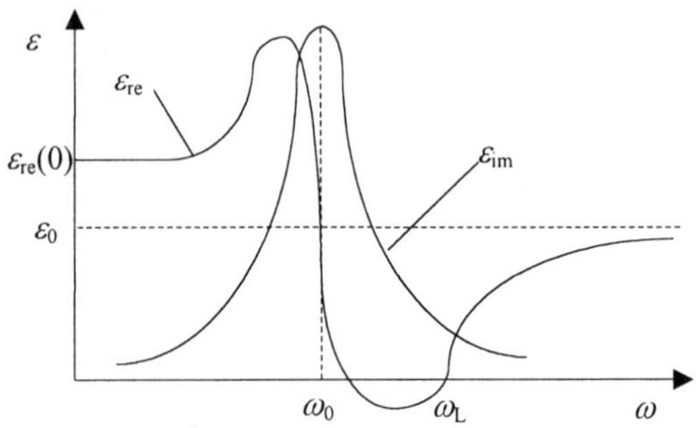

Fig. 2.1. The dependence of ε_{re}, ε_{im} on ω

$\varepsilon_{\text{im}}(\omega)$ peaks at $\omega = \omega_0$ and has a *Lorentzian shape* around this value with a halfwidth determined by δ:

$$\varepsilon_{\text{im}} \cong (N_0 e^2 \delta/m\omega_0)/[4(\omega_0 - \omega)^2 + \delta^2]. \quad (2.64)$$

For frequencies far from ω_0, $\varepsilon_{\text{im}} \cong 0$, ε is real and the wave is propagating *without* absorption. ω_0 is the resonance frequency of the system at which the electromagnetic field strongly excites the electronic oscillators.

$\varepsilon_{\text{re}}(\omega) = \varepsilon_0$ at $\omega = \omega_0$ and has a minimum and maximum shifted from ω_0 with about δ. Inside the region between the maximum and the minimum values, ε *decreases* with frequency – *anomal dispersion region*, and outside this region it *increases* with frequency – *normal dispersion region*. ε_{re} can even be negative, case in which no wave can propagate. In the region where $\varepsilon_{\text{re}} < 0$, $\varepsilon_{\text{im}} \cong 0$, n is purely imaginary, a *total reflection* of the incident wave taking place. This region extends between ω_0 and ω_L. At ω_L, $\varepsilon_{re} \cong 0$ and only *longitudinal* waves can propagate (the transversal wave condition $\mathrm{div}\boldsymbol{E} = 0$ is not satisfied since the first Maxwell's equation $\mathrm{div}(\varepsilon \boldsymbol{E}) = 0$ is automatically satisfied for any $\boldsymbol{E}$). These longitudinal waves are plasmon waves (see Sect. 1.5). The region between ω_0 and ω_L decreases with the increase of δ such that for sufficiently large δ values waves with any frequency can propagate. The formation of the forbidden band can be seen from the plot in Fig. 2.2 of the frequency as a function of $n_{re}\omega$ (n_{re} is obtained solving the equation $n_{re}^2 - (\varepsilon_{im}/2\varepsilon_0 n_{re})^2 = \varepsilon_{re}/\varepsilon_0$).

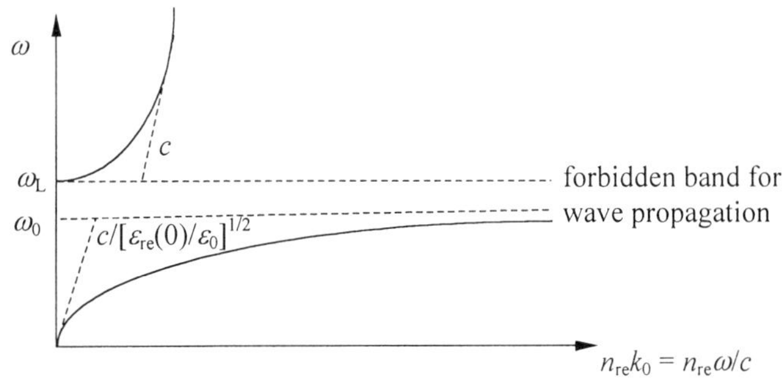

Fig. 2.2. Allowed and forbidden bands for wave propagation as a function of $n_{re}\omega$

For small ω the light propagates with constant speed $c/n(0) = c/[\varepsilon_{re}(0)/\varepsilon_0]^{1/2}$, while at high frequencies $\omega >> \omega_0$, $\varepsilon_{re}(\infty) \cong \varepsilon_0$, and the light propagates with a speed $\cong c$.

The expressions for ε_{re} and ε_{im}, for the free-electron case ($\omega_0 = 0$), are:

$$\varepsilon_{re}(\omega) = \varepsilon_0 - (N_0 e^2/m)/(\omega^2 + \delta^2), \quad \varepsilon_{im}(\omega) = (N_0 e^2 \delta/m)/[\omega(\omega^2 + \delta^2)]. \quad (2.65)$$

The contribution of free electrons at ε_{re} is *negative* because the electrons are moving in antiphase with the electromagnetic field due to the inertial mass and the lack of a restoring force. In this case, the plasma frequency introduced previously through $\pi\varepsilon_0\omega_p^2/2 = \int_0^\infty \omega\varepsilon_{im}(\omega)\mathrm{d}\omega$ is found to be $\omega_p^2 = N_0 e^2/m\varepsilon_0$, a relation identical to that introduced in Chap. 1 for plasmons.

Taking into account that at low frequencies $\omega \to 0$ for free electrons $\sigma = \varepsilon_{im}\omega \to (N_0 e^2/m\delta)$ the significance of δ can be inferred as $\delta = 1/\tau$ where τ is the relaxation time. This identification follows by comparing σ with the expression obtained for free-electron transport $\sigma_0 = N_0 e^2 \tau/m$.

The Drude–Lorentz theory, despite being a phenomenological one, is in agreement with experimental observations; it gives ω_p as a function of N_0 and m and is especially suitable for studying the behavior of metals, or the contribution of free carriers to the dielectric constant of doped nonpolar semiconductors. Moreover, a Drude-like theory has been proposed for a single critical-point transition of electrons from the valence to the conduction band, as well as for collective oscillations in semiconductors and insulators (Chen et al., 1993).

In this case $\int_0^\infty \omega'\varepsilon_2(\omega')\mathrm{d}\omega' = (\pi/2)\omega_\mathrm{p}^2$, $\int_0^\infty \omega'\alpha(\omega')\mathrm{d}\omega' = (\pi/4)\omega_\mathrm{p}^2$, and $\int_0^\infty \omega'\mathrm{Im}[-1/\varepsilon(\omega')]\mathrm{d}\omega' = (\pi/2)\omega_\mathrm{p}^2$. If there is more that one critical point which contributes to the spectrum, a sum over the separate contributions of each critical point is required.

2.3.3 Sum Rule

According to (2.63), denoting by $f_j = N_j/N_0$, the susceptibility in the presence of several oscillations with frequencies ω_j, at frequencies close enough to all ω_j can be written as

$$\chi(\omega) = \frac{e^2 N_0}{2m\varepsilon_0}\Sigma_j \frac{f_j}{\omega_j}\frac{1}{\omega_j - \omega - \mathrm{i}\delta_j/2}, \tag{2.66}$$

with $\Sigma_j f_j = 1$.

This form of the susceptibility remains valid also in a quantum mechanical treatment. However, a full quantum treatment is able to determine both the value of the parameter f_j and of the damping constants δ_j. More precisely, in a quantum theory the material medium is modeled as a collection of oscillators with different frequencies $\omega_{fi} = (E_f - E_i)/\hbar$, which are excited from their initial state i by the action of the electromagnetic field. The susceptibility of the material system is then obtained by adding the contribution of all oscillators, irrespective of their final state. The result is the expression

$$\chi(\omega) = \frac{e^2 N_0}{2m\varepsilon_0}\Sigma_f \frac{f_{fi}}{\omega_{fi}}\frac{1}{\omega_{fi} - \omega - \mathrm{i}\delta_{fi}/2}, \tag{2.67}$$

with

$$\Sigma_f f_{fi} = 1. \tag{2.68}$$

Relation (2.68) is called the sum rule for the oscillator strengths, where in quantum mechanics the dimensionless parameter $f_{fi} = (2m\omega_{fi}/e^2\hbar)|d_{fi}|^2$ is called the strength of the oscillator with frequency ω_{fi}, and d_{fi} are the matrix elements of the electric dipole moment between the initial and final state. The sum rule suggests that in quantum mechanics the total strength of the transition can be imagined as that due to an oscillator which is distributed over many partial oscillators, each of them with an oscillator strength f_{fi}. (In a complete quantum mechanical treatment $(\omega_{fi} - \omega - \mathrm{i}\delta_{fi}/2)^{-1}$ in (2.67) is in fact replaced with $(\omega_{fi} - \omega - \mathrm{i}\delta_{fi}/2)^{-1} - (-\omega_{fi} - \omega - \mathrm{i}\delta_{fi}/2)^{-1}$ (see Haug and Koch (1990)). Note that the strength of the interaction between the electromagnetic field and the oscillator is described classically by the dimensionless quantity S_j

$= f_j N_0 e^2 /(\varepsilon_0 m \omega_j^2)$, with f_j, ω_j having the meanings discussed in connection with expression (2.66) (Basu, 1997).

2.3.4 Microscopic Origin of Absorption

In order to find the relation between the absorption coefficient and the transition probability per unit time, we return to the definition of the absorption coefficient for a wave propagating along the z direction $\alpha = -(\mathrm{d}I/\mathrm{d}z)/I$ $= -(n/c)(\mathrm{d}I/\mathrm{d}t)/I$, where n is the refractive index of the medium with and $\mathrm{d}I/\mathrm{d}t$ the absorbed energy per unit time. This last quantity can also be written as

$$\mathrm{d}I/\mathrm{d}t = \hbar\omega(\mathrm{d}P_{\mathrm{tot}}/\mathrm{d}t), \tag{2.69}$$

where $\hbar\omega$ is the absorbed energy at the induced transition, taking into account only single-photon processes. P_{tot} is the total transition probability, expressed as the sum or integral over all states which can participate in the transition. More precisely,

$$\alpha = (n/c)\hbar\omega(2\pi/\hbar)\int |H_{\mathrm{int},fi}|^2 \delta[E_f(\boldsymbol{k}_f) - E_i(\boldsymbol{k}_i) - \hbar\omega]\mathrm{d}\boldsymbol{k}_f \mathrm{d}\boldsymbol{k}_i / N\hbar\omega, \tag{2.70}$$

where the integral is performed over all values of $\boldsymbol{k}_{f,i}$ in the first Brillouin zone and $N\hbar\omega$ is the incident photon energy. $H_{\mathrm{int},fi} = H_{A,fi}$ in the first-order approximation and $H_{AA,fi} + \sum_m H_{A,fm} H_{A,mi}/(E_m - E_i)$ in the second-order approximation of perturbation theory. We have not used the expression of the transition probability in which $\rho(E)$ appears because the integral is performed over $\boldsymbol{k}$, not over E. Also, (2.70) does not account for the occupation probability of different states in the initial and final energy bands.

In this way, the macroscopic parameter – the absorption coefficient, is related to the transition probabilities between different states. A more precise expression, and the selection rules for absorption can only be derived with a particular model for the specific type of transitions. For example, for transitions between electron states in energy bands in a crystalline solid

$$\alpha = (\alpha_0/\hbar\omega)\int |M_{fi}|^2 \delta(E_f - E_i - \hbar\omega)\mathrm{d}\boldsymbol{k}, \tag{2.71}$$

with $M_{fi} = [(2\pi)^3/\Omega]\hat{e}_s \delta(\boldsymbol{k}_i - \boldsymbol{k} - \boldsymbol{k}_f)\int_{\mathrm{cell}} u_i^*(\boldsymbol{k}_i, \boldsymbol{r})\nabla u_f(\boldsymbol{k}_f, \boldsymbol{r})\mathrm{d}\boldsymbol{r}$ and α_0 a constant (Callaway, 1991). In (2.71) the integral over the first Brillouin zone can be transformed into a surface integral

$$\alpha = (\alpha_0/\hbar\omega)\int_S \mathrm{d}S\, |M_{fi}|^2 / |\nabla_{\boldsymbol{k}}(E_f - E_i)|_{E_f - E_i = \hbar\omega}, \tag{2.72}$$

over the surface S of equal energy in the $\boldsymbol{k}$-space, defined by $E_f - E_i = \hbar\omega$.

In many cases M_{fi} has a slow variation with $\boldsymbol{k}$, so that α is determined by the total density of states $J_{fi}(\omega) = \int_S \mathrm{d}S / |\nabla_{\boldsymbol{k}}(E_f - E_i)|_{E_f - E_i = \hbar\omega}$. In a semiconductor, f = c (conduction) and i = v (valence), if we are referring at interband transitions. J_{cv}, and thus the absorption coefficient, has a strong variation with ω for those frequencies for which $\nabla_k E_c(\boldsymbol{k}) = \nabla_k E_v(\boldsymbol{k}) = 0$ or $\nabla_k E_c(\boldsymbol{k}) = \nabla_k E_v(\boldsymbol{k}) \neq 0$. The first relation defines the *critical points* in $\boldsymbol{k}$-space, which we have discussed in Chap. 1.

2.3.5 Absorption Mechanisms

2.3.5.1 Band-to-Band Light Absorption: Direct and Allowed Absorption

This is the most important absorption mechanism, predominant in crystals with direct band structure. The dependence of the absorption coefficient on the frequency of the electromagnetic radiation can be calculated with (2.72). For nondegenerate semiconductors (semiconductors which are not strongly doped) where the valence band is practically fully occupied and the conduction band empty, the occupation probability of the bands need not be explicitly considered, and we can take M_{cv} as slowly varying on the surface of constant energy. Moreover, if the conduction and valence bands are both isotropic and parabolic, and have their extrema at $\boldsymbol{k} = 0$ (see Fig. 2.3), i.e. if $E_c = E_g + \hbar^2 k^2/(2m_c)$, $E_v = -\hbar^2 k^2/(2m_v)$, then $E_c - E_v - \hbar\omega = 0 = E_g + \hbar^2 k^2/2m_r - \hbar\omega$, where $m_r = m_c m_v/(m_c + m_v)$ is the reduced mass.

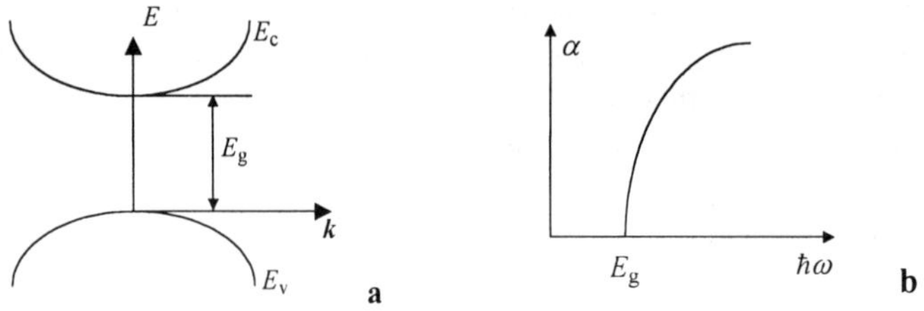

Fig. 2.3. (**a**) Schematic representation of energy bands and (**b**) the behavior of the absorption coefficient at band-to-band absorption

The net result after performing the integral in (2.72) is

$$\alpha(\omega) \approx (\hbar\omega)^{-1}(\hbar\omega - E_g)^{1/2}, \tag{2.73}$$

which has a threshold at $\hbar\omega_{th} = E_g$ (see Fig. 2.3). Typical values for $\hbar\omega > E_g$ are $\alpha \cong 10^4$–10^5 cm^{-1}. This value is about 100–1000 times larger than for indirect transitions.

When the effective masses in the conduction and valence bands are not isotropic, the above integral must be performed over an ellipsoid, and the absorption coefficient is anisotropic. There are also situations when the effective mass takes negative values in some or in all three directions.

2.3.5.2 Band-to-Band Direct, Forbidden Transitions

This kind of transition appears when direct transitions between the conduction and valence bands are forbidden by a selection rule at the extreme point $\boldsymbol{k} = 0$, but can take place in the neighborhood of the extreme. In this case the interaction Hamiltonian is written as a series around $\boldsymbol{k} = 0$

$$|H_{\text{int,cv}}(\boldsymbol{k})| \cong \boldsymbol{k}(\partial |H_{\text{int,cv}}(\boldsymbol{k})| / \partial \boldsymbol{k})\Big|_{\boldsymbol{k}=0} + \ldots, \tag{2.74}$$

in which $|H_{\text{int,cv}}(0)|$ is zero. In the calculation of the absorption coefficient only the first term in the expansion is retained; integrating (2.72) over the spherical surface of radius $k = 2m_{\text{r}}(\hbar\omega - E_{\text{g}})^{1/2} / \hbar$ the absorption coefficient is

$$\alpha(\omega) \approx (\hbar\omega)^{-1}(\hbar\omega - E_{\text{g}})^{3/2}. \tag{2.75}$$

It has again a threshold for $\hbar\omega_{\text{th}} = E_{\text{g}}$, but the value of the absorption coefficient is smaller than for the allowed case (the density of states in points other than $\boldsymbol{k} = 0$ is generally smaller).

So, by measuring $\alpha(\omega)$ around the absorption threshold, one can obtain information about the symmetry of the wavefunctions in the conduction and valence bands (if the wavefunctions have the same parity at $\boldsymbol{k} = 0$, forbidden transitions occur). From measurements of $\alpha(\omega)$ the most exact values of E_{g} can be obtained, for both allowed and forbidden direct transitions.

2.3.5.3 Band-to-Band Indirect, Allowed Transitions

These processes have a much weaker intensity than direct transitions since they are second-order processes in H_{int}. They appear when the conduction and valence bands have their extreme points localized in different directions in the $\boldsymbol{k}$-space (see Fig. 2.4)

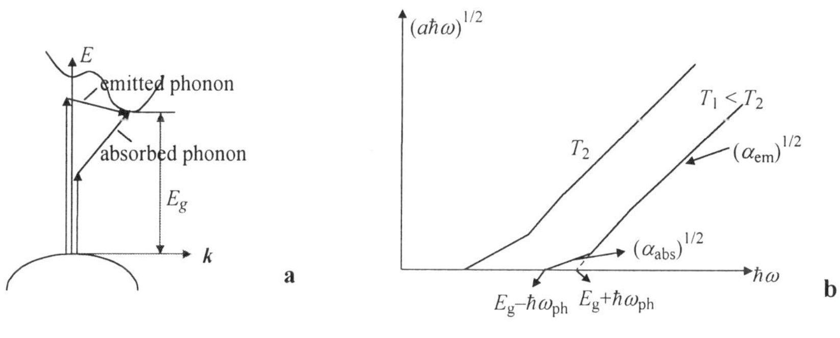

Fig. 2.4. (**a**) Indirect band-to-band allowed transitions and (**b**) the energy dependence of the absorption coefficient

The transitions take place with the emission or absorption of phonons in order to conserve the total momentum. In this case the transition probability is not determined by $H_{\text{int},fi}$, which is equal to zero (because the moment is not conserved at transitions in which only the photon and electrons take place), but by $\sum_m H_{\text{int},fm} H_{\text{int},mi} /(E_{0i} - E_{0m})$, where the sum is performed over the intermediate states m. We suppose that the interaction takes place in two steps: i) absorption of the photon, represented by a vertical transition, ii) absorption or emission of a phonon, represented by an oblique transition. The inverse processes: the absorption/emission first of phonons and then the absorption of photons are also possible, but have a much lower probability since $E_{0i} - E_{0m}$ is in general much larger than in the case when the photon is absorbed first (Basu, 1997).

Thus, $H_{\text{int},mi}$ describes the direct transitions and has the same expression as for direct transitions, whereas $H_{\text{int},fm}$ describes the electron-phonon coupling. This last term is generally difficult to calculate; however, in the simplified case of a semiconductor in which the transition takes place with the absorption of a phonon, and in which the reference level for the energy is taken at the minimum of the conduction band, energy conservation requires that $E_c = E_v + \hbar\omega + \hbar\omega_{\text{ph}}$ where ω_{ph} is the frequency of the phonon. In the intermediate state the total momentum is conserved, but the conservation of energy is not necessary if the lifetime of the intermediate state is short. Assuming that the density of states in the conduction and valence bands are $\rho(E_c) \approx (E_c)^{1/2}$ and $\rho(E_v) \approx (-E_g - E_v)^{1/2}$, respectively, and that $H_{\text{int},fm} H_{\text{int},mi} /(E_{0i} - E_{0m})$ depends slowly on energy, $\alpha_{\text{abs}}(\omega) \approx [\bar{n}(\omega_{\text{ph}})/\hbar\omega] \int \delta(E_c - E_v - \hbar\omega - \hbar\omega_{\text{ph}}) \rho(E_v) \rho(E_c) \mathrm{d}E_v \mathrm{d}E_c$. The contribution of transitions with phonon emission is obtained by changing the sign of $\hbar\omega_{\text{ph}}$ and by replacing $\bar{n}(\omega_{\text{ph}})$ with $1 - \bar{n}(\omega_{\text{ph}})$. The total absorption coefficient is thus

$$\alpha = \alpha_{\text{abs}} + \alpha_{\text{em}} \approx \frac{1}{\hbar\omega} \left[\frac{(\hbar\omega + \hbar\omega_{\text{ph}} - E_g)^2}{\exp(\hbar\omega_{\text{ph}} / k_B T) - 1} + \frac{(\hbar\omega - \hbar\omega_{\text{ph}} - E_g)^2}{1 - \exp(-\hbar\omega_{\text{ph}} / k_B T)} \right]. \tag{2.76}$$

At low temperatures α_{abs} is small. The curve $(\alpha\hbar\omega)^{1/2} = f(\hbar\omega)$ is approximately a straight line for $E_g - \hbar\omega_{\text{ph}} < \hbar\omega < E_g + \hbar\omega_{\text{ph}}$, and at small energies, where phonon absorption is much more probable than phonon emission, we have again another straight line for $\hbar\omega > E_g + \hbar\omega_{\text{ph}}$ (see Fig. 2.4). More than one phonon branch can be present in this process; in this case there are more 'knees' in the absorption curve where one reaches the threshold for the new phonons to participate.

In the case of indirect transitions α gives information on the value and $\boldsymbol{k}$-space position of the energy gap E_g as well as on its temperature dependence; by comparing these data with other experimental data (obtained from vibration spectra) it is possible to determine the separation in the $\boldsymbol{k}$-space between the extremes of the conduction and valence bands.

2.3.5.4 Band-to-Band Indirect, Forbidden Transitions

This type of transition occurs if, as for forbidden direct transitions, the matrix element of H_{int} is equal to zero for a given $\boldsymbol{k}_0$, but different from zero in general.

Again, the interaction Hamiltonian is replaced with its first derivative multiplied by $\boldsymbol{k}-\boldsymbol{k}_0$, which determines the appearance of a term to the third power: $(\hbar\omega \pm \hbar\omega_{\rm ph} - E_{\rm g})^3$ instead of the corresponding term to the second power in the expression of α. These different forms of the dependence of α on ω are helpful to experimentally distinguish between different transition types.

The absorption coefficient for indirect transitions is smaller than for direct transitions: $\alpha \cong 10$–$20\ \text{cm}^{-1}$, and the dependence of α on $\hbar\omega$ as well as on temperature is more pronounced than for direct transitions.

2.3.5.5 Effects That Can Appear at Band-to-Band Absorption

There are several effects that accompany the band-to-band absorption (Basu, 1997). One of them is the Burstein–Moss shift, which is observed as a shift of the absorption threshold from $E_{\rm g}$ to $E_{\rm g}+\xi$ in heavily doped semiconductors for which the Fermi level $E_{\rm F}$ is situated at a depth ξ inside the conduction band. This effect is stronger when the effective mass is smaller and can be used to determine ξ from absorption measurements (and from it one can then determine the carrier concentration). The Burstein–Moss effect is not only a shift of the fundamental absorption in (2.79); the shape of the absorption coefficient is different from (2.79) because we must now take into account $f_{\rm c}$, $f_{\rm v}$, which were neglected in nondegenerate semiconductors.

Another effect present in heavily doped dielectrics is the exponential decrease of the energy bands into the energy gap, which produces an exponential absorption dependence on energy $\alpha(\omega) \approx \exp[g(\hbar\omega - E_{\rm g})]$. This Urbach law is explained by the appearance of disorder-induced density of levels below the conduction and valence bands. The coefficient g can depend on temperature, as for example $g = 1/k_{\rm B}T$ in KBr, or can be practically independent of temperature, as in compensated GaAs.

When an external electric field is applied on an absorbing probe, the absorption coefficient is modified due to the Franz–Keldysh effect, which is in fact a photon-assisted tunneling phenomenon.

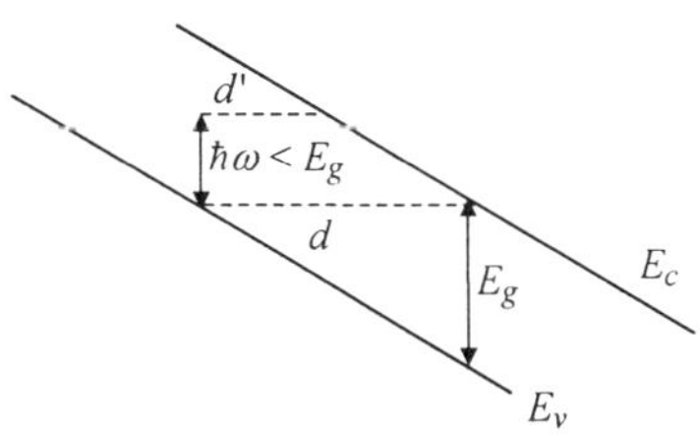

Fig. 2.5. The 'tilted' conduction and valence bands due to an applied electric field

In the presence of an electric field, the conduction and valence bands are 'tilted' in space, as in Fig. 2.5. The electron wavefunction decreases exponentially in the forbidden band, so that to pass from the valence to the conduction band the electron must tunnel over a length $d = E_{\rm g}/(eF)$, where F is the applied electric field. If photons with energy $\hbar\omega < E_g$ are simultaneously absorbed, the length of the tunneling path decreases to $d' = (E_g - \hbar\omega)/(eF)$ and hence the tunneling probability increases.

If the Franz–Keldysh effect is associated with the fundamental direct absorption, it shows as a shift of the absorption threshold towards lower energies as well as in a change of its dependence on the photon energy

$$\alpha(\omega) \approx [F^{1/3} / \hbar\omega(E_g - \hbar\omega)]\exp[-(4/3)(E_g - \hbar\omega)^2 /(edF)^2]. \tag{2.77}$$

So, in the presence of the Franz–Keldysh effect, the absorption coefficient depends exponentially on the energy of incident photons and on the electric field intensity F, the absorption being possible for radiation with energy lower than the bandgap. This effect can be used to determine E_g in materials for which there is no available radiation source near their bandgap energy.

2.3.5.6 Intraband Absorption

Intraband absorption takes place between subbands of the conduction or valence bands, for $\hbar\omega << E_g$. The most common are the transitions between subbands of the valence band, separated due to the spin-orbit interaction.

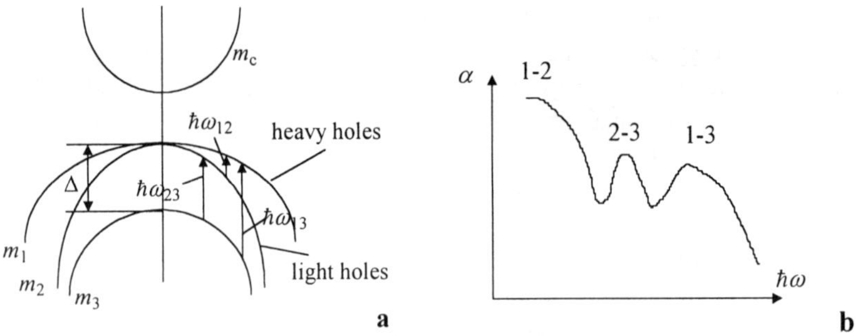

Fig. 2.6. (**a**) Possible transitions at intraband absorption and (**b**) the corresponding absorption coefficient behavior with energy

Three direct transitions are possible, with corresponding energies $\hbar\omega_{12}$, $\hbar\omega_{13}$, $\hbar\omega_{23}$ (see Fig. 2.6). The transitions at $\boldsymbol{k} = 0$ are forbidden so that the absorption is quite weak (these are forbidden direct transitions) and the transition probabilities are proportional to $|\boldsymbol{k}|^2$. Since the density of states for holes decreases with $|\boldsymbol{k}|$, the absorption due to the possible transitions have maxima as a function of energy. From absorption data it is possible to determine the spin-orbit split energy Δ.

Intraband indirect transitions can also take place through scattering of charge carriers on phonons or impurities in order to ensure the simultaneous conservation of energy and total momentum. The contribution of intraband transitions at ε_{re} can be identified in $\varepsilon_{re}(0)$ (in the limit of very small frequencies).

For intraband transitions α is proportional to the hole density. If the doping or the temperature are modified, both intensity and position of the absorption peaks change. More precisely, when $E_g < \Delta$, moves deeper inside the valence band, the 1-3 and 1-2 peaks shift towards higher energies, and the 2-3 peak shifts towards lower energies, since in the valence band the states *above* E_F are occupied by holes and those *below* E_F are empty. In semiconductors for which $E_g < \Delta$, the 1-3 and 2-3 transitions can be masked by the fundamental band-to-band absorption.

2.3.5.7 Absorption on Free Carriers

This absorption mechanism is characteristic of metals, especially alkaline metals, but is also encountered in highly doped semiconductors. In metals, the absorption on free carriers is very well described by the Drude–Lorentz model with $\omega_0 = 0$, $\delta = 1/\tau$ (with τ the relaxation time), N_0 the concentration of free carriers, and the mass m replaced by the optical mass m_{opt}. The absorption in metals, including the definition of the optical mass, is discussed in more detail in Chap. 4. The optical mass can be identified with the effective mass for isotropic Fermi surfaces; as mentioned in Abelès (1972), the infrared reflectivity for crystals with different orientations of the surface with respect to the crystallographic axes can be used to obtain the Fermi surface. Also, τ can be anisotropic and frequency dependent, its parabolic decrease with frequency being caused by interelectronic collisions. The interaction of free electrons with the surface of the metal produces an increase of the relaxation time and of the absorption.

As follows from (2.62), in semiconductors α is larger for larger N_0, and smaller m_{eff} (instead of m) and ω; it becomes important at frequencies less than ω_{th} for the fundamental band-to-band absorption. At high free-carrier concentrations, the threshold of the fundamental absorption can even be masked by the absorption on free carriers; in this case the absorption curve looks like that in Fig. 2.7.

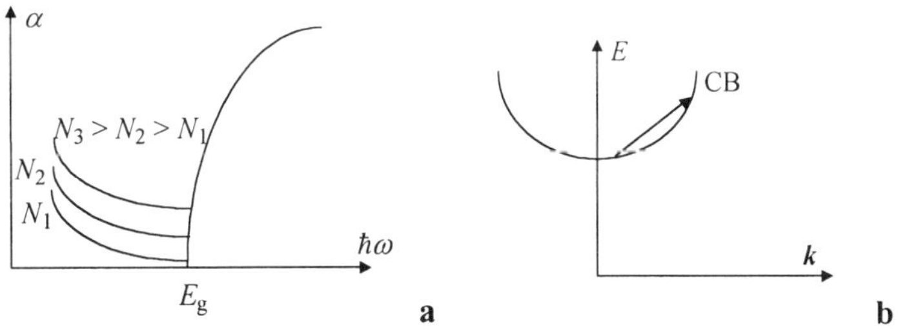

Fig. 2.7. (**a**) Energy dependence of the absorption on free carriers in highly doped direct semiconductors and (**b**) the corresponding transitions

Since free carriers move inside a band, photon absorption is accompanied by a transition to a state with higher energy inside the same band; therefore, this absorption mechanism has no threshold. However, an additional interaction (with phonons, ionized impurities, etc.) is necessary in general to conserve the total momentum – see Fig. 2.7. The scattering mechanism that ensures the conservation of momentum influences the frequency dependence of α through $\tau(\omega)$: $\alpha_{ac} \approx 1/\omega^{3/2}$ – for scattering on acoustic phonons, $\alpha_{op} \approx 1/\omega^{5/2}$ – for scattering on optical phonons, and $\alpha_{ion} \approx 1/\omega^{7/2}$ – for scattering on ionized impurities. In the latter case the absorption coefficient is also dependent on the impurity type. Also, scattering on complex defects (growth defects, defects due to irradiation or mechanical processing) can induce an absorption α_{cd}, which has different values for each crystal. In general, all these mechanism can contribute simultaneously, in which case their contributions add. The order of magnitude of the absorption coefficient depends on the concentration of free carriers. For example, in Ge, $\alpha = 4$ cm^{-1} for $N_0 = 10^{17}$ cm^{-3}, whereas $\alpha = 400$ cm^{-1} (comparable with the fundamental absorption) for $N_0 = 10^{19}$ cm^{-3}.

2.3.5.8 Absorption on Wannier–Mott Excitons

For excitons, the sum or integral over $\boldsymbol{k}$ in the absorption coefficient $\alpha \approx \sum_{\boldsymbol{k}} |H_{\text{int,cv}}(\boldsymbol{k})|^2 \, \delta(E - \hbar\omega)$ must be replaced with a sum or an integral *over n* – the exciton levels. Since on optical absorption only electron-hole pairs with $\boldsymbol{K} = 0$ are created, the absorption coefficient can be written as (Basu, 1997)

$$\alpha = (\alpha_0 / \hbar\omega) \sum_n |f_n(0)|^2 \, \delta(\hbar\omega - E_g + E_b / n^2), \tag{2.78}$$

where we have assumed that $H_{\text{int},\omega(n)} \approx f_n(\boldsymbol{r} = 0)$ with $f_n(\boldsymbol{r}) = R_{n,e}(r) Y_e^n(\theta, \phi)$ the eigenfunctions of the hydrogen atom and $|f_n(0)|^2 = 1/\pi a_B^3 n^3$. When $\hbar\omega \to E_g$ we can consider that n varies continuously, i.e. the sum in (2.78) is replaced with an integral (inside the forbidden band), the net result after making the change of variables $x = \hbar\omega - E_g + E_b / n^2$ being

$$\alpha = (\alpha'_0 / \hbar\omega) \int_{\hbar\omega - E_g}^{\hbar\omega - E_g + E_b} \mathrm{d}x \delta(x) / E_b. \tag{2.79}$$

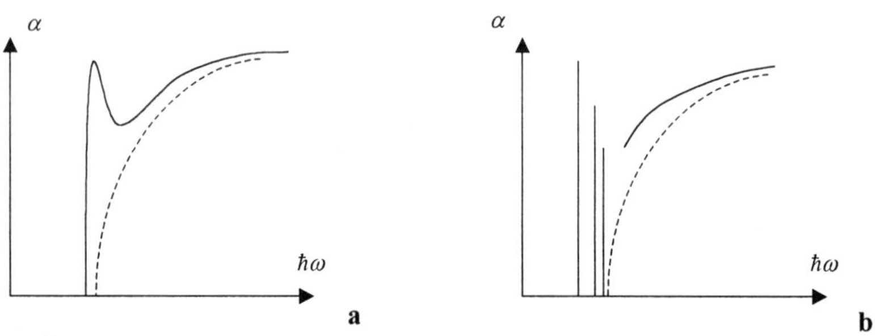

Fig. 2.8. (**a**) Typical experimental absorption curve in the presence of excitons and (**b**) ideal excitonic absorption curve in direct semiconductors (the *dashed curve* is for fundamental absorption)

For $\hbar\omega > E_g$ the exciton states are continuous and $|f_n(0)|^2 = z\exp z/\sinh z$, with $z = \pi / ka_{ex}$ where $k = |\boldsymbol{k}|$ is the momentum of the electron-hole relative motion $\boldsymbol{k} = (m_c\boldsymbol{k}_c - m_v\boldsymbol{k}_v)/M$. The absorption coefficient in this case is

$$\alpha = \frac{\alpha_0}{\hbar\omega}\int dk \frac{4\pi k^2 \delta(\hbar^2 k^2/2m_r + E_g - \hbar\omega)}{1-\exp(-2\pi/ka_{ex})} \approx \frac{1}{\hbar\omega}\frac{2\pi\sqrt{E_b}}{1-\exp[-2\pi\sqrt{E_b/(\hbar\omega - E_g)}]}. \tag{2.80}$$

When $\hbar\omega >> E_g$, (2.80) approaches $\alpha \approx (\hbar\omega)^{-1}(\hbar\omega - E_g)^{1/2}$, i.e. (2.80) has the same expression as at direct band-to-band transitions.

2.3.5.9 Indirect Excitonic Absorption

In semiconductors with an indirect bandgap, as in Ge and Si, excitons are also created below the indirect gap energy with nonzero translational kinetic energy, the absorption curve being no longer characterized by a series of lines but by a series of 'knees' (see Fig. 2.9). The 'knees' in the absorption curve appear because more than one phonon can take place in the transition; these phonons can be absorbed or emitted in any combinations.

Fig. 2.9. Excitonic absorption spectrum in indirect semiconductors

In indirect bandgap semiconductors, if $\boldsymbol{k}_0$ is the position of the conduction band minimum, the dispersion relations for electrons and holes become $E_c(\boldsymbol{k}_c) = E_g + \hbar^2(\boldsymbol{k}_c - \boldsymbol{k}_0)^2/(2m_c)$, $E_v(\boldsymbol{k}_v) = -\hbar^2\boldsymbol{k}_v^2/(2m_v)$, and the energies of the relative exciton motion and of the center-of-mass motion are, respectively, $E_2(\boldsymbol{k}_0) = E_g - E_b/n^2$ and $E_1 = \hbar^2(\boldsymbol{K} - \boldsymbol{k}_0)^2/(2M)$ (see Chap. 1). The absorption coefficient is defined in this case as (Basu, 1997)

$$\alpha \approx (\hbar\omega)^{-1} |\textstyle\sum_{n'} H_{\text{int},n'i} H_{\text{int},fn'}/(E_{n'} - \hbar\omega)|^2 \, \delta[\hbar\omega \pm \hbar\omega_{ph} - E_2(\boldsymbol{k}_0) - \hbar^2\boldsymbol{K}^2/2M], \tag{2.81}$$

where $H_{\text{int},n'i}$ describes the transition between the initial and intermediary states, and $H_{\text{int},fn'}$ between the intermediary and the final excitonic state of label n and center-of-mass vector $\boldsymbol{K}$. The transition can only take place with the participation of a phonon with wavevector $\boldsymbol{q} = \boldsymbol{K}$. If we consider only the process with phonon absorption and $n = 1$, (2.81) becomes

$$\alpha_{\mathrm{abs}} \approx (\hbar\omega)^{-1} \mid f_1(0)\mid^2 \int \mathrm{d}\boldsymbol{K}\delta(\hbar\omega + \hbar\omega_{\mathrm{ph}} - E_2(\boldsymbol{k}_0) - \hbar^2\boldsymbol{K}^2/2M), \tag{2.82}$$

where the sum must be performed over the excitonic states $\boldsymbol{K}$ which conserve the energy and the momentum. Transforming the integral over $\boldsymbol{K}$ into an integral over energy through $E_1 = \hbar^2(\boldsymbol{K} - \boldsymbol{k}_0)^2/(2M)$, the final result is

$$\alpha_{\mathrm{abs}} \approx (\hbar\omega)^{-1}\sqrt{\hbar\omega - E_{\mathrm{g}} + E_{\mathrm{b}} + \hbar\omega_{\mathrm{ph}}}\,\overline{n}(\omega_{\mathrm{ph}}) \mid f_1(0)\mid^2 H(\hbar\omega - E_{\mathrm{g}} + E_{\mathrm{b}} + \hbar\omega_{\mathrm{ph}}), \tag{2.83}$$

where $H(x)$ is the Heaviside function. The absorption coefficient depends on the energy of incident photons as $(\hbar\omega - E')^{1/2}$, whereas for the indirect fundamental absorption the dependence is $(\hbar\omega - E')^2$.

Analogously, the absorption term with phonon emission is

$$\alpha_{\mathrm{em}} \approx (\hbar\omega)^{-1}\sqrt{\hbar\omega - E_{\mathrm{g}} + E_{\mathrm{b}} - \hbar\omega_{\mathrm{ph}}}\,[\overline{n}(\omega_{\mathrm{ph}}) + 1] \mid f_1(0)\mid^2 H(\hbar\omega - E_{\mathrm{g}} + E_{\mathrm{b}} - \hbar\omega_{\mathrm{ph}}), \tag{2.84}$$

where $\mid f_1(0)\mid^2$ is the probability that an electron and a hole exist in the same position in space (i.e. with a relative separation $\boldsymbol{r} = 0$).

For transitions into continuous excitonic states

$$\alpha \approx (\hbar\omega)^{-1}\int \mathrm{d}\boldsymbol{K}\mathrm{d}\boldsymbol{k}(z\exp z/\sinh z)\delta(E_{\mathrm{g}} + \hbar^2\boldsymbol{K}^2/2M + \hbar^2\boldsymbol{k}^2/2m_{\mathrm{r}} \pm \hbar\omega_{\mathrm{ph}} - \hbar\omega). \tag{2.85}$$

At $\hbar\omega \cong E_{\mathrm{g}}$ (2.85) can be approximated with $\alpha \approx (\hbar\omega)^{-1}(\hbar\omega - E_{\mathrm{g}} \pm \hbar\omega_{\mathrm{ph}})^{3/2}$, whereas for $\hbar\omega >> E_{\mathrm{g}}$, $\alpha \approx (\hbar\omega)^{-1}(\hbar\omega - E_g \pm \hbar\omega_{ph})^2$ has the same dependence as for fundamental transitions. The absorption coefficient in the presence of excitons always lies above that without excitons, but at high energies the two curves are almost parallel.

2.3.5.10 Biexciton Production Through Photon Absorption

At absorption, it is also possible to generate directly a pair of bound excitons (a biexciton) with $\boldsymbol{k} = 0$ by the absorption of two photons with the same energy $\hbar\omega_{\mathrm{a}} = E_{\mathrm{X}} - (1/2)\Delta E_{\mathrm{XX}}$. Biexcitons can also be produced by free-exciton generation through irradiation in the intrinsic exciton line, followed by bonding of two excitons. Another possibility of biexciton formation is by absorption of a photon of energy $\hbar\omega = E_{\mathrm{X}} - \Delta E_{\mathrm{XX}}$ by a single exciton; this process is called 'induced absorption'. Here, E_{X} is the energy of a single exciton at $\boldsymbol{k} = 0$ and ΔE_{XX} is the binding energy of the biexciton. The total biexciton energy is $2E_{\mathrm{X}} - \Delta E_{\mathrm{XX}} + \hbar^2k^2/[4(m_{\mathrm{e}} + m_{\mathrm{h}})]$; it decays radiatively by emission of a photon of momentum $\boldsymbol{q}$, process in which one exciton is annihilated and the other stays as a free exciton of momentum $\boldsymbol{k} - \boldsymbol{q}$ (Curie, 1979).

2.3.5.11 Light Absorption on Phonons

Light absorption on phonons can be observed in far infrared (10^{11}–10^{13} Hz) and is characterized by a series of absorption maxima whose intensity and position do not change with the modification of the concentration and type of impurities and which remains the same after irradiation, thermal treatment or even small deformations. The number of absorption bands is not larger than the number of *optical* branches of oscillation. Typical values for the absorption coefficient are $\alpha \cong 1\text{–}10\ \text{cm}^{-1}$.

Two or more phonons can participate in this type of transition; we consider only one here. The absorption coefficient of the electromagnetic radiation on the oscillations of the crystalline lattice is calculated with the interaction Hamiltonian

$$H_{\text{int}} = -\textstyle\sum_{m,n} (e_m / M_m) \boldsymbol{A}(\boldsymbol{r}_{mn}) \cdot \boldsymbol{p}_{mn}, \tag{2.86}$$

where e_m, M_m are the charge and mass of the mth atom in the nth elementary cell. Denoting the number of phonons with wavevector $\boldsymbol{q}$ in branch r by $n_r(\boldsymbol{q})$, their energy is $E_{r,\boldsymbol{q}} = [n_r(\boldsymbol{q}) + 1/2]\hbar\omega_r(\boldsymbol{q})$. $\boldsymbol{p}_{mn} = M_m \dot{\boldsymbol{u}}_{mn}$, the moment associated to the atom m in the cell n, can be obtained from (1.50) and (1.53), as

$$\begin{aligned} \boldsymbol{p}_{mn} = &-\mathrm{i}(\hbar M_m N/2)^{1/2} \textstyle\sum_{r,\boldsymbol{q}} [\omega_r(\boldsymbol{q})]^{1/2} [\hat{\boldsymbol{e}}_m^r(\boldsymbol{q}) b_r(\boldsymbol{q}) \exp(\mathrm{i}\boldsymbol{q}\cdot\boldsymbol{R}_n) \\ &- \hat{\boldsymbol{e}}_m^r{}^*(\boldsymbol{q}) b_r^+(\boldsymbol{q}) \exp(-\mathrm{i}\boldsymbol{q}\cdot\boldsymbol{R}_n)], \end{aligned} \tag{2.87}$$

where N is the number of elementary cells and $\hat{\boldsymbol{e}}_s$ the polarization versor.

Considering the process of absorption of one photon and emission of one phonon, and taking into account that

$$\begin{aligned} &\langle \varphi_{n-1}^s(\boldsymbol{k}) \varphi_{n_{\text{ph}}+1}^r(\boldsymbol{q}) \,|\, a_{s'}^+(\boldsymbol{k}') b_{r'}(\boldsymbol{q}') \,|\, \varphi_{n_{\text{ph}}}^r(\boldsymbol{q}) \varphi_n^s(\boldsymbol{k}) \rangle \\ &\quad = \sqrt{n_{\text{ph}}(\boldsymbol{q})+1}\sqrt{n(\boldsymbol{k})}\,\delta(\boldsymbol{k}-\boldsymbol{k}')\delta(\boldsymbol{q}-\boldsymbol{q}')\delta_{ss'}\delta_{rr'}, \end{aligned} \tag{2.88}$$

where $\varphi_n^s(\boldsymbol{k})$ is the wavefunction of the mode of the electromagnetic field with polarization s, propagation vector $\boldsymbol{k}$ and n photons in the mode ($\varphi_{n_{\text{ph}}}^r(\boldsymbol{q})$ has the same meaning for phonons) one obtains that

$$H_{\text{int},fi} \approx \sqrt{n_{\text{ph}}(\boldsymbol{q})+1}\sqrt{n(\boldsymbol{k})}\sqrt{\omega_r(\boldsymbol{q})/\omega_s(\boldsymbol{k})} \textstyle\sum_m (e_m/\sqrt{M_m}) \hat{\boldsymbol{e}}_s(\boldsymbol{k}) \hat{\boldsymbol{e}}_m^r(\boldsymbol{q}) \delta(\boldsymbol{k}-\boldsymbol{q}), \tag{2.89}$$

where, as in (2.87),

$$\boldsymbol{A}(\boldsymbol{r}_{mn}) = A_0 \textstyle\sum_{s,\boldsymbol{k}} [\hat{\boldsymbol{e}}_s(\boldsymbol{k})/\sqrt{\omega_s(\boldsymbol{k})}][a_s(\boldsymbol{k})\exp(\mathrm{i}\boldsymbol{k}\cdot\boldsymbol{r}_{mn}) + a_s^+(\boldsymbol{k})\exp(-\mathrm{i}\boldsymbol{k}\cdot\boldsymbol{r}_{mn})]. \tag{2.90}$$

For acoustic phonon modes all particles, in every elementary cell, move in phase, with equal amplitudes, i.e. $\boldsymbol{u}_{mn}$ = const. and $e_m/(M_m)^{1/2}$ = const. If there are two atoms in the elementary cell of a ionic crystal such that $e_m = +e$ and $e_m = -e$, then $H_{\text{int},fi} \equiv 0$. On the contrary, for the optical phonon modes in the same ionic crystal $H_{\text{int},fi} \neq 0$, since the two ions in the elementary cell move in antiphase such that $\sum_m M_m \boldsymbol{u}_{mn} = 0$. So, only optical phonons take place in absorption processes. Moreover, since $\hat{e}_m^r(\boldsymbol{q})$ is perpendicular on $\boldsymbol{q}$, *only transversal optical modes contribute to the absorption.*

The absorption coefficient, obtained by integrating $|H_{\text{int},fi}|^2$ over $\boldsymbol{k}$ and transforming the integral over $\boldsymbol{q}$ into an integral over ω_r is

$$\alpha \approx I(\omega)[1 + \coth(\hbar\omega / 2k_{\text{B}}T)], \tag{2.91}$$

where $\omega = \omega_r = \omega_s$ and $I(\omega_s)$ is the incident light intensity, proportional to the integral over $\boldsymbol{k}$ of $|H_{\text{int},fi}|^2$. This formula describes the optical absorption with the participation of a single phonon; analogously one can deduce the formula for the optical emission of a single phonon.

2.3.6 Optical Properties of Solids in External Magnetic Fields

The effect of the electric field on the band-to-band fundamental absorption is described by the Franz–Keldysh effect. The presence of a magnetic field has a greater influence on electrons than on the collective oscillations of the lattice, such that the magneto-optic properties of the crystal offer information on the electron motion in the solid.

2.3.6.1 Energy Levels in Magnetic Field

The wavefunction of a free electron with effective mass m_{c} in the conduction band of a solid is given by $\psi(\boldsymbol{r}) = u(\boldsymbol{r})\varphi(\boldsymbol{r})$ where $u(\boldsymbol{r}) = u_{jk}(\boldsymbol{r})\exp(\mathrm{i}\boldsymbol{k}\cdot\boldsymbol{r})$ is a Bloch function (denoted by $u_j(\boldsymbol{r})$ for $\boldsymbol{k} = 0$) and $\varphi(\boldsymbol{r})$ satisfies the equation

$$[(\boldsymbol{p} + e\boldsymbol{A})^2 / (2m_{\text{c}})]\varphi(\boldsymbol{r}) = E_{\text{c}}\varphi(\boldsymbol{r}). \tag{2.92}$$

If the magnetic field $\boldsymbol{B} = \nabla \times \boldsymbol{A}$ with amplitude $B = |\boldsymbol{B}|$ is applied along the z direction, $\boldsymbol{A}$ can be chosen as $\boldsymbol{A} = (0, Bx, 0)$ and (2.92) has solutions of the form $\varphi(\boldsymbol{r}) = g(x)\exp[\mathrm{i}(k_y y + k_z z)]$, where

$$-\frac{\hbar^2}{2m_{\text{c}}}\frac{\mathrm{d}^2 g}{\mathrm{d}x^2} + \left[\frac{1}{2m_{\text{c}}}(\hbar k_y + eBx)^2 + \frac{\hbar^2 k_z^2}{2m_{\text{c}}}\right] g = E_{\text{c}} g, \tag{2.93}$$

(the motion along z is independent). Equation (2.93) is identical to the equation of motion of a 1D harmonic oscillator with equilibrium position $x_0 = -\hbar k_y / eB$ and frequency $\omega_{\text{cc}} = eB / m_{\text{c}}$ – the cyclotron frequency. So, the eigenenergies are

$$E_{\text{c}} = (n + 1/2)\hbar\omega_{\text{cc}} + \hbar^2 k_z^2 / 2m_{\text{c}}, \tag{2.94}$$

with $n = 0,1,2,...$ and $g(x) = \psi_n(x - \hbar k_y / eB)$ with

$$\psi_n(x) = \sqrt{2\sqrt{\pi}/(2^n n!)}\sqrt{eB/\hbar}H_n(x\sqrt{eB/\hbar})\exp[-x^2 eB/(2\hbar)], \tag{2.95}$$

where H_n are Hermite polynomials. In a similar manner, for a hole in the valence band with effective mass m_v the total energy is $E_v = -E_g - (n'+1/2)\hbar\omega_{cv} - \hbar^2 k_z^2 / 2m_v$, where $\omega_{cv} = eB / m_v$. (The c, v labels in the cyclotron frequency will be dropped whenever no confusion in possible between conduction and valence bands.) The continuous, 3D parabolic energy band for the electron has been transformed into a discrete energy spectrum (Landau levels), associated with electron motion in a plane perpendicular to the direction of the magnetic field; the motion of the electron parallel to the magnetic field is not modified. Since the energy separation between Landau levels $\hbar eB / m_{c,v} = 1.1577 \times 10^{-4}\,\mathrm{eV} \times B/(m_{c,v} / m_0)$ is small compared to the thermal energy, the quantization of energy levels is ignored except in the cases of very low temperatures, strong magnetic fields or materials with $m_{c,v} / m_0 << 1$ (in InSb, Bi, for example, this ratio is about 10^{-2}). Moreover, the cyclical character of electron motion on a close orbit appears only if the period of electron rotation is smaller than the duration of collisions with phonons, impurities, etc., a condition that can be met at low temperatures and strong magnetic fields.

The Landau energy levels are strongly degenerate, the degeneracy being proportional to B. For example, for a system inside a rectangular box with dimensions L_x, L_y, L_z the number of possible values for $k_{x,y,z}$ in an interval $\Delta k_{x,y,z}$ is $L_{x,y,z}\Delta k_{x,y,z} / 2\pi$, the allowed values for k_y being those for which the center of the orbit is in the box, i.e. for which $-L_x / 2 \le x_0 \le L_x / 2$, or $-eBL_x / 2\hbar \le k_y \le eBL_x / 2\hbar$. So, the number of states with k_z fixed, on a Landau level, is $L_y \Delta k_y / 2\pi = eBL_x L_y / 2\pi\hbar$ (Basu, 1997).

2.3.6.2 Interaction of the Electromagnetic Radiation with a System in a Magnetic Field

The Hamiltonian that describes the interaction of an incident electromagnetic field $\boldsymbol{A}_{\mathrm{rad}}$ with an electron in a magnetic field is

$$H_{\mathrm{int}} = (\boldsymbol{p} + e\boldsymbol{A} + e\boldsymbol{A}_{\mathrm{rad}})^2 / 2m_c \cong e\boldsymbol{A}_{\mathrm{rad}} \cdot (\boldsymbol{p} + e\boldsymbol{A}) / m_c, \tag{2.96}$$

the transition probability of an electron from an initial state i to a final state f being proportional to

$$\langle \psi_f \mid H_{\mathrm{int}} \mid \psi_i \rangle = \int u_f{}^*(\boldsymbol{r})\varphi_f{}^*(\boldsymbol{r})(e/m_c)\boldsymbol{A}_{\mathrm{rad}} \cdot (\boldsymbol{p} + e\boldsymbol{A})u_i(\boldsymbol{r})\varphi_i(\boldsymbol{r})\mathrm{d}\boldsymbol{r}. \tag{2.97}$$

Supposing that $\boldsymbol{A}$ and $\varphi_{i,f}(\boldsymbol{r})$ are slowly variable in comparison with the Bloch functions, such that they are constant in an elementary cell in the crystal, (2.97) can be rewritten as (see Sect. 2.2.1)

$$\langle\psi_f \mid H_{\text{int}} \mid \psi_i\rangle = \int_{\text{cell}} u_f{}^*(\boldsymbol{r}) u_i(\boldsymbol{r}) \mathrm{d}\boldsymbol{r} \int_{\text{cryst}} \varphi_f{}^*(\boldsymbol{r})(e/m_{\text{c}}) \boldsymbol{A}_{\text{rad}} \cdot (\boldsymbol{p}+e\boldsymbol{A}) \varphi_i(\boldsymbol{r}) \mathrm{d}\boldsymbol{r}$$
$$+ \int_{\text{cell}} u_f{}^*(\boldsymbol{r})(e/m_{\text{c}}) \boldsymbol{A}_{\text{rad}} \cdot \boldsymbol{p} u_i(\boldsymbol{r}) \mathrm{d}\boldsymbol{r} \int_{\text{cryst}} \varphi_f{}^*(\boldsymbol{r}) \varphi_i(\boldsymbol{r}) \mathrm{d}\boldsymbol{r}. \tag{2.98}$$

The expression of the matrix elements of the Hamiltonian of interaction depends of the type of transition:

i) Intraband transitions (cyclotron resonance)

Since the Bloch functions are orthonormal, the first term in $H_{\text{int},fi}$ is different from zero only if $u_f = u_i$. These transitions are the well-known *Landau transitions*; for them $H_{\text{int},fi} \cong \text{const}.\sqrt{n+1}$ (as at the phonon or photon absorption or emission). The selection rules for cyclotron transitions are $\Delta n = \pm 1$, $\Delta k_z = 0$, $\Delta k_y = 0$ such that $\Delta E = \pm\hbar\omega_{\text{cc}}$ or $\pm\hbar\omega_{\text{cv}}$ – resonant cyclotron absorption/emission, the first rule $\Delta n = \pm 1$ being valid only if $\boldsymbol{A}_{\text{rad}}$ is uniform over the orbit. If the radiation field is not uniform over the orbit the transitions can take place for any integer value of Δn, and resonant cyclotron absorption can be observed for subharmonics of the fundamental frequency. The other selection rules, $\Delta k_z = 0$, $\Delta k_y = 0$ imply vertical transitions in which $\boldsymbol{k}$ is conserved.

The absorption is proportional to $\sum_{i,f} | H_{\text{int},fi} |^2 \, \delta(\hbar\omega - E_f - E_i)$ where ω is the frequency of the radiation field and $E_{f,i}$ are the energies on the initial and final Landau levels such that (in the conduction band) $E_f - E_i = \hbar\omega_{\text{cc}}$. It follows that the absorption is resonant for $\hbar\omega = \hbar\omega_{\text{cc}}$.

The cyclotron frequency of the electron in the crystal is determined from measurements of the absorption or reflection of circularly polarized electromagnetic waves with frequency ω, which propagate in the same direction as the magnetic field. The absorption or reflection increases strongly when $\omega = \omega_{\text{c}}$. The width of the resonance, γ_{c}, is determined by the relaxation time of the cyclotronic motion $1/\gamma_{\text{c}}$ (proportional to the average time between two electron-phonon collisions), if $1/\gamma_{\text{c}}$ is greater than $2\pi/\omega_{\text{c}}$.

Up to now we have neglected the effect of the spin. If we take it into account, $E_{\text{c}} = (n+1/2)\hbar\omega_{\text{cc}} + \hbar^2 k_z^2/2m_{\text{c}} + S_z \mu_{\text{B}} g B$, where $S_z = \pm 1/2$ is the spin component along the direction of the magnetic field, $\mu_{\text{B}} = e\hbar/2m_{\text{c}}$ is the Bohr magneton and g is the gyromagnetic factor. In the presence of the magnetic field each Landau level splits into two, an effect known as Zeeman splitting, the level separation being equal to $\mu_{\text{B}} g B$. The cyclotronic resonance corresponds then to transitions between levels with the same S_z but with $\Delta n = 1$.

ii) Direct interband transitions

In this case u_i, u_f are orthogonal (are situated in different – conduction and valence – bands) and the first term in (2.98) vanishes. Then

$$H_{\text{int},fi} = (e/m_{\text{c}})(\boldsymbol{p} \cdot \boldsymbol{A}_{\text{rad}})_{\text{cv}} \int \varphi_{\text{c}}{}^*(\boldsymbol{r}) \varphi_{\text{v}}(\boldsymbol{r}) \mathrm{d}\boldsymbol{r}, \tag{2.99}$$

with $(\boldsymbol{p} \cdot \boldsymbol{A}_{\text{rad}})_{\text{cv}} = \int_{\text{cell}} u_{\text{c}}{}^*(\boldsymbol{r}) \boldsymbol{p} \cdot \boldsymbol{A}_{\text{rad}} u_{\text{v}}(\boldsymbol{r}) \mathrm{d}\boldsymbol{r}$. u_{c}, u_{v} must have different parities in order to have $(\boldsymbol{p} \cdot \boldsymbol{A}_{\text{rad}})_{\text{cv}} \neq 0$. The selection rules are in this case

$\Delta k_y = \Delta k_z = 0$, $\Delta n = 0$, the transition energy being $E_c - E_v = E_f - E_i = E_g + \hbar^2 k_z^2 / 2m_r + (n + 1/2)\hbar(\omega_{cc} + \omega_{cv})$ where $m_r = m_c m_v / (m_c + m_v)$ is the reduced mass. An additional term $(S_{zc} g_c - S_{zv} g_v)\mu_B B$ must be included in the transition energy to account for the spin; the inclusion of the spin determines additional selection rules $\Delta S_z = 0, \pm 1$, which give rise to polarization effects.

In the absorption coefficient expression, the term $\sum_{i,f} \delta(\hbar\omega - E_f + E_i)$, which corresponds to $2Vd^3\mathbf{k}/8\pi^3$ for spherical bands in the absence of the magnetic field, now becomes $(V/2\pi^2)(2m_r/\hbar^2)^{3/2}(\hbar\omega - E_g)^{1/2}$ as for the case of direct band-to-band fundamental absorption. However, here the density of states does not depend on k_x since ψ does not depend on k_x, and k_y is restricted to $-eBL_x/2\hbar \le k_y \le eBL_x/2\hbar$. So, the transition probability per unit volume, for a fixed n, is

$$\begin{aligned} P_{fi}^n &= (eB/\pi\hbar^2)\int dk_z \,|H_{\text{int},fi}|^2\, \delta[E_g + \hbar(n+1/2)(\omega_{cc} + \omega_{cv}) + \hbar^2 k_z^2/2m_r - \hbar\omega] \\ &= (2eB/\pi\hbar^3)\sqrt{m_r/2}\,|H_{\text{int},fi}|^2 / \sqrt{\hbar\omega - [E_g + \hbar(n+1/2)(\omega_{cc} + \omega_{cv})]}. \end{aligned} \tag{2.100}$$

The total absorption coefficient is obtained by summing over terms for which the function under the square root at the denominator in (2.100) is positive:

$$\alpha = \alpha_0 (eB/\omega)\sqrt{2m_r}\,\sum_n 1/\sqrt{\hbar\omega - [E_g + \hbar(n+1/2)(\omega_{cc} + \omega_{cv})]}. \tag{2.101}$$

Unlike the direct band-to-band transition, where there is an absorption threshold, in a magnetic field there are *a series of absorption thresholds* corresponding to electron transitions between the nth Landau level in the valence band and the nth Landau level in the conduction band (Basu, 1997). The absorption spectrum looks like that in Fig. 2.10.

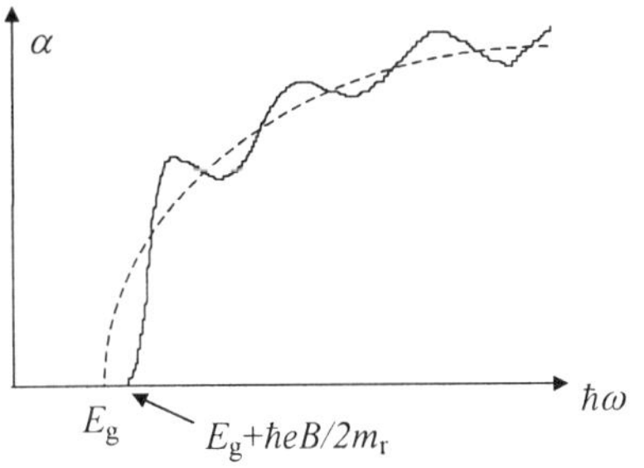

Fig. 2.10. Absorption in the absence (*dashed line*) and in the presence of the magnetic field (*solid line*) in direct semiconductors

In particular, the absorption threshold is shifted from E_g to $E_g + \hbar eB/2m_r$ and α has an oscillatory function on both B and $\hbar\omega$. A series of peaks in α can be obtained by either fixing $\hbar\omega$ at a value larger than E_g and varying B or if B is kept constant and $\hbar\omega$ is varied (the peaks in this case are equally spaced in ω).

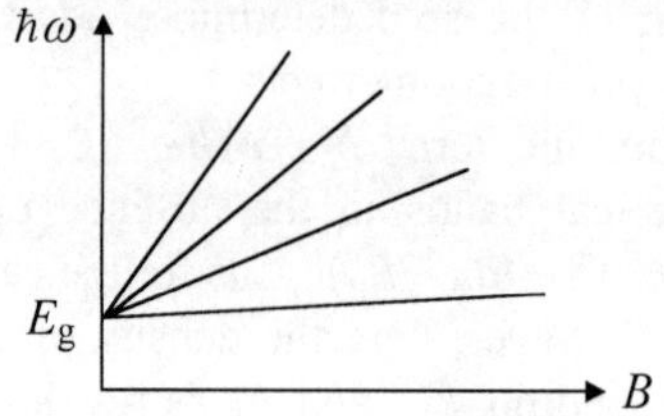

Fig. 2.11. Peak shifts in the absorption spectrum in the presence of a magnetic field

If we represent by straight lines in Fig. 2.11 the condition for obtaining the peaks in the absorption spectrum, the reduced mass can be determined from the slope of the lines. With m_c determined from the cyclotron resonance and m_r from magnetoabsorption, one can then determine m_v. If one takes also into account the spin, each of the lines in the above figure will be split into two; the spin-orbit interaction factor g can be obtained from these split lines.

The selection rule $\Delta n = 0$ is broken when an additional electric field is applied normal to the magnetic field. The theory of one-phonon resonant Raman scattering (see Sect. 2.5.1) in diamond- and zincblende-type semiconductors, developed by de la Cruz and Trallero-Giner (1998), which includes the deformation potential and Fröhlich interactions for electron-one-phonon coupling, shows that in this case the Raman intensity decreases exponentially with increasing electric field for fixed laser frequency. The degeneracy of magneto-electron-hole pairs is partially reduced when an electric field is applied normal to the magnetic field, making it possible to explore different electron-hole transitions and even the electron-phonon interaction itself.

The direct forbidden transitions can be treated similarly to the case of direct band-to-band forbidden transitions, with the difference that the density of states is no longer 3D but 1D. The result is that, as for allowed direct transitions, the absorption coefficient has an *oscillatory behavior* around the curve which corresponds to the $B = 0$ case.

If the crystal has an anisotropic effective mass (as in Ge and Si), such that in the absence of the magnetic field

$$E(\boldsymbol{k}) = (\hbar^2 / 2m_t)(k_x^2 + k_y^2) + (\hbar^2 / 2m_l)k_z^2, \tag{2.102}$$

with m_t, m_l the effective transversal and longitudinal masses, respectively, the effective cyclotron mass which would be measured is

$$1 / m_{eff}^2 = \cos^2\theta / m_t^2 + \sin^2\theta / m_t m_l, \tag{2.103}$$

where θ is the angle in a plane perpendicular on y between the magnetic field and the z axis ($A_x = A_z = 0$, $A_y = B(x\cos\theta + z\sin\theta)$). So, if the magnetic field is parallel to the z axis, the electron trajectory is in the xy plane and its effective mass is m_t, while if the field is parallel to the x axis, the trajectory is in the yz plane and the effective mass is $m_{eff} = (m_t m_l)^{1/2}$.

2.3.6.3 Light Interaction with Metals in Magnetic Field

In metals, the number of free carriers is much greater than in semiconductors (Mavroides, 1972) and the Fermi level is inside the conduction band such that optical transitions, unlike in semiconductors, depend on its position (in particular on the position in the **k**-space). The cyclotron frequency in metals cannot be determined through absorption or reflection measurements as above since i) the electromagnetic field is totally reflected for $\omega < \omega_p$, ii) high-frequency electromagnetic waves do not penetrate inside the metal, and iii) the Fermi surface, even for cubic crystals, is not ellipsoidal. So, another method should be used, based on the fact that (see Fig. 2.12) when the magnetic field is parallel to the metal surface, the axis of the *helicoidal* electron trajectory is in the surface plane.

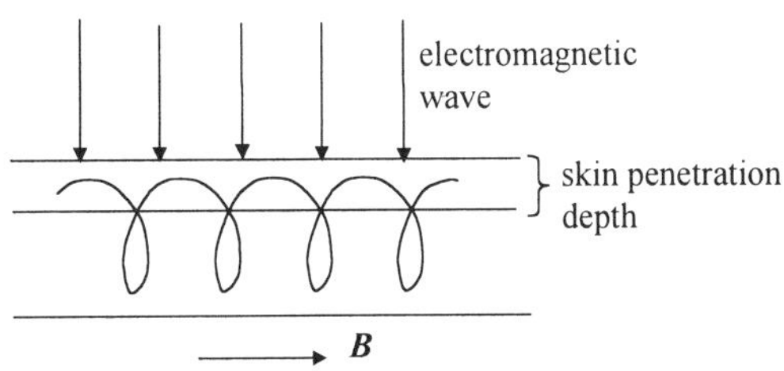

Fig. 2.12. Helicoidal electron trajectory in illuminated metals placed in magnetic fields parallel to the metal's surface

For $B = 10^3$–10^4 Oe, the diameter of the orbit is about 10^{-3} cm, much larger than the skin depth of 10^{-6} cm, and the electron is not influenced by the electromagnetic wave along most part of its trajectory. If the period of the electromagnetic wave is equal to, or is a multiple of, the rotation period of the electron, each time the electron passes through the skin-depth region, it is accelerated or delayed. Since $\omega_c(k_z)$ depends on the projection k_z of the wavevector on the direction of $\boldsymbol{B}$, cyclotron resonance appears whenever $n\omega_c = neB/m_{eff} = \omega$, with an integer n. This phenomenon, called *Azbel–Kaner cyclotron resonance*, is different from the usual cyclotron resonance in which the electron interacts continuously with the uniform electric field and in which $n = 1$. The absorption in metals has *several peaks* at the cyclotronic resonance, corresponding to different n. Also, the conductivity σ_{zz} for magnetic fields parallel to the z axis has an oscillatory behavior with the frequency. Moreover, in thin metal layers, the resonance can only be observed if the layer width is larger than the diameter of the orbit (otherwise the synchronism is destroyed). So, by decreasing B the dimension of the orbit increases until it becomes equal to the layer width, a further decrease of B

making the resonance disappear. In this way one can either measure the thickness of thin films, or obtain information about the form of the Fermi surface with layers with different orientations of the crystallographic axes.

The reflection coefficient R at normal incidence in metals, at high frequencies, can be approximated by

$$R = (n_{re} + in_{im} - 1)^2 / (n_{re} + in_{im} + 1)^2 \cong (n_{re} - 1)^2 / (n_{re} + 1)^2, \tag{2.104}$$

since (see Sect. 2.3.2) $n_{im} \cong 0$ and $n_{re}^2 = \varepsilon_{re}(\omega)/\varepsilon_0 = \varepsilon_r(1 - \omega_p^2/\omega^2)$ with $\omega_p^2 = Ne^2/(m_{eff}\varepsilon_0)$ the frequency of the free-electron plasma. The maximum value of R, $R = 1$, is attained at $\omega_{max} = \omega_p$, whereas R takes a minimum value, equal to 0, when $n_{re}^2 = 1$, i.e. for a frequency of the incident wave $1 - \omega_p^2/\omega^2 = 1/\varepsilon_r$, or $\omega_{min} = \omega_p/(1 - 1/\varepsilon_r)^{1/2}$. m_{eff} can thus be determined from ω_{min} if N is known, the latter being obtained from the frequency variation at $\omega_{max} = \omega_p$. When ε_r is large, the extreme values of R, 1 and 0, are attained over a narrow frequency range; the frequency of this jump in the reflectivity value is called the plasma reflection edge. (Note that the same behavior of the reflectivity is also encountered in highly doped semiconductors.) If the reflectivity does not reach the zero value, m_{eff} can be determined from the variation of the slope of n_{re} as a function of $1/\omega^2$. An alternative method to determine the effective mass in metals, which uses only the frequency dependence of R around ω_{min}, involves the application of a magnetic field, which changes the value of ω_{min} with an amount dependent on m_{eff} and B.

To find the variation of the refractive index with the magnetic field, we start from the motion of a damped free harmonic oscillator in the magnetic field, described by an equation

$$m_{eff}(\mathrm{d}^2\boldsymbol{r}/\mathrm{d}t^2) - m_{eff}\delta(\mathrm{d}\boldsymbol{r}/\mathrm{d}t) - e(\mathrm{d}\boldsymbol{r}/\mathrm{d}t) \times \boldsymbol{B} = -e\boldsymbol{E}_0 \exp(\mathrm{i}\omega t), \tag{2.105}$$

with $1/\delta = \tau$ the relaxation time. Looking for solutions of $\boldsymbol{r}$ proportional to $\exp(\mathrm{i}\omega t)$ and considering $\boldsymbol{B}$ parallel to the z direction, we obtain for the components of the conductivity tensor, defined as $\boldsymbol{J} = Ne\boldsymbol{v} = Ne(\mathrm{d}\boldsymbol{r}/\mathrm{d}t) = \hat{\sigma}\boldsymbol{E}$ the following nonzero values

$$\sigma_{xx} = \sigma_{yy} = \frac{\sigma_0(1 + \mathrm{i}\omega\tau)}{(1 + \mathrm{i}\omega\tau)^2 + \omega_c^2\tau^2}, \quad \sigma_{yx} = -\sigma_{xy} = \frac{\sigma_0\omega_c\tau}{(1 + \mathrm{i}\omega\tau)^2 + \omega_c^2\tau^2}, \quad \sigma_{zz} = \frac{\sigma_0}{(1 + \mathrm{i}\omega\tau)}, \tag{2.106}$$

with $\sigma_0 = Ne^2\tau/m_{eff}$ and $\omega_c = eB/m_{eff}$.

If the direction of the incident light is parallel to $\boldsymbol{B}$, the right and left polarized components of light have different effective refractive indices. This phenomenon, called the Faraday effect, can be demonstrated starting from Maxwell's equations with currents in non-magnetic materials, with ε constant in time

$$\nabla \times \boldsymbol{E} = -\partial \boldsymbol{B} / \partial t = -\mu_0 \partial \boldsymbol{H} / \partial t, \quad \nabla \times \boldsymbol{H} = \boldsymbol{J} + \partial \boldsymbol{D} / \partial t = \hat{\sigma} \boldsymbol{E} + \varepsilon \partial \boldsymbol{E} / \partial t. \tag{2.107}$$

Applying the rotor operator on the first equation in (2.107) and supposing that $\nabla \boldsymbol{E} = 0$ with $\boldsymbol{E} = \boldsymbol{E}_0 \exp[\mathrm{i}(\omega t - n\boldsymbol{k}_0 \cdot \boldsymbol{r})]$, it follows that

$$\boldsymbol{E}_0 (k_0^2 n^2) = \boldsymbol{E}_0 (-\mathrm{i}\mu_0 \hat{\sigma} \omega + \mu_0 \varepsilon \omega^2) = \boldsymbol{E}_0 \omega^2 \mu_0 \varepsilon (1 - \mathrm{i}\hat{\sigma} / \omega\varepsilon), \tag{2.108}$$

with $k_0 = |\boldsymbol{k}_0| = \omega / c$. The effective refractive indices for right or left circularly polarized waves $\boldsymbol{E}_0 = \boldsymbol{E}_{0x} \pm \mathrm{i}\boldsymbol{E}_{0y}$ are $n_\pm^2 = \varepsilon_\mathrm{r}(1 - \mathrm{i}\sigma_\pm / \omega\varepsilon)$, with $\varepsilon_\mathrm{r} = \varepsilon / \varepsilon_0$, $\sigma_\pm = \sigma_{xx} \pm \mathrm{i}\sigma_{xy}$. So,

$$n_\pm^2 = \varepsilon_\mathrm{r}\{1 - \omega_\mathrm{p}^2 / [\omega(\omega \pm \omega_\mathrm{c} - \mathrm{i}/\tau)]\} \cong \begin{cases} \varepsilon_\mathrm{r}\{1 - \omega_\mathrm{p}^2 / [\omega(\omega \pm \omega_\mathrm{c})]\}, & \omega\tau >> 1, \\ \varepsilon_\mathrm{r}\{1 - \omega_\mathrm{p}^2 / [(\omega \pm \omega_\mathrm{c}/2)^2]\}, & \omega_\mathrm{B} << \omega, \end{cases} \tag{2.109}$$

and the reflectivity becomes minimum for a frequency ω_min displaced with $\pm\omega_\mathrm{c}/2$ from the $\boldsymbol{B} = 0$ case, depending on the direction of the circular polarization. In particular, the difference between the minima for the two circular polarizations is equal to $\omega_\mathrm{c} = eB / m_\mathrm{eff}$, from which one can obtain m_eff. The relaxation time τ can also be determined from the minimum of R as a function of ω, if the complete expressions of R are considered.

In a similar manner it can be shown that if the electric field of the incident radiation $\boldsymbol{E}$ is parallel to $\boldsymbol{B}$ and $\boldsymbol{k}_0$ is normal to $\boldsymbol{B}$, the refractive index is not changed, while if both $\boldsymbol{k}_0$ and $\boldsymbol{E}$ are perpendicular to $\boldsymbol{B}$, for $\omega\tau >> 1$, $n^2 = \varepsilon_\mathrm{r}\{1 - \omega_\mathrm{p}^2(\omega^2 - \omega_\mathrm{p}^2) / [\omega^2(\omega^2 - \omega_\mathrm{p}^2 - \omega_\mathrm{c}^2)]\}$.

2.3.6.4 Phenomena at Low Frequencies

Especially in semimetals the positive ions are also moving from their equilibrium positions under the action of a magnetic field, so that if $\boldsymbol{k}_0$ is parallel to $\boldsymbol{B}$ and $\omega\tau >> 1$ (Mavroides, 1972),

$$n_\pm^2 = \varepsilon_\mathrm{r}\{1 - \omega_\mathrm{pe}^2 / [\omega(\omega \pm \omega_\mathrm{ce})] - \omega_\mathrm{ph}^2 / [\omega(\omega \pm \omega_\mathrm{ch})]\}, \tag{2.110}$$

where ω_ce, ω_ch and ω_pe, ω_ph are, respectively, the cyclotron and plasma frequencies of the electrons and holes, which have, respectively, effective masses m_e, m_h and densities N_e, N_h. Since in semimetals $N_\mathrm{e} = N_\mathrm{h}$, ω_ce, ω_ch have opposite signs and (2.110) becomes

$$n_\pm^2 = \varepsilon_\mathrm{r}\{1 - \omega_\mathrm{pe}^2 / [\omega(\omega \pm \omega_\mathrm{ce})] - \omega_\mathrm{ph}^2 / [\omega(\omega \mp \omega_\mathrm{ch})]\}. \tag{2.111}$$

In this plasma of electrons and holes two types of waves can propagate: helicoidal and Alfven waves.

The helicoidal waves are circularly polarized and can be observed even at room temperatures in thin semimetal layers as stationary waves that produce variations of the transmitted and/or reflected power. For an n-doped semimetal for which $\omega_{ph} \cong 0$, if $\omega_{ce} >> \omega_{ch}$ ($m_e << m_h$) and $\omega << \omega_{ce}$,

$$n_{\pm}^2 = \varepsilon_r[1 \mp \omega_{pe}^2 /(\omega\omega_{ce})]. \tag{2.112}$$

For low frequencies the medium can be transparent only for one type of helicoidal wave, n_- ($n_-^2 > 0$) and not for the other type, n_+, for which $n_+^2 < 0$. The waves with refractive index n_- propagate with a small phase velocity $v_- = c/n_- \cong c(\omega\omega_{ce})^{1/2}/\omega_{pe}$ proportional to $B^{1/2}$ and $N_e^{-1/2}$, and independent of the effective mass. Therefore, the transmission coefficient dependence on B has two peaks in thin layers of semimetals (InSb), corresponding to a variation of the effective thickness with $\lambda/2$ (as in Fabry–Perot interferometers), as can be seen from Fig. 2.13. From these data the electron concentration can be determined.

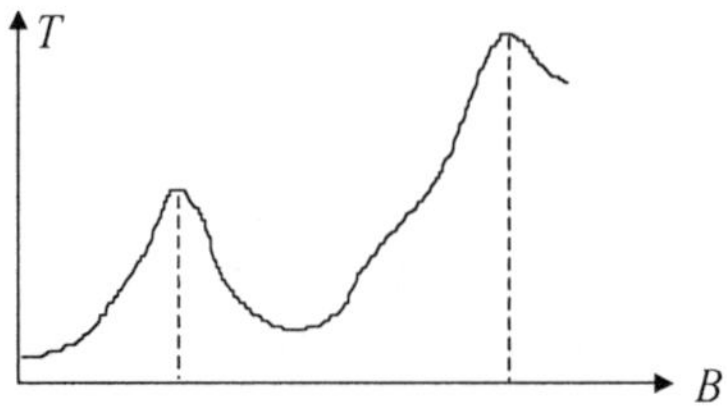

Fig. 2.13. The transmission coefficient dependence on B in thin layers of semimetals

When the plasma contains both electrons and holes with $N_e = N_h = N$ (as in semimetals or intrinsic semiconductors) and $\omega << \omega_{ce}, \omega_{ch}$, in the same approximation as above,

$$n_{\pm}^2 = \varepsilon_r\left(1 \mp \frac{\omega_{pe}^2}{\omega\omega_{ce}}\left(1 \mp \frac{\omega}{\omega_{ce}}\right) \pm \frac{\omega_{ph}^2}{\omega\omega_{ch}}\left(1 \pm \frac{\omega}{\omega_{ch}}\right)\right) = \varepsilon_r\left(1 + \frac{N}{\varepsilon_0 B^2}(m_e + m_h)\right). \tag{2.113}$$

This expression describes the effective refractive index of Alfven waves; these waves are linearly polarized since the circularly polarized components have the same velocity. For Alfven waves the equal densities of positive and negative charges oscillate in phase, with equal amplitudes, both modes propagating with the phase velocity

$$v \approx B/\sqrt{N(m_e + m_h)/\varepsilon_0}, \tag{2.114}$$

independent of ω. Again, the optical thickness of a thin layer (the product of its thickness and the refractive index of the propagating waves) can be varied by modifying B – from the corresponding peaks in the transmission coefficient one can determine the concentration or the sum of the electron and hole masses.

2.3.6.5 The Shubnikov–de Haas Effect

When the electric field is parallel to the magnetic field ($\boldsymbol{E} \parallel \boldsymbol{B}$), there is no Lorentz force acting on the free carriers, but still a periodic variation with B can be observed in the resistivity (conductivity). The Shubnikov–de Haas (SdH) effect can be explained by the passing of a Landau level through the Fermi surface and is caused by the discontinuous density of states in the magnetic field. At low temperatures all energy levels below the Fermi surface are occupied, and all levels above it are empty, so that when a Landau level is moving above E_F the electrons on it are redistributed on other Landau levels below the Fermi level. This electron redistribution produces an oscillation of the density of states (of the conductivity or resistivity) with the magnetic field, the oscillation condition $E_F = (n+1/2)\hbar\omega_c$ being periodic in B. Unlike in the Azbel–Kaner effect, the oscillations are independent of ω, but the strength of the effect depends on frequency, since at frequencies $\omega < \omega_p$ the reflectivity is large and any small changes in the refractive index can be easily observed. On the contrary, for $\omega > \omega_p$ the effect of the conduction electrons upon the refractive index is negligible so that the amplitude of the reflectivity (conductivity, resistivity) oscillations is maximum for $\omega \cong \omega_p$. Also, when passing from $\omega < \omega_p$ to $\omega > \omega_p$ a shift of π in the phase of the oscillations is observed, the relative minima at $\omega < \omega_p$ transforming to relative maxima for $\omega > \omega_p$.

2.4 Photoluminescence

Luminescence is defined as a nonequilibrium emission of radiation, i.e. the light emitted by a material system in excess of thermal radiation. This can be excited by irradiation with photons, in which case it is called photoluminescence (PL), by bombardment with cathode rays, when it is called cathodoluminescence, by chemical reaction (chemiluminescence), by applying an electric field – electroluminescence, etc. In this section we refer only to PL, sometimes also called fluorescence or phosphorescence.

After absorbing a photon, the material system can be excited to a virtual or real state; the restoration of the equilibrium state is achieved through several processes, as depicted in Fig. 2.14a, b. If the material system emits a photon directly from the excited state (processes 1, 2, 3) the process is called scattering, and is treated in the next section; the scattering can be elastic or Rayleigh, when the emitted photon has the same energy as the incident one, or inelastic, or Raman, in the opposite case. PL is usually understood as the incoherent radiative de-excitation from an excited state different from that reached by light absorption; this excited state can be reached by either radiative or nonradiative transitions (accompanied or not by emission of photons).

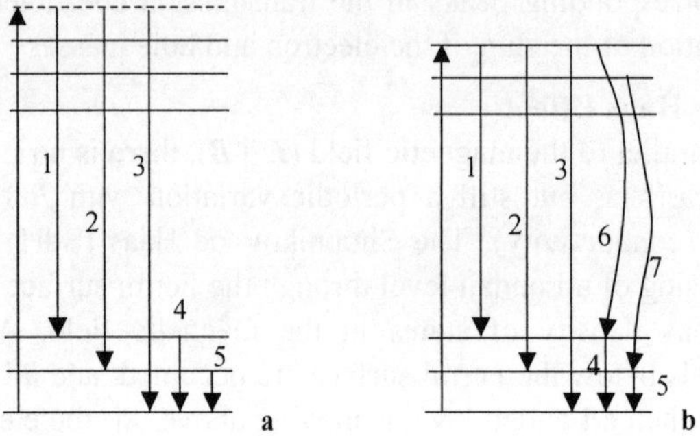

Fig. 2.14. Types of radiative recombinations at excitations of a virtual (**a**) and real (**b**) excited state: 1,2-inelastic scattering; 3-elastic scattering, 4,5-photoluminescence, 6,7-associated nonradiative intermediate transitions

The nonradiative decay is generally a multiphonon process, due to the interaction of the luminescent center with the vibrating crystal field. The stronger the coupling with the lattice, the larger the probability of nonradiative decay. In nonradiative relaxation processes between two levels, the number of emitted phonons is given by the energy required to bridge the gap. The nonradiative transition probability decreases with increasing phonon number, so it decreases with increasing separation between levels. A review of nonradiative transitions in semiconductors can be found in Stoneham (1981). On the other hand, the radiative transition probability increases with increasing separation, the net result being that PL occurs generally from the first excited state, since the energy gap between it and the ground state is the largest.

The rate of luminescence emitted at a transition between two levels separated by an energy $\hbar\omega_{ji}$

$$W_{ji}^{\text{lum}} = [\Delta N_j - (g_j / g_i)\Delta N_i \exp(-\hbar\omega_{ji} / k_{\text{B}}T)]A_{ji}\hbar\omega_{ji} / [1 - \exp(-\hbar\omega_{ji} / k_{\text{B}}T)] \tag{2.115}$$

is defined as the difference between the total rate of spontaneous emission and the rate of absorption of thermal radiation (Stepanov and Gribkovskii, 1963). In (2.115) A_{ji} is the integral Einstein coefficient $A_{ji} = (2/\pi)B_{ji}\hbar\omega_{ji}^3/(c/n)^3$ and ΔN_j, ΔN_i are the departures from equilibrium population of the upper level *j* and lower level *i*, respectively, between which the transition occurs. So, when the system reaches thermodynamic equilibrium, the luminescence vanishes.

One can also define positive luminescence as that associated with the increase in the population of the upper level or the reduction in the population of the lower level, while negative luminescence corresponds to a reduction in the population of the upper level or an increase in the population of the lower level. At normal temperatures it is much easier to observe positive luminescence, but negative luminescence has been observed at high temperatures or low frequencies

$\hbar\omega < k_B T$, for $\hbar\omega << k_B T$ the intensities of positive and negative luminescence being of the same order of magnitude (Stepanov and Gribkovskii, 1963). In solid-state physics, however, at least for all the experiments described in this book, we deal with positive PL, called simply PL. Luminescence of most inorganic solids occurs by transitions between electronic states.

The PL intensity depends on both the intensity of the incident light beam and the strength of interaction between the system and the incident radiation (the PL is weak, for example, if the radiation is poorly absorbed). After the state of nonequilibrium is reached, the equilibrium state tends to be restored not only by PL, but also by nonradiative transitions, which can lead to quenching of luminescence. PL quenching can be general, when the PL intensity is simultaneously reduced at all frequencies, or selective. A mechanism of general PL quenching is, for example, the increase in temperature, which favors nonradiative transitions.

Especially when the sample is a plane-parallel layer, secondary luminescence can become important. Defining the primary luminescence as that produced by the radiation reaching the layer from outside, secondary luminescence is produced by the absorption of radiation from primary luminescence; higher-order luminescence can also be defined in a similar manner. Of course, in order to unambiguously interpret PL spectra, secondary as well as higher-order luminescence must be negligible.

PL emission can be polarized when the medium has natural or induced anisotropy, as is the case in crystals, and depolarized otherwise. Generally, the PL anisotropy is much less than that of the exciting radiation, the degree of polarization of the PL being defined as $\rho = (I_{\parallel} - I_{\perp})/(I_{\parallel} + I_{\perp})$ where $I_{\parallel,\perp}$ are the PL intensities polarized parallel and perpendicular, respectively, to the linear polarization of the excitation radiation. The degree of polarization depends on the exciting and luminescent electronic transitions, the nonradiative decay of excited carriers, and in confined structures or microcrystals, on the size and shape of the confined domain.

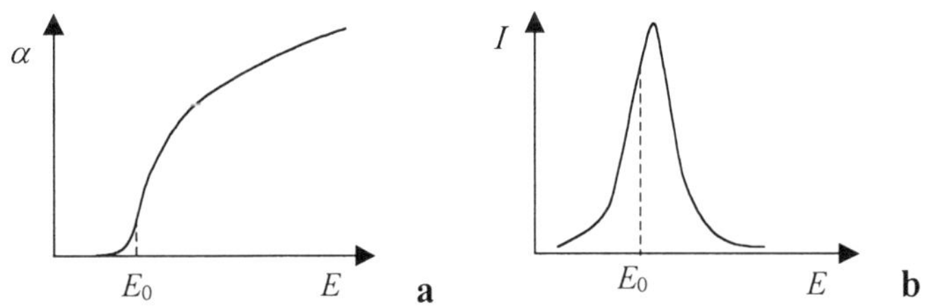

Fig. 2.15. The connection between the (**a**) absorption and (**b**) PL spectra

For processes that have an energy threshold E_0, the shape of the PL emission spectrum is related to the absorption coefficient by means of the principle of detailed balance, through

$$I_{PL}(\hbar\omega) \cong \alpha(\hbar\omega)\exp[-(\hbar\omega - E_0)/k_B T]. \tag{2.116}$$

A typical relation between absorption and ideal (dotted line) and real (solid line) emission curves is represented in Fig. 2.15. For free carriers or free excitons, (2.116) must be modified to $I_{PL}(\hbar\omega) \cong \alpha(\hbar\omega)\exp[-\hbar\omega / k_B T_c]$ where T_c is the effective temperature of carriers, related to the conduction electron temperature T_e and hole temperature T_h by $T_c^{-1} = T_e^{-1} + (m_e / m_h)T_h^{-1}$, with $m_e << m_h$. T_c can also be related to the pump intensity I_p by $I_p \approx \exp(-\hbar\omega_{LO} / k_B T_c)$ (hot photogenerated carriers lose kinetic energy mainly by longitudinal optic (LO) phonon interaction) (Curie, 1979).

In PL one can sometimes observe lines not revealed in absorption. For example, in Ge the radiative annihilation of the free exciton assisted by emission of a longitudinal acoustic (LA) phonon is not observed in absorption, but produces a PL spectrum $I_{PL}(\omega) \approx (\hbar\omega - E_g - \hbar\omega_{LA})^{1/2} \exp[-(\hbar\omega - E_g - \hbar\omega_{LA}) / k_B T]$ with T the bath temperature. PL spectroscopy is also the most suitable method of characterization and quality control of heterostructures (see, for example, Vasko and Kuznetsov (1999)). The nature of line broadening can also be determined in PL experiments, when the excitation energy is tuned continuously into resonance with the PL peak. Inhomogeneous broadening is then characterized as a decrease in linewidth, since all contributions to the PL signal above excitation energy disappear (Woggon et al. 1997).

Apart from the common PL, there is also hot luminescence (HL), which is the radiative de-excitation that takes place before the excited system relaxes with respect to the interaction modes, i.e. before the phonon-assisted emission. HL appears when the excitation has much higher energy than the bandgap. One method to distinguish HL from resonant Raman scattering (RRS) is based on the fact that for a sharp discontinuity in the incident beam, RRS has an instantaneous response, while the HL response is exponential, due to the time needed for the buildup or decay of electrons in the intermediate state. Generally, the HL spectrum has a broad background on which the RRS peaks are superimposed. Mathematically, RRS and HL are described by two separate terms in the scattering probability. One is proportional to the number of electrons in the excited state, $\rho_{nn}(0)$, while the other term is proportional to an off-diagonal element of the density matrix ρ_{ij} between the initial and final state. The term proportional to electron concentration in the excited state is HL, that proportional to ρ_{ij}, which represents a process independent on the number of excited electrons, is RRS (Shen, 1974).

2.5 Inelastic Light Scattering

In a scattering process, the system composed from the electromagnetic field and the solid-state material evolves from an initial state in which the matter is in the state a_i in the presence of a photon with wavevector $\boldsymbol{k}_i$ and polarization $\hat{e}_i$, in the final state in which the matter is in the state a_f in the presence of another photon with wavevector $\boldsymbol{k}_f$ and polarization $\hat{\boldsymbol{e}}_f$. Light scattering can be imagined as taking place in three different ways:

(i) the material system absorbs a photon with $\boldsymbol{k}_i$, $\hat{\boldsymbol{e}}_i$, jumps to an excited state b and then emits a photon with $\boldsymbol{k}_f$, $\hat{\boldsymbol{e}}_f$;

(ii) the material system emits a photon with $\boldsymbol{k}_f$, $\hat{\boldsymbol{e}}_f$ and arrives in the state b where it absorbs a photon with $\boldsymbol{k}_i$, $\hat{\boldsymbol{e}}_i$;

(iii) the absorption of the photon with $\boldsymbol{k}_i$, $\hat{\boldsymbol{e}}_i$ and the emission of the other photon with $\boldsymbol{k}_f$, $\hat{\mathbf{e}}_f$ take place simultaneously, without involving another state b of the material system.

Processes (i) and (ii) involve two steps, as the indirect transitions, without the necessity of conserving the energy of the system in the intermediate state. Therefore, as for indirect transitions, $H_{\text{int},fi}$ for the three processes mentioned above is given, respectively, by

$$H_{\text{int},fi} = \sum_b \frac{\langle a_f;\boldsymbol{k}_f,\hat{\boldsymbol{e}}_f \mid H_A \mid b;0\rangle\langle b;0 \mid H_A \mid a_i;\boldsymbol{k}_i,\hat{\boldsymbol{e}}_i\rangle}{E_{a_i} + \hbar\omega_i - E_b}, \tag{2.117}$$

$$H_{\text{int},fi} = \sum_b \frac{\langle a_f;\boldsymbol{k}_f,\hat{\boldsymbol{e}}_f \mid H_A \mid b;\boldsymbol{k}_i,\hat{\boldsymbol{e}}_i,\boldsymbol{k}_f,\hat{\boldsymbol{e}}_f\rangle\langle b;\boldsymbol{k}_i,\hat{\boldsymbol{e}}_i,\boldsymbol{k}_f,\hat{\boldsymbol{e}}_f \mid H_A \mid a_i;\boldsymbol{k}_i,\hat{\boldsymbol{e}}_i\rangle}{E_{a_i} - \hbar\omega_f - E_b}, \tag{2.118}$$

$$H_{\text{int},fi} = \langle a_f;\boldsymbol{k}_f,\hat{\boldsymbol{e}}_f \mid H_{AA} \mid a_i;\boldsymbol{k}_i,\hat{\boldsymbol{e}}_i\rangle. \tag{2.119}$$

The intermediate states in inelastic light scattering are virtual excitations, in contrast to interband absorption and luminescence, where only real excitations are involved, and in which the energy is conserved in all individual acts of absorption or emission (Vasko and Kuznetsov, 1999). Since the frequencies and wavevectors of incident and scattered photons can be controlled independently, the energy and wavevector of various elementary excitations, in particular their dispersion relations can be directly measured through inelastic light scattering. The interactions described by the H_A term are predominant in phonon, polariton, magnon, spin-flip scattering, as well as in electronic transitions in atoms, whereas H_{AA} describes scattering processes on free electrons. Both terms contribute in scattering on plasmons and conduction electrons in crystals (Hayes and Loudon, 1978).

As can be seen from (2.117) and (2.118), a singularity can appear in the denominator of $H_{\text{int},fi}$ when $\hbar\omega_i = E_b - E_{ai}$ in process i) or when $\hbar\omega_f = E_{a_i} - E_b$, in process ii). In this case the frequency of the incident or the scattered photon is in resonance with the transition frequency of the crystal from state b to state a_i or vice-versa – a resonant scattering process takes place, the scattering signal being much stronger than in the nonresonant situation. It is even possible that these two resonant processes take place simultaneously; the situation corresponds to a double resonance first-order scattering phenomenon, with an even stronger signal. The double-resonance condition can be achieved in bulk crystals by applying a uniaxial stress that splits the valence band states or by applying a magnetic field and tuning the Landau energy levels. In semimagnetic semiconductors, where there is a strong exchange interaction between magnetic impurities, this condition can be achieved either by applying a magnetic field or by varying the temperature. Such a tuning of exciton energy levels of a semimagnetic semiconductor in magnetic field was demonstrated by Gubarev et al. (1991). The most important application of resonant scattering is to elucidate the scattering mechanism and to extract the correspondent interaction constants. We would like to mention that at resonant scattering the interaction with light cannot be described as a perturbation; one consequence is the observation of vibrational overtones with intensities comparable to the fundamental line in scattering spectra, due to the breakdown in the selection rules obtained in the harmonic oscillation approximation.

As for any transition, the energy of the total system must be constant, i.e. $E_{ai} + \hbar\omega_i = E_{af} + \hbar\omega_f$. If during the scattering process the state of the material system does not change, i.e. if $E_{ai} = E_{af}$, then $\omega_i = \omega_f$ and we have elastic Rayleigh scattering (this denomination is used for energies of the incident photon lower than the ionization energy E_{ion} of the atoms in the crystal; otherwise it is called Thomson scattering and appears at X-ray scattering on atoms). If $\omega_i \neq \omega_f$ the scattering is inelastic and the variation of photon energy $\hbar(\omega_f - \omega_i)$ reflects the variation of the energy of the material system. The inelastic scattering is called Raman scattering if the energy of the incident photon is lower than E_{ion}; otherwise it is called Compton scattering. The inelastic scattering on very low-frequency excitations, for example on acoustic phonons, is called Brillouin scattering.

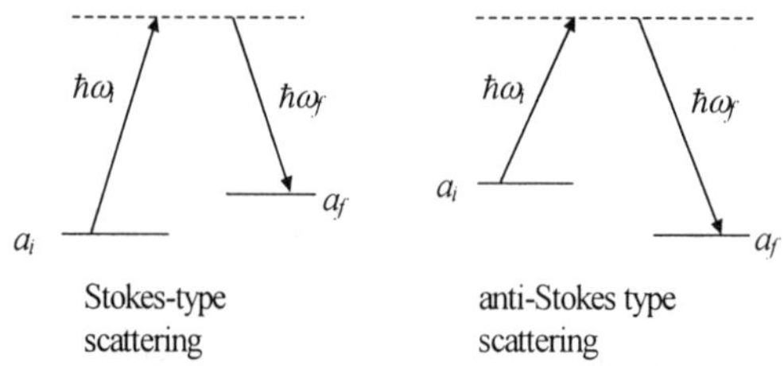

Fig. 2.16. Types of inelastic scattering processes

If $E_{ai} < E_{af}$ $(\omega_i > \omega_f)$ the scattering processes are called *Stokes*, whereas if $E_{ai} > E_{af}$ $(\omega_i < \omega_f)$ they are *anti-Stokes* processes (see Fig. 2.16). Anti-Stokes scattering can take place only if initially the system is in an excited state. On the contrary, if the material system is initially in the fundamental state with $E_{ai} = 0$, *only Stokes processes* are allowed. The energy conservation law implies $\hbar\omega_i = \hbar\omega_f + E_{af}$ where E_{af} is the energy of the excited state of the material system. Light scattering, being a *second-order process* in the electric field, can only be observed with high-power radiation sources, in particular lasers for Raman scattering.

2.5.1 Raman Scattering

The Raman spectrum is formed from Stokes and anti-Stokes lines, the latter being generally much weaker in intensity than the former and symmetrically displaced with respect to the Rayleigh line, which has the same frequency as the incident radiation. For vibrational spectra, for example, if v_j is the quantum vibration number on the vibrational level *j*, Rayleigh scattering corresponds to transitions with $\Delta v_j = 0$, and Raman spectra correspond to transitions with $\Delta v_j = \pm 1$. The distribution of the total N_0 population on the vibrational levels with energies $E(v_j) = \hbar\omega_j(v_j + 1/2)$ is made in agreement with the Boltzmann distribution $N(v_j) = N_0 \exp(-\hbar\omega_j v_j / k_B T)/[1 - \exp(-\hbar\omega_j / k_B T)]$, so that the intensity ratio of the Stokes and anti-Stokes Raman lines in this case (no resonance effects included) is given by

$$I_{AS} / I_S = [(\omega_i + \omega)/(\omega_i - \omega)]^4 \exp(-\hbar\omega_j / k_B T). \tag{2.120}$$

The anti-Stokes lines are much weaker than the Stokes lines.

In crystals transparent both to the incident and scattered light, with elementary excitations of infinite lifetime, both *energy and momentum conservation laws*:

$$\omega = \omega_i - \omega_f = \pm\omega_j, \tag{2.121}$$

$$\boldsymbol{k} = \boldsymbol{k}_i - \boldsymbol{k}_f = \pm\boldsymbol{q}_j \tag{2.122}$$

are satisfied, where ω_j, $\boldsymbol{q}_j$ are the frequency and the wavevector, respectively, of the *j*-type excitation in crystal. The plus (minus) sign in the equations above refer to Stokes (anti-Stokes) processes. Whether the energy conservation law is always satisfied, the wavevector is *not conserved if*:

i) the scattering medium has no translational symmetry, as for example in amorphous materials, or materials with high concentrations of impurities or defects. In this case, scattering by modes with $\boldsymbol{k} \neq \boldsymbol{q}_j$ is allowed;

ii) the scattering volume is small. Then, light scattering is due to excitations with wavevectors in a range $\Delta q \cong 2\pi / L$ with L the characteristic length in the scattering volume;

iii) the incident and scattered waves are damped in the scattering volume, as for example in metals or small-gap semiconductors that are opaque to visible light. In these cases $\boldsymbol{k}_i$, $\boldsymbol{k}_f$ are complex and inelastic scattering involves excitations with a range of wavevectors $\Delta\boldsymbol{q} = \mathrm{Im}(\boldsymbol{k}_i) + \mathrm{Im}(\boldsymbol{k}_f)$ around $\boldsymbol{q} = \mathrm{Re}(\boldsymbol{k}_i - \boldsymbol{k}_f)$.

In crystalline media the value of the scattered wavevector is determined by the scattering geometry (see Fig. 2.17):

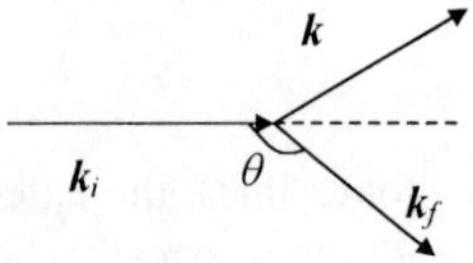

Fig. 2.17. Raman scattering geometry

The minimum value of the wavenumber $k = |\boldsymbol{k}|$ is obtained in the forward scattering direction, when $\theta = 180°$, and is given, in isotropic media, by $k_{\min} = [n(\omega_i)\omega_i - n(\omega_f)\omega_f]/c$ where $n(\omega)$ is the refractive index of the crystal at frequency ω. The maximum value of k is obtained in backscattering, where $\theta = 0$: $k_{\max} = [n(\omega_i)\omega_i + n(\omega_f)\omega_f]/c$.

Raman scattering can take place on phonons, on free electrons or any other collective or individual oscillations of atoms in the crystal, any spatial or temporal fluctuations of the electrons or lattice vibrations that determine the electric polarizability. Which of these states are involved in Raman scattering, and thus are denoted as a_i, a_f, depends on the incident and scattered photon frequencies; so, by measuring $\omega_i - \omega_f$ one can obtain information about the excitations of the material system at low energies. Scattering can also take place due to coupling of the electromagnetic wave with the magnetic moments of the crystal; this type of scattering is very weak and can be neglected in non-magnetic solids, but becomes important for solids with magnetic properties, where the electromagnetic field interacts with magnons. There are a large number of excellent reviews on inelastic light scattering in different materials. For example, the electronic light scattering in semiconductors is reviewed in (Klein, 1975), the light scattering on phonons in bulk semiconductors and low-dimensional semiconductor heterostructures is reviewed in (Ruf, 1998), whereas light scattering on materials which contain scattering centers able to freely rotate in space is treated in (Stepanov and Gribkovskii, 1963). This type of material is encountered in solid-state physics in the form of crystalline powders. Raman scattering in semiconductors and insulators is well reviewed in (Pinczuk and Burstein, 1975). Light scattering on other excitation types such as magnons, interfaces, metals, etc. can be found in the excellent series *Light Scattering in Solids*, edited by Cardona, which has now reached the eighth volume.

For material systems initially in the ground state, only the first term in the expression of $H_{\text{int},fi}$ can produce resonant scattering. In general, the energy values E_b and the eigenfunctions of the material system in the intermediate states are not known; approximations are therefore used. However, the intermediate states $|b\rangle$ have generally a finite lifetime τ_b, directly related to the delay between the absorption of the initial photon ω_i and the emission of the scattered photon

ω_f. This means that E_b has a small imaginary part $\mathrm{Im}(E_b) = \hbar\gamma_b = \hbar / \tau_b$, the resonant scattering being the process in which $\hbar\omega_i \cong \mathrm{Re}(E_b)$ and

$$H_{\mathrm{int},fi} \cong \sum_{b,\mathrm{Re}(E_b)\cong\hbar\omega_i} \langle a_f, \omega_f \mid H_{\mathrm{int}} \mid b,0\rangle\langle b,0 \mid H_{\mathrm{int}} \mid 0,\omega_i\rangle / [\hbar\omega_i - \mathrm{Re}(E_b) - \mathrm{i}\hbar\gamma_b], \tag{2.123}$$

the sum being restricted to energies near $\hbar\omega_i$. So, only some (very few) intermediate states contribute to resonant scattering.

In typical light scattering experiments $0 \le k \le 10^6\ \mathrm{cm}^{-1}$. This means that for first-order scattering processes, the accessible range of $q_j = |\boldsymbol{q}_j|$ is small compared to reciprocal lattice vectors. However, this is an important range for many excitations that may not be otherwise accessible, scattering measurements providing information about the energy, lifetime and symmetry properties of the excitation. First-order Raman scattering implies that the final state is reached through only one intermediate state. However, more than one intermediate state can be involved in the process. For example, at inelastic light scattering on phonons in semiconductors, the intermediate states are electron-hole pair states separated energetically through the energy of a phonon. In this case

$$H_{\mathrm{int},fi} = \sum_{b_1,b_2} M_{b_1 i} \mathfrak{M}_{b_2 b_1} M_{f b_2} / [\hbar^2 (\omega_i - \omega_{b_1} - \mathrm{i}\gamma_{b_1})(\omega_i - \omega_{b_2} - \mathrm{i}\gamma_{b_2})], \tag{2.124}$$

where $M_{b_1 i} = \langle b_1 \mid H_{\mathrm{int}} \mid 0\rangle$, $M_{fb_2} = \langle a_f \mid H_{\mathrm{int}} \mid b_2\rangle$, $\mathfrak{M}_{b_2 b_1} = \langle b_2 \mid H_m \mid b_1\rangle$, and H_m is the Hamiltonian which couples the intermediate states $|b_1\rangle$ and $|b_2\rangle$. If there are three intermediate states, analogously,

$$H_{\mathrm{int},fi} = \sum_{b_i} \frac{M_{b_1 i} \mathfrak{M}_{b_2 b_1} \mathfrak{M}_{b_3 b_2} M_{f b_3}}{\hbar^3 (\omega_i - \omega_{b_1} - \mathrm{i}\gamma_{b_1})(\omega_i - \omega_{b_2} - \mathrm{i}\gamma_{b_2})(\omega_i - \omega_{b_3} - \mathrm{i}\gamma_{b_3})}. \tag{2.125}$$

In higher-order scattering processes, the individual excitation wavevectors $\boldsymbol{q}_j$ can range from zero to a reciprocal lattice vector since now $\boldsymbol{k} = \sum \boldsymbol{q}_j$; in particular second-order scattering gives information about the two-magnon or two-phonon density of states, or about one-phonon density of states in defect-induced scattering experiments.

If electrons with electric charge $-e$ and mass m are scattered by the radiation field with vector potential $\boldsymbol{A}$, $H_{\mathrm{int}} = H_A + H_{AA}$ where

$$H_A = -(e/m) \sum_{j,\boldsymbol{k}_i,\omega_i} \boldsymbol{p}_j \cdot \boldsymbol{A}(\boldsymbol{k}_i, \omega_i), \tag{2.126a}$$

$$H_{AA} = (e^2 / 2m) \sum_{\boldsymbol{k}_i,\omega_i,\boldsymbol{k}_f,\omega_f} \boldsymbol{A}(\boldsymbol{k}_i, \omega_i) \cdot \boldsymbol{A}(\boldsymbol{k}_f, \omega_f). \tag{2.126b}$$

Taking into account that the vector potentials of the incident and scattered photons at the position $\boldsymbol{r}_j$ of the electron are

$$A_i(\boldsymbol{k}_i,\omega_i)=(A_0/\sqrt{\omega_i})\hat{\boldsymbol{e}}_i[a_i\exp(\mathrm{i}\boldsymbol{k}_i\cdot\boldsymbol{r}_j)+a_i^+\exp(-\mathrm{i}\boldsymbol{k}_i\cdot\boldsymbol{r}_j)], \tag{2.127a}$$

$$A_f(\boldsymbol{k}_f,\omega_f)=(A_0/\sqrt{\omega_f})\hat{\boldsymbol{e}}_f[a_f\exp(\mathrm{i}\boldsymbol{k}_f\cdot\boldsymbol{r}_j)+a_f^+\exp(-\mathrm{i}\boldsymbol{k}_f\cdot\boldsymbol{r}_j)], \tag{2.127b}$$

and that the creation and annihilation operators of the electromagnetic field act only on the part of the total wavefunction that represents the state of the electromagnetic field, $H_{\mathrm{int},fi}=\langle f\,|\,H_{\mathrm{int}}\,|\,i\rangle$ can be written as

$$\begin{aligned}H_{\mathrm{int},fi}=(e^2A_0^2/m^2\sqrt{\omega_i\omega_f})\Big[&m\langle a_f\,|\,\textstyle\sum_j\exp[\mathrm{i}(\boldsymbol{k}_i-\boldsymbol{k}_f)\cdot\boldsymbol{r}_j]\hat{\boldsymbol{e}}_i\cdot\hat{\boldsymbol{e}}_f\,|\,a_i\rangle\\&+\sum_b\frac{\langle a_f\,|\,\sum_j\exp(-\mathrm{i}\boldsymbol{k}_f\cdot\boldsymbol{r}_j)\boldsymbol{p}_j\cdot\hat{\boldsymbol{e}}_f\,|\,b\rangle\langle b\,|\,\sum_j\exp(\mathrm{i}\boldsymbol{k}_i\cdot\boldsymbol{r}_j)\boldsymbol{p}_j\cdot\hat{\boldsymbol{e}}_i\,|\,a_i\rangle}{E_{a_i}+\hbar\omega_i-E_b}\\&+\sum_b\frac{\langle a_f\,|\,\sum_j\exp(\mathrm{i}\boldsymbol{k}_i\cdot\boldsymbol{r}_j)\boldsymbol{p}_j\cdot\hat{\boldsymbol{e}}_i\,|\,b\rangle\langle b\,|\,\sum_j\exp(-\mathrm{i}\boldsymbol{k}_f\cdot\boldsymbol{r}_j)\boldsymbol{p}_j\cdot\hat{\boldsymbol{e}}_f\,|\,a_i\rangle}{E_{a_i}-\hbar\omega_f-E_b}\Big].\end{aligned} \tag{2.128}$$

In (2.128) $|\,i\rangle=|\,\phi_i\rangle\,|\,a_i\rangle$, $|\,f\rangle=|\,\phi_f\rangle\,|\,a_f\rangle$, where ϕ_i, ϕ_f represent the state of the electromagnetic field; we have considered in (2.128) that there is one initial phonon and one scattered phonon and we have accounted for all possible scattering mechanisms.

In general, $|\,H_{\mathrm{int},fi}\,|^2$ is obtained by multiplying the square modulus of the expression in (2.128) with $N_i(N_f+1)$, where N_i, N_f are the mean number of photons of the incident and scattered radiation. The term proportional to N_i corresponds to the usual, spontaneous Raman scattering, whereas that proportional to N_iN_f describes the stimulated Raman process, which is a nonlinear scattering mechanism.

We can define a Raman tensor $\hat{R}$ that determines the matrix element for any polarization direction of the incoming and outgoing photons as

$$H_{\mathrm{int},fi}=\hat{\boldsymbol{e}}_i\hat{R}\hat{\boldsymbol{e}}_f. \tag{2.129}$$

The Raman tensor depends only on the frequencies ω_i, ω_f and on the properties of the initial and final crystal states. When $a_i\neq a_f$ the scattering tensor is not symmetric and its nine independent components are, in general, complex and different one from another, whereas for $a_i=a_f$, i.e. for coherent Rayleigh scattering, the scattering tensor is hermitic. Moreover, if the elements of the scattering tensor are also real, it is symmetric. If the material system is initially in the ground state, the possible nonzero components of $\hat{R}$ correspond to different symmetries of the excited $|\,a_f\rangle$ states. The Raman tensor for phonon scattering is symmetric away from any resonant electronic energy, while near resonance antisymmetric components can appear. In scattering on magnetic excitations (magnons) the antisymmetric components of the Raman tensor are important.

The displacements of ions in the crystal during the scattering process determine a modification of the polarization of the elementary cell in the crystal. Since the polarization is a symmetric tensor of second order, it is different from zero for the vibration branches (the optical ones) for which the transformation of coordinates is symmetric. This means that only some vibration branches are active. In particular, in crystals with a center of symmetry the vibration branches that are active in the Raman spectrum are inactive in the IR absorption spectrum on phonons and vice-versa.

Since for light scattering $\boldsymbol{k}_i \cdot \boldsymbol{r}_j << 1$, $\boldsymbol{k}_f \cdot \boldsymbol{r}_j << 1$, the matrix elements can be evaluated by multimode expansion, i.e. the electric dipole, electric quadrupole and magnetic dipolar interactions can be considered separately. In the electric dipole approximation (2.129) becomes

$$H_{\text{int},fi} \cong (A_0^2 \sqrt{\omega_i \omega_f}) \left[\sum_b \frac{\langle a_f \,|\, \boldsymbol{P} \cdot \hat{\boldsymbol{e}}_f \,|\, b \rangle \langle b \,|\, \boldsymbol{P} \cdot \hat{\boldsymbol{e}}_i \,|\, a_i \rangle}{E_{a_i} + \hbar\omega_i - E_b} \right.$$
$$\left. + \sum_b \frac{\langle a_f \,|\, \boldsymbol{P} \cdot \hat{\boldsymbol{e}}_i \,|\, b \rangle \langle b \,|\, \boldsymbol{P} \cdot \hat{\boldsymbol{e}}_f \,|\, a_i \rangle}{E_{a_i} - \hbar\omega_f - E_b} \right], \tag{2.130}$$

where $\boldsymbol{P} = \sum_j e_j \boldsymbol{r}_j$ is the induced electric dipolar moment of the scattering system. Then, by decomposing the scattering tensor into an isotropic, a symmetric and an antisymmetric part, respectively, as

$$\begin{aligned} \langle a_f \,|\, R^0 \,|\, a_i \rangle &= \langle a_f \,|\, R_{xx} + R_{yy} + R_{zz} \,|\, a_i \rangle / 3, \\ \langle a_f \,|\, R^s_{\rho\sigma} \,|\, a_i \rangle &= (1/2) \langle a_f \,|\, R_{\rho\sigma} + R_{\sigma\rho} \,|\, a_i \rangle - \langle a_f \,|\, R^0 \,|\, a_i \rangle \delta_{\rho\sigma}, \\ \langle a_f \,|\, R^a_{\rho\sigma} \,|\, a_i \rangle &= (1/2) \langle a_f \,|\, R_{\rho\sigma} - R_{\sigma\rho} \,|\, a_i \rangle, \end{aligned} \tag{2.131}$$

we can identify the different parts with different multipole contributions. The isotropic component has spherical symmetry, so that the selection rules for isotropic scattering are the same as for a scalar quantity, i.e. only transitions between states with the same symmetry are allowed. The symmetric part is governed by the same selection rules as the spontaneous emission of radiation due to the electric quadrupole moment, whereas the antisymmetric part corresponds to magnetic dipole scattering. The total Raman scattering tensor is $\langle a_f \,|\, R_{\rho\sigma} \,|\, a_i \rangle = \langle a_f \,|\, R^0 \delta_{\rho\sigma} + R^s_{\rho\sigma} + R^a_{\rho\sigma} \,|\, a_i \rangle$.

The selection rules for Raman scattering are derived from $H_{\text{int},fi}$. These selection rules are especially easy to obtain in the dipolar approximation, where the interaction Hamiltonian is proportional to the displacement operators. As we have mentioned in Chap. 1, in order to determine the selection rules, the group theoretical approach is very useful, since the symmetry of the Hamiltonian must be retrieved in the symmetry of the wavefunction. A very intuitive example of the derivation of selection rules for Raman and IR absorption can be found in (Hirata et al., 1996). In the $Ti_{1-x}Sn_xO_2$ crystal, where both TiO_2 and SnO_2 have tetragonal symmetry D_{4h} with two molecular units per primitive cell, the irreducible representations of the point group of the crystal are $\Gamma = A_{1g}(\mathrm{R}) + A_{2g} + A_{2u}(\mathrm{IR}) + B_{1g}(\mathrm{R}) + B_{2g}(\mathrm{R}) + 2B_{1u} + E_g(\mathrm{R}) + 3E_u(\mathrm{IR})$ where we have used the standard notations for the irreducible representations (see any textbook about applications of group theory in the solid state). The irreducible representations/symmetries of motion that have R in brackets are Raman active, those which have IR in brackets are active in IR absorption, and those which are not followed by a bracket are visible in neither Raman or IR absorption. The atom displacements viewed along the c axis, for the IR and Raman active modes are represented in Fig. 2.18, where the black and white circles represent the Ti and O atoms, respectively.

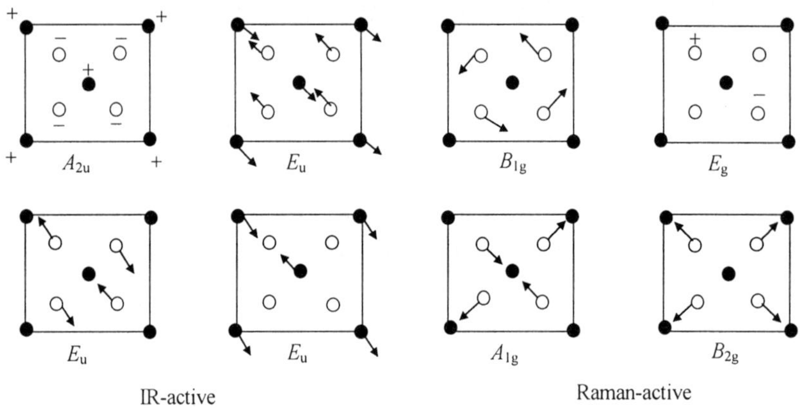

Fig. 2.18. Atomic displacements in a TiO_2 crystal corresponding to different types of IR and Raman active modes

It should be mentioned that the selection rules found by applying group theoretical methods, for example, or simple calculations of the interaction matrix elements break down when electric or magnetic fields, as well as any other perturbation with lower symmetry, are applied on the system.

Analogously to (2.19), the differential scattering cross-section $\mathrm{d}\sigma$ of an initial photon $(\boldsymbol{k}_i, \hat{\boldsymbol{e}}_i)$ in a scattered photon with a frequency between ω_f and $(\omega_f + \mathrm{d}\omega_f)$, and a polarization $\hat{\boldsymbol{e}}_f$ such that $\boldsymbol{k}_f$ is within a solid angle $\mathrm{d}\Omega$ around the direction θ is defined as

$$\mathrm{d}\sigma(\boldsymbol{k}_i, \hat{\boldsymbol{e}}_i; \omega_f, \hat{\boldsymbol{e}}_f, \theta) = (2\pi/\hbar c)\Sigma_{\boldsymbol{k}_f, a_f} |H_{\mathrm{int},fi}|^2 \,\delta(\hbar\omega_f + E_{af} - \hbar\omega_i). \qquad (2.132)$$

The sum in $\mathrm{d}\sigma$ is performed for $\boldsymbol{k}_f$ over a fraction $\mathrm{d}\Omega/4\pi$ of the surface of the k^2 sphere, the result being

$$\mathrm{d}\sigma(\boldsymbol{k}_i,\hat{\boldsymbol{e}}_i;\omega_f,\hat{\boldsymbol{e}}_f,\theta)=(\mathrm{d}\Omega/4\pi^2\hbar^4c^4)\Sigma_{af}\,|\,H_{\mathrm{int},fi}\,|^2\,(\hbar\omega_i-E_{af})^2. \tag{2.133}$$

The remaining sum over a_f must be performed over states for which $\hbar\omega_f<\hbar\omega_i-E_{af}<\hbar\omega_f+\hbar\mathrm{d}\omega_f$. When the final state E_{af} belongs to the discrete part of the spectrum, or when E_{af} is in the continuum spectrum but the selection rules are such that $H_{\mathrm{int},fi}$ is different from zero for one (or a finite number) of final states, the scattered photon has a very narrow, δ-type frequency band around $\hbar\omega_f=\hbar\omega_i-E_{af}$. On the contrary, if E_{af} belongs to the continuum energy spectrum, such that for a continuum subset $H_{\mathrm{int},fi}\neq 0$, the differential scattering cross-section is proportional to the density of states $\rho(E_{af}=\hbar\omega_i-\hbar\omega_f)$, and the scattered photon has a spectral frequency band for any direction θ.

In resonant Raman scattering, if the discrete initial and final states $|\,0,\omega_i\rangle$, $|\,a_f,\omega_f\rangle$ are connected through only one intermediate state $|\,b,0\rangle$,

$$|\,H_{\mathrm{int},fi}\,|^2\approx|\,M_{fb}M_{bi}\,|^2\,/\,\hbar^2[(\omega_i-\omega_b)^2+\gamma_b^2], \tag{2.134}$$

where $\mathrm{Re}(E_b)=\hbar\omega_b$, $\langle a_f,\omega_f\,|\,H_{\mathrm{int}}\,|\,b,0\rangle=M_{fb}$ and $\langle b,0\,|\,H_{\mathrm{int}}\,|\,0,\omega_i\rangle=M_{bi}$. Very near resonance, for $\Delta\omega=|\,\omega_i-\omega_b\,|<<\gamma_b$, the scattering cross-section is dominated by slow processes, whereas far from resonance, for $\Delta\omega>>\gamma_b$, fast processes dominate. At intermediate frequencies fast and slow processes coexist. For a continuum of final states and a one-to-one correspondence between the final and resonant intermediate states, a sum of terms of the form (2.134) (multiplied by $(\hbar\omega_i-E_{af})^2$) appear in $\mathrm{d}\sigma$, the sum being performed over either the states $|\,b\rangle$ or $|\,a_f\rangle$, due to the one-to-one correspondence. *Interference* effects can however appear between the terms in $H_{\mathrm{int},fi}$ if there is more than one discrete state in the resonant region, which corresponds to the same a_f. In this case the sum must be performed *before* the square modulus is applied:

$$|\,H_{\mathrm{int},fi}\,|^2=|\,\Sigma_b M_{fb}M_{bi}\,/[\hbar(\omega_i-\omega_b)-\mathrm{i}\hbar\gamma_b]\,|^2\,. \tag{2.135}$$

The form of the resonance is determined by the phase and the functional dependence of the matrix elements, as well as by the position of the poles. The sum in (2.135) must be replaced by an integral if there is a continuum of states $|\,b\rangle$ that resonate with any initial $|\,0,\omega_i\rangle$ and final $|\,a_f,\omega_f\rangle$ state. A more detailed discussion of resonant Raman scattering can be found in (Martin and Falicov, 1975).

The Raman scattering efficiency per unit length and solid angle, in the dipolar approximation, is

$$\frac{\mathrm{d}S}{\mathrm{d}\Omega_f}=\frac{V}{(2\pi)^2}\,\frac{\omega_f^3\omega_i n^3(\omega_f)n(\omega_i)}{c^4(\hbar\omega_i)^2}\,|\,H_{\mathrm{int},fi}\,|^2, \tag{2.136}$$

and the Raman scattering intensity is obtained by multiplying it by the energy of scattered photons and integrating over the solid angle.

In A_3B_5 bulk semiconductor materials the scattering events that conserve the spin and those which involve spin-flip transitions from the spin-split-off band have different polarization dependencies, proportional to $(\hat{e}_i \cdot \hat{e}_f)^2$ and $|\hat{e}_i \times \hat{e}_f|^2$, respectively. In spin-flip scattering H_A is the dominant term, whereas for plasmon scattering or single-electron non-flip transitions the H_{AA} term dominates. Detailed calculations show that $dS/d\Omega_f \approx \Delta^2$ for $\Delta < E_g$ and $\approx \Delta^4$ for $\Delta > E_g$, where Δ is the spin-orbit split-off energy. For electron spin-flip $dS/d\Omega_f \approx (2 - g_e)^2$ if $\Delta > E_g$, whereas for hole spin-flip scattering, the transition rate is proportional to the square of $\mu_B gH/(E_g - \hbar\omega)^2$, the hole spin-flip scattering being weaker than electron spin-flip by a factor of $[\mu_B gH/(E_g - \hbar\omega)^2]^2$. Except for very close to resonance, where this factor can approach unity, the hole spin-flip scattering is very difficult to detect. The lineshape of the spin-flip scattering line is highly asymmetric, with a tail on the low-energy side of the peak. Spin-flip Raman scattering can be used to determine the g value of the carrier from spin-flip energy shifts, to distinguish between free and bound electrons (through linewidths), between electrons and holes, between effective-mass and non-effective-mass acceptors (for non-effective-mass acceptors, g is different from zero for $\boldsymbol{H}$ perpendicular to the c axis). For a review of Raman spin-flip transitions for free holes, holes bound to acceptors, photoexcited electrons, excitons, multiple and stimulated spin-flip scattering, see Scott (1980).

Sometimes, it is useful to work with reduced Raman spectra, defined as $I_{red}(\omega) = \omega(\omega_L - \omega)^{-4}[\bar{n}_{ph}(\omega) + 1]^{-1} I(\omega)$ for the Stokes part, or $I_{red}(\omega) = \omega(\omega_L + \omega)^{-4} \bar{n}_{ph}(\omega)^{-1} I(\omega)$ for the anti-Stokes part. Here, ω_L is the frequency of the incident laser radiation and $\bar{n}_{ph}$ the phonon occupation number. The reduced spectra depend only on the density of states and the frequency-dependent transition matrix element, which in many cases are slowly varying, and in some cases calculable.

2.5.2 Brillouin Scattering

Brillouin scattering denotes light scattering on very low-frequency excitations, with frequency shift $|\omega_i - \omega_f|$ within the GHz frequency range, as for example, on acoustical phonons. Brillouin scattering on phonons is mainly used for the mechanical and acoustical characterization of thin films, multilayers, heterostructures, and superlattices. As for all inelastic scattering processes, the energy and momentum are conserved, so that the frequency and wavenumber of the acoustic phonon, labeled by a, are given, respectively, by $\omega_f = \omega_i \pm \omega_a$, $\boldsymbol{k}_f = \boldsymbol{k}_i \pm \boldsymbol{k}_a$, where $k_i = n(\omega_i)\omega_i / c$, $k_f = n(\omega_f)\omega_f / c$. The frequency shift in the Brillouin scattering spectrum

$$\omega_A = \pm\omega_i (v_a / c)[n^2(\omega_i) + n^2(\omega_f) - 2n(\omega_i)n(\omega_f)\cos\theta]^{1/2}, \quad (2.137)$$

where $\theta = \cos^{-1}(\hat{\boldsymbol{k}}_i \cdot \hat{\boldsymbol{k}}_f)$ is the scattering angle, is directly proportional to the sound velocity v_a. The + sign corresponds to an absorbed acoustical phonon (anti-Stokes shift) and the – sign signifies an emitted acoustical phonon (Stokes shift). (2.137) is analogous to the formula describing Bragg reflection from a grating of period $2\pi / k_a$ moving with velocity v_a. As for Raman scattering, only excitations that change the polarizability of the medium couple to light and are observed in Brillouin scattering. These changes in polarizability for long-wavelength acoustic excitations can be related to macroscopic properties such as the elasto-optic constants, which govern the changes in the refractive index due to strain, or magneto-optic coefficients for magnetic excitations. So, Brillouin spectra provide information on elastic and magnetic constants, acoustic-velocity anisotropy, relaxation processes, phase transitions and acoustic-phonon interactions with other low-frequency excitations.

In materials transparent to the incident light wavelength, scattering is caused by thermally excited acoustic phonons, whereas in opaque materials light scattering is produced by surface ripple mechanisms, called surface waves. Surface acoustic waves have penetration depths that decrease with increasing frequency. Any phonon that corrugates the surface and satisfies the parallel-to-surface wavevector conservation is observed in light scattering. The sharp features in Brillouin scattering are due to surface waves, which are solutions of the elastic wave equation in a semi-infinite solid subject to boundary conditions of the free surface. These waves have amplitudes that decrease exponentially away from the surface, so that the wavevector component normal to the surface is limited by the inverse of the penetration depth. Their velocity is generally closely related to that of a bulk transverse wave polarized perpendicular to the free surface. The Brillouin spectrum also has broad features, due to phonons in the bulk reflected at the free surface. Since the phonon wavevector normal to the surface is not subject to any restriction, the features corresponding to this process are broad.

Light scattering in opaque materials can occur on several types of surface waves in thin supported layers or multilayers, depending on the geometrical characteristics of the layer and overlayer. We have Rayleigh surface acoustic waves (SAW), pseudo-SAW, Sezawa surface waves and pseudo-Sezawa waves, surface wave effects due to bulk-wave contributions (Lamb shoulder) and Stonely interface waves (Zhang et al., 1998). All these types of waves as well as their dispersion curves can be identified with Brillouin spectroscopy. In particular, multilayer structures with special characteristics are tested with Brillouin spectroscopy to check the existence and propagation conditions of different surface waves.

The determination of elastic constants using Brillouin scattering is based on the fact that the quantity $X = \rho v_a^2$ is a combination of elastic moduli constants c_{ij}. In order to assign measured velocities to elastic constants we have to know the crystallographic orientation of the sample with respect to the scattering geometry. The specific combination of elastic constants that determines the sound velocity in a given crystallographic direction is determined from the Christoffel equation and is measured by taking Brillouin scattering spectra for different scattering configurations. For example, in a crystal of cubic symmetry, as in diamond, only three independent elastic moduli c_{11}, c_{12} and c_{44}, exist. c_{11} is obtained in a backscattering geometry, for a phonon wavevector direction along [100], from longitudinal phonon spectra, whereas $(c_{11}+c_{12}+2c_{44})/2c_{44}$ and c_{44} are obtained in the 90° scattering geometry, with the phonon wavevector direction along [110], from the LA and TA (transverse acoustic) phonons, respectively. Other combinations of elastic moduli are determined in other scattering geometries (Zouboulis and Grimsditch, 1998). The typical Brillouin spectrum looks like that displayed in Fig. 2.19. The resonances of TA and LA phonons are clearly visible; the vertical axis is the intensity of the Brillouin scattering represented in arbitrary units. The background is due to luminescence phenomena.

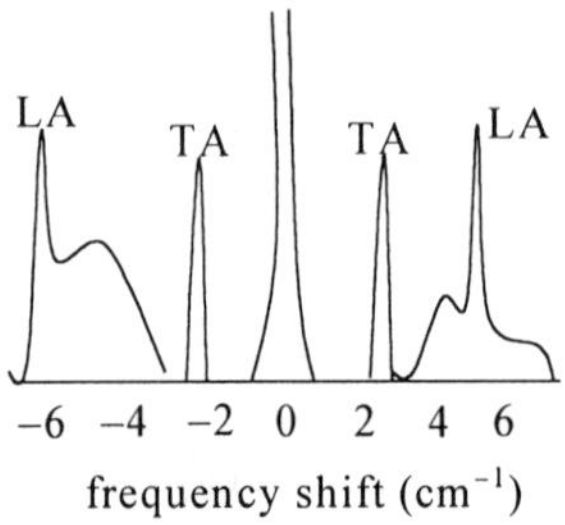

Fig. 2.19. Typical Brillouin spectrum for scattering on acoustic phonons

As in Raman spectroscopy, the Brillouin scattering signal increases – there is resonant Brillouin scattering – when the incident or scattered light frequency coincides with different types of excitations in the medium.

The spectrum of acoustic phonons with a lifetime τ_a has a Lorentzian shape

$$S(\omega_f,\omega_a)=(\Gamma_a/\pi)[(\omega_f-\omega_i\mp\omega_a)^2+\Gamma_a^2], \tag{2.138}$$

where $\Gamma_a=\tau_a^{-1}=\alpha_a v_a$ with α_a the sound attenuation (Pine, 1975). Different damping mechanisms due to the interaction of light with acoustic phonons can be identified and characterized by measuring the lineshape of Brillouin spectra. In opaque crystals the broadening is enhanced by the absorption of incident and scattered light, the halfwidth of the spectrum being proportional to the absorption coefficient $\Delta\omega_{FWHM}\cong\alpha v_a$. So, knowing v_a, the real and imaginary part of the refractive index of scattering media can be accurately estimated from measurements of this excess width.

The power P_B and cross-section σ_B of Brillouin scattering are determined from the following equations (Pine, 1975)

$$\partial^2 P_B/\partial\Omega\partial\omega_f=\sigma_B P_i LS(\omega_f,k_a)/2, \tag{2.139a}$$

$$\sigma_B = \pi^2 n(\omega_f)\mu\hbar\omega_a |\boldsymbol{m}_d|^2 \sin^2\varphi / \rho v_a n_i \lambda_f^4, \qquad (2.139b)$$

where L and ρ are the length and density, respectively, of the scattering region, and $\mu = \bar{n}_a + 1$ for Stokes shifts and $\mu = \bar{n}_a$ for anti-Stokes shifts, with $\bar{n}_a$ the phonon occupation number, which in the case of thermal scattering becomes $\bar{n}_a + 1 \cong \bar{n}_a \cong [\exp(k_B T / \hbar\omega_a) - 1]^{-1}$ for $k_B T >> \hbar\omega_a$. The dipole moment $\boldsymbol{m}_d$ in (2.138b) is given by $\boldsymbol{m}_d = \hat{\varepsilon}\,\hat{p}\,\hat{\varepsilon}\,\boldsymbol{E}_{inc} \cdot \boldsymbol{k}_a \cdot \hat{\boldsymbol{u}}_a$, where $\hat{\varepsilon}$ is the optical dielectric tensor, $\hat{p}$ the Pockels tensor, $\hat{\boldsymbol{u}}_a$ is the unit acoustic displacement vector and $\varphi = \arccos(\hat{\boldsymbol{m}}_d \cdot \hat{\boldsymbol{k}}_f)$. When light couples only to density fluctuations of longitudinal acoustic (LA) phonons $|\hat{\varepsilon}\,\hat{p}\,\hat{\varepsilon}| = (\varepsilon - 1)(\varepsilon + 2)/3$.

Another important application of Brillouin spectroscopy is the determination of the main properties of magnetic thin films, magnetic multilayers or magnetic superlattices (Grünberg, 1989). In this case the light is inelastically scattered on surface spin waves (magnons), with a relatively large range of wavevectors ($q \cong 10^5$–10^6 cm^{-1}). For comparison, microwave absorption can only excite spin waves with $q \cong 0$. Since magnetic excitations do not produce ripples on the surface (magnetostriction effects are negligible), coupling to light in this case occurs only through polarizability changes in the bulk. As in the case of scattering on phonons, the magnon wavevector parallel to the surface, $q_{\|} = k_0(\sin\theta_i + \sin\theta_f)$, must be conserved, with k_0 the wavenumber of the incident radiation. If scattering is on bulk magnetic materials, the normal wavevector component must only satisfy $0 \le q_{\perp} \le 2\pi / l_0$, where l_0 is the light penetration depth, since the radiation is strongly attenuated in the direction normal to the surface. The frequency of bulk magnons of wavevector $\boldsymbol{q}$ propagating perpendicular to $\boldsymbol{H}$ in an isotropic ferromagnet is $\omega_b = \gamma[(H + Dq^2)(H + 4\pi M + Dq^2)]^{1/2}$, where γ is the gyromagnetic ratio, M the saturation magnetization and D the spin-wave stiffness. The Brillouin lines in this case are broad and asymmetric, since $q_{\perp}$ is not well defined. On the other hand, the Brillouin lines for surface magnons or magnetic excitations in thin films are sharp. The dynamics of spin waves in this case is described by

$$\mathrm{d}\boldsymbol{M} / \mathrm{d}t = \gamma \boldsymbol{M} \times \boldsymbol{B}_{eff}, \qquad (2.140)$$

where $\boldsymbol{M}$ is the magnetization vector, $\gamma = -g\mu_B / \hbar$ is the gyromagnetic ratio and $\boldsymbol{B}_{eff} = \boldsymbol{B} - \nabla_u E_a / M + (2A / M^2)\nabla_{\hat{u}}^2 \boldsymbol{M}$ is the effective magnetic induction, with $\boldsymbol{B} = \mu_0(\boldsymbol{H} + \boldsymbol{M})$, E_a [J/m^3] the volume anisotropy energy and A[J/m] the exchange stiffness parameter. Due to the exchange interaction, an additional boundary condition appears $\boldsymbol{M} \times (\nabla_{\hat{u}} E_{surf} - D\partial\boldsymbol{M} / \partial n) = 0$, where E_{surf} [J/m^2] is the surface anisotropy energy density, which can be perpendicular to the magnetic film surface $E_{surf}^{\perp}$ or in-plane uniaxial anisotropy $E_{surf}^{\|}$.

The spin waves consistent with these general boundary conditions are called Damon–Esbach (DE) modes or magnetostatic surface waves when $\boldsymbol{q}_{\|} \perp \boldsymbol{J}$, where $\boldsymbol{J} = \mu_0 \boldsymbol{M}$ is the magnetic polarization. Light scattering on DE modes can occur when the incidence plane of light is perpendicular on $\boldsymbol{J}$. The dispersion relation of these modes is

$$\omega = \gamma\{[1-\exp(-2q_{\parallel}d)](J/2) + B^2 + B^2 J\}^{1/2}, \tag{2.141}$$

where $B = B_0 + 2K/M$, with K the bulk or surface anisotropy constant, depending on the geometry and the type of the magnetic film of thickness d. If the exchange constant $D = 2A/M$ is included in the treatment, the dispersion relation for DE modes becomes $\omega = \gamma[(B + Dq^2)(B + Dq^2 + J)]^{1/2}$, where $q^2 = q_{\parallel}^2 + q_{\perp}^2$ and $q_{\perp} = n\pi/d$. A typical Brillouin scattering spectrum for a metallic film is represented in Fig. 2.20. The surface modes (DE) are identified.

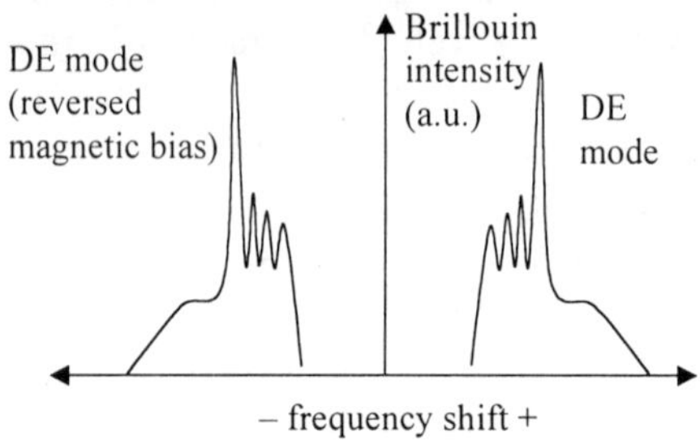

Fig. 2.20. Magnetic Brillouin scattering

The determination of DE mode frequencies allows us to determine the magnetic parameters of the films: J, g, D, $E_{\text{surf}}^{\perp}$ and/or $E_{\text{surf}}^{\parallel}$, and the anisotropy bulk constant K_{b}, where $E_{\text{a}} \approx K_{\text{b}}(\boldsymbol{u}\cdot\boldsymbol{u})^2$. In a similar manner, the parameters of magnetic multilayers or magnetic superlattices can be determined. The incident light wavevectors necessary to induce surface modes are given by $q_{\parallel} = 2k_i \sin\theta$, where θ is the angle between the incident light and the normal to the sample surface.

2.6 Nonlinear Optical Response of Solids

In the previous sections it was assumed that the optical constants, which characterize the interaction between the solid state and the electromagnetic field, are independent of the intensity of the incident radiation. One of the consequences of this assumption is that the polarization of the matter oscillates with the same frequency as the electromagnetic radiation.

The nonlinear optical response refers to the situations when this assumption is not valid. In this regime, the dielectric susceptibility tensor $\hat{\chi}$, defined as $\boldsymbol{P} = \varepsilon_0 \hat{\chi} \boldsymbol{E}$ is no longer independent of $\boldsymbol{E}$, a situation which can be dealt with by expanding the polarizability as a power series in the electric field (Klingshirn, 1997):

$$P_i / \varepsilon_0 = \sum_j \chi_{ij}^{(1)} E_j + \sum_{j,k} \chi_{ijk}^{(2)} E_j E_k + \sum_{j,k,l} \chi_{ijkl}^{(3)} E_j E_k E_l, \tag{2.142}$$

$\chi^{(1)}$ is the linear susceptibility. The tensor $\chi^{(2)}$, as well as all other even-order terms, vanishes in crystals with inversion symmetry, for which $\boldsymbol{E} \to -\boldsymbol{E}$ corresponds to $\boldsymbol{P} \to -\boldsymbol{P}$. If the different field amplitudes in (2.142) have different frequencies, this equation can describe the so-called parametric oscillations, which include the second-harmonic, sum- and difference-frequency generation, etc. In general $\hat{\chi}^{(k)}$ is a tensor of rank $k+1$, the kth order susceptibility describing the generation of the kth harmonic, or the nonlinear mixing between $k+1$ waves.

The different order susceptibility tensors are in general complex, their imaginary part being related to the absorption coefficient and n_{im}, whereas their real part is closely related to n_{re}. These real and imaginary parts are correlated through the Kramers–Krönig relations. An expression for the susceptibility can be obtained by assuming a mechanism through which the susceptibility is dependent on the frequency of the incident radiation. Namely, it is supposed that one photon is absorbed in a virtually excited intermediate state $|m\rangle$, and after some delay, another photon with the same frequency and polarization is spontaneously emitted. The interaction of the electromagnetic radiation with matter is therefore a two-photon process, the linear susceptibility for frequencies far away from the resonant frequencies of the intermediate states being

$$\chi^{(1)} = \sum_m \langle i | H_{\text{int}} | m \rangle \langle m | H_{\text{int}} | i \rangle / [\hbar\omega - (E_i - E_m) + \mathrm{i}\delta], \tag{2.143}$$

where we considered that $H_{\text{int}} \cong H_A$. In a similar manner, the second-order susceptibility characterizes an interaction involving two virtually excited intermediate states $|m_1\rangle$, $|m_2\rangle$. Supposing that the different field amplitudes have different frequencies, and neglecting the damping constant δ,

$$\chi^{(2)} = \sum_{m_1, m_2} \frac{\langle i | H_{\text{int}}(-\omega_1 - \omega_2) | m_2 \rangle \langle m_2 | H_{\text{int}}(\omega_2) | m_1 \rangle \langle m_1 | H_{\text{int}}(\omega_1) | i \rangle}{[\hbar(\omega_1 + \omega_2) - (E_{m_2} - E_i)][\hbar\omega_1 - (E_{m_1} - E_i)]} + \text{c.p.}, \tag{2.144}$$

where c.p. denotes cyclic permutation. Another contribution comes from the H_{AA} term. In particular for the second-harmonic generation, this contribution is

$$\chi^{(2)} = \sum_{m_2} \langle i | H_{\text{int}}(-2\omega) | m_2 \rangle \langle m_2 | H_{AA}(\omega) | i \rangle / (2\hbar\omega - E_{m_2}). \tag{2.145}$$

The $\chi^{(2)}$ term describes the two-photon absorption, second-harmonic generation, sum- and difference-frequency generation. Note that the absorption coefficient for the two-photon absorption is defined phenomenologically as $-\mathrm{d}I / \mathrm{d}z = \gamma I^2$.

For the third-order susceptibility tensor, denoting by ω_1, ω_2, ω_3 the incident, absorbed and stimulating frequencies, and by ω the frequency which results from the process,

$$\chi^{(3)}(\omega;-\omega_3,\omega_2,\omega_1) = \sum_{m_1,m_2,m_3} \frac{\langle i|H_{\text{int}}(\omega)|m_3\rangle\langle m_3|H_{\text{int}}(-\omega_3)|m_2\rangle\langle m_2|H_{\text{int}}(\omega_2)|m_1\rangle\langle m_1|H_{\text{int}}(\omega_1)|i\rangle}{[\hbar(\omega_1+\omega_2-\omega_3)-(E_{m3}-E_i)][\hbar(\omega_1+\omega_2)-(E_{m2}-E_i)][\hbar\omega_1-(E_{m1}-E_i)]} + \text{c.p.}\,. \tag{2.146}$$

$\chi^{(3)}$ effects include four-wave mixing (FWM), coherent anti-Stokes Raman scattering (CARS), hyper-Raman scattering, etc. As an example of nonlinear optical phenomena we present in the next section nonlinear Raman effects.

In semiconductors, for example, nonlinear effects appear when the field-induced momentum eF/ω where $F = |\boldsymbol{F}|$ is the electric field, exceeds the interband momentum E_g/P with P the characteristic interband velocity in the Kane model. Then, the high-frequency contribution to the Hamiltonian leads to mixing of the conduction and valence band states and nonlinear phenomena appear. Although the susceptibilities of all orders, including the linear susceptibility, are by construction independent of the driving field $\boldsymbol{F}$, they depend on the density matrix that describes the distribution of carriers over electronic states. However, the excitation field may affect the electron distribution and thus parametrically modify the linear and nonlinear susceptibility; this indirect nonlinear response is called dynamic nonlinearity. In semiconductors, the sources of dynamic nonlinearities are (Vasko and Kuznetsov, 1999)

(i) Pauli-blocking, or phase-space-filling, determined by the population difference between the initial and final states. In a noninteracting isotropic system Pauli blocking reduces the low-density absorption coefficient by $1 - f^c_{nn}(\hbar\omega - E_n) - f^h_{n'n'}(\hbar\omega - E_{n'})$, where $f^{c,v}$ are the Fermi–Dirac distribution functions in the conduction and valence bands, respectively. At high excitation densities, this population difference factor can become negative at certain frequencies and optical gain is possible.

(ii) Coulomb renormalization of single-particle energies manifests itself as a decrease of the energy bandgap with increase of carrier density, and implicitly as a redshift of the absorption edge. Coulomb renormalization effects are treated as a rigid shift of the whole band in bulk materials, whereas in heterostructures different subbands renormalize in a different way, the energy splittings between subbands becoming density dependent.

(iii) screening of the Coulomb potential suppresses excitonic effects by reducing the strength of the electron-hole attraction.

(iv) collisional broadening of optical transitions washes out the excitonic peaks at high excitation densities and is responsible for the absorption tail that extends below the band edge.

These effects are strongly interdependent and occur in the same range of densities, so that their separate treatment is impossible. In low-dimensional semiconductor heterostructures the importance of screening is reduced compared to bulk semiconductors, while the importance of phase-space filling is increased due to quantum confinement.

2.6.1 Nonlinear Raman Effect

The linear (spontaneous) Raman effect depends linearly on the intensity of the electric field of the incident radiation. If this approximation is not valid the scattering must be studied in terms of higher-order Raman tensors. Note that due to the similarities of (2.128), (2.129) and (2.145), the higher-order Raman tensor can be defined similarly as the higher-order optical susceptibility (2.146). The linear Raman scattering corresponds to a second-order process, since it involves two-photons.

2.6.1.1 Stimulated Raman Effect

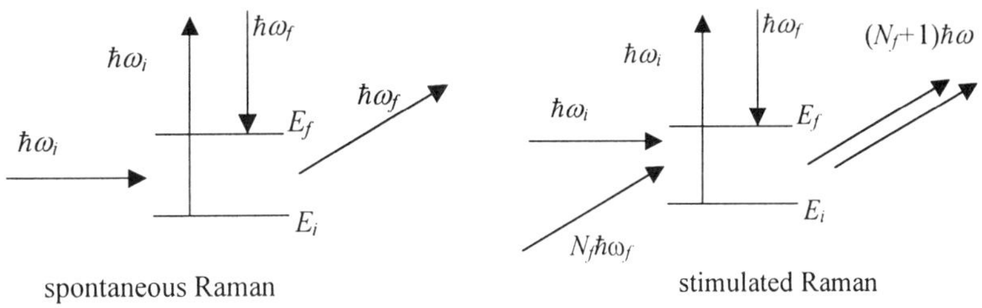

Fig. 2.21. Scattering processes in spontaneous and stimulated Raman effects

The stimulated Raman effect appears if, simultaneously with the radiation with frequency ω_i, a radiation containing N_f photons of frequency ω_f is also incident on the system. Then, the scattered radiation contains N_f+1 photons with frequency ω_f. The scattering probability is the same as in the spontaneous Raman effect, multiplied with $N_i(N_f+1)$, or with N_iN_f for $N_f >> 1$ (Shen, 1975). The variation of the number of scattered photons with respect to the distance of propagation in the scattering medium is determined by the transitions $i \rightarrow f$ characterized by the probability P_{fi} and by the transitions $f \rightarrow i$ with probability P_{if}. If N_i is the population of the level i, the evolution of the number of scattered photons is described by

$$\mathrm{d}N_f / \mathrm{d}z = (P_{fi}N_i - P_{if}N_f)[n(\omega_f)/c] - \alpha_f n_f = (GI_0 - \alpha_f)n_f, \tag{2.147}$$

where α_f is the absorption coefficient for the scattered radiation and G the corresponding amplification factor. By integrating this relation one obtains

$$N_f = N_f(0)\exp[(GI_0 - \alpha_f)L]. \tag{2.148}$$

So, for the stimulated Raman effect the number of scattered photons increases exponentially with the surface density of the energy flux of the incident radiation, I_0. In contrast, in the spontaneous Raman effect the increase is linear.

From (2.148) it follows that the stimulated Raman effect has a threshold for which $G = 0$. This threshold is in practice less than the calculated value due to self-focalization of radiation, which appears for high-power excitations. Other differences between the spontaneous and stimulated Raman radiation are:

(i) the spontaneous Raman effect is incoherent and the scattered radiation propagates in all directions, having a very small intensity compared to the incident radiation (their ratio is about 10^{-6}); the stimulated Raman effect is coherent;

(ii) the stimulated Raman effect appears for powers of the incident radiation greater than a threshold, or for propagation lengths inside the medium greater than a minimum one;

(iii) the shift of the frequencies of the Stokes and anti-Stokes components in the stimulated Raman effect are multiples of the fundamental frequencies of vibrations associated to the most intense transitions in the medium; they coincide with the corresponding shifts in the spontaneous Raman effect (but the number of lines are smaller since not all transitions appear in the stimulated effect);

(iv) the linewidth of the stimulated Raman radiation is much narrower than in the spontaneous scattering process;

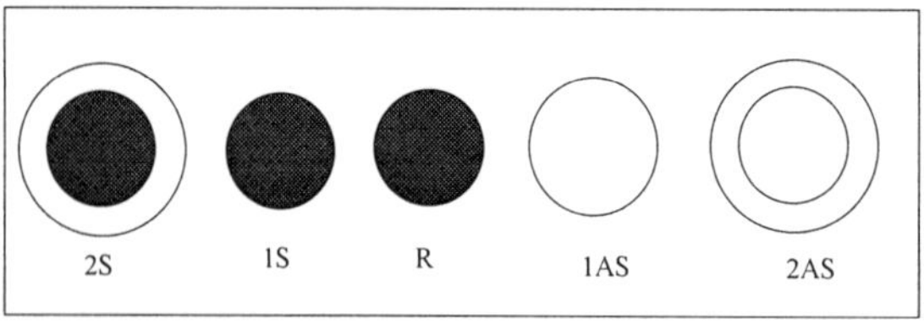

Fig. 2.22. Transverse patterns of scattered radiation in the stimulated Raman effect

(v) the angular distribution of the scattered radiation in the stimulated Raman effect is very complicated. The scattered radiation has a sharp maximum in the direction of the incident radiation corresponding to the first Stokes component. The higher Stokes components and anti-Stokes components have intensity maxima along directions that make well-determined angles with the direction of the incident radiation. These angles are determined by the nature of the medium. On a photographic plate perpendicular to the direction of the incident radiation one would observe a central blob for the Rayleigh and first Stokes line surrounded by concentric circles corresponding to the intersections of the photographic plate with the surface of the cones along which the higher-order Stokes (associated with $\omega_i - n\omega$, $n \geq 2$) and anti-Stokes components (with frequencies $\omega_i + n\omega$, $n \geq 1$) are propagating (see Fig. 2.22).

2.6.1.2 Inverse Raman Scattering

The inverse Raman scattering is the process of absorption of the radiation with frequency ω_f and stimulated emission of radiation with ω_i. It can be described with the same formulae as the stimulated Raman process, with indices i and f interchanged. In the inverse Raman process the sample is irradiated with a monochromatic beam with frequency ω_i and another beam with a broad spectrum. As a result, the medium absorbs the frequency $\omega_f = \omega_i \pm \omega$ from the incident radiation and stimulatedly emits the radiation ω_i. The amplification of the radiation with ω_i (attenuation of the radiation with ω_f) is proportional to the population N_{a_i} of the state a_i (population N_{a_f} of the state a_f). If $N_{a_i} > N_{a_f}$ the crystal has a net gain for the electromagnetic radiation of frequency ω_f. A laser oscillator on frequency ω_f can be obtained by introducing the crystal in a resonant cavity – this type of laser is called *laser with stimulated Raman effect.*

2.6.1.3 Hyper-Raman Effect

The hyper-Raman effect is a third-order process in which two photons, with the same or different frequencies, are absorbed, and one photon with a different frequency from that of the absorbed photons is emitted. More interesting is the case when the two absorbed photons have the same frequency. Then, the corresponding Raman tensor is given, in the dipolar approximation, by

$$\langle a_f | R_{\rho\sigma\tau} | a_i \rangle \approx \sum_{r,p} \left[\frac{\langle a_f | P_\rho | p \rangle \langle p | P_\sigma | r \rangle \langle r | P_\tau | a_i \rangle}{(E_{a_i} - E_r - \hbar\omega_i)(E_{a_i} - E_p - 2\hbar\omega_i)} \right.$$

$$\left. + \frac{\langle a_f | P_\sigma | r \rangle \langle r | P_\rho | p \rangle \langle p | P_\tau | a_i \rangle}{(E_{a_f} - E_r + \hbar\omega_i)(E_{a_i} - E_p - \hbar\omega_i)} + \frac{\langle a_f | P_\tau | r \rangle \langle r | P_\sigma | p \rangle \langle p | P_\rho | a_i \rangle}{(E_{a_f} - E_r + \hbar\omega_i)(E_{a_f} - E_p + 2\hbar\omega_i)} \right]. \tag{2.149}$$

The selection rules for the hyper-Raman scattering are different from those for spontaneous Raman scattering. Many modes that are inactive in the Raman and IR spectra are active in hyper-Raman spectra. However, the intensity of the hyper-Raman radiation is approximately six orders of magnitude lower than for the spontaneous Raman effect. The hyper-Raman effect can also be resonant; in this case there are nine cases of resonances, depending of the way in which the denominators vanish. In particular, the double-resonance case appears if $\hbar\omega_i = E_{a_i} - E_r$ and $E_{a_i} - E_p = 2\hbar\omega_i$ or when $\hbar\omega_i = -(E_{a_f} - E_r)$ and $E_{a_i} - E_p = \hbar\omega_i$, or when $\hbar\omega_i = -(E_{a_f} - E_r)$ and $-(E_{a_f} - E_p) = 2\hbar\omega_i$. In the double-resonance case, the energetic flux is about 12 orders of magnitudes higher than in the usual hyper-Raman effect. However, due to the high absorption of the scattered radiation, it is difficult to observe even in this case.

In contrast to the intensity-independent spontaneous Raman effect, the scattering efficiency of the hyper-Raman effect is $\mathrm{d}S_{\mathrm{HR}} / \mathrm{d}\Omega_f \approx I_i$, where I_i is the intensity of the incident light with frequency ω_i. Therefore, we can define a normalized hyper-Raman scattering efficiency

$$\frac{d\widetilde{S}_{HR}}{d\Omega_f}=\frac{1}{I_i}\frac{dS_{HR}}{d\Omega_f}=\frac{V^2}{(2\pi)^2}\omega_i\omega_f^3\frac{n^2(\omega_i)n^3(\omega_f)}{c^5}\frac{N+1}{(\hbar\omega_i)^3}|H_{\text{int},fi}|^2, \qquad (2.150)$$

where N is the number of quanta of the material excitation with frequency $\hbar\omega = 2\hbar\omega_i - \hbar\omega_f$. A theoretical model for the resonant hyper-Raman scattering on longitudinal optical (LO) phonons in semiconductors, which accounts for the excitonic effects and is valid for energies below and above the absorption edge can be found in Gárcia-Cristóbal et al. (1998). Resonant hyper-Raman scattering occurs when $2\hbar\omega_i$ or $\hbar\omega_f$ approaches a particular electronic transition. It was found that, in contrast to one-phonon Raman scattering, the one-phonon resonance hyper-Raman process mediated by the deformation potential interaction is dipole forbidden, but it becomes allowed when induced by the Fröhlich interaction.

2.6.1.4 Coherent Anti-Stokes Raman Scattering (CARS)

The stimulated Raman effect is not very useful for the study of the excitations of a material system since its spectrum is poor – it usually contains only one line which corresponds to the most intense and narrowest line of the spectrum of the spontaneous scattering, and since very intense laser beams must be used, which generate other nonlinear effects (self-focalization, for example).

Let us irradiate the sample with two monochromatic beams of frequencies ω_1, ω_2 such that their difference $\omega_1 - \omega_2$ is resonant to one of the vibration frequencies of the medium, and the powers of the two beams are under the threshold for the stimulated effect. We then study the excitation of the vibrational modes in the sample with a probe radiation of frequency Ω. In principle both Stokes radiation with frequency $\omega_S = \Omega - (\omega_1 - \omega_2)$ and the anti-Stokes mode with frequency $\omega_{AS} = \Omega + (\omega_1 - \omega_2)$ can appear. If one of the exciting waves, say that with frequency ω_1, is used as probe, the CARS is the anti-Stokes radiation with frequency $\omega_{AS} = \omega_1 + (\omega_1 - \omega_2) = 2\omega_1 - \omega_2$. CARS is thus a third-order nonlinear process, described by a second-order Raman tensor, as in the hyper-Raman effect. Since the probe radiation can be used to register the phase relations between elementary excitations in different points of the sample, the process is called coherent anti-Stokes Raman scattering or active coherent spectroscopy. The intensity of the scattered radiation is much larger than that of the spontaneous Raman scattering and the shape of the line, which can be modified by varying the difference $\omega_1 - \omega_2$ around the vibrational frequencies of the medium, differs from the spontaneous lineshape. The spatially coherent CARS radiation propagates in the direction $\boldsymbol{k}_{AS} = 2\boldsymbol{k}_1 - \boldsymbol{k}_2$.

As inverse Raman scattering is the inverse effect of the stimulated Raman scattering, CSRS (coherent Stokes Raman spectroscopy) is the inverse of CARS. In particular, if the frequency of CARS is given by $2\omega_1 - \omega_2$ (the implicit assumption was made that $\omega_1 > \omega_2$), the frequency of CSRS is $2\omega_2 - \omega_1$ (two photons of frequency ω_2 and one photon of frequency ω_1 participate in the process). The formalisms for CARS and CSRS are symmetric in ω_1 and ω_2. One of the advantages of coherent Raman scattering over the spontaneous Raman effect is the fact that the spectral resolution depends only on the linewidth of the exciting laser, which can be as low as 0.05 cm^{-1}. This property, as well as the possibility of detecting small frequency shifts and of generating photoexcited free carriers that can be captured by ionized impurities, makes coherent Raman spectroscopy quite a useful tool (see, for example, Rupprecht et al. (1998) for its applications in the analysis of diluted magnetic semiconductors).

2.7 Optical Methods for the Characterization of Light-Matter Interaction

In the following we will present briefly the experimental methods of measuring the parameters that characterize light-matter interaction. Usually, the processes that can appear at the interaction between the electromagnetic field and the solid state are divided into one-photon or multiphoton processes, or into linear and nonlinear phenomena. We think that such a classification is sometimes misleading; there are experimental methods that cannot be unambiguously put into either such category. For example, pump-probe reflection spectroscopy can be characterized as either a one-photon process when we refer only to the probe beam or a two-photon method when we refer to both pump and probe which can generally have different wavelengths. Therefore, we will not attempt to classify the physical processes or experimental methods, although we will specify, whenever appropriate, their linear or nonlinear character, or the number of photons involved.

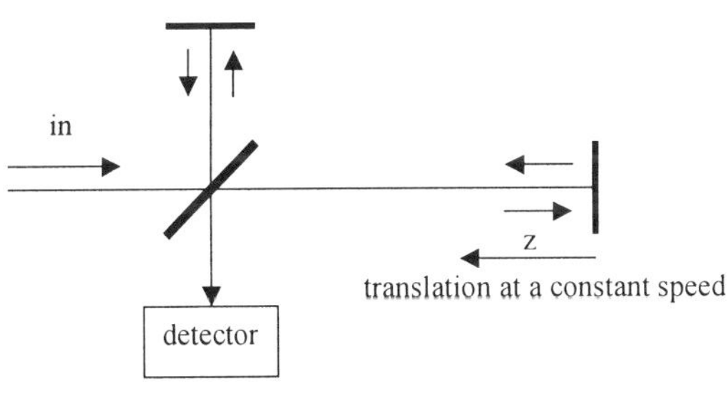

Fig. 2.23. Set-up for Fourier transform spectroscopy

All optical characterization methods presented in this section can be performed either as a function of wavelength, in which case we talk about frequency-resolved or spectrally resolved spectroscopy methods, or in time, when the sample is illuminated with light pulses. In the latter case we deal with time-resolved characterization methods. A special category of optical characterization methods is the Fourier transform method, used mostly in the IR spectral region (Klingshirn, 1997). When it is not possible to find the frequency dependence of the transmitted or reflected beam in a large bandwidth due, for example, to the lack of a spectrometer with the desired bandwidth, the frequency dependence can be transformed, with a Michelson interferometer, into the spatial dependence of the beam intensity via a Fourier transform. A schematic set-up is presented in Fig. 2.23. In Fourier transform spectroscopy one measures the Fourier transform of $I(\omega)$, i.e. the dependence of the intensity as a function of z, $I(z)$. If $I(z)$ is found to be periodic, then $I(\omega)$ is monochromatic and its frequency can be easily determined from the period in z.

Frequency-resolved spectroscopy needs high spectral resolution, and is used mainly for probing the static features of spectra and the effects of the environment on different energy levels, whereas time-resolved spectroscopy probes the dynamic interactions in solids; a high temporal resolution is needed in the latter type of measurements.

Time-resolved optical processes need pulsed illumination. Since the energies of optical transitions are broadened due to several mechanisms, a pulse with duration τ_p is indistinguishable from a monochromatic excitation as along as its energy width $\hbar / \tau_\text{p}$ is smaller than the linewidth of the transition. A special category of time-resolved spectroscopy is ultrafast optical spectroscopy, which uses pulses with smaller durations than the relevant scattering times, usually in the 10–100 fs range (see Shah (1999) and Vasko and Kuznetsov (1999)). When an ultrashort laser pulse hits a bulk semiconductor coherent excitations are produced, the photoexcited carriers having a well-defined phase relation with the laser pulse. If the excitation intensity is high enough, a nonequilibrium carrier distribution is formed, whose coherence is subsequently destroyed, the nonthermal carriers relaxing to a thermalized hot electron plasma that cools to the lattice temperature by interacting with phonons. If the central frequency of excitation is highly detuned from the transition frequency, the excitation regime is called off-resonant, the carriers created in this regime being called virtual carriers since they are only present when the excitation is on. On the contrary, real carriers remain in the sample after the excitation pulse is over. The generation rate of carriers is not described by the semiclassical Fermi golden rule even under resonant excitation conditions, since this rule is valid only when the dephasing time T_d is much shorter than the pulse duration. The dephasing time is the time it takes for two dipolar oscillators to get out of phase with each other by one radian under the influence of some disturbance. On excitation with ultrashort pulses, virtual carrier behavior of the coherent regime gradually changes into the real carrier behavior when T_d becomes comparable to the pulse duration. In the dephasing-dominated regime, the width of the carrier distribution function after excitation, $\gamma = 1/T_\text{d}$, is much larger than the spectral width of the pulse $1/\tau_\text{p}$ that determines the width of the distribution in the coherent regime. The dephasing limit is the only case when the interband polarization follows adiabatically the electric field of the pulse, the polarization persisting generally for times of the order of $1/\gamma$ after the excitation pulse is switched off.

The distribution function of carriers in ultrafast excitation is very different from the equilibrium Fermi–Dirac shape, the initial nonequilibrium carrier distribution relaxing towards equilibrium through various scattering mechanisms. During this process the distribution function passes through several physically distinct, but sometimes overlapping stages. The initial coherent stage of carrier relaxation lasts from the moment of carrier creation to the first scattering event that destroys coherent coupling between electrons and holes. The duration of this process is extremely short. In the following thermalization stage, the carriers (generated in a relatively narrow range of energies) redistribute over the whole $\boldsymbol{p}$-space, the carrier distribution function evolving from the initial peak at the generation energy into a Fermi–Dirac distribution. The process is dominated by electron-electron collisions that do not change the total energy of the electronic system, so that the temperature T_e of the thermalized distribution is of the order of the initial generation energy. This process lasts from a few tens of fs to a few tens of ps, depending on the density of electrons and the geometry of the structure. The next cooling stage is controlled by the energy relaxation time, determined by the coupling of electrons to various phonon branches. During this process the carriers, often referred to as hot electrons, cool down to the lattice temperature T_l by emitting acoustic and optical phonons within a few ps to a few ns. Ultrafast optical excitation can also produce a nonequilibrium population of phonons (the optical phonon modes are not in equilibrium with the rest of lattice), this effect being known as the 'hot phonon effect'. After the electronic temperature reaches the lattice temperature, the photogenerated electrons can still relax by recombining with holes. The recombination stage lasts a few tens of ns for radiative recombination in direct-gap materials, the duration being shortened to a few hundred ps in the presence of nonradiative recombinations.

The early stage of carrier relaxation is usually described by the Boltzmann kinetic equation that includes the carrier-carrier scattering, interaction with various phonon branches, scattering on impurities and imperfections. During the coherent stage of relaxation, the Boltzmann equation must be replaced by a quantum kinetic equation that accounts for the presence of coherent interband polarization, memory effects due to rapidly changing distribution functions, and nonequilibrium screening. After thermalization, the carrier distribution in each subband is completely characterized by only two parameters: the total density (or the chemical potential) and the electronic temperature.

2.7.1 Transmission/Reflection Spectroscopy

The absorption coefficient is usually measured via the transmission coefficient of the electromagnetic radiation through the sample. When the electromagnetic field with intensity I_0 hits the sample, only part of it, TI_0, is transmitted, the other part RI_0 being reflected (see Fig. 2.24).

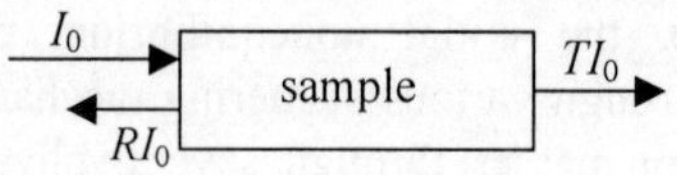

Fig. 2.24. Light transmission and reflection from a sample

On propagation through bulk materials, the transmitted intensity of the radiation with frequency ω is given by

$$I(\omega) = T(\omega)I_0(\omega) = [1 - R(\omega)]I_0(\omega)\exp[-\alpha_{tot}(\omega)d], \tag{2.151}$$

where d is the thickness of the sample, $[1 - R(\omega)]I_0(\omega)$ is the fraction of the incident radiation intensity which enters the sample, and the total absorption (called sometimes extinction) is $\alpha_{\text{tot}} = \alpha_{\text{abs}} + \alpha_{\text{scat}}$. The transmission coefficient is defined as $T(\omega) = I(\omega)/I_0(\omega)$. The two terms in the absorption represent distinct physical mechanisms: the first term is responsible for the transformation of the electromagnetic field into heat, chemical energy or incoherent electromagnetic field emission (photoluminescence) frequency shifted compared to the incident radiation, while the second term is responsible for the attenuation due to scattering phenomena on inhomogeneities (Klingshirn, 1997). Generally, the scattering contribution to the extinction coefficient is much weaker than the absorption, and can be neglected. Neglecting or taking into account light reflection, one can determine the absorption coefficient, and implicitly the imaginary part of the dielectric function by measuring the intensity of light transmitted through the probe. The phase of the transmitted light can then be calculated with the Kramers–Krönig relations.

When the light propagates in a plane-parallel slab of width d instead of a bulk material, the transmission coefficient has an oscillatory behavior as a function of d:

$$T = (1 - R_{\text{f}})(1 - R_{\text{b}})\exp(-\alpha d)/[(1 - R_\alpha)^2 + 4R_\alpha \sin^2(n\omega d/c)], \tag{2.152}$$

where R_{f}, R_{b} are the reflectivities of the front and back surface of the slab, and $R_\alpha = (R_{\text{f}} R_{\text{b}})^{1/2} \exp(-\alpha d)$ (Klingshirn, 1997). In this case great care must be used to extract the absorption coefficient.

The reflection coefficient measurement is used whenever the sample is highly absorbing, or when it is opaque to the incident radiation, i.e. when the radiation is not transmitted through the sample, as in the case of metals. Since it is very sensitive to the nature of the surface on which reflection takes place, measurements of reflection coefficients are used for surface characterization. The reflection coefficient defined as above is a real quantity; the phase of the reflected light can be determined from measurement data by (2.99).

2.7.2 Modulated Reflectance and Transmittance

The reflection or the transmission coefficient of a sample can be modified by applying electric or magnetic fields, by stressing the material, heating it or by employing any combination of these external factors. The spectra obtained under these conditions are called modulated reflectance or transmittance, the reason for modulating the spectra being either to increase the resolution of some spectral lines otherwise hidden in the unmodulated optical spectrum or to study the effect of the application of different external factors.

Differentiating the reflection coefficient for example with respect to ε_1 and ε_2 we obtain (Seraphin, 1972)

$$\Delta R / R = \alpha(\varepsilon_1, \varepsilon_2)\Delta\varepsilon_1 + \beta(\varepsilon_1, \varepsilon_2)\Delta\varepsilon_2, \tag{2.153}$$

where α and β are the Seraphin coefficients, which can be calculated from the Fresnel formula at both normal and oblique incidence. Their magnitude and phase in different spectral regions tell us the behavior of the reflectance at increasing or decreasing modulation. A typical dependence on energy of the Seraphin coefficients for bulk semiconductors such as Si, Ge or GaAs is schematically represented in Fig. 2.25.

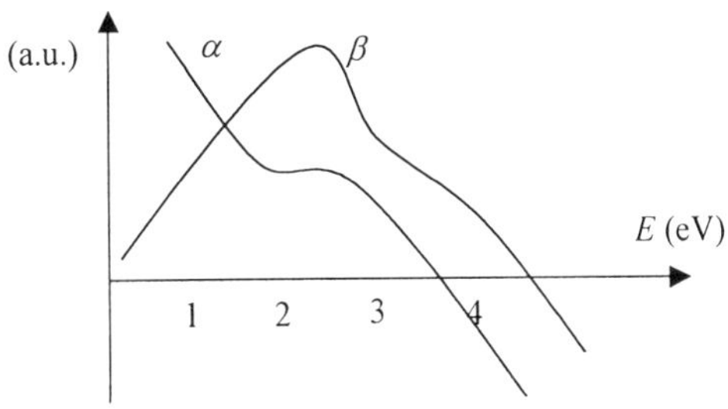

Fig. 2.25. Dependence of Seraphin coefficients on energy

These coefficients are the macroscopic part of $\Delta R / R$ since they are common for any type of modulation reflectance, while $\Delta\varepsilon_1$ and $\Delta\varepsilon_2$ are the microscopic part, which change with the modulation type. Different types of modulation induce specific and identifiable line shapes of $\Delta\varepsilon_1$ and $\Delta\varepsilon_2$ at different types of critical points of the material. At critical points discontinuities in both the density of states and ε_2 occur. For a modulation that modifies the energy gap E_g, for example,

$$\Delta\varepsilon_2 / \Delta E_g \approx (1/\omega^2)\Delta\rho / \Delta E_g, \tag{2.154}$$

where $\Delta\rho$ is the change in the density of states and ΔE_g the change in the energy gap. $\Delta\rho / \Delta E_g$ has a singularity at the critical points, which is responsible for the selectivity and enhancement of contrast lines at these points and the vanishing of the modulated spectrum far from their neighborhoods. The change of ε_2 produces a change in ε_1 which is given by the Kramers–Krönig relation

$$\Delta\varepsilon_1 / \Delta E_g \approx \int_0^{\infty} \Delta\rho \, d\omega / [\Delta E_g \omega(\omega_0^2 - \omega^2)]. \tag{2.155}$$

The principal types of modulated reflectance are electroreflectance (Franz–Keldysh effect) when an electric field is applied perpendicular on the sample, photoreflectance, if the modulation is due to an optical field, piezoreflectance, when a static uniaxial stress is applied on the sample, or thermoreflectance when the temperature is the modulating parameter. When a magnetic field is applied, the resulting spectrum is called magnetoreflectance. All these modulation types can be encountered in solid-state spectroscopy. The modulated transmittance can be described in a similar way.

2.7.3 Pump-Probe Spectroscopy

Pump-probe spectroscopy (PPS), when referring to reflection or transmission measurements, can be considered as a particular case of modulated reflection or transmission spectroscopy. PPS can be either degenerate or nondegenerate, depending on whether the pump and probe pulses originate from the same source or not. In degenerate PPS an ultrashort laser pulse is split into two beams. One beam, called *pump*, excites the sample under test. The other beam, called *probe*, monitors the changes produced in the sample due to its interaction with the pump beam. The probe beam is delayed by a known amount τ with respect to the probe beam and has, in general, a much weaker intensity. In the nondegenerate PPS the pump and probe beams are generated by two distinct sources, for example two synchronized lasers or one laser and a broadband light source.

The probe beam is further analyzed to extract information about the reflectivity, transmission, absorption, PL or Raman scattering processes of the sample under study. A simple set-up for time-resolved transmission/reflection PPS is schematically presented in Fig. 2.26.

The changes in the transmission or reflection of the pump pulse are measured as a function of τ, the results being commonly displayed as differential transmission or reflection spectra, defined as

$$\Delta T / T = (T - T_0) / T, \quad \Delta R / R = (R - R_0) / R, \tag{2.156}$$

where the index 0 denotes the transmission or reflection in the absence of the pump pulse. More sophisticated variants of the method are sometimes encountered. For example, the electro-optic sampling (EOS) technique is used to detect the difference in reflection (REOS) or transmission (TEOS) of two perpendicularly polarized components of the same time-delayed optical pulse. In this method, the real-space movement of charges induced by the pump pulse creates a macroscopic screening electric field, which can break the cubic symmetry, for example, of the lattice and makes the system birefringent. The refractive indices for electromagnetic waves propagating parallel and perpendicular to the screening field become different, the signal being directly proportional to the instantaneous value of the screening electric field, and thus to the dipole moment in the material (Shah, 1999).

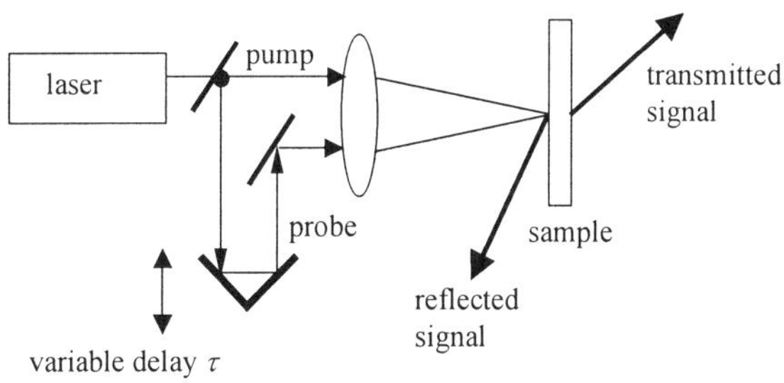

Fig. 2.26. Time-resolved transmission/reflection PPS

In DTS (differential transmission spectroscopy) the pulses are usually linearly and perpendicular polarized in order to suppress coherent coupling of the two beams. Since it is not a background-free method, it needs stronger pulses than four-wave-mixing (see Sect. 2.7.7). DTS is commonly used to study carrier kinetics in samples with medium-to-high carrier density, where it measures completely the distribution function and its evolution, giving information about the scattering mechanism and electron structure. However, the electron and hole distributions cannot be determined independently; the measured distribution function is generally their sum, and it can be obtained only at the energy of the laser pulse. Moreover, if the dephasing time of polarization is long compared to the probe pulse duration, an averaged distribution function is measured over the dephasing time.

2.7.4 Amplitude and Phase Spectroscopy

In this type of spectroscopy both amplitude and phase of the transmitted or reflected light is measured. The simultaneous knowledge of amplitude and phase is crucial for a complete characterization of the interaction between light and matter, and it is particularly used in the study of quantum beats in four-wave-mixing spectroscopy, Fano interferences in low-dimensional semiconductors, microcavity modes, or for the determination of the complex refractive index (Tignon et al., 1999).

Amplitude and phase spectroscopy can be performed with continuous or pulsed excitation; in the first case the method is usual known as ellipsometry. Spectroscopic ellipsometry is an advantageous technique to measure the optical constants of solids because it can directly determine $\varepsilon(\omega)$ on a wavelength-by-wavelength basis, without the need to use Kramers–Krönig relations, which imply measurements over large wavelength intervals. It is also a surface-sensitive technique. The ellipsometric variables measured in ellipsometry are ψ and Δ, the ratio of amplitudes and the phase difference of the reflectance for s- and p-polarized states (perpendicular and parallel polarized light with respect to the plane of incidence, respectively). These parameters are sometimes united in $\rho = (\chi_i / \chi_r) = \tan\psi \exp(i\Delta)$ where $\chi_{i,r}$ are the incident and reflected light polarizations; in an isotropic material $\rho = r_p / r_s$ is the ratio between the Fresnel reflection coefficients for p- and s-polarized light. The dielectric function is related to ρ as $\varepsilon / \varepsilon_a = [(1-\rho)/(1+\rho)]^2 \tan^2\varphi \sin^2\varphi + \sin^2\varphi$, where φ is the angle of incidence and ε_a the dielectric function of the ambient. The ellipsometric method cannot be used with sufficient accuracy if $\psi \to 0$ (Azzam and Bashara, 1977). Ellipsometry measurements can also be performed at different wavelengths *and* incidence angles, to obtain more data about the sample. In this case the method is called VASE (variable-angle spectroscopic ellipsometry) and is useful to detect the anisotropy of the dielectric constant. In the generalized VASE (GVASE) method, all Jones matrix elements, in transmission or reflection are measured, and p to s wave conversions, or vice-versa, can be detected (Schubert et al., 1996a, 1996b). In a related method, 2-MGE (two-modulator generalized ellipsometry) (Jellison Jr. and Modine, 1997a, 1997b), a single measurement is necessary to obtain the optical function from appropriately aligned uniaxial crystals. In this method two photoelastic modulator-polarizer pairs are used, one in the polarization-state generator arm of the ellipsometer, the other in the polarization-state-detector arm, which operate at different resonant frequencies, such that eight different elements of the reduced sample Mueller matrix can be measured. If the two modulators are aligned respectively at (0,|45|), or (|45|,0) with respect to the plane of incidence, the light reflected from the non-depolarizing sample is completely determined (its Jones matrix is obtained).

Amplitude and phase spectroscopy with pulsed excitation is based on spectral interferometry. The working principle of time-resolved spectral interferometry is presented in Fig. 2.27. The unknown signal and the reference pulse, delayed by τ, interfere, the Fourier transform of the interference pulse being given by (Iaconis et al., 1998)

$$I_i(\omega) = I_s(\omega) + I_{ref}(\omega) + 2[I_s(\omega) I_{ref}(\omega)]^{1/2} \cos[\phi_s(\omega) - \phi_{ref}(\omega) - \omega\tau]. \tag{2.157}$$

Knowing the intensity and phase of the reference pulse, the intensity and phase of its Fourier transform, $I_{ref}(\omega)$ and $\phi_{ref}(\omega)$, respectively, can be calculated and then the amplitude and phase of the Fourier transform of the unknown signal, $I_s(\omega)$ and $\phi_s(\omega)$, can be determined. Finally, the signal itself is obtained by using a Fourier transform. The amplitude and phase of the reference signal can be determined, for example, using the FROG (frequency-resolved optical gating) method (Kwok et al., 1998). In this method two delayed parts of the reference signal reach a nonlinear crystal coming from different directions and self-diffract on it. From the self-diffracted signal the amplitude and phase are determined using an iterative algorithm. The amplitude and phase spectroscopy can analyze signals with very weak intensities, of even 10^{-21} J. The combination of spectral interferometry and FROG is also known as TADPOLE (temporal analysis by dispersing a pair of electric light fields).

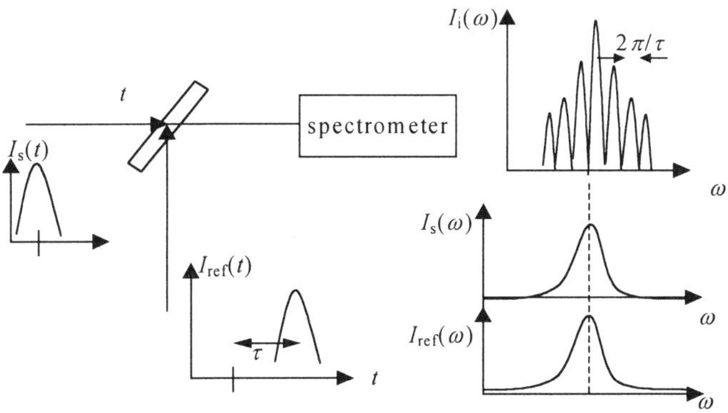

Fig. 2.27. The principle of amplitude and phase spectroscopy

When the reference signal is a fraction of the signal itself, the method is called self-referenced interferometry (Chu et al., 1995). A simple set-up is presented in Fig. 2.28.

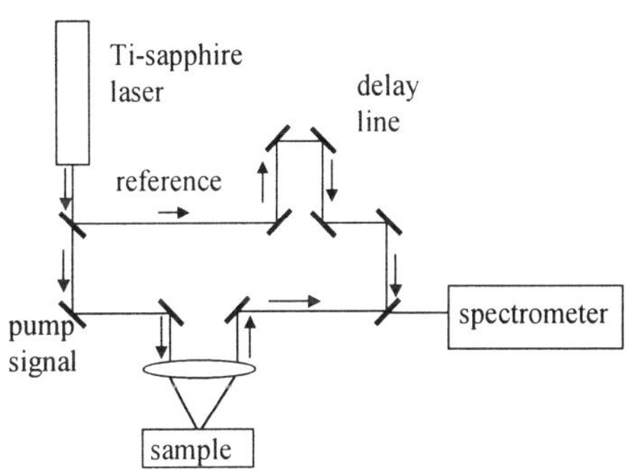

Fig. 2.28. Schematic principle of self-referencing interferometry

The self-interferometer in Fig. 2.28 works at a fixed time delay and does not need nonlinear elements; it also has a high sensitivity.

Measuring the transmitted signal through a sample of length L using amplitude and phase spectroscopy, the complex refractive index of the sample can be recovered if we note that:

$$E_s(\omega) = E_0(\omega)\exp[-\alpha(\omega)L/2]\exp[ik_0 n_{re}(\omega)L] = \sqrt{I_s(\omega)}\exp[i\phi_s(\omega)]. \tag{2.158}$$

This method allows the determination of both $\alpha(\omega)$ and the real part of the index of refraction $n(\omega)$ without using Kramer–Krönig relations. It is extremely useful when there are sharp resonances in the absorption spectrum due to excitons in a GaAs semiconductor, for example (Tignon et al., 1999).

2.7.5 Photoluminescence

The PL can be separated in practice from scattering by observing some of their characteristic features. One of these is the presence of an afterglow in PL with a longer decay time than the period of electromagnetic vibrations. The afterglow consists of the natural lines of the material system whose positions and profiles are independent of the properties of the radiation, the spectral composition and intensity of the excitation affecting only the intensity of the afterglow lines. In contrast to this, scattering spectra, although having some lines in common with afterglow spectra, are characterized by lines with positions and profiles depending not only on the properties of the system but also on the spectral composition of the exciting radiation. Scattering spectra retain more evidence of the condition of the incident radiation, for example, memory of the state of polarization. The afterglow is caused by the fact that transition probabilities between levels are finite and the PL exhibits time inertia. PL excitation begins immediately after the source of excitation is switched on and continues until the stationary state is reached, the afterglow, or the attenuation of PL, beginning immediately after the exciting radiation is switched off. The lifetime of the PL is determined solely by the lifetime of spontaneous transitions, when nonradiative transitions are absent, the latter leading to a reduction in lifetime, and thus to PL quenching. The measurement of PL lifetime for a particular system under different external excitations gives information also on the magnitude of *nonradiative* transition probabilities.

Since the lines emitted in PL are natural lines of the system, part of the emitted radiation is reabsorbed; therefore, the backward geometry is the most suitable since the distortion of the PL spectrum is minimized in this way. In a typical backward geometry the excitation and PL radiation propagate in almost opposite directions; a PL spectrum $I_{PL}(\omega)$ is always taken for a fixed frequency ω_{ex} of the incident radiation with intensity I_{ex} (see Fig. 2.29). The spectral content of the PL emission is determined with a spectrometer. Since the PL intensity depends on the geometry of the sample and that of the experimental set-up, its absolute value is of little practical interest, and PL spectra are usually given in arbitrary units. On the contrary the frequency of PL lines, their angular and polarization dependencies provide useful information about the sample.

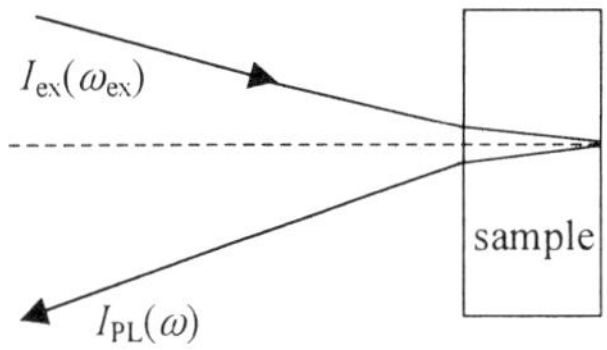

Fig. 2.29. PL backscattering geometry

One interesting application of PL is in measuring the absorption of thin films. The absorption of very thin films cannot be measured with sufficient accuracy due to the small values of the absorption coefficient. Therefore, in these situations one can use the set-up in Fig. 2.30. The laser is normal to the thin film; if the refractive index of the substrate is lower than that of the film, the excited PL is guided along the film and propagates along it. By monitoring the variation in the PL intensity at a change in the lateral position of the laser, the absorption coefficient can be detected with quite good resolution (Taylor, 1989).

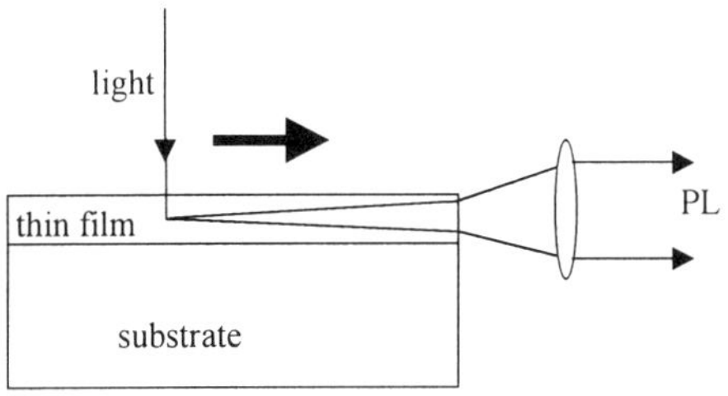

Fig. 2.30. Principle of absorption measurement through PL in thin films

If instead of a thin film, we have a graded-index planar optical waveguide, by coupling the laser light to the guide by a prism, in order to excite one of the TM or TE modes, the PL spectrum obtained by exciting the waveguide in different modes characterizes the guide at different depths. This is possible since different guided modes penetrate to different depths in the film, the depth reached by the light increasing with the mode number m. So, PL spectra of different modes, detected by collecting the emitted light from the front of a planar surface, probe different structures and different ion concentration or crystal-field distributions (assuming that the graded-index profile is obtained by ion doping). The spectrum measured for excitation in one mode is the sum of contributions from different depths, weighted by the local value of the squared electric field. Raman spectra instead of PL can be also used to this end (Ferrari et al., 1999).

In another related method, PLE (photoluminescence excitation), the input excitation wavelength is varied and takes different values $\lambda_{ex1},\ldots,\lambda_{exN}$, whereas the output is detected at a single, fixed wavelength $\lambda_d > \lambda_{exi}$, $i = 1,2\ldots N$. The output intensity of the laser must be kept constant when its wavelength is varied. The PLE spectrum is represented by the curve $I_{\mathrm{PLE}} = f(\lambda_{\mathrm{ex}i})$. A set-up for PLE in a forward geometry is represented in Fig. 2.31.

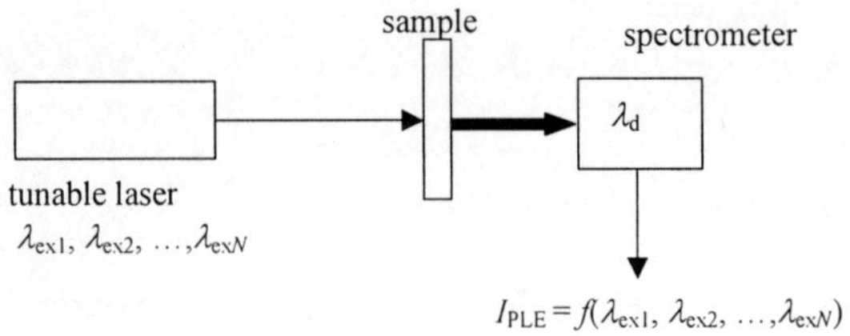

Fig. 2.31. PLE set-up

With PLE one can excite material excitations not visible, or poorly visible in PL spectra. In particular, PLE is used for the study of hot-exciton processes, which differ from resonant Raman scattering in that they possess an 'energy memory' of the monochromatic excitation, but not a phase memory. The transformation of the exciton to the free-hole luminescence under a transverse electric field that gradually ionizes the excitons, for example, is easily observed in PLE; when the electric field inside the structure grows, the PLE intensity of the corresponding peaks decreases, the excitonic peak decreasing at a faster rate compared to the carrier-acceptor peak. This behavior reflects the different origin of the peaks; the position of both peaks slightly shifts with the applied field (Stark shifts). PLE measurements can also be used to determine the shape of the absorption spectrum. Since PLE intensity is proportional to the number of photogenerated electron-hole pairs, which in turn is proportional to $\alpha(\omega)$ at a given frequency, the PLE spectrum is directly proportional to the absorption, but is easier to measure.

2.7.6 Light Scattering

One of the most common experimental geometries in light scattering is backscattering, when the incident light propagates almost perpendicular to the surface of the sample, although forward scattering is also used. The scattering geometry is denoted by $\boldsymbol{k}_i(\hat{\boldsymbol{e}}_i, \hat{\boldsymbol{e}}_f)\boldsymbol{k}_f$, and the modes observed in Raman spectra are denoted after the irreducible representations of the point group of the crystal that become active in the particular scattering geometry. Two types of light scattering spectra are measured: polarized spectra when incident and scattered light polarizations are parallel, and depolarized spectra when they are orthogonal.

Since the scattered light intensity is proportional to $|H_{\mathrm{int},fi}|^2$, only the absolute values of Raman tensor components are usually determined in scattering experiments. However, it is possible to obtain also the corresponding phases using interference phenomena produced when the scattering polarizations are not parallel along crystal axes (Strach et al., 1998). Such measurements were made on the tetragonal crystal $SmBa_2Cu_3O_{7-x}$, for which the Raman tensor in the backscattering configuration on the (110) surface, in the A_{1g} symmetry can be written as:

$$A_{1g} = \begin{pmatrix} a & 0 & 0 \\ 0 & a & 0 \\ 0 & 0 & c \end{pmatrix} = \begin{pmatrix} |a|\exp(\mathrm{i}\varphi_a) & 0 & 0 \\ 0 & |b|\exp(\mathrm{i}\varphi_b) & 0 \\ 0 & 0 & |c|\exp(\mathrm{i}\varphi_c) \end{pmatrix}, \qquad (2.159)$$

where the incident and scattered polarizations are referred to the tetragonal axes

$$\hat{e}_{i,f} = \begin{pmatrix} -(2)^{-1/2}\sin\theta_{i,f} \\ +(2)^{-1/2}\sin\theta_{i,f} \end{pmatrix}. \qquad (2.160)$$

$\theta_{i,f}$ is the angle between $\hat{e}_i$, $\hat{e}_f$ and the z axis. The Raman intensity $I \cong |\hat{e}_f A_{1g} \hat{e}_i|$ for the parallel ($\theta_i = \theta_f$) and perpendicular ($\theta_f = \theta_i + 90°$) polarizations is, respectively,

$$I_{\|}(\theta) \approx |a|^2 (\sin\theta)^4 + |c|^2 (\cos\theta)^4 + 2|a||c|\cos\varphi_{ac}(\sin\theta\cos\theta)^2, \qquad (2.161a)$$

$$I_{\perp}(\theta) \approx (|a|^2 + |c|^2 - 2|a||c|\cos\varphi_{ac})(\sin\theta\cos\theta)^2, \qquad (2.161b)$$

where $\theta = \theta_i$ and $\varphi_{ac} = \varphi_a - \varphi_c$. Thus $|a|$, $|c|$ and φ_{ac} can be determined from $I_{\|}$ by varying θ, and the results can be checked using $I_{\perp}$.

Vibrational Raman scattering is used, for example, to determine the chemical composition of a system, since vibrational energies depend on the bonds between different atoms. For the same reason, vibrational Raman scattering is also sensitive to the large strain fields at the interface between two materials, which perturb the pre-existing bonds and shift the vibrational frequencies. As will be seen in subsequent chapters, vibrational Raman scattering can even monitor in situ chemical reactions on the surface of the solid. Raman scattering in bulk crystals cannot, however, show the dispersion of optical phonon branches even if $\boldsymbol{k}$ differs somewhat from zero, since the dispersion curves are usually flat for $\boldsymbol{k} \cong 0$. These dispersion curves are obtained, for example, through inelastic neutron scattering at wavelengths of a few Å (Bilz and Kress, 1979). The acoustic phonon dispersion relation, on the other hand, is linear in frequency, and the dependence of ω on $\boldsymbol{k}$ can be measured in Brillouin spectroscopy.

Light scattering on magnons is a consequence of either direct coupling between the spin system of a ferromagnet and the radiation field through the magnetic dipole interaction, or is based on the existence in magnetic crystals of the spin-orbit interaction, which cause inelastic light scattering by phonons. Scattering of light on magnons has a much smaller cross-section than that of vibrational excitations, and appears in paramagnetic materials or ordered magnetic structures at low temperatures. Except for antiferromagnetic materials, where the frequencies of magnons can be detected by Raman spectroscopy, in other materials the frequencies of magnons are much lower than those of phonons and therefore are analyzed through Brillouin spectroscopy.

There are three basic configurations for Brillouin scattering represented in Fig. 2.32.

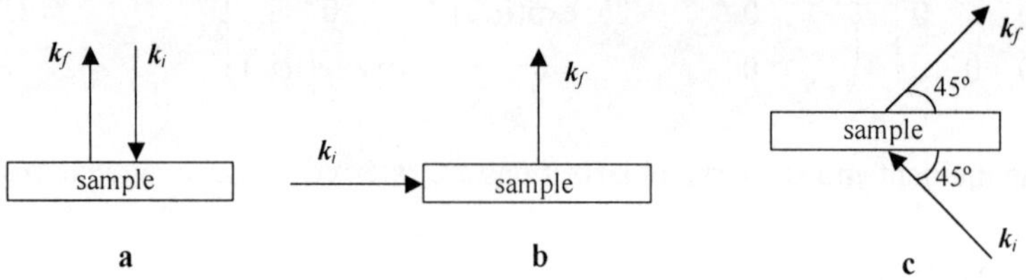

Fig. 2.32. Basic configurations of Brillouin scattering

These correspond to backscattering (Fig. 2.32a), 90° scattering (Fig. 2.22b) and platelet-type Brillouin scattering (Fig. 2.32c), their corresponding frequency shifts being given, respectively, by

$$\omega_a = \pm\omega_i (v_a / c)[n(\omega_i) + n(\omega_f)], \tag{2.162a}$$

$$\omega_a = \pm\omega_i (v_a / c)[n^2(\omega_i) + n^2(\omega_f)]^{1/2}, \tag{2.162b}$$

$$\omega_a = \pm\omega_i (v_a / c)\sqrt{2}. \tag{2.162c}$$

The phase velocity for surface acoustic phonons is $v_{s,a} = \omega_a / k_i(\sin\theta_i + \sin\theta_f)$, while for bulk acoustic phonons the corresponding formula is $v_a = \omega_a / 2n(\omega_f)k_i$.

2.7.7 Four-Wave Mixing

This nonlinear optical method is based on the superposition of three coherent optical beams in a medium, the fourth wave being emitted in a phase-matched direction. The phenomenon is described by a third-order susceptibility tensor. In contrast to other optical methods, where the optical source need only have a high spectral resolution and a sufficient intensity, four-wave mixing (FWM) relies on the coherence of the optical source (laser) to produce the desired effect. In this method the laser initially prepares a coherent state of the system, and the decay of coherence due to dephasing is monitored through various methods as a function of time. FWM measurements are easiest to conduct when the relaxation rates are long or, equivalently, the homogeneous line is sharp. Several variants of the FWM method exist, usually distinguished by the number of incident beams.

For example, in the three-beam degenerate FWM, three pulses with generally non-collinear wavevectors $\boldsymbol{k}_1$, $\boldsymbol{k}_2$, $\boldsymbol{k}_3$ arrive at the sample at different times, t_1, t_2, t_3 (see Fig. 2.33) and the diffracted pulse propagating in the phase-matched direction $\boldsymbol{k} = \boldsymbol{k}_1 - \boldsymbol{k}_2 + \boldsymbol{k}_3$ is measured for various delays between the incident pulses. In this configuration the two time delays between the three beams can be individually controlled. In the most common variant of the three-beam degenerate FWM, the first two pulses arrive simultaneously ($\tau_{12} = t_2 - t_1 = 0$), a transient grating being formed in the sample due to their interference. The grating moves in the lateral direction with a speed $v = (\omega_1 - \omega_2)/|\boldsymbol{k}_1 - \boldsymbol{k}_2|$. The third pulse is then diffracted on this grating, the diffracted signal as a function of the delay τ_{13} providing information about the population decay of the system. When the two laser beams that write the grating have equal frequencies, the interference pattern is stationary and the period of the induced grating is given by $\Lambda = \lambda_1/[2\sin(\theta/2)]$, where θ is the angle between the two laser beams.

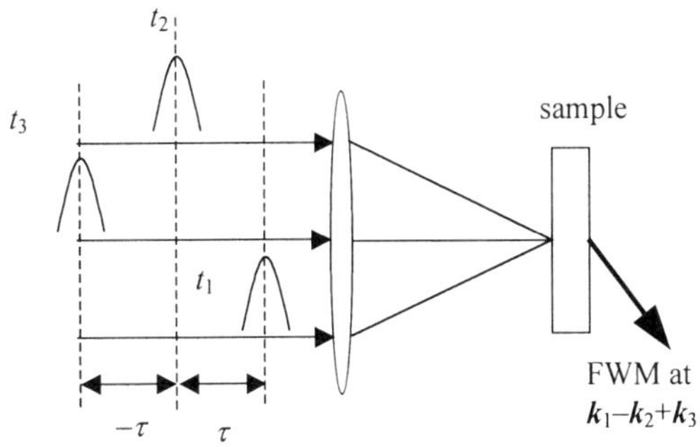

Fig. 2.33. Degenerate FWM with three beams

The method is called degenerate because we consider only one of the diffracted beams. In principle, the beam can be diffracted in a direction corresponding to any of the combinations $\pm\boldsymbol{k}_1 \pm \boldsymbol{k}_2 \pm \boldsymbol{k}_3$.

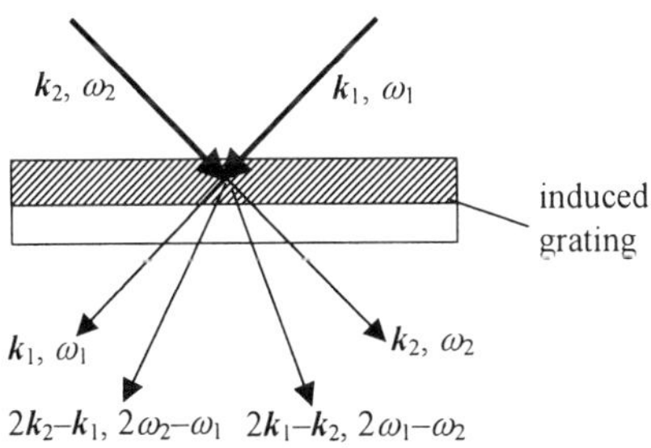

Fig. 2.34. Nondegenerate FWM with two beams

FWM spectroscopy can be performed with only two beams, in which case it is called two-beam FWM. In this case pump photons with wavevector $\boldsymbol{k}_1$ generate a coherent polarization in the sample, the probe beam with wavevector $\boldsymbol{k}_2$ arriving after a delay τ_{21}. It then self-diffracts on the interference grating and exits the sample in the phase-matching direction $\boldsymbol{k} = 2\boldsymbol{k}_2 - \boldsymbol{k}_1$. If only this beam is observed, we have degenerate two-beam FWM; in a nondegenerate two-beam FWM, also the diffracted beam in the direction $2\boldsymbol{k}_1 - \boldsymbol{k}_2$ is considered (see Fig. 2.34).

In any of the variants of the FWM method, different combinations of linear or circular polarizations of the incident beams can be employed to study different excitations in the materials. Since there is no background signal in the FWM method, it can be used for the investigation of polarization kinetics in a low-to-medium carrier density regime.

To quantitatively describe the FWM investigation procedure, we introduce the optical Bloch equations, which explain many results of FWM experiments.

2.7.7.1 Optical Bloch Equations

These equations describe the evolution of the polarization and population of ensembles of independent two-level systems, which interact with a radiation field, as in Fig. 2.35 (Shah, 1999).

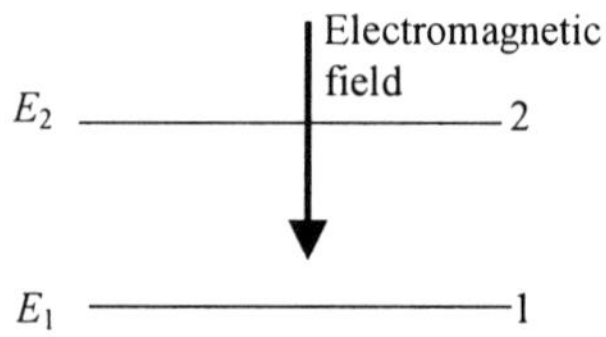

Fig. 2.35. A two-level system in interaction with an electromagnetic field

The density matrix of the two-level system is given by

$$\rho = \begin{bmatrix} \rho_{22} & \rho_{21} \\ \rho_{12} & \rho_{11} \end{bmatrix}, \tag{2.163}$$

where the diagonal elements represent the normalized populations of the two energy levels so that $\rho_{11} + \rho_{22} = 1$, and the off-diagonal elements describe the coherence of the superposition of states.

The time evolution of the density matrix is described by

$$\mathrm{i}\hbar(\mathrm{d}\rho / \mathrm{d}t) = [H, \rho], \tag{2.164}$$

where the total Hamiltonian is $H = H_0 + H_{\mathrm{int}} + H_{\mathrm{r}}$, with H_0 the unperturbed Hamiltonian with energy eigenvalues E_1, E_2. The interaction with the electromagnetic field is described by the electric-dipole Hamiltonian

$$H_{\mathrm{int}} = \begin{bmatrix} 0 & \Delta_{21} \\ \Delta_{21}{}^* & 0 \end{bmatrix}, \tag{2.165}$$

where $\Delta_{21} = -\boldsymbol{d}_{21} \cdot \boldsymbol{E}$ with $\boldsymbol{E}$ the electric field of light and $\boldsymbol{d}_{21}$ the dipole moment operator, whereas the relaxation Hamiltonian towards thermal equilibrium is introduced by

$$[H_{\mathrm{r}}, \rho]_{22} = -\rho_{22} / T_1, \quad [H_{\mathrm{r}}, \rho]_{21} = -\rho_{21} / T_2. \tag{2.166}$$

Here, T_1 is the lifetime of state 2, and T_2 the lifetime of the coherent superposition of states, related to T_1 through $1/T_2 = 1/T_1 + 1/T_{\mathrm{d}}$, with T_{d} the dephasing time. This form of H_{r} is valid if the medium has either a very slow or a very fast response in comparison with the interaction time with the electromagnetic field.

The time evolution of the density matrix is finally expressed as

$$\begin{aligned} &\partial n / \partial t + \tau_1 u + \mathrm{i}(\Delta_{21} v^* - v \Delta_{21}^*) / \hbar = 0 \\ &\partial v / \partial t + \tau_2 v + \mathrm{i}\Delta_{21}(1 - 2u) / \hbar = 0, \end{aligned} \tag{2.167}$$

where $u = \rho_{22} = 1 - \rho_{11}$, $v = \rho_{21}$, $\tau_1 = 1/T_1$ and $\tau_2 = \mathrm{i}\Omega + 1/T_2$, with $\Omega = (E_2 - E_1)/\hbar$. Equations (2.167) are the optical Bloch equations, which must be generally solved numerically to model different types of time-resolved spectroscopy results. However, analytical expressions for the FWM signal can be found if the excitation pulses in the two-beam FWM are modeled with Dirac functions, in which case the electric field is $E(\boldsymbol{r},t) = [E_1 \delta(t) \exp(\mathrm{i}\boldsymbol{k}_1 \cdot \boldsymbol{r}) + E_2 \delta(t - \tau_{21}) \exp(\mathrm{i}\boldsymbol{k}_2 \cdot \boldsymbol{r})] \exp(\mathrm{i}\omega t)$. Two cases are of importance: a homogeneously broadened system where the linewidths of the energy levels are $\Gamma_{\mathrm{hom}} = 2\hbar / T_2$, and inhomogeneously broadened systems where the inhomogeneous broadening $\Gamma_{\mathrm{inh}} >> \Gamma_{\mathrm{hom}}$ (Shah, 1999).

In the first case, where the homogeneous broadening can be due to phonon interaction, lifetime broadening, or carrier-carrier interaction, the FWM signal has a free polarization decay for $\tau_{21} > 0$. For Dirac-like pulses the time-resolved FWM has a maximum at $t = \tau_{21}$ and decays with a time constant $T_2/2$. The same decay constant is observed for pulses of width τ_{p}, for which the FWM signal peaks, however, at $\tau_{21} + \tau_{\mathrm{p}}$. The time-integrated FWM signal is proportional to $\exp(-2\tau_{21}/T_2)$, from which the dephasing time T_2 can be obtained if the nature of the line broadening is known.

For an inhomogeneously broadened system, the macroscopic polarization created by the first pulse decays rapidly to zero since the phases of different frequency components evolve at different rates. If, however, τ_{21} is sufficiently short compared to T_2, the phase evolution of the different frequency components can be reversed, and a photon echo is observed at a time τ_{21} after the arrival of he second pulse, described by $\exp[-(t - 2\tau_{21})^2 / 2\tau_{\mathrm{inh}}^2]$. The width of the photon echo is determined by the inhomogeneous linewidth Γ_{inh}, and its height is proportional to $\exp(-4\tau_{21}/T_2)$. The time-integrated FWM signal has the same decay form in time, $\exp(-4\tau_{21}/T_2)$, with a decay constant half that for a homogeneously broadened system.

So, the inhomogeneous or homogeneous nature of the spectral line can be determined from time-resolved FWM, and then T_2 can be determined form its decay; these considerations are, however, not valid in general, because in many cases the homogeneous and inhomogeneous parts of the broadening are comparable. Another possibility to identify the nature of broadening is to use three-beam FWM, for which the time-integrated signal is symmetric for a homogeneous broadened system and asymmetric for inhomogeneous broadening (Shah, 1999).

A variant of the FWM is the three-photon echo spectroscopy. In the three-photon echo method, a sequence of three pulses hits the sample; the first generates a coherent polarization in the material, the second interacts with the polarization forming a population grating and the third is scattered off the grating in a phase-matched direction. This method can be used to study the systems in which the electron dephasing is comparable to the vibrational period in the medium. In this case the contribution of vibronic quantum beats (see Sect. 2.7.7.2) to echo decay can be suppressed by a proper tuning of the third-pulse delay with respect to $2\pi / \omega_V$, where ω_V is the frequency of the coupled vibrational mode. Then, the vibrational beats modulate the two-pulse photon echo without mode suppression, inducing an extremely rapid decay (Banin et al., 1997). Both electronic and vibrational dephasing times can be probed in this way.

The two-pulse Stark-modulated photon-echo method can be used also to measure the small Stark shift ($\leq$ 1 kHz) induced by the static electric field in impurity-ion-doped crystals. In this case the photon-echo intensity I_e is measured when a static electric pulse E_S (Stark pulse) with width τ_S is applied either during the dephasing period (between the first and second pulse) or during the rephasing period (between second pulse and echo). The Stark-shift frequencies are obtained from the modulation pattern of $I_e = f(\tau_S)$ (Graf et al., 1997). The resolution of the method is limited by the pulse separation. The modulation pattern becomes quite complicated when several crystal sites participate in generating photon echos, due to interference of inequivalent ions. For example, for N ions, each having an eigenfrequency ω_j, the normalized echo intensity is given by $I_e = |\sum_{j=1}^{N} A_j \exp[\mathrm{i}\int_0^T \omega_j(t)\mathrm{d}t - \mathrm{i}\int_\tau^{2\tau} \omega_j(t)\mathrm{d}t]|^2$, where the coefficients A_j depend on the angle between the transition dipole moment and laser polarization. If the Stark pulse is applied in the dephasing period, the induced Stark shift of the kth ion is $\Omega_k = (\delta\boldsymbol{\mu}_k \cdot \boldsymbol{E}_S)/\hbar$ with $\delta\boldsymbol{\mu}_k$ the difference between ground and excited-state dipole moments, and $I_e(\tau_S) = |(1/n_S)\sum_{k=1}^{n_S} A_k \exp(\mathrm{i}\Omega_k\tau_S)|^2$, where n_S is the number of sites (of possible Stark shifts). Ω_k are obtained by Fourier transforming the signal I_e.

The simple optical Bloch equations introduced above do not include the Coulomb attraction between electron-hole pairs; the many-body Coulomb interaction can be accounted for considering a system of coupled two-level systems, known as the semiconductor Bloch equations. In this treatment, it is possible to introduce both renormalized electron and hole energies due to Coulomb interaction, and excitons. Also, the interaction strength between matter and the electromagnetic field is renormalized (see Shah (1999) and Haug and Koch (1990) for details). Semiconductor Bloch equations are, however, difficult to solve, the numerical solutions being extremely computer time consuming.

2.7.7.2 Quantum Beats and Polarization Beats

The simplified model of a two-level system cannot account for the experimentally observed phenomena of quantum beats and polarization beats, which manifest in an oscillatory behavior of the FWM. These effects can only be described in a three-level model, such as that in Fig. 2.36.

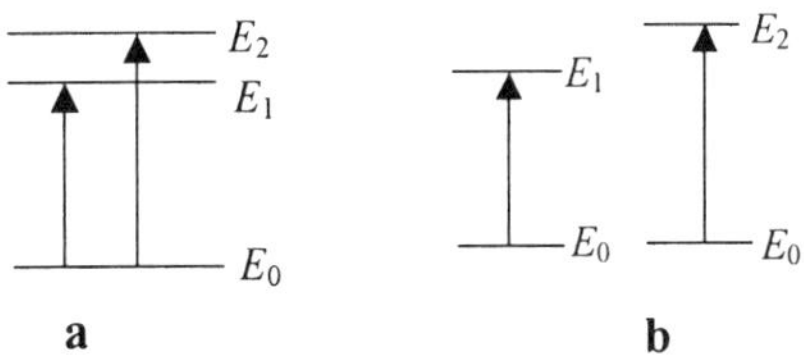

Fig. 2.36. Three-level models for (**a**) quantum beats and (**b**) polarization beats

The temporal evolution of the polarization components, p_{01} and p_{02} is described by the optical Bloch equations, which in the low-excitation limit are

$$\begin{aligned} i\hbar(\partial p_{01} / \partial t) &= (E_1 - E_0) p_{01} - d_{01} E(t), \\ i\hbar(\partial p_{02} / \partial t) &= (E_2 - E_0) p_{02} - d_{02} E(t), \end{aligned} \tag{2.168}$$

where d_{ij} are the dipole matrix elements for the two transitions (Vasko and Kuznetsov, 1999). Quantum beats (see Fig. 2.36a) occur when both transitions from the lowest level to the upper levels are excited simultaneously by the same spectrally broad laser pulse, so that a linear coherent superposition of the states of the upper energy levels is formed, described by the time-dependent Schrödinger equation. Spectrally broad pulses correspond to short excitation pulses that can be approximated by a δ-function ($E(t) = E_0 \tau_p \delta(t)$). In this case the FWM intensity, proportional to the squared modulus of the total optical polarization at $\tau_{21} > 0$ is

$$\begin{aligned} I_{\text{FWM}}(\tau_{21}) &\approx | d_{01} p_{01}(\tau_{21}) + d_{02} p_{02}(\tau_{21}) |^2 \\ &= (E_0^2 \tau_p^2 / \hbar^2)\{d_{01}^4 + d_{02}^4 + 2 d_{01}^2 d_{02}^2 \cos[\tau_{21}(E_2 - E_1)/\hbar]\}. \end{aligned} \tag{2.169}$$

As seen from (2.169), the FWM signal has an oscillatory behavior, extremely useful in resolving the presence of closely spaced levels separated by $E_2 - E_1$. Quantum beats can also be observed in time-resolved pump-probe experiments and in PL.

However, such an oscillatory behavior of the FWM characterizes not only the excitation of a coherent superposition of wavefunctions in a three-level system, but also the case presented in Fig. 2.36b, of two two-level systems with close transition energies, but no common state. The time-evolution of the dipole moment in this case has the same form as (2.169), i.e.

$$D(t)=(E_0^2\tau_p^2/\hbar^2)\theta(t)\{d_{01}^2D_{11}+d_{02}^2D_{22}+2d_{01}d_{02}D_{12}\cos[t(E_2-E_1)/\hbar]\}, \quad (2.170)$$

where D_{ij} is the matrix element of the dipole moment between states i and j, produced by these real-space charge oscillations. This interference phenomenon involves dipole moments and not superposition of wavefunctions; it is not a quantum interference, as in the previous case. In order to distinguish it from quantum beats, the interference of dipole moments is called polarization beats.

Although quantum and polarization beats have different origins, it is not always easy to distinguish between them in practice. They can, however, be identified by spectrally resolved FWM or time-resolved FWM techniques. These two methods are presented in Shah (1999) (see also the references therein). The spectral shape of the FWM signals is shown to be different in the two cases when the transitions with frequencies $\omega_1=(E_1-E_0)/\hbar$ and $\omega_2=(E_2-E_0)/\hbar$ are excited by pulses with spectral width much smaller than ω_1 and ω_2, but larger than $|\omega_1-\omega_2|$, for τ_{21} larger than the pulse widths. In the time-resolved FWM method, the signal for the quantum beats case, for a given τ_{21}, is maximum when $t=\tau_{21}+2\pi n\hbar/(E_2-E_1)$, whereas for polarization beats it is maximum for $t=2\tau_{21}\pm 2\pi n\hbar/(E_2-E_1)$, if $t>\tau_{21}$. Moreover, when there is no dephasing, the maximum in the quantum beat signal oscillates as a function of τ_{21}, while the polarization interference signal does not depend on the delay τ_{21}.

2.7.8 Spectral Hole Burning

Spectral hole burning is a method of recovering the homogeneous width of an optical transition from an inhomogeneously broadened line of width Γ_{inh}, which is a sum of separate, independent lines of width $\Gamma_{hom} \ll \Gamma_{inh}$, each centered at a different frequency within the inhomogeneous line profile. The method consists in tuning an excitation laser beam of frequency ω_L and bandwidth $\Gamma_L<\Gamma_{hom}$ over an inhomogeneously broadened absorption band, so that the absorption of laser radiation depletes only that subassembly of excited centers with energies within Γ_{hom} of the laser frequency. The absorption signal is thus determined selectively by only some of the atoms in the sample, a hole of width $2\Gamma_L$ being burned in the inhomogeneous broadened line (see Fig. 2.37). If the line is homogeneously broadened, the whole line is burned down at illumination with the excitation laser beam, and no hole appears. To resolve the homogeneous width, it is necessary that $\Gamma_L<\Gamma_{hom}\ll\Gamma_{inh}$. The narrow laser linewidth and high power used in hole burning spectroscopy makes it possible to maintain a significant fraction of the subassembly with transition frequency ω_L in the excited state, where it no longer contributes to the absorption at this frequency. A large enough depletion of the ground state population is thus created, so that a subsequent probe beam can detect the difference in absorption between pumped and unpumped sites.

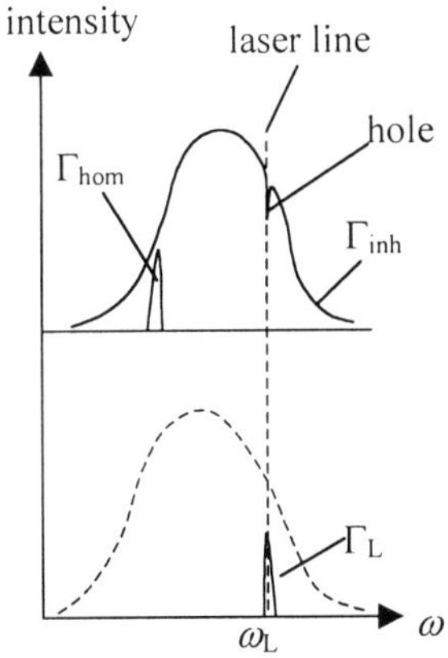

Fig. 2.37. The principle of spectral hole burning method

The width of the spectral hole Γ_L gives information about the lifetime of the transition τ, based on the relation $\Gamma_L\tau \cong h$, the resolution of the method being limited by the laser linewidth. Since the very narrow hole is burned at the frequency of the excitation laser, the need for a high-resolution spectrometer is eliminated. The probe beam can be constituted, for hole widths less than 20 MHz, by the sideband which is created at the laser modulation frequency, so that scanning the intensity laser modulation rate creates a probe beam which scans across one side of the hole (Selzer, 1981). The hole burning signal can be defined as $-\Delta OD = (I_{wp} - I_p)/I_{np}$, where I_{wp} is the intensity of the probe signal going through the sample with the pump, I_p the contribution of the pump alone, and I_{np} the intensity of the probe signal through sample without pump. The signal is linear in the pump energy, its spectral distribution being given by $-\Delta OD(\omega) = C\int_{-\infty}^{\infty} H(\omega - \omega')H(\omega_L - \omega')g(\omega' - \omega_0)\mathrm{d}\omega$, with $H(\omega)$ the homogeneous absorption lineshape, $g(\omega - \omega_0)$ the inhomogeneous distribution of absorption centers, centered at ω_0, and C a constant proportional to the excitation intensity. After burning a spectral hole, the behavior of the narrow hole under external perturbations can be studied instead of the broad line, with much higher resolution.

A technique closely related to hole burning is polarization spectroscopy (Selzer, 1981; Weber, 1981). In this method, a probe laser detects the nonlinear birefringence in the sample produced by the pump. In general, the pump and probe beams are linearly polarized and at 45° with respect to one another. Any frequency-dependent birefringence induced in the sample by the pump produces a frequency-dependent polarization component orthogonal to the original probe polarization. Tuning the probe beam frequency through the pump beam frequency, a Lorentzian curve is obtained, with a width twice the homogeneous linewidth of the transition under examination. This technique is more sensitive than hole burning, since it involves deviation-from-null measurements.

2.7.9 Fluorescence Line Narrowing

Fluorescence line narrowing (FLN) is also a site-selective excitation technique, in which from an inhomogeneously broadened distribution of luminescent centers, only a few, with a given frequency, are excited (see Fig. 2.38) (Selzer, 1981, Weber, 1981, Morgan and Yen, 1989). Their subsequent luminescent emission is narrower than the width of the inhomogeneous broadened distribution. The homogeneous PL linewidth $\Delta\omega_{\text{hom}}$ of the selected subset of ions can be measured in this way in the presence of inhomogeneous broadening. The homogeneous linewidth is related to the relaxation time T_2 by $1/T_2 = \Delta\omega_{\text{hom}}$. If there is a mechanism of energy transfer between different luminescent centers, then the excitation is transferred to other luminescent centers as well, which are not initially in resonance with the excitation. Therefore, in this case the luminescence line broadens in time, becoming eventually identical to the inhomogeneously broadened distribution of centers. Time-resolved FLN experiments allow then the direct observation of energy migration, the fluorescence wavelength within the inhomogeneous profile labeling the impurities that are excited after some time delay by nonresonant interactions with the initial pumped sites.

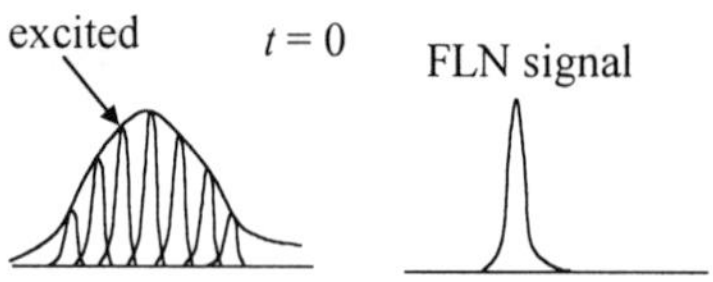

Fig. 2.38. Fluorescence line narrowing

FLN and spectral hole burning are to some extent complementary spectroscopic methods. Spectral hole burning is more suitable for resolving narrow linewidths in systems with no dynamics or with weakly fluorescing transitions, while FLN is preferred for the study of the hyperfine or superhyperfine structure. Hole burning is susceptible to 'dynamic inhomogeneous broadening' which can occur during the creation of the hole, causing it to spread. Also, the greater laser power requirement in a hole-burning experiment may produce power broadening of the line, which occurs when the induced transition rate is comparable to the dephasing rate of the system. In systems where slow spectral dynamics take place, pulsed hole burning using a delayed probe is more difficult and less effective than time-resolved FLN.

2.7.10 Magneto-Optical Kerr Effect

In crystals with at least threefold rotational symmetry (hexagonal, tetragonal, trigonal, cubic) with a magnetic field applied along the z axis, the complex dielectric constant is

$$\hat{\varepsilon} = \begin{pmatrix} \varepsilon_0 & \mathrm{i}\varepsilon_1 & 0 \\ -\mathrm{i}\varepsilon_1 & \varepsilon_0 & 0 \\ 0 & 0 & \varepsilon_z \end{pmatrix}. \tag{2.171}$$

Its eigenvalues are $\varepsilon_{\pm} = \varepsilon_0 \pm \varepsilon_1$, and the normal modes for light propagating with $\boldsymbol{k} \parallel z$ are circularly polarized $\hat{\varphi}_+ = (-\mathrm{i}/\sqrt{2})(\hat{x} + \mathrm{i}\hat{y})$, $\hat{\varphi}_- = (1/\sqrt{2})(\hat{x} - \mathrm{i}\hat{y})$. If the incoming beam is linearly polarized at an angle α, measured counterclockwise to x, i.e. if $\boldsymbol{E}_{\mathrm{in}} = E_0(\hat{x}\cos\alpha + \hat{y}\sin\alpha) = E_0\hat{r}_\alpha = E_+\hat{\varphi}_+ + E_-\hat{\varphi}_-$, then the reflected beam is $\boldsymbol{E}_{\mathrm{out}} = r_+E_+\hat{\varphi}_+ + r_-E_-\hat{\varphi}_-$, where $r_{\pm} = (1 - \sqrt{\varepsilon_{\pm}})/(1 + \sqrt{\varepsilon_{\pm}})$ are the reflection coefficients for the left and right circularly polarizations. Then, $\boldsymbol{E}_{\mathrm{out}}$ $= (1/2)(r_+ + r_-)[\hat{r}_\alpha + \mathrm{i}\hat{r}_\beta(r_+ - r_-)/(r_+ + r_-)]$, where $\hat{r}_\beta = -\hat{x}\sin\alpha + \hat{y}\cos\alpha$ is the direction normal on $\hat{r}_\alpha$. The rotation of the polarization angle is called complex Kerr rotation and is defined as $\Psi = \mathrm{i}(r_+ - r_-)/(r_+ + r_-) = \theta + \mathrm{i}\psi$, where θ is the Kerr angle and $\mathrm{i}\psi$ the ellipticity. For small rotations, $\Psi = \theta + \mathrm{i}\psi$ $\cong \mathrm{i}\varepsilon_1/[\sqrt{\varepsilon_0}(\varepsilon_0 - 1)]$. The change in polarization is called the Faraday effect when it refers to the transmitted light, and the Kerr effect when the reflected light is considered. The Faraday rotation can be enhanced by placing the sample in a microcavity, due to multiple round-trips of light between mirrors. This enhancement of Faraday rotation at a quantum well exciton resonance is accompanied by a resonant redistribution of the signal between right- and left-circular polarizations (Kavokin et al., 1997).

For normal incident light and a magnetic field applied along the z direction, the complex magneto-optic Kerr (MOKE) rotation is the same for s- and p-polarizations and is a measure of the size of the off-diagonal component of the dielectric tensor $\Psi_{\mathrm{K}} \approx \sigma_{yz}/\sigma_{zz}$. There are two basic configurations in magneto-optic experiments: the Faraday configuration, when the magnetic field is parallel to the optical excitation, and the Voigt configuration, when the magnetic field is perpendicular to the optical excitation. Each of them probes different excitations of the solid, with different selection rules.

MOKE is used, for example, in surface magnetic analysis; see for example Moog and Bader (1985). The ac- and dc-MOKE in polar geometry is sensitive to the out-of-plane component of magnetization, whereas Kerr loops in longitudinal geometry record the in-plane component. A recent theory for single-ion electric-dipole contribution to the dielectric constant, between ground and excited states, predicts that the polar MOKE can distinguish between left-circular polarized (LCP) and right-circular polarized (RCP) light (Fontijn et al., 1997). This is possible due to the optical anisotropy in MOKE caused by spin-orbit coupling, which induces different oscillator strengths for LCP and RCP. When the oscillator strength for transitions caused by LCP and RCP are equal the dielectric tensor has a type I, diamagnetic lineshape, for which the real part of ε_{xy} has an extremum, and its imaginary part an oscillatory behavior. On the contrary, when the oscillator strength for transitions caused by LCP and RCP differ, the off-diagonal components of the dielectric tensor have a type II, paramagnetic lineshape for which the real part of ε_{xy} has an oscillatory behavior, and the imaginary part has an extremum as a function of frequency.

The nonlinear MOKE (NLMOKE) effect, which describes the rotation, at reflection from a ferromagnetic surface, of the polarization plane for the second-harmonic generation reflects the symmetry of the surface and its magnetism. It can be used as an ultrafast spectroscopic probe of the 2D magnetism or for the measurement of spin-lattice relaxation times in transition metals. The peak positions and peak height ratios of MOKE and NLMOKE spectra are mainly determined by the electron band structure, the NLMOKE spectrum being almost the derivative of the linear MOKE spectrum for Fe, for example (Pustogowa et al. 1993). A new type of MOKE rotation, the surface-induced transverse MOKE (SITMOKE) was predicted to appear at reflection from a magnetized surface in transverse geometry, at normal incidence (Petukhov et al., 1999). SITMOKE vanishes at most bulk-terminated low-index faces of cubic media, but is allowed at lower-symmetry planes and is linear in the surface magnetization. It can therefore be used to study the magnetism of vicinal and reconstructed surfaces. In order to detect it, the usual MOKE effect must be eliminated; this can be achieved at normal incidence, if the light propagates along the direction normal to $\boldsymbol{M}$.

2.8 Line Identification

We have seen that several optical methods can be used to demonstrate the excitations of a sample. However, the nature of these excitations is not always easy to identify. A possible way to identify them is to compare their spectral position, linewidth and strength with theoretical calculations. The result is not always unambiguous since theoretical calculations often use parameters that are not known with great precision for the sample under study, or the theoretical models do not always reflect all the processes in the sample, or several theoretical models give different results. Therefore, experimental ways to identify the nature of the excitations are of great help.

For example, one way to distinguish a homogeneously broadened line from an inhomogeneous one is to perform temperature-dependent measurements. The homogeneously broadened line is strongly temperature dependent, while the inhomogeneous line is practically temperature independent as long as its linewidth is broader than that of the homogeneous constituents. By applying magnetic fields it is possible to distinguish between electron and vibrational Raman lines, since electron lines have characteristic splittings and/or shifting of their spectral position, whereas vibrational lines are not affected by the magnetic field. Host phonon modes are sensitive to an externally applied hydrostatic pressure, whereas the energy position of defect lines does not change under pressure. Different lines have, in general, different behaviors under hydrostatic pressure. A model for the pressure-optic coefficient in optical materials and its dispersion, including contributions of isothermal compressibility, isobaric bandgaps, and pressure coefficients of excitonic and lattice resonance absorption bandgaps can be found in Ghosh (1998). Different types of interactions give rise to different temperature and frequency dependence of electron scattering lines, for example. Electrons can scatter on polar optical phonons, ionized impurities, neutral impurities, acoustic-phonon deformation potential and acoustic-phonon piezoelectric potential, or interface roughness. The rates of these electron scattering mechanisms in 2D structures are given in Kozhevnikov et al. (2000); see also the references therein.

Bound exciton PL lines in II-VI semiconductor materials can be distinguished, for example, through their binding energies. In these materials the binding energies ΔE_1, ΔE_2 of excitons responsible for PL lines caused by an exciton trapped to a neutral acceptor, and trapped to a neutral donor, respectively, satisfy quite well the approximate formulas (Curie, 1979) $\Delta E_1 / E_A \cong 0.1$, $\Delta E_2 / E_D \cong 0.2$ where E_A, E_D are the binding energy of a hole on acceptor, and an electron on donor, respectively. The PL lines of free excitons differ from those of bound excitons through their temperature dependence. Unlike free exciton lines, all bound exciton lines follow the gap variation with temperature, shifting along almost parallel curves when the temperature is increased. The linewidths of bound excitons, however, increase only slightly with temperature in, contrast to the considerable broadening of free exciton lines. The PL lineshapes of free and bound excitons are also different: free excitons show a temperature broadening of the high-energy part of the line and a sharp cut of the low-energy part, whereas bound excitons are characterized by either a symmetric or an asymmetric broadened line, with a more pronounced broadening towards the low-energy part. Free and bound excitons also behave differently under hydrostatic and uniaxial pressures, or external magnetic and electric fields.

Up to now we have referred mainly to optical excitations of electronic, or quasiparticle states. However, most transitions are accompanied by phonon replica, caused by the interaction of electrons or excitons with the lattice, as well as by the Franck–Condon principle. According to this principle, in optically induced transitions between electronic levels, the lattice is almost frozen. Lattice relaxation after excitation of electron levels, occurs through emission of phonons. The number of emitted phonons during lattice relaxation is measured by the Huang–Rhys factor *S*, defined as $S = E_r / \hbar\omega_{ph}$, where E_r is the relaxation energy and ω_{ph} is the frequency of the emitted phonons. So, in PL or Raman experiments the electronic lines are almost always accompanied by phonon replica. Modeling phonon vibrations as harmonic oscillators, the phonon factor that scales the electronic transition probability from a ground electronic state with no phonons to an excited electronic state with *n* phonons in the Franck–Condon approximation is

$$| M_{0n} |^2 = \exp(-S) S^n / n!. \tag{2.172}$$

In PL experiments, S can be obtained by fitting the intensity ratio of the *n*th phonon replica to the zero-phonon line with $I_n / I_0 = S^n / n!$. Spectrally, the *n*th phonon replica is located at an energy lower by $n\hbar\omega$ than the zero-phonon line. The probability of transition with creation of *n* phonons is $W_n = \exp[nz - S\coth(z)] I_n[S\text{csch}(z)]$, where $z = \hbar\omega_{\text{ph}} / 2k_B T$, and $I_n(z)$ is the modified Bessel function (Davies, 1981).

The Huang–Rhys factor can also be determined from the Stokes shift between the absorption and PL edges, caused by the lattice relaxation between the excited electronic state, from which the PL emission starts, and the ground electronic state on which the photon is absorbed. In terms of electron-phonon coupling strength γ_j, the Stokes shift and the Huang–Rhys factor can be expressed as $\Delta = 2\sum_j \gamma_j^2 / \hbar\omega_j$ and $S = \sum_j \gamma_j^2 / (\hbar\omega_j)^2$, respectively, where the sums must be performed over all phonon modes. Typical values for the Stokes shift are several meV and the *S* factor is usually 0.3–0.7. The *S* factor can also be determined from the ratio of the fundamental to overtone resonant Raman intensities; for a more detailed discussion of how the Huang–Rhys factor is obtained in this case see, for example, Krauss and Wise (1997).

Phonon replicas have a more pronounced temperature dependence than the zero-phonon line, and also a different pressure dependence. For example, in Be-doped Si, the zero-phonon line and phonon replicas decrease in energy with increasing pressure with different rates, in contrast to peaks with different origin that increase with pressure (Kim et al., 1995).

Not only phonon sidebands but also magnon sidebands can appear in magnetic materials. In this case the lattice relaxation takes place with the emission of magnons. For example, the exciton line can have a magnon sideband, this exciton-magnon pair production occurring via the electric-dipole interaction. The magnon sidebands are characteristically more intense than the no-magnon line because of their spin-allowed, electric-dipole nature; the exciton and magnon may appear on the same or different sublattices.

References

Abelès, F. (1972) in *Optical Properties of Solids*, ed. F. Abelès, North-Holland, Amsterdam, p.95.

Azzam, R.M.A. and N.M. Bashara (1977) *Ellipsometry and Polarized Light*, North-Holland, Amsterdam.

Banin, U., G. Cerullo, A.A. Guzelian, C.J. Bardeen, A.P. Alivisatos, and C.V. Shank (1997) Phys. Rev. B **55**, 7059.

Basu, P.K. (1997) *Theory of Optical Processes in Semiconductors*, Clarendon Press, Oxford.

Bilz, H. and W. Kress (1979) *Phonon Dispersion Relations in Insulators*, Springer, Berlin, Heidelberg.

Callaway, J. (1991) *Quantum Theory of the Solid State*, Academic Press, Boston, 2nd edition.

Chen, Y.F., C.M. Kwei, and C.J. Tung (1993) Phys. Rev. B **48**, 4373.

Chu, K.C., J.P. Heritage, R.S. Grant, K.X. Liu, A. Dienes, W.E. White and A. Sullivan (1995) Opt. Lett. **20**, 904.

Curie, D. (1979) in *Luminescence of Inorganic Solids*, ed. B. di Bartolo, Plenum Press, New York, p.337.

Davies, G. (1981) Rep. Prog. Phys. **44**, 787.

de la Cruz, G.G., and C. Trallero-Giner (1998) Phys. Rev. B **58**, 9104.

di Bartolo, B. (1968) *Optical Interactions in Solids*, Wiley, New York.

Ferrari, M., M. Montagna, S. Ronchin, F. Rossi, and G.C. Righini (1999) Appl. Phys. Lett. **75**, 1529.

Fontijn, W.F.J., P.J. van der Zaag, M.A.C. Devillers, V.A.M. Brabers, and R. Metselaar (1997) Phys. Rev. B **56**, 5432.

Gárcia-Cristóbal, A., A. Cantarero, C. Trallero-Giner, and M. Cardona (1998) Phys. Rev. B **58**, 10443.

Ghosh, G. (1998) Phys. Rev. B **57**, 8178.

Graf, F.R., A. Renn, U.P. Wild, and M. Mitsunaga (1997) Phys. Rev. B **55**, 11225.

Grünberg, P. (1989) in *Light Scattering in Solids V*, eds. M. Cardona and G. Güntherodt, Springer, Berlin, Heidelberg, p.303.

Gubarev, S.I., T. Ruf, and M. Cardona (1991) Phys. Rev. B **43**, 1551.

Haug, H. and S.W. Koch (1990) *Quantum Theory of the Optical and Electronic Properties of Semiconductors*, World Scientific, Singapore.

Hayes, W. and R. Loudon (1978) *Scattering of Light by Crystals*, Wiley, New York.

Hirata, T., K. Ishioka, M. Kitajima, and H. Doi (1996) Phys. Rev. B **53**, 8442.

Iaconis, C., V. Wong, and I.A. Walmsley (1998) IEEE J. Selected Topics in Quantum Electron. **4**, 285.

Jellison Jr., G.E. and F.A. Modine (1997a) Appl. Opt. **36**, 8184.

Jellison Jr., G.E. and F.A. Modine (1997b) Appl. Opt. **36**, 8190.

Kavokin, A.V., M.R. Vladimirova, M.A. Kaliteevski, O. Lyngnes, J.D. Berger, H.M. Gibbs, and G. Khitrova (1997) Phys. Rev. B **56**, 1087.

Kim, S., I.P. Herman, K.L. Moore, D.G. Hall, and J. Bevk (1995) Phys. Rev. B **52**, 16309.

Klein, M.V. (1975) in *Light Scattering in Solids*, ed. M. Cardona, vol.8 in Topics in Applied Physics, Springer, Berlin, Heidelberg, p.147.

Klingshirn, C.F. (1997) *Semiconductor Optics*, Springer, Berlin, Heidelberg.

Kozhevnikov, M., E. Cohen, A. Ron, and H. Shtrikman (2000) Phys. Rev. B **60**, 16885.

Krauss, T.D. and F.W. Wise (1997) Phys. Rev. B **55**, 9860.

Kwok, A., L. Jusinski, M.A. Krumbügel, J.N. Sweetser, D.N. Fittinghoff, and R. Trebino (1998) J. Selected Topics in Quantum Electron. **4**, 271.

Martin, R.M. and L.M. Falicov (1975) in *Light Scattering in Solids*, ed. M. Cardona, vol.8 in Topics in Applied Physics, Springer, Berlin, Heidelberg, p.79.

Mavroides, J.G. (1972) in *Optical Properties of Solids*, ed. F. Abelès, North-Holland, Amsterdam.

Moog, E.R. and S.D. Bader (1985) Superlattices Microstruct. **1**, 543.

Morgan, G.P. and W.M. Yen (1989) in *Laser Spectroscopy of Solids II*, ed. W.M. Yen, vol.65 in Topics in Applied Physics, Springer, Berlin, Heidelberg, p.77.

Petukhov, A.V., A. Kirilyuk, and Th. Rasing (1999) Phys. Rev. B **59**, 4211.

Pinczuk, A. and E. Burstein (1975) in *Light Scattering in Solids*, ed. M. Cardona, vol.8 in Topics in Applied Physics, Springer, Berlin, Heidelberg, p.23.

Pine, A.S. (1975) in *Light Scattering in Solids*, ed. M. Cardona, vol.8 in Topics in Applied Physics, Springer, Berlin, Heidelberg, p.253.

Pustogowa, U., W. Hübner, and K.H. Bennemann (1993) Phys. Rev. B **48**, 8607.
Ruf, T. (1998) *Phonon Raman Scattering in Semiconductors, Quantum Wells and Superlattices*, Springer, Berlin, Heidelberg.
Rupprecht, R., B. Müller, H. Pascher, I. Miotkowski, and A.K. Ramdas (1998) Phys. Rev. B **58**, 16123.
Schubert, M., B. Rheinländer, J.A. Woollam, B. Johs, and C.M. Herzinger (1996a) J. Opt. Soc. Am A **13**, 875.
Schubert, M., B. Rheinländer, C. Cramer, H. Schmiedel, C.M. Herzinger, B. Johs, and J.A. Woollam (1996b) J. Opt. Soc. Am. A **13**, 1930.
Scott, J.F. (1980) Rep. Prog. Phys. **43**, 951.
Selzer, P.M. (1981) *Laser Spectroscopy of Solids*, eds. W.M. Yen, P.M. Selzer, vol.49 in Topics in Applied Physics, Springer, Berlin, Heidelberg, p.113.
Seraphin, B.O. (1972) in *Optical Properties of Solids*, ed. F. Abelés, North-Holland, Amsterdam, p.163.
Shah, J. (1999) *Ultrafast Spectroscopy of Semiconductors and Semiconductor Nanostructures*, Springer, Berlin, Heidelberg, 2nd edition.
Shen, Y.R. (1974) Phys. Rev. **9**, 622.
Shen, Y.R. (1975) in *Light Scattering in Solids*, ed. M. Cardona, vol.8 in Topics in Applied Physics, Springer, Berlin, Heidelberg, p.275.
Stepanov, B.I. and V.P. Gribkovskii (1963) *Theory of Luminescence*, Iliffe Books Ltd., London.
Stoneham, A.M. (1981) Rep. Prog. Phys. **44**, 1251.
Strach, T., J. Brunen, B. Lederle, J. Zegenhagen, and M. Cardona (1998) Phys. Rev. B **57**, 1292.
Taylor P.C., (1989) in *Laser Spectroscopy of Solids II*, ed. W.M. Yen, Springer Verlag, Berlin, Heidelberg, p.257.
Tignon, J., M.V. Marquezini, T. Hasche, and D.S. Chemla (1999) IEEE J. Quantum Electron. **35**, 510.
Vasko, F.T. and A.V. Kuznetsov (1999) *Electronic States and Optical Transitions in Semiconductor Heterostructures*, Springer, Berlin, Heidelberg.
Weber, M.J. (1981) *Laser Spectroscopy of Solids*, eds. W.M. Yen, P.M. Selzer, vol.49 in Topics in Applied Physics, Springer, Berlin, Heidelberg, p.189.
Woggon, U., W. Petri, A. Dinger, S. Petillon, M. Hetterich, M. Grün, K.P. O'Donnell, H. Kalt, and C. Klingshirn (1997) Phys. Rev. B **55**, 1364.
Zhang, X., J.D. Comins, A.G. Every, P.R. Stoddart, W. Pang, and T.E. Derry (1998) Phys. Rev. B **58**, 13677.
Zouboulis, E.S. and M. Grimsditch (1998) Phys. Rev. B **57**, 2889.

3. Optical Properties of Impurities in Solids

In this chapter we analyze the optical properties of doped solid-state materials. When the doping concentration is small, we expect the appearance in optical spectra of sharp lines due to transitions between different discrete levels of the dopants or between the level of the dopants and the bands of the host material. The first problem is to calculate the energy levels of a dopant atom or ion inside a host. From the point of view of group theory, the normal modes of vibration of an atom/ion in a real crystal are found as follows: first we identify the symmetry group G_m of the free atom/ion. The normal vibrations of the free atom/ion can be classified after the irreducible representations of G_m. If we introduce the atom/ion into the crystal, equivalent from a physical point of view to placing the atom/ion in an external field with the same symmetry as the crystal, it would have the symmetry of position G_s. This symmetry is in general lower than that of G_m; G_s is therefore a subgroup of G_m, any irreducible representation of G_m being a (generally reducible) representation of G_s. Decomposing this reducible representation after the irreducible representations of G_s we can find the relation between the normal vibrational modes of the free atom/ion and of the same atom/ion in the host crystal. One can also introduce in the description the coupling between the identical normal vibrations of the neighboring molecules, which introduce a new group G_f – the factor group of the crystal: $G_m \to G_s \to G_f$.

3.1 Electronic Structure of Isolated Atoms

A neutral atom with atomic number Z is composed from a nucleus with positive charge Ze and Z electrons. Since the dimension of the nucleus is extremely small, and its mass much larger than that of the electron, it is modeled as a fixed point. The movement of the system of Z electrons is very difficult to treat; the simplest approximation is to consider each electron in an effective electrostatic potential due to the nucleus and the interaction with all other $Z-1$ electrons, and modeled as an electric cloud with spherical symmetry. This effective, radial potential, depends only on the distance r of the electron from the nucleus, varying from $V(r) \cong -Ze^2/(4\pi\varepsilon_0 r)$ for the electron closest to the nucleus (unscreened potential) to $V(r) \cong -e^2/(4\pi\varepsilon_0 r)$ for the furthest electron.

The solution of the Schrödinger equation for the electron in a radial potential, called the orbital of the electron, can be separated in

$$\psi_{nlm} = R_{nl}(r)Y_{lm}(\theta,\varphi), \tag{3.1}$$

where n, l, m, are respectively the principal, azimuthal and magnetic quantum numbers ($0 \le l \le n-1$, $-l \le m \le l$), the radial function $R_{nl}(r)$ depends on the specific form of the radial potential, and

$$Y_{lm}(\theta,\varphi) = \frac{(-1)^{l+m}}{2^l l!}\sqrt{\frac{(l-m)!}{(l+m)!}}\sqrt{\frac{2l+1}{4\pi}}\frac{\mathrm{d}^{l+m}(1-\cos^2\theta)^l}{(\mathrm{d}\cos\theta)^{l+m}}\sin^m\theta\exp(\mathrm{i}m\varphi), \tag{3.2}$$

are the normalized spherical functions. The electron energy E_{nl} does not depend, in general, on the magnetic quantum number m. In spectroscopy, the energy levels are denoted by $n\xi$, where n is the principal quantum number and $\xi = \text{s, p, d, f}\ldots$ is a notation for the quantum number l = 0, 1, 2, 3… These discrete energy levels are grouped in shells denoted by K, L, M…, each of them containing the following energy levels (in increasing order)

K: 1s
L: 2s, 2p
M: 3s, 3p
N: 4s, 3d, 4p
O: 5s, 4d, 5p
P: 6s, 4f, 5d, 6p
Q: 7s, 5f, 6d …

The energy levels are occupied by electrons in increasing order of their values. The maximum number of electrons that can occupy an orbital characterized by a quantum number l is $2(2l+1)$ (2 accounts for the spin, $2l+1$ are the possible m values for a given l); the total degeneracy of the nth level, including spin degeneracy is $2n^2$. The properties of different elements, including the optical properties, depend on the electronic configuration. Thus we can distinguish the rare gases, whose electronic configuration consists of closed shells, halogens with an electron less than a closed-shell configuration, alkaline metals with an electron more than a closed-shell configuration. Transition metals are elements in which the d orbitals are partially occupied, whereas lanthanides and actinides (rare earths) are the elements for which the 4f and 5f orbitals, respectively, are partially occupied. In f orbitals the electrons move in a region close to the nucleus, so that rare earths (RE) are quite insensitive to the effects of the surrounding medium; all elements in the lanthanide or actinide families thus have similar properties.

When each electron is moving in a central potential, not only the electron orbital kinetic momentum, but also the total orbital kinetic momentum operator $\boldsymbol{L} = \sum_i \boldsymbol{l}_i$ is conserved, due to the invariance of the total Hamiltonian at rotations around the nucleus. Another invariant is the spin operator $S = \sum_i s_i$; L and S can serve as quantum numbers for the isolated system of Z electrons.

When there is spin-orbit coupling the operators L and S are no longer conserved, but the total kinetic momentum $J = L + S$ is. Due to this interaction, considered as a perturbation, the $(2L+1)(2S+1)$ degenerate energy levels split; this splitting reveals the fine structure of energy levels. In an external magnetic field a new perturbation characterized by a Hamiltonian $-(\hbar e/2m)(J+S)\cdot B = -g(\hbar e/2m)J \cdot B$, with $\mu_B = \hbar e/2m$ the Bohr magneton, acts on the atom and an additional splitting of each level – the Zeeman splitting – occurs. The gyromagnetic or Landé factor g takes the value 2 when the magnetic moment is caused only by the spin contribution.

3.2 Ions in Crystals

If the atom/ion is introduced into a crystal, with generally lower symmetry than the free space, two major effects appear: the effective field introduced by the surrounding ions (ligand field) can displace the energy levels and/or can split the levels by removing some degeneracies. The exact effect can be calculated using the character table of groups. For example, the $(2l+1)$ degeneracy is completely lifted for a low symmetry C_{2v}, whereas in cubic point symmetry, d orbitals are split into e_g and t_{2g}, but p orbitals are not split. The effect of site symmetry on the optical spectrum of ions, can be illustrated by Cr^{3+} transitions between ^2E and $^4A_2$ levels. These are single energy levels in octahedral sites, but split in lower symmetry (see Fig. 3.1). The PL intensity ratio of these two lines varies with temperature as the Boltzmann factor $N_1/N_2 = \exp(-E/k_BT)$ (Imbusch, 1979a).

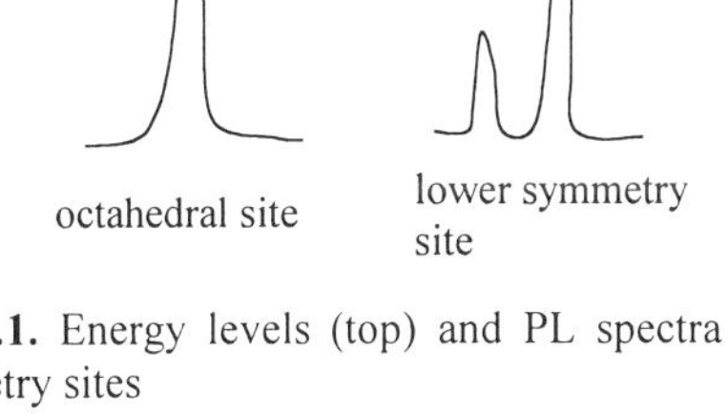

Fig. 3.1. Energy levels (top) and PL spectra (bottom) of Cr^{3+} in octahedral and lower symmetry sites

The optical properties of dopant atom/ions are generally defined by the electron configuration of the partially occupied shell. This can be an outer shell, or an inner shell (as for RE elements), shielded from the effect of the surrounding media by the outer s or p orbitals.

In either case, the strength of the interaction between the electromagnetic radiation and these optically active electrons is characterized by $S(ab) = \Sigma_{a,b} |\langle b | \boldsymbol{D} | a \rangle|^2$ (Imbusch and Kopelman, 1981) where the radiative transition (emission or absorption) between the initial level a and the final level b is induced by either the electric or the magnetic field of the radiation, described respectively by the operator $\boldsymbol{D} = -\boldsymbol{p} \cdot \boldsymbol{E}$ or $\boldsymbol{D} = -\boldsymbol{m} \cdot \boldsymbol{B}$. Here $\boldsymbol{p} = \Sigma_i e\boldsymbol{r}_i$, where the sum is taken over all optically active electrons, is the electric dipole momentum, whereas $\boldsymbol{m} = \Sigma_i (e/2m)(\boldsymbol{l}_i + 2\boldsymbol{s}_i)$ denotes the magnetic dipole momentum. The strength $S(ab)$ of the transition vanishes, i.e. $S(ab) = 0$, if the states a and b differ in their total spin; in the absence of spin-orbit coupling we thus have the spin selection rule $\Delta S = 0$. When the spin-orbit coupling is present, the total spin is no longer a valid quantum number and the spin selection rule is therefore not rigorous. In general, however, spin-allowed transitions with $\Delta S = 0$ are much stronger than spin-forbidden transitions. It is useful sometimes to define the dimensionless oscillator strength

$$f(ab) = 2m\omega S(ab)/(g_a 3\hbar e^2), \tag{3.3}$$

where $\hbar\omega = |E_b - E_a|$ and g_a is the statistical weight of state a. For spin-allowed electric dipole transitions $f_{\text{ED}} \cong 1$, whereas for spin-allowed magnetic dipole transitions $f_{\text{MD}} \cong 10^{-6}$, these values decreasing with about two orders of magnitude for spin-forbidden transitions. The integral absorption coefficient $\alpha(\omega)$ obtained by averaging over all polarizations relative to the crystal axes of the host material, for the case when electric dipole processes take place is

$$\int \alpha(\omega)\mathrm{d}\omega = N\pi\omega[(n^2+2)^2/9n]S_{\text{ED}}(ab)/(\varepsilon_0 3\hbar c g_a), \tag{3.4}$$

where N is the number of absorbing centers per unit volume, and n the refractive index of the material. The term in square brackets accounts for the polarization of the material by the electric field. Einstein's spontaneous transition probability for a radiative transition between the same levels a and b is defined as

$$A_{\text{ED}}(\omega) = \omega^3[(n^2+2)^2/9n]S_{\text{ED}}(ab)/(\pi\varepsilon_0 3\hbar c^3 g_a), \tag{3.5}$$

and is equal to the inverse of the radiative decay time $A_{\text{ED}}(ab) = 1/\tau_{\text{R}}$ if $A_{\text{ED}}(ab)$ is the only radiative process from a. Formulae analogous to (3.4) and (3.5) can be defined also for magnetic dipolar transitions if $[(n^2+2)/3]^2$ is replaced by n^2.

Taking into account the different possible interactions of the optically active electrons (the electronic center) in an optically inert crystalline material, their total Hamiltonian can be written as

$$H = H_{\text{isol}} + H_{\text{el-stat}} + H_{\text{el-dyn}} + H_{\text{dyn}}. \tag{3.6}$$

H_{isol} describes the isolated electronic center disregarding the influence of the surrounding solid. $H_{\mathrm{el-stat}}$, the Hamiltonian of the electrostatic lattice, characterizes the effect of the average static environment, i.e. the crystal field (or ligand field) on the electronic center. There are different models to estimate the effects of the crystal field, depending on the relation between its strength and the strengths of the spin-orbit coupling of electrons in atom, and the multiplet splitting between atomic energy levels. Namely, when the crystal field is weak in comparison to spin-orbit coupling, it can be treated as a perturbation of the atomic states, which are characterized by the values of $\boldsymbol{J}^2$. This is the case of RE ions. When the crystal field is strong compared to spin-orbit coupling but weak compared to the multiplet splitting, as in the case of transition metals, the crystal potential is treated as a perturbation on states with definite L and S values. Finally, for crystal fields stronger than multiplet splitting, the atom states have no definite L and S values and the multiplet splitting is treated as a perturbation of the level structure determined including the crystal potential (Callaway (1991)). H_{dyn} in (3.6) is the Hamiltonian of the dynamic crystal lattice, the vibrational lattice energy being described in terms of lattice phonon modes. Finally, $H_{\mathrm{el-dyn}}$ characterizes the effects of lattice modes on the electronic center, effects which include a modification of the shape and strength of optical transitions of ions in crystalline solids, as well as the occurrence of nonradiative processes. The transfer of optical excitation energy between adjacent centers is also affected by $H_{\mathrm{el-dyn}}$ when the concentration of electronic centers is high enough so that adjacent centers interact.

The effect of impurities on crystals is studied with different mathematical methods depending on the short- or long-range changes in the crystal potential due to the impurity (Callaway, 1991).

The short-range case is encountered for example in metals, where any unbalanced charge is screened by the redistribution of electrons. Also, electrically neutral impurities in semiconductors, the effect of substituted atoms of different mass on lattice vibrations, the effect of magnetic impurities on spin excitations in ferromagnets, etc. can sometimes be included in this category. These problems are treated with the theory of electron scattering by a static potential U. The transition probability per unit time for an electron in band n with wavevector $\boldsymbol{k}$ to a state in band l with wavevector $\boldsymbol{q}$ is defined as

$$P_{nk\to lq} = (2\pi)^7 \, |\langle l\boldsymbol{q} \,|\, t \,|\, n\boldsymbol{k}\rangle|^2 \, \delta[E_l(\boldsymbol{q}) - E_n(\boldsymbol{k})], \tag{3.7}$$

where $t = U + U[1/(E^+ - H)]U$ with $H = H_0 + U$ the total Hamiltonian. H_0 is a time-independent part of the total Hamiltonian that usually contains a periodic potential, and $E^+ = E + \mathrm{i}\varepsilon$ with E the continuous spectrum of H_0 and ε a positive infinitesimal part introduced for computational reasons and which is set to zero after the calculations are performed. The scattering amplitude introduced by assuming the asymptotic form for the wavefunction

$$\psi^{+}(r) \to \exp(\mathrm{i}kz) + f\exp(\mathrm{i}kr)/r, \tag{3.8}$$

is proportional to the t matrix:

$$f_{lj}(\boldsymbol{q},\boldsymbol{k}) = -(2\pi^2/\gamma_l)\langle l\boldsymbol{q}|t|j\boldsymbol{k}\rangle, \tag{3.9}$$

with γ_l the reciprocal effective mass at the top of the lth band ($E_l(k) = E_m - \gamma_l k^2$ with E_m the highest energy in the band). The imaginary part of the forward scattering amplitude is proportional to the total cross-section – this is a result known as the optical theorem. More precisely, it says that

$$\mathrm{Im}\langle n\boldsymbol{k}|t|n\boldsymbol{k}\rangle = -\pi\textstyle\sum_l\int \mathrm{d}S_q(E_l)/|\nabla_q E_l|\,|\langle n\boldsymbol{k}|t|l\boldsymbol{q}\rangle|^2. \tag{3.10}$$

For a single parabolic band this reduces to the well-known result $\sigma_{\text{tot}} = (4\pi/k)\,\mathrm{Im}\, f(0)$. A more detailed discussion about the calculation of the forward scattering amplitude and the associated phase shift, as well as their effects on optical spectra for different potential strengths, can be found in Callaway (1991).

For long-range impurities, encountered mainly in semiconductors, the effective mass representation is used. Starting with the $\boldsymbol{k}\cdot\boldsymbol{p}$ Hamiltonian, with the assumption that $U(\boldsymbol{r})$ is sufficiently slowly varying such that terms with $\boldsymbol{K}_s \neq 0$ are not involved, we have the following equation

$$\begin{aligned}&[E_n + (\hbar^2/2m)(\boldsymbol{k}^2 - \boldsymbol{k}_0^2) - E]A_n(\boldsymbol{k}) + (\hbar/m)(\boldsymbol{k}-\boldsymbol{k}_0)\cdot\textstyle\sum_l \boldsymbol{p}_{nl}(\boldsymbol{k}_0)A_l(\boldsymbol{k})\\ &\quad + \int \mathrm{d}\boldsymbol{q}\,U(\boldsymbol{k}-\boldsymbol{q})A_n(\boldsymbol{q}) = 0,\end{aligned} \tag{3.11}$$

where $U(\boldsymbol{k}-\boldsymbol{q}) = [1/(2\pi)^3]\int \exp[\mathrm{i}(\boldsymbol{q}-\boldsymbol{k})\cdot\boldsymbol{r}]U(\boldsymbol{r})\mathrm{d}\boldsymbol{r}$. This equation can be simplified in the second order in $\delta\boldsymbol{k} = \boldsymbol{k}-\boldsymbol{k}_0$ to

$$[E_n(\boldsymbol{k}) - E]C_n(\boldsymbol{k}) + \int \mathrm{d}\boldsymbol{q}\,U(\boldsymbol{k}-\boldsymbol{q})C_n(\boldsymbol{q}) = 0, \tag{3.12}$$

with the change of variables $A_n(\boldsymbol{k}) = C_n(\boldsymbol{k}) - (\hbar\delta\boldsymbol{k}/m)\sum_l[\boldsymbol{p}_{nl}/(E_n - E_l)]C_l(\boldsymbol{k})$. In ordinary space the last equation reads as

$$[E_n(1/\mathrm{i}\nabla) - E]F_n(\boldsymbol{r}) + U(\boldsymbol{r})F_n(\boldsymbol{r}) = 0, \tag{3.13}$$

where $F_n(\boldsymbol{r}) = \int \exp(\mathrm{i}\delta\boldsymbol{k}\cdot\boldsymbol{r})C_n(\boldsymbol{k})\mathrm{d}\boldsymbol{k}$ and $E_n(1/\mathrm{i}\nabla)$ means that we have to replace $(1/\mathrm{i})\partial/\partial x_j$ for δk_j.

For a single electron bound to a donor ion, the potential $U(\boldsymbol{r})$ at large distance is that of a point charge, screened by the dielectric function ε of the crystal. For a parabolic band with effective mass m_{eff} and zero energy at the band minimum, the above equation reduces to a hydrogenic problem with solutions

$E_n = -m_{\text{eff}} e^4 / 2\varepsilon^2 n^2 \hbar^2$. The effective Bohr radius for the lowest orbit, $a_{\text{B}} = \varepsilon\hbar^2 / m_{\text{eff}} e^2$ is about 100 Å for GaAs, for example, much larger than the lattice constant. For a more realistic approach, the anisotropy of the conduction band must be included in the treatment, as well as the degeneracy of the impurity state in the effective mass approximation. In Si, for example, near the face centers X, there are six equivalent conduction band minima along the (100) axis. Methods to deal with these situations are discussed in Callaway (1991).

Generally, not only do the impurity ions influence the properties of the host, but the influence is reciprocal: the hosts influence the linewidths of the impurity lines. For example, the temperature dependence of the linewidth of spectral holes burned in ${}^7F_0 \rightarrow {}^5D_0$ transitions in Eu^{3+} in insulating Eu_2O_3 nanoparticles is different from that in bulk Eu_2O_3. The bulk power-law temperature dependence $\approx T^7$, well known experimentally and theoretically, is weakened to T^3 for nanoparticles. This dependence can be deduced from two-phonon Raman scattering involving discrete phonon modes of homogeneous nanoparticles with stress-free boundary conditions. The linewidths of phonon modes broaden with $\approx \omega^2$ and the size dependence can be described as $\approx D^{-2.5}$ (Meltzer and Hong, 2000).

Analogously, there is a quantitative and qualitative difference in magnitude and behavior of the homogeneous linewidth of impurities in glassy and crystalline solids. Due to two-phonon Raman processes which couple the impurity level to the acoustic phonons in the crystal, the linewidth in crystalline solids has a T^2 dependence for temperatures above the Debye temperature of the glass, and a T^7 power law below it, whereas in amorphous materials the broadening is orders of magnitude larger at the same temperatures and has a T^α dependence, where $\alpha = 1$–2.6 depending on the material (García and Fernandez, 1997).

The main difference between the properties of ions in crystalline and amorphous materials is due to the lack of symmetry and periodicity of the latter. As a consequence, the local fields at individual ion sites vary in glasses, and so do the energy levels and the radiative and nonradiative transition probabilities of ions. In particular, optical spectra exhibit inhomogeneous broadening and the decays do not have a single exponential time dependence. Following pulse excitation, the fluorescence at frequency ω is $S(\omega,t) = \sum_i N_i(t) A_i g_i(\omega)$ where A_i is the spontaneous emission probability for the transition at the ith site, g_i the line-shape function ($\int g(\omega)\mathrm{d}\omega = 1$), and $N_i(t)$ the number of excited ions at time t. The population of each subset of sites i after excitation at $t = 0$ decays as $N_i(t) = N_i(0)\exp(-t/\tau_i)$ where τ_i is the fluorescence lifetime of an individual site i determined by both radiative and nonradiative contributions. Because excited ions in glass decay at different rates, the time dependence of the fluorescence decay has a non-single-exponential behavior, the contributions of longer-lived ions becoming increasingly prominent at later times. Moreover, a spectral shift can appear if the ions are located in sites with different emission frequencies, time-resolved and spectral-resolved fluorescence acting as methods of site selectivity. Polarized excitation can also play this role since each site in glass has a set of principal axes, although the orientation in space of a given geometry site is random. Another important difference between the spectra of ions in crystalline and glass environments concerns the selection rules for optical transitions; these rules are usually highly relaxed in glasses due to their lack of symmetry (Weber, 1981).

The ions introduced in a host occupy several crystal sites, the environmental symmetry and/or interaction strength being usually different. This can sometimes cause interferences among inequivalent ion sites, complicating the interpretation of experimental results. For example, when measuring the Stark shift induced by the static electric field in doped crystals with the two-pulse photon-echo method, the interferences between polarizations at different sites modulate the echo intensity as $I_e(\tau_s) = |(1/N_s)\sum_{k=1}^{N_s} A_k \exp(i\Omega_k \tau_s)|^2$. Here A_k are modulation amplitudes at different sites, whose number is N_s, τ_s is the width of the static electric field pulse $\boldsymbol{E}_s$ applied during the dephasing period (between the first and second optical pulses) and the Stark shifts are $\Omega_k = (\delta\boldsymbol{\mu}_k \cdot \boldsymbol{E}_s)/\hbar$ with $\delta\boldsymbol{\mu}_k$ the difference between the static dipole moments in the ground and excited states. The interferences in the modulated echo intensity are different from quantum beats in that they do not originate from a single quantum system, but from two or more sets of different oscillations with different frequencies. The Stark shifts are obtained after performing a Fourier transform on the echo intensity. If the crystal group has inversion symmetry, N_s is even, $N_s = 2N_i$, half of the sites being related by inversion to the other half so that $I_e(\tau_s) = |(1/N_i)\sum_{k=1}^{N_i} A_k \cos(\Omega_k \tau_s)|^2$ since $\Omega_{N_i+k} = -\Omega_k$. Experiments in $YAlO_3$ have revealed the existence of two inequivalent sites with two corresponding Stark split frequencies, the frequencies and the amplitudes A_k being determined from fitting (Graf et al., 1997).

If the dopant ion enters inequivalent sites in the crystal, such that luminescent transitions from different centers overlap, it is possible to see if there is more than one distinct center by measuring the decay times of various luminescence features. Distinct lifetimes suggest transitions from different types of luminescent centers. Alternatively, different temperature and/or pressure dependencies of luminescence lines suggest the existence of different centers. If PL spectra from different inequivalent centers overlap, but have different decay rates, the number and intensity signals from each center can be determined by phase-sensitive detection techniques. This can be done by pumping a square-wave modulated light and using a lock-in detector to record the sine-wave component of luminescence. Since PL signals from different inequivalent centers have different decay rates and their sine-wave components are out of phase with each other, the number of inequivalent centers is determined by beating the signal with a reference sine wave. By adjusting the phase of the reference signal to be in phase with luminescence emitted by a certain center, the output of the beating signal can be maximized for that center. Different PL components can be thus individualized and separated. This technique has been applied to study the PL from Cr^{3+} in MgO (Imbusch, 1979a).

The symmetry of different sites can also be determined optically. If the site has uniaxial symmetry (one direction is distinguished from the others), this is reflected in the luminescence spectrum. Even the dipole nature of the optical processes can be determined in this case from the PL (Imbusch, 1979a).

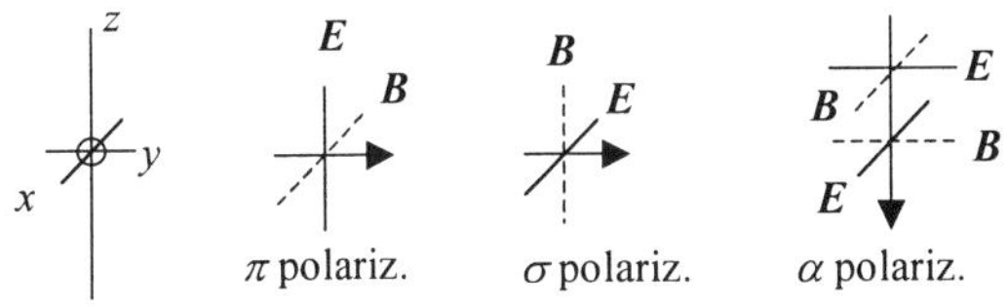

Fig. 3.2. Geometry for different polarizations in uniaxial symmetry sites

For the example illustrated in Fig. 3.2 the σ and α spectra should be identical, but different from π polarization for electric dipole processes, since in this case the dipole operator is zE for π polarization, whereas $\boldsymbol{E}$ is perpendicular to z in σ or α polarizations. The transition strength is different in the two situations due to the uniaxial symmetry of the site, which individualizes the z direction. On the contrary, for magnetic dipole processes π and α spectra should be identical, but different from the σ spectrum. This identification of the dipole nature can only be made if the site has one preferential axis. If this is not the case, such as for isotropic lines, the preferential axis can be imposed externally by applying a perturbation (pressure, electric or magnetic field). It is interesting to note that the dipole nature of the zero-phonon line and the phonon sidebands can be different. For example, for the ${}^6A_1 \rightarrow {}^4T_1$ transition of the Mn ion in $RbMnF_3$, the zero-phonon line has a magnetic-dipole character, while the one-magnon sidebands are electric-dipole in nature (see references in Imbusch (1979a)). If there are several equivalent sites in the crystal with different directions of their preferential axis, the overall crystal spectrum is isotropic, although the ion occupies a uniaxial site. This situation is encountered for Cr ions in $MgAl_2O_4$ of trigonal symmetry (lower than octahedral symmetry). Then, we can preferentially excite one type of site, if we polarize the pumping light along one of the trigonal directions, then we can apply the π, σ, α polarization test.

Analogously, one can use circular dichroism to study the symmetry of the ion and the dipole nature of the transitions. Another method to study the symmetry of the site is based on the dependence of PL on the incident polarization. For example, Cr^{3+} in MgO has three sites: (i) octahedral, (ii) tetragonal center, with a vacancy in the next Mg^{2+} site along the [001] direction, (iii) diagonal center, with vacancy in the nearest Mg^{2+} site along [110]. The PL from each tetragonal site for example is anisotropic, but is isotropic for the whole crystal. By applying a large enough uniaxial stress along [001], the energy levels of the center with [001] axis of symmetry are different from the others and thus correspond to different transition frequencies; splitting of the lines occurs under stress along [001]. On the contrary, by applying the stress along [111] all tetragonal sites are equally affected and no split of the transitions is observed. Analogously, the spectrum from each diagonal center is anisotropic, but isotropic for the whole crystal (Imbusch, 1979a). By optically pumping the sample with light polarized along a specific direction, the equilibrium population in the excited states of centers with axes pointing along that direction is different from other centers, the uniaxial direction of the centers being indicated by the anisotropy of the PL. For example, under pumping with plane-polarized light incident along $[\bar{1}00]$, the PL intensity I from diagonal centers varies with the angle θ of incident polarization as $I(\theta) = a + b\sin(2\theta)$ with a, b constants, whereas under similar conditions the luminescence intensity due to tetragonal centers is independent of θ.

The neutral-donor nature of isolated neutral donors made up of defect pair complexes can be determined from magnetic-field measurements and from two-electron transitions. If the PL lines associated with the isolated neutral donors are strongly polarized, it can be inferred that during growth defect complexes are preferentially incorporated in certain crystallographic orientations. The crystal is strained in the vicinity of the defect pairs, the strain being oriented in the direction of the pair (Reynolds et al., 1998). The PL lines result from the collapse of excitons bound to the neutral donor-complex.

A substitutional impurity occupies the lattice sites of atoms or ions of the host lattice when the size of the host ion is comparable with the effective size of the impurity. If the impurity ion is smaller than the host ion, it no longer occupies the center of symmetry of the cluster, i.e. it is situated 'off-center'. Whether or not a substituting ion is located 'off-center' can be determined experimentally from the dependence of the oscillator strength of the transitions on temperature, hydrostatic pressure, or on an applied electric field. The displacement of the impurity can be evaluated by comparing the total energy of the impurity-doped crystal, which includes electronic, repulsive and electronic polarization contributions, with that of a perfect crystal.

Localized impurities destroy the translational invariance of the crystal, perturbing the normal lattice modes in a way specific to the impurity type, as well as inducing the polarization of normal modes, independent of the nature of impurities. This latter effect can make active normal lattice modes, which are otherwise inactive, as for example the normal modes of the homopolar crystals of the diamond type. For these crystals, formed from two face-centered cubic lattices displaced by a quarter of the volume diagonal along the volume diagonal, there are no electric moments due to the special crystal symmetry. Therefore, the normal vibration modes are not optically active in the infrared spectrum.

In a perfect lattice the acoustic and optical frequency spectra are separated by a forbidden frequency band. By introducing localized defects, allowed frequencies can appear in the forbidden band, as well as above the maximum frequency of the unperturbed system. These narrow-width frequencies are localized, in the sense that the corresponding eigenvector does not have a wave-like dependence in space, but is strongly peaked at the impurity atom and falls off rapidly a few lattice sites away. Localized modes can exist also within the band of normal frequencies, or within the energy gap of the crystal, in which case they are called gap modes. Moreover, impurities can smooth the density of states in the crystal, removing sharp features in the optical response. Numerical simulations in the 1D diatomic chain which include interaction only between nearest neighbors, show that the defect modes become more localized as the mass of the substitutional impurity is smaller, the odd symmetry modes appearing in the IR spectrum, whereas the even-symmetry modes become Raman active. The localization of the defect mode is less strong for modes close to the band edge (Barker Jr. and Sievers, 1975).

3.3 Franck–Condon Principle for Defects with Large Orbits

If the orbits of electronic particles trapped by defects are large compared to the interionic distance, the particles interact with many ions or, in terms of phonons, with many LO lattice vibrations with different $\boldsymbol{q}$ vectors. In this case the Franck–Condon (FC) principle can be extended as follows (DiBartolo, 1979).

According to FC, at light absorption, the transition occurs in two steps: in the first step the electrons undergo transitions to excited states in a short time compared to the period of lattice oscillations for LO modes, so that the lattice has no time to relax and the polarization remains adjusted to the initial state. Then, the ions of the lattice begin to move to new equilibrium positions, corresponding to the polarization adapted to the new charge distribution. During this relaxation process the wavefunctions of electronic particles follow adiabatically the evolution of the lattice until the system reaches a relaxed excited state with energy E_{R} , lower than the energy of the unrelaxed state. The excess energy is dissipated by phonon emission, the absorption occurring not only at the frequency $\omega_{\mathrm{r}} = (E_{\mathrm{R}} - E_0)/\hbar$ but also at $\omega_{\mathrm{r}} = (E_{\mathrm{R}} - E_0 + n\hbar\omega_{\mathrm{LO}})/\hbar$ where ω_{LO} is the frequency of the LO phonons. By extension of the FC, the oscillator strength of these phonon sidebands is proportional to the square of the transition matrix element

$$M = \sum_i z_i \langle l_{\mathrm{R}} \mid l_0 \rangle \int \mathrm{d}^3 r_1 ... \mathrm{d}^3 r_n \boldsymbol{E}\boldsymbol{r}_i \phi_{\mathrm{u}}^* (\boldsymbol{r}_1 .. \boldsymbol{r}_n) \phi_0 (\boldsymbol{r}_1 ,.., \boldsymbol{r}_n), \tag{3.14}$$

where $\boldsymbol{E}$ is the electric field of the incident radiation, $l_{0,\mathrm{R}}$ the lattice wavefunctions in the initial and final relaxed states, and $\phi_{0,\mathrm{u}}$ the wavefunctions of the initial and unrelaxed excited electron states. So, the relative intensities of different phonon replica are governed by the behavior of $\langle l_{\mathrm{R}} \mid l_0 \rangle$ as a function of the number of emitted phonons in the final state l_{R}. The final states of the relaxed lattice constitute a complete orthonormal set of lattice wavefunctions.

In the opposite case, the lattice distortion follows adiabatically the orbital motion of electrons, Fröhlich polarons being formed, which move in the field of defects. Since the transitions between different orbital states do not affect the distortion cloud in the polaron, no strong phonon sidebands are expected. In particular, for isolated acceptors or donors the energy levels are $E_n = -\alpha\hbar\omega - R/n^2$, the first term in E_n being the self-energy of the charge carrier in the field of ionic lattice polarization. Here R is the Rydberg constant for polarons and $\alpha = e^2(1/\varepsilon_\infty - 1/\varepsilon_0)\sqrt{m}/\sqrt{2\omega}\hbar^{3/2}$ is the Fröhlich coupling constant. In GaP for example, by appropriate doping and selective excitation of the donor-acceptor (DA) particle pair the transition from adiabatic to non-adiabatic lattice motion should be observed.

3.4 Jahn–Teller Effect

As we have already mentioned, lower symmetry environments, e.g. hexagonal, trigonal or tetragonal symmetries, induce splittings of degenerate energy levels. It is to be expected that in this case the system occupies a lower energy state that in the original degenerate configuration. If u is a parameter that measures the distortion of the degenerate system caused by the lower symmetry environment, the total energy should decrease with an amount proportional to u, let's say with au. On the other hand, the distortion tends to be compensated by elastic restoring forces that increase the total energy with a term quadratic in u, bu^2, so that the energy change due to distortions is $E(u) = -au + bu^2$. The new position equilibrium of the system, i.e. the position of minimum energy with respect to small distortions u, is attained for a finite distortion $u_0 = a/2b$, the minimum energy value being $E_{JT} = -a^2/4b$. This static effect, which occurs for electrons sufficiently weakly bound to inner atomic shells in order to sense the crystal environment, is called the Jahn–Teller effect. The dynamic Jahn–Teller effect occurs when E_{JT} is smaller than the zero-point energy of phonon modes, in which case the vibrational motion of the electronic center is not localized on a single stable configuration, so that the vibrations carry the system between different distorted configurations (Callaway, 1991).

3.5 Interactions Between Ions

With increasing ion concentration, the interactions between ions become increasingly important. The strength of these interactions is expressed in terms of the exchange and/or superexchange integrals. The electrostatic exchange operator, for example, can be represented as the scalar product of spin variables $J_{12}\boldsymbol{S}_1 \cdot \boldsymbol{S}_2$. The pair interaction in the fundamental level of RE and iron-group ions is generally studied by EPR and it was evidenced by optical spectroscopy in the excited levels, with very different characteristics for the RE and iron-group ions. Since the pair interaction for RE ions is small, it is very difficult to distinguish in optical spectra lines corresponding to singly excited pairs even for nearest neighbors, double-excited pairs, however, being more easily observed. On the contrary, for iron-group ions, the pair interaction is sufficiently large to allow a direct study of energy levels for both singly and doubly excited pairs (see Imbusch (1979b) and the references therein).

One manifestation of interaction between dopant ions is energy transfer between them. Energy transfer was mostly studied in dielectric materials doped with RE ions (Morgan and Yen, 1989). Excitation transfer in disordered systems has been well reviewed by Holstein et al. (1981). The energy transfer process involves the initial photoexcitation of a set of ions called donors (D), followed by energy transfer to other, acceptor (A) ions. It can occur through several processes which include: i) resonant and ii) nonresonant transfer, iii) cross-relaxation and iv) up-conversion processes (see Fig. 3.3a, b, c, d, respectively).

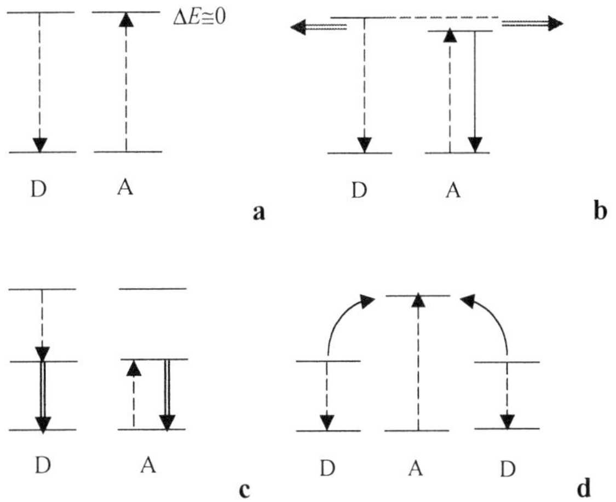

Fig. 3.3. Types of energy transfer process: (**a**) resonant D-A transfer, (**b**) nonresonant transfer, (**c**) cross-relaxation, (**d**) up-conversion (*solid vertical line* – radiative transitions, *double line* – nonradiative transitions, *dotted line* – D-A energy transfer processes)

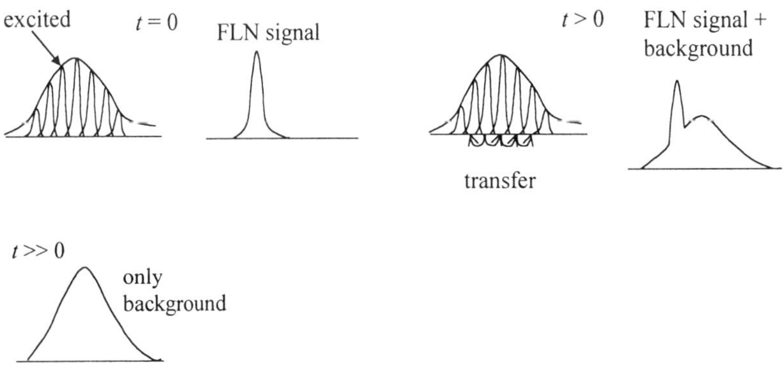

Fig. 3.4. Temporal development of the donor or acceptor fluorescence following the pulsed excitation of the donor system

The energy transfer can be studied by spectroscopic techniques, in particular by the temporal development of the donor or acceptor fluorescence following the pulsed excitation of the donor system. The donor system emits decaying fluorescence and the acceptor system both rising and decaying fluorescence components, due to transfer dynamics and intrinsic decay. If the pulsed laser source has a frequency bandwidth much less than the inhomogeneous transition profiles of the optically active ions, it becomes possible to excite and study a subset of ions and to monitor the transfer of optical excitation to other ions within the inhomogeneous feature (see Fig. 3.4) by time-resolved fluorescence line narrowing (TRFLN) (Huber et al., 1977). This spectral diffusion process requires an energy-shifting mechanism among ions placed in slightly different environments, with slightly different local crystalline fields. TRFLN cannot provide, however, information about the spatial transfer process, or about coherent resonant energy dynamics. Such information can only be gained through FWM and hole-burning techniques.

In TRFLN the temporal evolution of the narrow fluorescence I_{N} normalized to the total fluorescence (plus background) $I_{\mathrm{N}} + I_{\mathrm{B}}$ is measured. This ratio $R(t) = I_{\mathrm{N}}(t)/[I_{\mathrm{N}}(t) + I_{\mathrm{B}}(t)]$ represents the possibility that an initially excited ion remains excited at time t in the absence of radiative decay. $R(t)$ represents an ensemble average of ions with various configurations of neighboring dopant ions and is typically a nonexponential function of time. The (microscopic) dynamics of the system is described by ion-ion energy transfer probabilities in increasing orders of perturbations, with different assumptions. If intraline transfer processes are dominated by a two-site nonresonant interaction (one phonon, second-order process), as for example in $\mathrm{LaF_3{:}Pr^{3+}}$, the transfer rate is predicted to be independent of ion-ion energy mismatch and has a T^3 dependence on temperature (Morgan and Yen, 1989). Any asymmetry in transfer rates from a FLN signal on the high-energy side compared to that on the low-energy side of the inhomogeneous profile is caused by thermal population effects of phonons, which mediate the transfer process. On the other hand, for a one-site resonant process with an electric dipole-electric dipole ion pair interaction, as for example at ${}^4\mathrm{F}_{3/2} \to {}^4\mathrm{I}_{9/2}$ transition in $\mathrm{Nd}_x\mathrm{La}_{1-x}\mathrm{P}_5\mathrm{O}_{14}$, for $x = 0.2$ and $17\ \mathrm{K} < T < 24\ \mathrm{K}$, the transfer rate has an exponential temperature dependence $\exp(-\Delta / k_{\mathrm{B}}T)$, characteristic of energy activation processes. In still another example, the nonresonant donor-donor transfer rate from the R_1 line of samples with Cr^{3+} concentrations in the range 0.025–0.9 at.% at low temperatures (5–50 K) increases linearly with T, indicating a direct one-phonon process, due to the dominant effects of weakly exchange-coupled ion pairs which were folded into the main R_1 line. A more detailed discussion of these energy transfer mechanisms, as well as the corresponding references can be found in Morgan and Yen (1989).

To obtain the macroscopic quantities such as fluorescence lifetime and R in terms of microscopic interactions, various averaging procedures over microscopic parameters must be performed. Generally, a set of coupled rate equations $\mathrm{d}P_n(t)/\mathrm{d}t = -(\gamma_{\mathrm{r}} + X_n + \sum_{n' \neq n} W_{nn'})P_n(t) + \sum_{n' \neq n} W_{n'n}P_{n'}(t)$ is employed, where

$P_n(t)$ is the probability that an optically active ion at site n is in an excited state at time t, γ_r is the reciprocal of the excited state lifetime, $W_{nn'}$ the energy transfer rate from an excited ion at lattice site n to an unexcited ion at n' and $W_{n'n}$ the back transfer rate to n from n', whereas X_n is the total rate at which energy is removed from the level of interest to other species which do not allow any back transfer to occur. For $X_n = 0$, the initial decay of R is almost exponential: $\mathrm{d}R(t)/\mathrm{d}t \cong \mathrm{d}P_0(t)/\mathrm{d}t = -c\sum_n W_{0n}$, where W_{0n} is the transfer rate from an initially excited ion at site 0 to an ion at site n and c is the dopant concentration. Approximate expressions for the temporal evolution of $R(t)$ as a function of dopant concentration can be found in Morgan and Yen (1989).

Energy-transfer processes can be near resonant, or intraline, when occurring inside the same line (as considered in the discussion above) and nonresonant, or interline, if they take place between ions/particles in two different lines. Resonant processes are also possible in which the energy transfer occurs in only a small frequency interval around the excitation frequency, the FLN being no longer (or only slightly) broadened at large times (the background is practically absent and the FLN has approximately the same width and shape at all times). The probability of resonant energy transfer increases with the dopant concentration, and thus with the coupling strength between ions. Coherent coupling of resonant ions in stoichiometric crystals, for example, leads to Frenkel excitons. As the dopant ion concentration increases, the migration of optical excitation progresses from no transfer, to temperature-dependent incoherent hopping, to coherent, and then excitonic propagation. At very low temperatures, the transition from no propagation to rapid coherent propagation (Lyo, 1971) occurs abruptly at a critical dopant concentration.

Interline spectral transfer occurs usually between two distinct optically active centers. When donors and acceptors are composed of different ionic species, interline spectral transfer manifests itself as trapping of donor fluorescence or sensitization of acceptor emission. Sensitized luminescence is the phenomenon of observing luminescence from a type of ions A, if another type S is excited by illumination. The two types of ions with which the crystal is codoped are called activator and sensitizer, respectively. This phenomenon is usually observed when the A type of ions cannot be directly excited.

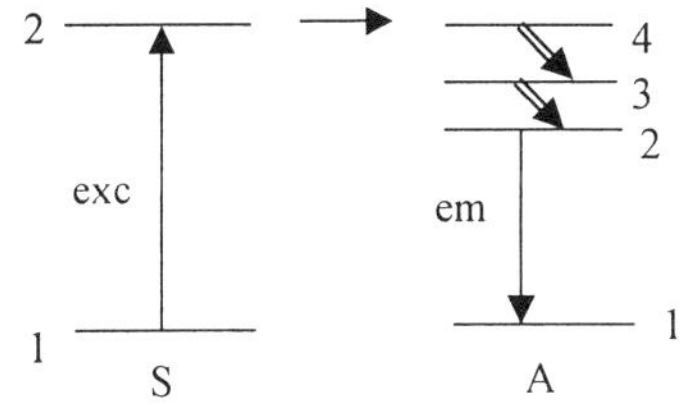

Fig. 3.5. Interline spectral transfer (*solid line* – radiative transitions, *double line* – nonradiative transitions; exc – excitation, em – emission)

An energy transfer usually occurs from S to an excited level of ion A (level 4 in Fig. 3.5), the sensitized luminescence being emitted from a lower level (level 2), which is reached by nonradiative decay. The energy transfer takes place if the upper energy level of A is in resonance with the excited level of S, the interaction between the two ions being of a Coulomb type if they are separated by a sufficiently large distance such that their charge clouds do not overlap, or is an exchange interaction. The transfer of excitation energy through the lattice from one host ion to another can also be modeled as exciton diffusion, the exciton migration being described as nearest-neighbor random hopping. During energy migration 'killer sites' can be reached where the energy is lost nonradiatively, leading to luminescence quenching.

Energy transfer between different levels of similar ions, also an interline spectral transfer mechanism, leads to cross-relaxation or concentration quenching. The time evolution of fluorescence is also characterized by a set of coupled rate equations (see Morgan and Yen (1989) for details).

The resonant energy transfer that involves migration of optical energy among ions with excited states within a homogeneous and possibly line-narrowed profile, can be probed with TRFLN especially when the process is incoherent. The FLN signal would then broaden (as the energy is transferred to ions with partially overlapping spectral profiles) and would eventually fill the inhomogeneous broadened feature. Two-photon techniques can directly probe the resonant energy transfer in absence of spectral transfer (Strauss et al., 1981). In this method one low-power narrowband laser beam line-narrows the $^3P_0 \to (^3H_4)_1$ fluorescence profile of LaF_3:Pr^{3+}, for example, the second narrow but more powerful laser beam, tuned to the $^3P_0 \to (^3H_6)_1$ transition, burning a hole in the broad emission. Refilling the hole is a clear manifestation of resonant energy transfer.

3.6 Spatial Migration

As we have already mentioned, TRFLN does not probe any spatial features of the dynamic optical system, the spatial migration of energy being demonstrated through other techniques, as for example the transient grating technique based on FWM. The transient grating with period $\Lambda = \lambda / 2\sin(\theta/2)$ – a spatially modulated excited state population – is formed when two coherent laser beams with the same frequency (in resonance with an optical transition in the material) cross in the material at angle θ. A third beam can then be Bragg scattered from the grating, its intensity $I_S \approx I_P(\Delta N)^2$ depending on the grating depth $\Delta N = N_{peak} - N_{valley}$, where N is the excited state population and I_p the probe beam intensity. If one of the writing beams is amplitude modulated, the scattered beam can probe the temporal behavior of the grating and in particular the spatial energy migration, which tends to increase the rate of decay after turning off the writing beam, as the grating contrast ΔN decreases with time due to energy diffusion from peaks to valleys. The energy diffusion constant can be measured in this way (Morgan and Yen, 1989).

3.7 Optical Properties of Rare-Earth Ions

In RE ions the f electrons are shielded by the outer s and p electrons, so that $H_{\text{el-stat}}$ is weak and acts as a perturbation to H_{isol}. In the crystal field theory the ions of the host matrix are treated as point charges and the interaction between RE ions and the ligand is assumed to be purely electrostatic, the bond covalency, the spatial extent of electron clouds and the orbital overlap being neglected. The crystal field Hamiltonian is then $H_{\text{CF}} = \sum_{km} B_{km}(r)C_{km}(\theta,\phi)$ where the spherical tensor $C_{km}(\theta,\phi)$ is related to $Y_{km}(\theta,\phi)$ and can be analytically calculated, whereas B_{km} are empirical parameters calculated from fitting the experimental energy levels to the calculated ones (Imbusch and Kopelman, 1981). Theoretically $B_{km} = A_{km}\langle r^k\rangle(1-\sigma_k)$, where the averages $\langle r^k\rangle$ are related to the radial wavefunctions, and σ_k are corrections for the shielding effects of outerlying $5s^2$, $5p^6$ electrons. $A_{km} = -(e^2/4\pi\varepsilon_0)\sum_j Z_j C_{km}(\theta_j,\phi_j)/R_j^{k+1}$ where the sum is taken over all ligands in glass, or ions in the host lattice, and R_j is the radial distance between the RE ion and the point charge Z_j. For the C_{2v} symmetry, the highest symmetry that allows the complete splitting of all crystal-field components, under the assumption that J mixing is negligible, the overall scalar crystal field strength is defined as $S_{\text{CF}} = \{(1/3)\sum_k[1/(2k+1)][B_{k0}^2 + 2\sum_{m>0}(\text{Re}\,B_{km}^2 + \text{Im}\,B_{km}^2)]\}^{1/2}$.

Racah (1942, 1943) has shown that the calculation of energy levels and the effect of interactions within the f^n configuration is simplified for many-electron configurations, if in addition to *LSJ*, the states are labeled by additional quantum numbers describing the transformation between two different coupling schemes of any three angular momenta. Generalized Racah coefficients can be defined in a similar way as the transformation functions between two different coupling orders in pairs of any four angular momenta (Arima et al. 1954).

In RE-doped crystals, both broad and relatively featureless inter-configurational transitions $4f^n \to 4f^{n-1}5d$, and sharp intra-configurational transitions $4f^n \to 4f^n$ within a partially filled 4f shell can occur, the latter being parity allowed for two-photon transitions. The sharpness of $4f^n \to 4f^n$ transitions is caused by the screening from broadening influences of the crystalline environment of both initial and final states by the $5s^2 5p^6$ shield. These transitions are, however, quite weak, because the initial and final states have the same parity; they are thus electric dipole forbidden, and magnetic dipole and electric quadrupole parity-allowed. They should therefore obey the selection rules $\Delta L, \Delta J \le 1$ and $\Delta L, \Delta J \le 2$, respectively, but in practice values as large as six are observed for $\Delta L, \Delta J$. This fact was explained (van Vleck, 1937) by assuming that in crystals where RE ions occupy noncentrosymmetric sites, odd parity components of the crystal field mix states from opposite parity configurations into

$4f^n$ wavefunctions. The electric dipole transitions become thus allowed in second order, and transitions with $\Delta L, \Delta J \leq 6$ are possible.

3.7.1 Judd–Ofelt Theory of One- and Two-Photon Transitions

The energy levels of RE ions are calculated using the Judd–Ofelt formalism (Judd, 1962; Ofelt, 1962). This theory is based on the fact that in first order the perturbative effect of odd-parity components of the crystal field upon the $4f^n$ ground $\langle g|$ and final $|f\rangle$ state wavefunctions of an RE ion in noncentrosymmetric crystalline sites is described by $\langle g'| = \langle g| + \sum_i E_{ig}^{-1} \langle g | H_{CF}^{\mathrm{odd}} | i \rangle\langle i|$, $|f'\rangle = |f\rangle + \sum_i E_{if}^{-1} |i\rangle\langle i | H_{CF}^{\mathrm{odd}} | f\rangle$ where the sum extends over all levels of the excited configurations with different parity than $4f^n$ and E_{ig} is the energy difference between states *i* and *g*. In the intermediate coupling case, *g* and *f* can be written as $\langle f^n \psi J M|$ and $|f^n \psi' J' M'\rangle$ respectively where ψ, ψ' are other quantum numbers than the total angular momentum *J* and its azimuthal component *M*. The integrated intensity (line strength) of a single photon transition from $\langle g'|$ to $|f'\rangle$ is proportional to

$$\begin{aligned}\sum_{M,M'} |\langle g' | \boldsymbol{E} \cdot \boldsymbol{D} | f' \rangle|^2 = \sum_{M,M'} | -\sum_i [E_{if}^{-1} \langle g | \boldsymbol{E} \cdot \boldsymbol{D} | i \rangle\langle i | H_{CF}^{\mathrm{odd}} | f \rangle \\ + E_{ig}^{-1} \langle g | H_{CF}^{\mathrm{odd}} | i \rangle\langle i | \boldsymbol{E} \cdot \boldsymbol{D} | f \rangle] |^2,\end{aligned} \tag{3.15}$$

in the electric dipole approximation, where $\boldsymbol{E}$ is the electric field vector and $\boldsymbol{D}$ is the sum $\sum_j \boldsymbol{r}_j$ of position vectors for all electrons in the f^n configuration. The eventual contributions of the magnetic dipole and electric quadrupole interactions can also be included in the formalism. The sum in (3.15) cannot be calculated for RE ions due to the enormous number of levels in intermediate configurations, unless the simplification of a slight variation of E_{ig}, E_{if} over *i* is made. This approximation is valid at least in trivalent RE ions where the excited configurations lie much higher in energy than the $4f^n$ initial and final state levels, for divalent RE ions the approximation being less accurate. The energy denominators can then be considered constant and the *closure approximation* $\sum_n |n\rangle\langle n| = 1$ can be used, the single photon line strength obtained after summing over all components *M*, *M'* of the ground and final states and over all polarizations being

$$\boldsymbol{S} = \sum_{\lambda=2,4,6} T_\lambda |\langle f^n \psi J \| \boldsymbol{U}^{(\lambda)} \| f^n \psi' J' \rangle|^2, \tag{3.16}$$

where $\boldsymbol{U}^{(\lambda)}$ is a sum of single particle unit tensor operators $\sum_n \boldsymbol{u}_n^{(\lambda)}$ which satisfy $\langle f \| \boldsymbol{u}^{(\lambda)} \| f \rangle = 1$. The reduced matrix elements $\langle f^n \psi J \| \boldsymbol{U}^{(\lambda)} \| f^n \psi' J' \rangle$ can, in principle, be rigorously calculated since they involve only angular parts of the initial and final state wavefunctions. On the other hand the coefficients T_λ are difficult to calculate from first principles and are therefore regarded as phenomenological parameters, being determined from fitting experimental oscillator strengths. Other mechanisms such as coupling of 4f electrons with transient dipoles induced in ligands by the radiation field can be incorporated for an extension of the standard Judd–Ofelt formalism. Also, third-order contributions due to the static crystal-field and spin-orbit interaction between levels of the perturbing configuration $4f^{n-1}n'l'$ can be included (Downer, 1989), but the second-order contributions in the standard Judd–Ofelt theory are able to explain the vast majority of spectra.

The Judd–Ofelt method can also describe two-photon transitions if a more sophisticated treatment of intermediate states is used (Downer, 1989). Such intermediate states as $4f^{n-1}d$ and $4f^{n-1}g$, opposite in parity to the initial and final $4f^n$ states, belong now to high energy excited configurations that extend radially far into the crystalline environment, in contrast to the localized $4f^n$ states. The second-order contributions to the probability of direct two-photon transitions are

$$|-\sum_{i,j}[\Delta_{i1}^{-1}\langle g|\boldsymbol{E}_1\cdot\boldsymbol{D}|i\rangle\langle i|\boldsymbol{E}_2\cdot\boldsymbol{D}|f\rangle+\Delta_{j2}^{-1}\langle g|\boldsymbol{E}_2\cdot\boldsymbol{D}|j\rangle\langle j|\boldsymbol{E}_1\cdot\boldsymbol{D}|f\rangle]|^2, \tag{3.17}$$

where $\Delta_{i1,2}=E_i-h\nu_{1,2}$ is the energy of the intermediate state above the single-photon energy. An explicit reference to the crystalline environment is absent in the second-order description of two-photon processes in RE, the second dipole operator playing the role of the noncentrosymmetric part of the crystal field in the analogous expression for one-photon transitions. If by convention positive ν_i frequencies describe the absorption of a photon, and negative frequencies the emission, (3.17) can be applied for either two-photon absorption (TPA) if both photon frequencies are positive, or Raman scattering when one frequency is positive, and the other negative. After applying the closure approximation and performing the sums

$$S_{\mathrm{TPA}}=A_2[\langle\boldsymbol{E}^{(1)}\boldsymbol{E}^{(1)}\rangle^{(2)}]^2\,|\langle f^N\psi J\,\|\,\boldsymbol{U}^{(2)}\,\|\,f^N\psi J'\rangle|^2, \tag{3.18}$$

where A_2 is a coefficient which includes contributions from d and g orbitals and $\boldsymbol{E}^{(1)}$ is the electric field expressed in spherical tensor form. Only a single term, instead of the three in one-photon-absorption, appears in (3.18). No phenomenological parameters are needed to determine, for example, the ratio of line strengths of two transitions, or the polarization dependence of line strengths. The average $[\langle\boldsymbol{E}^{(1)}\boldsymbol{E}^{(1)}\rangle^{(2)}]^2$ equals $(2/3)E^4$ for two equivalent linearly polarized photons, and is equal to E^4 for two equivalent circularly polarized photons, where E is the electric field amplitude. Additional higher-order perturbative terms involving radiative or nonradiative Hamiltonians, in particular spin-orbit H_{so}, crystal field H_{CF}, central field and interelectronic Coulomb potentials, must sometimes be included to explain all experimental data, in particular anomalously strong intensity and polarization anisotropy. The inclusion of these additional terms also modifies the selection rules. Typical values of crystal-field splitting in RE are 15 000 cm^{-1} for $4f^{n-1}5d$ levels and 100 cm^{-1} for

$4f^n$ levels; a typical value of the spin-orbit splitting is 1000 cm^{-1}. The higher value of the crystal-field splitting for $4f^{n-1}5d$ levels is due to the extended nature of intermediate state radial wavefunctions.

The applications of TPA include the study of formation of precipitated phases of RE in and their diffusion rate through the lattice, and the measurement of narrow homogeneous linewidths of RE ions within a broad inhomogeneous profile (see references in Downer (1989)). TPA in RE-doped materials allows access to higher-energy regions than in linear absorption, whereas Raman spectroscopy allows the study of levels very near to the ground state with visible light sources (since photon energies are subtracted). Another advantage of TPA compared to linear absorption is an easier identification of the symmetry of initial and final states because of the two independently variable polarizations. Moreover, energy levels of the same parity as the ground state can be explored, the transitions between them being forbidden in single-photon electric dipole processes.

RE ions can be divided into two classes: lanthanides and actinides, the active electrons belonging respectively to the 4f and 4f levels. Since the absorption and emission lines exhibited by lanthanide ions in crystals are very narrow, they can act as probes of the solid-state environment. The perturbing crystalline environment slightly depresses the $4f^n$ free-ion levels of lanthanides. The effects of the crystal field and other interactions on lanthanides and trivalent actinides are similar, so that they can be treated mathematically in an analogous manner. In particular, the environmental perturbation on f electrons can be described through a crystal potential with almost the same symmetry as the crystallographic site.

The difference between lanthanide and actinide configurations makes the crystal field strength in trivalent actinides approximately twice as large as in the corresponding trivalent lanthanides in the same host matrix, since the 5f shell is less efficiently shielded from its interaction with the surrounding ions than the 4f shell. Also, the spin-orbit coupling constant is generally larger in trivalent actinides by a factor of about two for $n \leq 7$, indicating marked deviations from the LS-coupling scheme, valid for lanthanides. The ion-lattice coupling is also stronger in actinides, reducing the emission efficiency due to faster nonradiative relaxation rates. The reduced average energy separation between crystal levels belonging to contiguous free ion *J*-levels in actinides also favors the nonradiative relaxation of the excitation energy in the 5f configuration, at the expense of radiative processes (Pappalardo, 1979).

3.7.2 Optical Spectra of Rare-Earth Ions

Optical experiments on materials doped with RE ions have focused on the study of the environment of these ions. The Judd–Ofelt parameters Ω_2, Ω_4, Ω_6 are obtained from best-fit values of line strengths in absorption spectra if $\| U^t \|^2$ is considered independent of the ion environment. Ω_2 is higher for samples with more covalent bonding character, and reflects the symmetry of the local environment of the RE ion. A smaller value of this parameter describes a more centrosymmetric coordination environment. Studies of the effect of host lead borate glasses on these parameters (Saisudha and Ramakrishna, 1996) show that the variation of Ω_2 with the PbO content is due to the change in the ligand field asymmetry at the site of the RE ion R caused by structural changes, and due to the change in R-O covalency. The change in Ω_6 was related to the variation of R-O covalency. All Ω_2, Ω_4, Ω_6 characterize a given R ion in a given host and are related to the radial wavefunctions of the states $4f^n$, the admixing states $4f^{n-1}5d$, and the ligand field. These admixing states allow electric-dipole transitions in the host, although the electric-dipole transitions between states within the 4f configuration are forbidden for a free ion. The covalency increase of the Nd-O bond, for example, in glasses is observed as an increase of the ratio between the intensities of the long- and short-wavelength components I_L / I_S of the Stark splitting peaks.

The different defect types that can arise on RE doping of a certain host can be identified by a series of experiments made for different implantation and annealing parameters, different doping concentrations, different temperatures and excitation powers. For example, five different defect types have been found by these procedures in Er-implanted Si corresponding to isolated interstitial Er, axial symmetry Er complexes with oxygen, Er complex centers containing residual radiation defects. The interstitial Er defects increase with increasing Er concentration, while oxygen-related defects decrease with increasing Er. All these different defect types have different exciton binding energies and nonradiative quenching rates (Przybylinska et al. 1996).

Crystal-field parameters B_{kq} for Eu^{3+} in sodium borosilicate glasses have been identified from the phonon sidebands of the $^7F_0 \to {}^5D_0$ transition, which correspond to localized vibrational modes around Eu^{3+}. In particular, the crystal field energy of the $^7F_1$ multiplet is Stark split and forms a sequence of levels ε_0, ε_-, ε_+ with increasing energies. These energies are $E(\varepsilon_0) = E(^7F_1) + B_{20}/\sqrt{5}$, $E(\varepsilon_\pm) = E(^7F_1) - (B_{20} \pm \sqrt{6}B_{22})/10$ with $E(^7F_1)$ the barycenter of the $^7F_1$ multiplet (Pucker et al. 1996). The five nonzero crystal-field parameters B_{20}, B_{40}, B_{60}, B_{44} and B_{64} related to the C_{4v} symmetry of Nd^{3+} in Nd_2CuO_4 are determined by fitting the eigenvalues of the crystal field Hamiltonian $H_{CF} = \sum_{k,q} B_{kq} C_{kq}$ to the energies of the detected Raman transitions considering J mixing within the entire set of ^4I_J multiplets ($J = 9/2$, $11/2$, $13/2$, $15/2$, with $J = 9/2$ the ground state of the multiplet, and the other levels the excited state multiplets) and approximating the energies of the split Kramers doublets by their average value. The splitting of the Kramers doublets is due to the interaction between the RE sublattice and CuO_2 planes (Jandl et al., 1995).

Numerous models can be tested with optical experiments, and their parameters can be determined. For example, persistent spectral hole burning in β"-alumina doped with Eu^{3+} at high temperatures (300 K) showed that light induces rearrangement of the local structure around RE ions. The hole has a depth of 15–20%, and its relaxation can be described by a hopping model with a

Gaussian distribution $g(V) = (2\pi\sigma^2)^{-1/2} \exp[-(V - V_0)^2 / 2\sigma^2]$ of barrier heights V. Fitting the time dependence of the long-lived hole area with $A(t) = \int_{-\infty}^{\infty} g(V) \exp(-Rt) \mathrm{d}V$ where the hole relaxation is $R = R_0 \exp(-V / k_B T)$, the mean barrier height has found to be $V_0 \cong 0.33$ eV. This value suggests that the persistent spectral hole burning is due to optically activated motion of the residual Na^+ ion near Eu (Yugami et al., 1996).

Indications about the dynamics of impurity ions can be obtained from the temperature dependence of spectral hole burning. For example, both transient and persistent hole burning have been observed in Eu^{3+}-doped porous γ-aluminum oxide (Feofilov et al., 1996). The sharp increase in the temperature dependence of the transient holes for $T > 7$ K suggests a redistribution of the population among hyperfine ground-state sublevels when the homogeneous linewidth becomes smaller than the ground-state hyperfine splitting. The increase of the temperature dependence of transient holes from $T^{1.3}$ at low temperatures to T^2 at higher temperatures in porous hosts (as well as in glasses and disordered materials in general) is in contrast to the behavior in crystalline hosts where the respective dependencies are T^7 and T^2. A two-phonon Raman process involving interactions with localized vibrations in glasses agrees quite well with the experimental data on the transient hole burning temperature dependence for $T > 7$ K. The persistent hole burning is probably due to photophysical rearrangement of the local environment.

Parameters of phonon-ion interaction which modify the spectral widths and the positions of the sharp inter-Stark lines of RE ions in a certain host can also be determined by optical means. The temperature dependence of the spectral width can be modeled by

$$\Gamma(T) = \Gamma_0 + \sum_i \beta_i \frac{1}{\exp(\Delta E_i / k_B T) - 1} + \alpha \left(\frac{T}{\Theta_D} \right)^7 \int_0^{\Theta_D / T} \frac{x^6 \exp x}{(\exp x - 1)^2} \mathrm{d}x, \tag{3.19}$$

where the second term on the right-hand side is due to one-phonon absorption/emission processes, the sum being taken over transitions with energies ΔE_i, and the last term is the contribution of Raman multiphonon processes associated with phonon scattering by impurity ions. The parameters α, β_i which characterize the coupling of ion-phonon interactions, as well as the Debye temperature Θ_D are obtained from fitting. Analogously, the energy shift of the line i is given by

$$\Delta \nu_i(T) = \alpha_i \frac{T^4}{\Theta_D^4} \int_0^{\Theta_D / T} \frac{x^3}{\exp x - 1} \mathrm{d}x + \sum_{j \neq i} T_{ij} \beta_{ij} \frac{T^2}{T_{ij}^2} P \int_0^{\Theta_D / T} \frac{x^3}{\exp x - 1} \left[\frac{T_{ij}^2}{\Theta_D^2} - x^2 \right]^{-1} \mathrm{d}x, \tag{3.20}$$

where P denotes the principal part and $T_{ij} = (E_i - E_j) / k_B$ (Sardar and Stubblefield, 1999). Measuring the temperature-dependent line broadening of a large number of 4f transitions throughout the lanthanide series it was found that the ion-phonon (electron-phonon) coupling strength α is greater in the beginning (Ce^{3+}, Pr^{3+}, Nd^{3+}) and end (Er^{3+}, Tm^{3+}, Yb^{3+}) of the trivalent series, but smaller in the center (Eu^{3+}, Gd^{3+}, Tb^{3+}). This can be explained by lanthanide contraction and shielding of 4f electrons by the outer $5s^2$, $5p^6$ electrons (Ellens et al., 1997).

Energy transfer between shallow centers and RE ions can be evidenced by two-color optical spectroscopy. For example, by irradiating Si with a visible laser, the photogenerated carriers are trapped at shallow centers, the excitation energy being afterwards transferred to the 4f-electron core of the Er^{3+} ion by irradiation with a mid-IR pulse (Gregorkiewicz et al., 2000). This process can be observed in the PL decay (see Fig. 3.6) of the band-to-band excitation in Si. The second peak is due to Er emission caused by the energy transfer, which takes place when the mid-IR pulse is applied with some delay.

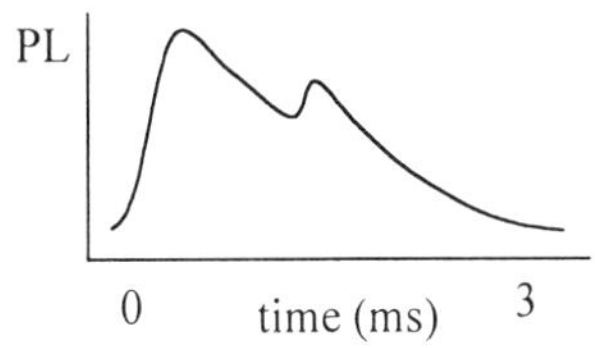

Fig. 3.6. PL decay of the band-to-band excitation in impurified Si

Energy transfer between RE ions can also be studied with TRFLN in inhomogeneously broadened absorption bands due to site-to-site variation of the crystalline field in RE-doped hosts. Such a study has been done for Yb^{3+}-doped fluoroindate glasses (Martín et al., 1998). For low concentration of Yb^{3+} the lowest Stark components of ${}^2F_{5/2} \rightarrow {}^2F_{7/2}$ transitions have a narrow spectrum that repeats the excitation profile, suggesting that energy migration between RE ions is negligible. At higher concentrations, however, (2.25 mol%) the narrow line emission of the ions initially excited by the pulse broadens as the time passes, due to energy transfer to other ions. TRFLN offers information about the different sites occupied by RE ions, and the homogeneous width of individual transitions. It can be modeled with a rate equation involving the two Yb^{3+} levels and their Stark components: $\mathrm{d}P_n(t)/\mathrm{d}t = -\sum_{n' \neq n} W_{nn'} P_n(t) + \sum_{n' \neq n} W_{n'n} P_{n'}(t)$ where $W_{nn'}$ is the transfer rate from ion n to n'. In thermal equilibrium, the probability that site L is occupied if the energy of the initially excited ions is E_0, is $W_{0L} \exp(-E_0 / k_B T) = W_{L0} \exp(-E_L / k_B T)$. The energy transfer is due to multipole interaction assisted by phonons and is described by $W_{0R} = W(E - E_0) C_{\mathrm{DA}}^{(s)} / R^s$ where $W(E - E_0)$ is the probability of emission or absorption of a phonon with energy $| E - E_0 |$ and $C_{\mathrm{DA}}^{(s)}$ is the parameter of energy transfer between a donor and an acceptor situated at a distance R due to dipole-dipole, dipole-quadrupole and quadrupole-quadrupole interactions, characterized respectively by $s = 6$, 8 and 10.

The persistent hole burning technique is the only way to measure simultaneously the homogeneous width and frequency shifts of inhomogeneously broadened transitions. Measurements of these parameters for the lowest one-photon allowed transitions in the $CaF^2:Sm^{2+}$ system, have shown a non-monotonous temperature dependence of the frequency shift, the dominant dephasing process being identified as the absorption of one acoustic phonon due to coupling of the system and its environment (Beck et al., 1996). When the homogeneous width is much smaller than the inhomogeneous width, the change in the absorption coefficient at the test frequency ω_t induced by the pump that generates the spectral hole burning at frequency ω_p is

$$\Delta\alpha(\omega_t;\omega_p) \approx \int \alpha_{hom}(\omega_p - \omega)\alpha_{hom}(\omega_t - \omega)d\omega, \tag{3.21}$$

indicating that the halfwidth of the hole Γ is the sum of the homogeneous halfwidths for pump and test pulses. A broadening of the hole linewidth occurs if there is saturation. By measuring the dependence of the width on the excitation intensity one can identify the nature of the process which takes place in the material: in linear, one-phonon processes the width increases linearly with the excitation intensity, this dependence becoming quadratic for two-photon processes.

The multipolar interaction responsible for energy transfer processes in yttria-stabilized zirconia crystals doped with Pr^{3+} was found by fitting the low-temperature decay curve of Pr^{3+} luminescence with

$$I(t) = I(0)\exp[-(t/\tau_r) - \Gamma(1-3/s)(N/N_{cr})(t/\tau_r)^{3/s}], \tag{3.22}$$

where τ_r is the radiative lifetime, N the concentration of luminescent centers, N_{cr} a critical concentration identical to the concentration at which an isolated donor-acceptor pair has the same transfer rate as the spontaneous decay rate of the donor. In this crystal Pr substitutes either Y^{3+} or Zr^{4+} cations, appearing as Pr^{3+} or Pr^{4+}, the Pr^{3+} to Pr^{4+} conversion being also observed. The parameter s = 6, 8 or 10 characterizes again dipole-dipole, dipole-quadrupole or quadrupole-quadrupole interactions. From fitting, the value of s (=6) and N/N_{cr} is obtained. Moreover, the activation energy ΔE of the nonradiative relaxation rate $W_{nr}(0)$ can be found from the temperature dependence of the $^3P_0$ lifetime $1/\tau = 1/\tau_r + W_{nr}(0)\exp(-\Delta E/k_BT)$ (Savoini et al., 1997).

The nature of the relaxation mechanism in Nd^{3+}-doped fluorophosphate glasses was studied by steady-state and time-resolved spectroscopy (Balda et al., 1996a). The stimulated emission cross-section and radiative lifetime of $^4F_{3/2} \rightarrow {}^4I_{11/2}$ transition was inferred from absorption spectra after calculating the Judd–Ofelt parameters. For low Nd^{3+} concentration and temperatures, the experimental lifetime is almost identical to the radiative lifetime calculated from Judd–Ofelt theory, so that the nonradiative decay rate by multiphonon emission can be neglected. This is not the case for higher Nd^{3+} concentrations when, although the decay remains single exponential, the lifetime $1/\tau_{exp} = 1/\tau_r + W_{Nd-Nd}(T)$ decreases due to nonradiative energy diffusion between Nd^{3+} ions.

The linear temperature dependence of the Nd-Nd nonradiative relaxation suggests a one-phonon assisted process; a T^3 dependence would imply a two-site nonresonant process. Fluorescence quenching is already present in samples with 1 wt % doping. Similar conclusions for Eu^{3+}-doped fluorophosphates glasses can be reached by studying the time-resolved PL after FLN selective excitation. In this case dipole-dipole energy transfer is characterized by the time dependence $\ln(1+I_B/I_N)=\gamma(E_L)t^{1/2}$ where I_B, I_N are, respectively, the intensities of the broad background and narrow luminescent component, and $\gamma(E_L)$ is the average rate of excitation transfer from donors to an inhomogeneous ensemble of acceptors. The small variation of $\gamma(E_L)$ with the excitation energy E_L suggests a small energy mismatch in the transfer process, assisted by one acoustic phonon (Balda et al., 1996b).

Up-conversion mechanisms in RE-doped materials are a subject of recent study. The power-law dependence of the emission intensity I with the excitation power P is an indication of the up-conversion mechanisms. The dependence $I \approx P^n$ with n = 1.5 corresponds to a cross-relaxation process, n = 2 to excited-state absorption, and n = 3 to energy transfer. The exponent n represents the number of photons absorbed per emitted up-converted photon. In experiments, however, deviations from these theoretical values are observed due to strong absorption, absorption of up-converted fluorescence, or nonradiative decay from higher-lying states to fluorescent states. For example, in Er^{3+}-doped fluoroindate glasses excited by an IR laser, up-converted emission is obtained in the green, blue and red regions, with n = 2.1, 2.9 and 1.6, respectively. The different intensities of the emitted lines (strong green emission, weaker blue and red) are explained by the different lifetimes (populations) of these transitions. For example, blue luminescence involves states with shorter lifetimes (lower populations) than green luminescence (Catunda et al., 1996).

Information about nonradiative processes can be inferred by optical means. For example, in Er-doped amorphous hydrogenated Si, it was found that the Er-induced PL decreased by 15 times in the temperature range 77–300 K, whereas the lifetime of excited Er ions is only a factor of 2.5 smaller. The discrepancy can be eliminated by the introduction of a defect-related Auger excitation, which should be larger by an order of magnitude compared to the radiative recombination process (Fuhs et al., 1997). This shows that although the host influences only slightly the energy positions of Er-induced PL, it has a pronounced effect on the relaxation mechanisms. Another example of the strong coupling between RE ions and the host are the anomalies in the elastic constants (Nipko et al., 1996). The elastic constants of $YbPO_4$ determined by Brillouin scattering, show a 20% softening of $(1/2)(c_{11}-c_{12})$ in the 10–300 K range (from about 0.8 at 10 K to 1 at 300 K) due to a Jahn–Teller coupling between the low-lying electron states of Yb^{3+} ions and the B_{1g} lattice strain. For extreme coupling a cooperative structural phase transition takes place, the tetragonal lattice being distorted to a lower symmetry, accompanied by a renormalization of the electronic states of RE ions.

The nuclear quadratic splitting in the ground state of Am^{3+} in $LaCl_3$ can also be studied by spectral hole burning (Liu et al., 1996). The transition $^7F_0 \rightarrow {}^5D_1$ at 17 173 cm^{-1} is excited by pumping with a dye laser, and the fluorescence emission at 14 455 cm^{-1} from $^5D_1$ to $^7F_1$ is monitored. In this broadened fluorescence line the spectral hole burning is accompanied by three equally spaced antiholes on each side (see Fig. 3.7), which measure the splitting between the nuclear energy levels $|I_z\rangle = \pm|1/2\rangle$, $\pm|3/2\rangle$, $\pm|5/2\rangle$ in the electronic ground state.

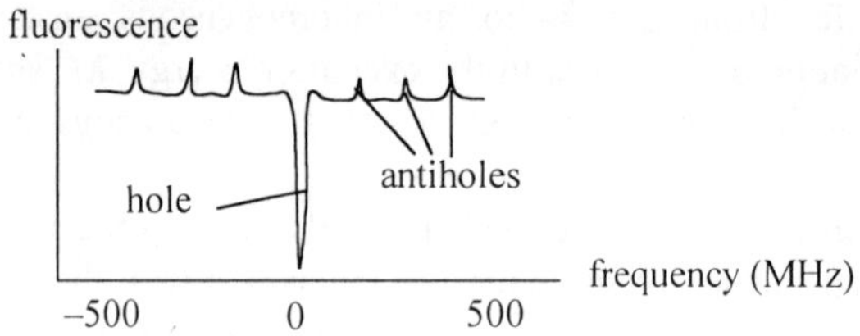

Fig. 3.7. Fluorescence spectrum of spectral hole burned ground state of Am^{3+} in $LaCl_3$

The antiholes are due to population accumulation in the nuclear quadrupole levels of the J = 0 ground state of Am^{3+}. The presence of antiholes indicates nuclear quadrupole splitting of the ground state, while the eventual presence of side holes, as in Eu^{3+}, would indicate the splitting of excited state splitting. The equal splitting between the antiholes, of Δf = 150 MHz, is consistent with an axial site of Am^{3+}, with the same symmetry as the site of host La^{3+}. The nuclear quadrupole coupling constant $P = \Delta f/2$ is defined through the nuclear quadrupole Hamiltonian $H_Q = P[I_z^2 - (1/3)I(I+1)]$ with I = 5/2 the nuclear spin and includes several contributions, the most important of which being the contribution of the lattice.

The isotope shifts and the hyperfine and superhyperfine interactions of Nd^{3+} ions in $YLiF_4$ have been also determined optically (Macfarlane et al., 1998). By monitoring the $^4I_{9/2} \rightarrow {}^4F_{3/2}$ transition of Nd^{3+} ions, it is possible to determine the isotope shifts of about 115 MHz/unit mass for the even mass number isotopes ^{142}Nd, ^{144}Nd, ^{146}Nd, ^{148}Nd, ^{150}Nd, which have very narrow linewidths of only 45 MHz. In the presence of an applied magnetic field each even isotope line is split by 30 MHz due to superhyperfine coupling between Nd^{3+} and its nearest-neighbor fluorine nucleus. The optical lineshapes of the even isotopes are obtained by summing over all transitions between the superhyperfine components of the ground and excited states, considering both magnetic dipole-dipole and contact interactions. Some hyperfine transitions of the odd isotopes are even narrower than those of even isotope lines and can reach 10 MHz. The hyperfine structure of ^{143}Nd and ^{145}Nd for example are resolved and hyperfine parameters determined in ground and electronically excited states. Since the transitions in this case occur within the f shell, the isotope shifts are due to the local lattice deformation and differential coupling to zero-point phonons whose frequencies are mass dependent. The spin Hamiltonian for the ground and excited states in a magnetic field is

$$H = g_{\parallel}\beta H_z S_z + g_{\perp}\beta(H_x S_x + H_y S_y) + AI_z S_z + B(I_x S_x + I_y S_y) + P\left[I_z^2 - \frac{I(I+1)}{3}\right], \tag{3.23}$$

where $\boldsymbol{S}$, $\boldsymbol{I}$ are the electron and nuclear spin operators, with $S = 1/2$, $I = 7/2$, A, B are hyperfine parameters and P the quadrupole constant. Assuming that the ground state hyperfine parameters are known, those for the excited state can be obtained from fitting. The splitting of even isotope lines due to superfhyperfine coupling is a result of the specific distribution of the electron-nuclear states of Nd^{3+} and its 8 nearest-neighbor ligand nuclei. The superhyperfine Hamiltonian is

$$H_{\text{shf}} = \sum_{i\alpha\beta} I_{i,\alpha} a_{\alpha\beta}(i) S_{\beta}, \tag{3.24}$$

where the sum is to be taken over the eight F nuclei with spin moments I_i (i = 1 to 8), a is the coupling constants due to magnetic dipole-dipole interactions between Nd^{3+} electron-spin moment and the surrounding fluorine nuclear moments, and the spin transfer from the Nd^{3+} ion onto 2s and 2p ligand orbitals.

3.7.3 Interaction Processes for Rare-Earth Ions

We refer in this subsection to interactions other than the energy transfer, in particular to interactions between pairs of RE ions (Parrot, 1979). The weak interaction between RE pairs makes it difficult to distinguish the energy levels of excited states of RE pairs from those of the isolated ions. A procedure to identify these states is to apply a magnetic field, which lifts the remaining Kramer's degeneracy and induces transitions between the Zeeman sublevels, the involved photons having an energy of about 1 cm^{-1}. The exchange interaction for a pair with axial symmetry is

$$H_{\text{ex}} = J[(\Lambda^2 - 1)/\Lambda]^2[g_{\parallel}^2 S_{iz} S_{jz} + g_{\perp}^2 (S_{ix} S_{jx} + S_{iy} S_{jy})], \tag{3.25}$$

where Λ is the ground state Landé factor, whereas the magnetic dipole contribution for pairs with axial symmetry is

$$H_{\text{dm}} = -2g_{\parallel}^2 \beta^2 S_{iz} S_{jz} / R^3 + g_{\perp}^2 \beta^2 (S_{ix} S_{jx} + S_{iy} S_{jy}) / R^3, \tag{3.26}$$

R denoting the interionic distance. Electric-multipole interactions can also be included through an effective Hamiltonian. All these contributions are bilinear in S_i, S_j, such that the effective spin Hamiltonian can be written as $H_{ij} = a_{ij} S_{iz} S_{jz} + b_{ij}(S_{ix} S_{jx} + S_{iy} S_{jy})$ in axial symmetry and $H_{ij} = \mathbf{S}_i \hat{K} \mathbf{S}_j$ for any symmetry, with $\hat{K}$ a symmetric second-rank tensor. $\hat{K}$ can be decomposed in an isotropic and a traceless tensor: $\boldsymbol{S}_i \hat{K} \boldsymbol{S}_j = \hat{J} \boldsymbol{S}_i \boldsymbol{S}_j + \boldsymbol{S}_i \hat{A} \boldsymbol{S}_j$, where $Tr(A) = 0$,

$Tr(K) = 3J$, the isotropic tensor $\hat{J}$ lifting the degeneracy of the singlet and triplet states, while $\hat{A}$ lifts the degeneracy of the triplet states. The remaining degeneracy of the triplet states is lifted in the presence of an external magnetic field described by the spin Hamiltonian $H_z = \beta H \hat{g}(S_i + S_j)$. For example, the interactions between Nd^{3+} ions in $NdCl_3$, $NdBr_3$ are mainly due to superexchange, the magnetic dipolar interactions having a small contribution. The superexchange interaction varies strongly with the level: the non-dipolar coupling is 0.53 cm^{-1} for B_2 level and 0.01 cm^{-1} for G_1 level. Interactions between non-identical RE ion pairs have also been observed, as for example between Ce-Pr and Ce-Nd, where the nearest-neighbor non-dipolar interaction is of the order 10^{-2} cm^{-1}, and between Ce-Gd, where the interaction is of a mainly magnetic dipolar nature.

3.8 Transition Metal Ions

Transition metal atoms occur in ionic states in which the electrons are not in closed shells; for example the positively charged transition metal ions have an incomplete outer 3d shell. Transitions between the initial and final states belonging to the same electronic configuration are parity forbidden in free transition metal atoms, but a slight relaxation of the parity selection rule is observed when the ions are embedded in a solid material.

The electric field of neighboring ions (the crystal or ligand field) splits the free-ion levels. When the environment has octahedral symmetry, the five-fold degenerate 3d state is split in two: a three-fold degenerate t_{2g} ground state and a two-fold degenerate e_g state with a higher energy. A tetrahedral environment also splits the 3d state into degenerate t_{2g} and e_g orbitals, but the first are higher in energy. If we write the splitting as 10 Dq then Dq (tetrahedral) $= -4/9\ Dq$ (octahedral). When more than one 3d electron is present, one must also account for the electrostatic interaction among them besides the effect of the crystal field. The Racah parameters A, B, C characterize the strength of the electron-electron interaction while the strength of the static octahedral crystal field is characterized by the value of Dq, 10 Dq being the crystal-field splitting between t_{2g} and e_g single-electron orbitals.

Spin-allowed transitions are the strongest, transitions involving a change in spin being not entirely forbidden due to spin-orbit coupling. The interaction of 3d electrons with the vibrating crystal field gives rise to broad absorption bands for some transitions, although the energy levels of the optically active ions should be narrow.

Since the interaction between the transition metal ions with the environment is stronger than for RE ions, optical spectra of transition metal ions are influenced by the host material in a much stronger manner. The wavelength of a transition can be shifted by stressing the material, different lines being shifted differently because they are generally formed from a mixture of e_g and t_{2g} orbitals with different sensitivities to the crystal field strength. This effect appears for a static change in an octahedral crystal field, which can be produced by a hydrostatic stress. Applying a uniaxial stress the symmetry of the crystal field changes from octahedral to a lower one, this perturbation splitting the octahedral crystal field levels. For example, the ^2E level in Cr^{3+} is usually split in two. If the crystal field contains a lower symmetry component that lacks inversion symmetry, as for Ga^{3+} site in $LiGa_5O_8$ where the arrangement of positively charged second-nearest-neighbor ions lacks charge symmetry and produces a small odd-parity perturbation, the energy levels are only slightly modified. On the contrary, the strengths of optical transitions are significantly changed. Odd-parity perturbations mix small amounts of $3d^{n-1}4p$ states into the even parity $3d^n$ states, the initial and final states of the optical transitions being no longer of exactly the same parity, the optical transition having a small amount of electrical dipole character that enhances the oscillator strength (Imbusch, 1979a).

The luminescence spectrum changes with the concentration of transition metal ions. For example, in diluted ruby, when Cr concentration increases, additional lines due to exchange-coupled pairs of Cr^{3+} ions appear, besides the two R lines emitted by isolated Cr^{3+} ions. The probability of pair formation increases as the square of Cr^{3+} ion concentration, the additional lines becoming as intense as the lines from single ions at 1% concentration. At still higher concentrations an additional broad band stretching from 0.74 μm to 0.84 μm is observed, due to centers involving three Cr^{3+} ions coupled together. More features due to larger clusters of ions appear progressively with increasing doping concentration, concentration quenching (no luminescence) being observed in the fully concentrated material Cr_2O_3. The excited states in fully concentrated material are collective excited states involving all ions (Imbusch, 1979b).

3.8.1 Optical Properties of Transition Metal Ions

Unlike RE, transition metal ions interact with the crystal field strength and lattice vibrations, having wide absorption spectra that differ from host to host. The dependence of optical properties on the host material is well illustrated by the temperature dependence of fluorescence lifetimes of Cr^{3+} (Zhang et al., 1993). In insulating crystals with low crystal field the transition between the lowest excited state ^4T_e and ground state $^4A_2$ has a radiative channel and a nonradiative one, the temperature dependence of the fluorescence intensity and lifetime being $I = I_0/[1+(\tau_i/\tau_q)\exp(-\Delta E_q/k_B T)]$ and $1/\tau = 1/\tau_i + (1/\tau_q)\exp(-\Delta E_q/k_B T)$, respectively, where $1/\tau_i$ is the intrinsic radiation rate, $1/\tau_q$ the thermal quenching rate, and ΔE_q the thermal activation energy up to the energy crossing point in the real space between the excited and ground states, from where nonradiative processes to the ground state dominate. On the contrary, in hosts with high crystal field, the fluorescence lifetime at low temperatures is dominated by thermally activated repopulation between ^2E and $^4T_2$ excited states, which are in quasi-thermodynamic equilibrium, the fluorescence lifetime being described by a two-level model $\tau = \tau_S[1+D_r\exp(-\Delta E/k_B T)]/[1+(\tau_S/\tau_i)\exp(-\Delta E/k_B T)]$, where

τ_l, τ_s are the lifetimes of the $^4T_2$ and ^2E levels, separated by an energy ΔE, and $D_r = 3$ is the ratio of degeneracy of the $^4T_2$ level to that of ^2E.

3d transition metal ions are used as probes to study the symmetry and strength of the crystal field of the host. In binary semiconductors, transition metals occupy cation sites; in ternary semiconductors such as I-III-VI_2 or II-III_2-VI_4 there are two cation sites that can be occupied by transition metal ions. In this case transition metal ions can sense the local site symmetry and local strength of the crystal field. One example of a ternary semiconductor is $AgGaS_2$, which has two species of cations corresponding to different types of distortion from cubic symmetry. In this material the length of the Ag-S bond is larger than that of Ga-S, all anions being displaced from their zincblende sites, and the whole crystal is compressed along the tetragonal direction. The Ga atom is surrounded by an almost regular tetrahedron of S atoms, whereas the symmetry of S atoms around an Ag atom is lowered to D_{2d}. Then, depending of the relative strength of the D_{2d} crystal-field and spin-orbit interaction, $^4T_1$ states are split in different numbers of levels. Optical measurements made in Co-doped crystals showed that all Raman modes are softened compared to the undoped crystals, which implies that Co^{2+} ions occupy Ga sites. Otherwise, if Co^{2+} would have had occupied Ag sites, some modes would have had larger frequencies compared to undoped crystals. Assuming that the lattice constants of Co-doped crystals are the same as in undoped samples, the frequency shifts of Raman modes are consistent with a Racah B parameter of 605 cm^{-1} and a crystal-field parameter $Dq = 359\,cm^{-1}$ (Park et al., 1996).

Many papers focus on the determination of the symmetry and crystal-field parameters in different sites and materials. One of the most recent from an innumerable series of such papers deals with Cr^{4+}-doped YAG and YGG (Riley et al., 1999) studied by selective luminescence and magnetic circular dichroism. Both materials have cubic structures, in which the cation sites are nominally dodecahedral, octahedral and tetrahedral. The luminescence of tetrahedral ions in the isoelectronic series Fe^{6+}, Mn^{5+}, Cr^{4+} is distinguished by a decreasing ligand field characterized by a decreasing value of Dq. The value of the ligand field determines which level is the lowest excited state and thus the emitting state in the PL. The value of the level splitting due to both crystal-field and spin-orbit coupling is compared with several theories, in particular with the ligand field model, in order to make a correct assignment of the different lines observed in absorption, PL or piezospectroscopy and to quantitatively relate them to microscopic quantities such as the degree of mixing of different types of orbitals depending on the symmetry of different sites.

In some cases Fano antiresonance lines can be observed in the absorption spectrum of materials doped with transition metal ions. For example, the absorption band corresponding to the transition from the $^4T_2$ band to $^4A_2$ ground state in the Cr^{3+}-doped pseudoternary system InF_3-GdF_3-GaF_3, looks like that in Fig. 3.8. The lineshape of the $^4T_2$ band is described by $\alpha(\omega) = \alpha_0 \exp(-S)S^p / \Gamma(p+1)$, where $p = (\omega - \omega_0)/\Omega$, from which one can determine the number and mean frequency Ω of emitted phonons by fitting. The two dips in the broad absorption band are due to Fano antiresonances caused by mixing with ^2E and $^2T_1$ lines. The antiresonances are caused by interference between the sharp, phononless transitions with t_2^3 configuration and the broad transition to the t_2^2e configuration in which phonons are emitted. Fano interference appears whenever a continuum of electronic excitations overlaps a discrete phonon, as shown in Fig. 3.9.

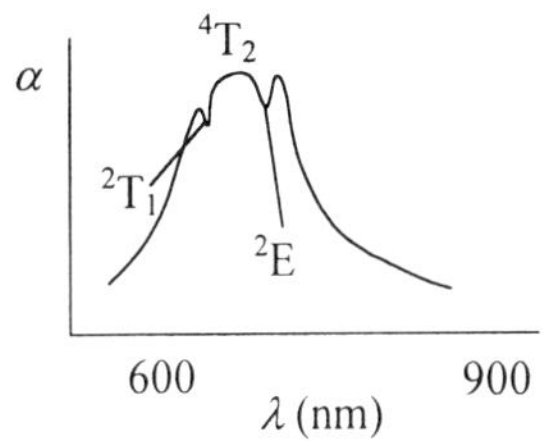

Fig. 3.8. Absorption band of the transition from the $^4T_2$ band to $^4A_2$ ground state in the Cr^{3+}-doped pseudoternary system InF_3-GdF_3-GaF_3

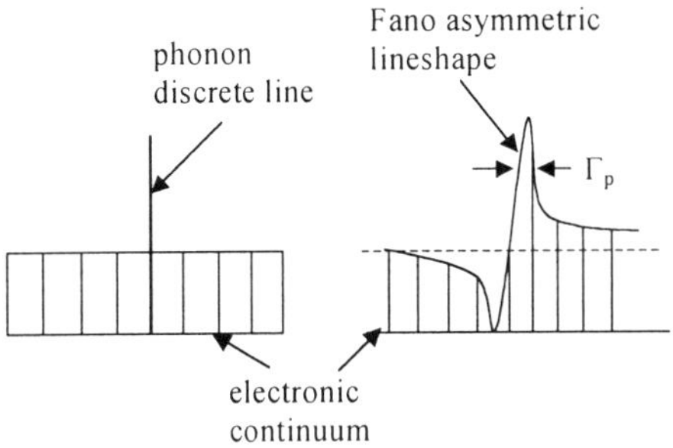

Fig. 3.9. Fano interference between an electronic continuum and a phonon discrete line

The antiresonance is formed at $^4A_2 \rightarrow {}^2E$ transitions of Cr^{3+} and is described (after subtracting the broad background) by $\alpha(\omega) = \alpha_0 (Q^2 + 2\xi Q - 1)/(1+\xi^2)$ where $\xi = (\omega - \omega_r)/\gamma$ with $1/2\gamma$ the lifetime of the sharp state against decay in the vibronic band. From fitting, it results that at the resonance frequency ω_r which corresponds to 657.7 nm, Q = 0.08, γ = 2.8×10^{12} rad/s. The nonlinear refractive index in this medium has been determined by the Z-scan technique. In this method a sample is translated through the focus of a Gaussian beam and the changes in the far-field intensity pattern are recorded. The light field induces a nonlinear Kerr contribution to the refractive index $n(I) = n_0 + n_2 I$ that modifies the far-field pattern depending on the Z position in the transverse plane along which the translation is made, and can be used to determine the induced nonlinear coefficient n_2. The experimental value, obtained from the difference between the peak and valley of the Z scan, is -6×10^{18} cm^2/kW at 608 nm (Mendonca et al., 1997).

3.8.2 Iron-Group Ion Pairs

When the concentration of transition metal ions is large enough, exchange interactions often dominant and are much stronger than for RE ions (Imbusch, 1979b). For example, the isotropic exchange interaction for the fundamental state of nearest-neighbor Cr^{3+} ions in Al_2O_3 is two orders of magnitude greater than the isotropic exchange for nearest-neighbor Nd^{3+} in $LaCl_3$. When both ions are orbitally nondegenerate, the exchange interaction is described to first-order by the Heisenberg Hamiltonian $J\boldsymbol{S}_1 \cdot \boldsymbol{S}_2$, where $\boldsymbol{S}_1$, $\boldsymbol{S}_2$ are spin operators of nondegenerate states of individual ions. The matrix elements of the Heisenberg Hamiltonian, for two identical spins, are diagonal in the coupled basis $|S_1S_2SM_S\rangle$:

$$\langle SM_S \,|\, J\boldsymbol{S}_1 \cdot \boldsymbol{S}_2 \,|\, S'M_{S'}\rangle = \delta(S,S')\delta(M_S,M_{S'})J[S(S+1) - 2S_1(2S_1+1)]/2, \quad (3.27)$$

where S is the total spin. Slight deviations from the energy levels given by the Heisenberg Hamiltonian for orbitally nondegenerate states can be caused by magnetic dipolar interactions and crystal-field effects (studied by electron paramagnetic resonance), or by biquadratic exchange terms $J(\boldsymbol{S}_1 \cdot \boldsymbol{S}_2)^2$ studied by optical methods.

The Heisenberg Hamiltonian is no longer appropriate when one or both ions are orbitally degenerate, the exchange interaction in this case being written in terms of equivalent tensor operators acting on LS or LSJ states. A considerable simplification of the theory is possible when there is one electron in each of the 3d orbitals and when the spin-orbit interaction acts as a small perturbation on the exchange interaction. In this case the total spin is a good quantum number and the dominant spin-dependent exchange interaction becomes

$$H_{ex} = \sum_{a=1}^{3}\sum_{b=1}^{3} J_{ab}\boldsymbol{S}_{a1} \cdot \boldsymbol{S}_{b2}, \quad (3.28)$$

where J_{ab} is the exchange parameter for an electron in orbital a on ion 1 and an electron in orbital b on ion 2.

3.9 Diluted Magnetic Semiconductors

Diluted magnetic semiconductors (DMS) are formed by replacing a small fraction of non-magnetic cations in (usually a II-VI) semiconductor with magnetic ions, such as Mn, Fe, Co, Cr. The magnetic properties, and magneto-optic response of such materials are therefore much enhanced. In particular, giant Faraday rotation and giant spin splitting are observed. DMS are characterized by a strong exchange interaction between delocalized s- and p-band electrons (in the conduction and valence band, respectively) and localized d- or f-electrons, as well as by the d-d interaction between magnetic ions, important at high magnetic ion concentrations.

The sp-d interaction leads to a splitting of the conduction and valence bands when an external magnetic field is applied, producing giant magneto-optical effects. The s-d exchange is a direct (potential) exchange between s and d one-electron orbitals centered on the same ion core; this exchange is ferromagnetic, characterized by a positive exchange constant, and is approximately independent of the host and the magnetic ion. On the contrary, the p-d exchange is a kinetic exchange, due to p-d hybridization, in which p (d) electrons virtually jump to orbitals already occupied by d (e) electrons. The exchange constant depends in this case on the energy difference between the p and d orbitals, and can be either negative or positive, corresponding to an antiferromagnetic and a ferromagnetic exchange, respectively. For II-VI DMS, the p-d exchange is ferromagnetic if the d-shell is less than half-filled (for example, for Cr(d^4)-based DMS), and antiferromagnetic for at least half-filled d-shell (Mn(d^5), Fe(d^6), Co(d^7)). The d-d interaction can induce magnetic ordering, and thus a transition from paramagnetic to antiferromagnetic phase transitions.

The magnetic properties of DMS influence the band structure of these materials. For example, in $Cd_{1-x}Mn_xTe$, the exchange interaction between free carriers and magnetic ions leads to an energy shift and an effective mass variation when a magnetic external field is applied. The two effects can be separated in the optical path modulation (OPM) method, where the transient photoreflectance is measured in a DMS layer, which is equivalent to a Fabry–Perot resonator. The interference phenomena produce features of the transient photoreflectance in the gap of the semiconductor where no optical transition exists. The contribution of the energy modulation to OPM can be suppressed by tuning the magnetic field, so that only the oscillator strength contribution remains (Farah et al., 1998). For samples glued on a glass plate and immersed in He the total reflectivity is $r=[r_1+r_2\exp(i\varphi)]/[1+r_1r_2\exp(i\varphi)]$, where r_1, r_2 are the reflectivities of the semiconductor/glass and He/semiconductor interfaces. From the four features present in photoreflectance, one is due to the excitonic line X and is caused by the change of r_1, the other three being due to interferences and coming from the modulation of the phase φ. To explain these features in the $\Delta R/R$ spectrum it must be assumed that the energy and effective mass averaged over the layer thickness vary due to thermomagnetic effects with ΔE, Δm, respectively. Whereas the renormalization of the effective mass under the influence of magnetic fluctuations always induce negative Δm values, ΔE is negative for zero applied magnetic field (corresponding to thermal modulation) and increases up to positive values with increasing H. The magnetic fluctuations induce a redshift of the energy gap as well as a modulation of the effective g factors. With increasing temperature ΔE increases, while Δm decreases.

Addition of magnetic ions also modifies the characteristic relaxation time of the material. For example, in $Hg_{1-x-y}Cd_xMn_yTe$, PL measurements at 5 K of band-to-band transitions showed splitting in both Faraday and Voigt geometries, followed by a decrease of the higher-energy component in magnetic fields larger than 2 T. This behavior is consistent with the following relation between the radiative lifetime τ_r, spin relaxation time τ_s and energy relaxation time τ_E: $\tau_r \geq \tau_s > \tau_E$. On the contrary, in a non-magnetic sample $Hg_{1-x}Cd_xTe$ with almost the same energy gap, $\tau_s \geq \tau_r > \tau_E$ (Hoerstel et al., 1998).

The optical properties of DMS are determined by the interaction between magnetic ions. One of the most studied DMS is $Cd_{1-x}Mn_xS$. In the absence of an applied magnetic field the energies of the A, B and C subbands of the valence bands are given by $E_{B,C} = E_A - (1/2)\{\Delta_1 + 3\Delta_2 \mp [(\Delta_1 - \Delta_2)^2 + 8\Delta_3^2]^{1/2}\}$, $E_A = \Delta_1 + \Delta_2$, where the crystal-field parameter Δ_1 and the anisotropy spin-orbit interaction constants $\Delta_{2,3}$ depend on the magnetic impurity (Mn) concentration. In the mean field approximation the exchange interaction of electrons with the system of Mn ions, which must be added to the Hamiltonian in the absence of the external field to calculate the giant spin splitting of the A, B, C lines, is given by $H_{ex} = J_{e(h)}x\langle \boldsymbol{S}\rangle \boldsymbol{S}_{e(h)}$ where $J_e = N_0\alpha$, $J_h = N_0\beta$ are the exchange constants for conduction and valence electrons, respectively, with spins $\boldsymbol{S}_e$, $\boldsymbol{S}_h$, N_0 is the cation concentration in the crystal and the thermodynamic average of magnetic ion spin momentum components in directions $\alpha = x, y, z$, induced by the magnetic field $\boldsymbol{H}$ is $\langle S_\alpha\rangle = (1/N_m)\mathrm{Tr}[\sum_{j=1}^{N_m} S_\alpha^j \exp(-H_m/k_BT)]/\mathrm{Tr}[\exp(-H_m/k_BT)]$. N_m is the total number of magnetic ions in the crystal, and the Hamiltonian H_m describes the spin momentum interaction with the external magnetic field $\boldsymbol{H}$, the crystal field influence on spins and the spin-spin interaction. This form of the exchange Hamiltonian is valid if $J_{e(h)}/W_{e(h)}$ is not small, where $W_{e(h)}$ is the width of the corresponding band; otherwise multiple scattering must be accounted for. Denoting by $Q = \Delta E^{(B)}(\boldsymbol{H} \parallel c)/\Delta E^{(e)}$, $R = \Delta E^{(B)}(\boldsymbol{H} \perp c)/\Delta E^{(e)}$ where $\Delta E^{(B)}$ is the giant spin splitting of the B-subband valence electrons, and $\Delta E^{(e)}$ the same parameter for conduction band electrons, for various orientations of the external field with respect to the crystal hexagonal axis c, all parameters of the band structure can be calculated from the giant spin splitting measured as a function of the applied external field. Namely, $\Delta_1 = \Delta E_{AB}\xi(1+\delta)/[2\xi - \xi\delta - \delta]$, $\Delta_2 = \Delta E_{AB}\xi(1-\delta)/[2\xi - \xi\delta - \delta]$, $\Delta_3 = \Delta E_{AB}\delta[(1-\xi^2)/2]^{1/2}/[2\xi - \xi\delta - \delta]$, $J_h = \eta J_e$, where $\Delta E_{AB} = E_A - E_B$, $\xi = 1/(1+2R/Q)$, $\eta = -Q/\xi$, and $\delta = 2\xi \times [\xi - (P+1)/(P-1)]^{-1}$, with $P = (E_A - E_B)/(E_A - E_C)$ (Semenov et al., 1997).

The s-d and p-d exchange interaction can be directly measured by free exciton spectroscopy. In a cubic lattice, as in $Zn_{1-x}Fe_xTe$, four excitons are visible in the Faraday configuration, two (A and B) in σ^+ polarization, the others (C and D) in σ^-, the energies of these excitons being $E_A = E_0 - 3b - 3a$, $E_B = E_0 + b + 3a$, $E_C = E_0 - b - 3a$, $E_D = E_0 - 3b + 3a$, where E_0 is the zero-field exciton energy, $a = -(1/6)N_0\alpha x\langle S\rangle$, $b = -(1/6)N_0\beta x\langle S\rangle$. The difference between A and D lines, which are three times stronger than B and C, provides the difference between exchange constants, $E_D - E_A = -(N_0\alpha - N_0\beta)x\langle S\rangle$, if the average spin $\langle S\rangle$ is known; the latter can be, for example, determined from measurements of the magnetization M per unit mass as $-x\langle S\rangle = Mm/g\mu_B$. Reflectance measurements of the sign of the splitting between σ^+ and σ^- reveal a positive $N_0\alpha - N_0\beta$, from which, since in all DMS $|\alpha| \ll |\beta|$, it follows that $N_0\beta$ is negative. This antiferromagnetic p-d exchange is typical for Fe-, Mn- and

Co-based DMS. The exciton splitting can also be determined from reflectance measurements as $\Delta E = E(\sigma^-) - E(\sigma^+) \cong -2P(E)/[\partial \ln I(E)/\partial E]$, where $P = [I(\sigma^-) - I(\sigma^+)]/[I(\sigma^-) + I(\sigma^+)]$ is the polarization degree, and $I = I(\sigma^-) + I(\sigma^+)$ (see Fig. 3.10). The lines which appear in σ^+ and σ^- polarizations, are roughly identified as A and D, although they are in reality a mixture of A, B and D, C, respectively. For strong exciton splitting, all four lines are resolved. Typical curves for reflectance and polarization degree for $Zn_{1-x}Fe_xTe$ are shown above; knowing from spin-flip Raman that $N_0\alpha = 0.22$ eV, it follows that for this material $N_0\beta = -1.9$ eV (Mac et al., 1996). The value $N_0\alpha$ of the s-d interaction, practically the same for all DMS, is obtained from the difference between the energy shifts of the spin-flip Raman lines from donor electrons with the applied magnetic field, for samples with and without magnetic impurities. A more detailed discussion of such measurements for $Cd_{1-x}Cr_xS$ can be found in Twardowski et al. (1996).

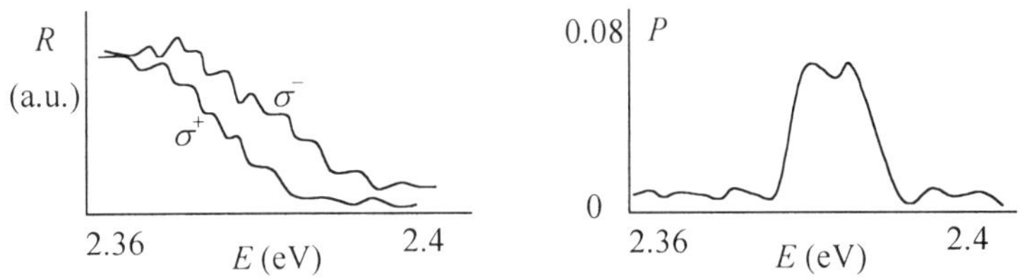

Fig. 3.10. Spectra of circularly polarized reflectance and polarization degree for $Zn_{1-x}Fe_xTe$

A study of the pressure effects in high magnetic fields for the $Cd_{1-x}M_xSe$ class of materials shows that the field-induced shift of the A exciton increases with the pressure for M = Co, $x = 0.012$ and M = Mn, $x = 0.05$, and decreases with the pressure for M = Mn, $x = 0.25$. These results reveal that the p-d exchange interaction is strengthened with increasing pressure, as is the antiferromagnetic coupling among Mn ions (Matsuda and Kuroda, 1996). The modification of the crystal-field levels in $Zn_{1-x}Co_xSe$ under hydrostatic pressure has been studied by Bak et al. (1996). The PL measurements between 2.24 and 2.38 eV revealed two zero-phonon lines, the energies of which vary quadratically with the pressure. The fitting procedure of the calculated energy levels with the experimental ones inferred that the Racah parameter B varies with pressure with a slope $dB/dP = -0.66$ meV/GPa, which suggests an increase of the hybridization between Co^{2+} and host states with the pressure, due to the proximity between the top of the valence band and the ground state of the Co^{2+} ion in the cubic crystal field. The crystal-field splitting parameter Dq did not influence the redshift of the energy levels under pressure, as would be expected for d-d transitions between levels that do not involve changes of the electron configuration.

The giant Faraday rotation in DMS is used to study optically the magnetization of Mn ions. For DMS with low concentration of magnetic ions the Faraday rotation is given by $\theta_F = \theta_0(\hbar\omega)B_S(B,T) + \chi(\hbar\omega)B$ where $\theta_0(\hbar\omega)$,

$\chi(\hbar\omega)$ are fitting parameters, dependent on the photon energy, corresponding to the paramagnetic contributions of Mn ions, and the diamagnetic contribution of the interband transition, respectively. $B_S(B,T)$ is the Brillouin function of magnetic ions with spin S in the magnetic field B and at the temperature T (S = 5/2, when the magnetic ion is Mn, and 3/2 when it is Co). The first term saturates in large magnetic fields, the saturation corresponding to a complete alignment of the spins of isolated Mn^{2+} ions. The diamagnetic contribution, present also in non-magnetic semiconductors, gives a Faraday rotation in the opposite direction to the magnetic component and it dominates at high fields, for example, for $B > 7$ T in $Zn_{1-x}Mn_xSe$, after the magnetic contribution saturates. θ_0 increases as the photon energy increases toward the bandgap, while $-\chi$ has a maximum at 2.4 eV as a function of the photon energy, which suggests that d-d transitions in Mn^{2+} system play also a role. For larger Mn concentrations, Mn pairs and clusters appear in the crystal, the Faraday rotation showing a stepwise increase corresponding to the magnetization steps of Mn ion pairs. The nearest-neighbor antiferromagnetic exchange constant in the same material, $J_{NN} = -13.1$ K, is obtained from these steps (Yasuhira et al., 2000). The contribution of nearest-neighbor pairs of ions becomes important for $x > 1\%$, and it introduces an additional term in $\langle S\rangle$. Denoting by P_1 the probability of the magnetic ion to occupy the cation site in isolation, and P_2 the probability of two magnetic ions to form a nearest-neighbor pair, $\langle S\rangle = SP_1B_S(Sg_M\mu_BH / k_B(T + T_0))$ $+(P_2/2)\Sigma_n 1/[1+\exp[g_M\mu_B(H_n - H)/k_BT]]$ where T_0 is an effective temperature representing the exchange mean field due to isolated magnetic ions and g_M the g factor of magnetic ions. The energy of the nearest-neighbor pair is quantized in states with total spin $S_p = 0,1,..,2S$, the ground state with $S_p = 0$ at $H = 0$ being successively replaced by the lowest Zeeman sublevel as the external magnetic field increases. At $H_n = 2n|J_{NN}|/g_M\mu_B$ the lowest Zeeman sublevel of the state $S_p = n$ becomes the fundamental state of the pair of magnetic ions, crossing downward over the lowest Zeeman sublevel of the state $S_p = n-1$. This is the origin of the steps in the Faraday rotation experiment discussed above (Matsuda and Kuroda, 1996).

When the magnetic ions are not uniformly distributed, but form clusters, specific effects can be observed (Weston et al., 1998). In clustered samples, the concentration of magnetic ions is larger in some regions relative to the mean concentration, and smaller in others. This can affect either $\langle S\rangle$ or T_0, or both, these quantities becoming position/concentration-dependent. Moreover, clustering reduces the paramagnetism due to the local accumulation of magnetic ions. The clustering can especially be observed in heterostructures between semimagnetic semiconductors. In this case the dependence on the growth direction coordinate of T_0 can be approximated as $T_0(z) = T_0^m + (T_0^b - T_0^m)(z/z_0)^2$, where T_0^b is the bulk value, T_0^m is the maximum value of this parameter at the interface to the barrier and z_0 is the spatial extent of the alloy clustering.

Direct determination of the components of the g tensor of electrons and holes in many-valley DMS can be performed with CARS in the mid-IR region.

Such measurements have been made on $Pb_{1-x}Mn_xSe$, a material with a small direct gap at L points of the Brillouin zone. In this case the exchange Hamiltonian calculated in the $\boldsymbol{k} \cdot \boldsymbol{p}$ model has a more complicated form due to spin-orbit mixing

$$H_{\text{ex}} = \frac{5\mu_{\text{B}} B S_0 B_{5/2}}{2k_{\text{B}}(T+T_0)} \begin{pmatrix} b_1\sigma_z\alpha_3 - b_2(\sigma_x\alpha_1 + \sigma_y\alpha_2) & 0 \\ 0 & a_1\sigma_z\alpha_3 + a_2(\sigma_x\alpha_1 + \sigma_y\alpha_2) \end{pmatrix}, \tag{3.29}$$

where a_i, b_i are exchange parameters for the valence and conduction band, σ_i are the Pauli matrices and α_i the direction cosines in the valley axis system. S_0, T_0 are determined from fitting CARS intensities as a function of B to the calculated values. The Landau levels are obtained by diagonalizing the total Hamiltonian. In the Voigt geometry a spin-flip resonance appears at $\hbar(\omega_{\text{L}} - \omega_{\text{s}}) = g_{\text{eff}}\mu_{\text{B}}B_{\text{res}}$, from which the effective g factor for the particular orientation of $\boldsymbol{B}$ is determined. ω_{L}, ω_{s} are here the frequencies of the incident laser and scattered radiation, respectively. The resonance magnetic field B_{res} is determined from the intensity maxima or from the points of inflexion of the CARS signal, depending on the strength of the resonant contribution to the CARS signal, which is dominated by interference of resonant and nonresonant parts of the nonlinear susceptibility (Geist et al., 1996).

Magnetic relaxation of DMS can be studied for example, with transient reflectivity. When excited with above-gap light, relaxation times of $Cd_{1-x}Mn_xTe$ materials are characteristic for spin-lattice relaxation, no phonon-bottleneck being observed. The time-dependent shift of the excitonic structure after excitation is directly connected with the temporal evolution of the magnetization, in particular with the spin temperature T of the magnetic subsystem. In Mn-based DMS, a small change in T for σ^+ polarization corresponding to the lowest exciton component (σ^- is often broader) creates a Zeeman shift proportional to the slope of the Brillouin function $\Delta E^+ = \Delta E_{\text{sat}} B_{5/2}(g\mu_{\text{B}}SH / k_{\text{B}}T_{\text{eff}})$ where $T_{\text{eff}} = T + T_0$. Transient reflectivity measurements show a temporal evolution during and after the pulse width τ_{p} as, respectively, $T_{\text{eff}}(t) = T^\infty - (T^\infty - T^0)\exp(-t/\tau_1)$, for $0 < t < \tau_{\text{p}}$, and $T_{\text{eff}}(t) = T^0 + (T^\infty - T^0)[1 - \exp(-\tau_{\text{p}}/\tau_1)]\exp[-(t - \tau_{\text{p}})/\tau_2]$ for $t > \tau_{\text{p}}$. The relaxation times τ_1, τ_2 correspond to different lattice temperatures and thus different spin-lattice relaxations, having a strong x dependence which rules out the thermalization of the spin and lattice as a whole, in which case the spin relaxation would be bottlenecked by lattice thermalization and would not show such a strong x dependence; usually $\tau_1 < \tau_2$ (Farah et al., 1996).

By exciting a DMS with intense fs pump pulses with circular polarization, and measuring the time-resolved MOKE rotation, Larmor precession of Mn^{2+} initiated by photoinjected carriers can be observed. The circularly polarized pump pulse, tuned to the heavy-hole (HH) exciton resonance, creates magnetic moments of electrons and HH along the growth direction in $CdTe/Cd_{1-x}Mn_xTe$ quantum wells (QWs), which is perpendicular to both the magnetic field and the sample surface (Voigt geometry). After pumping, conduction electron spins precess with

Larmor frequency amplified by the s-d exchange, while HH cannot precess, being quantized along the growth direction, and decay. The MOKE rotation, probed with a linearly polarized pulse, follows the time evolution of the magnetic moments of electrons, HH and Mn^{2+} ions. A typical signal as a function of the delay between pump and probe pulses is shown in Fig. 3.11.

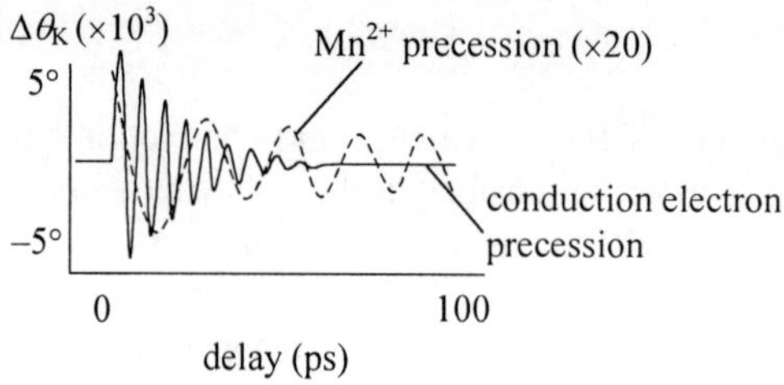

Fig. 3.11. Typical MOKE signal as a function of the delay in $CdTe/Cd_{1-x}Mn_xTe$ QWs

By fitting the Kerr rotation with a damped precession $\cos(2\pi\Omega_c t)\exp(-t/\tau_c)$, which describes the electron moments, where the Larmor frequency is $\Omega_c = \Delta E_c/\hbar$ with $\Delta E_c = g_c\mu_B H$, and a decaying component for HH, $\exp(-t/\tau_h)$, the spin relaxation times τ_c, τ_h and ΔE_c (the g factor for the conduction band electrons) are obtained as a function of B. Besides the rapid decaying oscillations of the electrons and HH, there is also a weaker, persistent oscillation modeled by $A_{Mn}\sin[2\pi\Omega_{Mn}(t-t_d)]\exp[-(t-t_d)/\tau_{Mn}]$, due to the precession of Mn^{2+} magnetization, where the delay time t_d describes the tipping of the Mn^{2+} moment away from the axis of the external field. $g_{Mn} \cong 2.01$, as well as the dependence on the external field of t_d, τ_{Mn} and A_{Mn} are obtained from fitting. Numerical simulations show that Mn^{2+} tipping is mainly initiated by the HH exchange field (Akimoto et al., 1998).

Besides the giant Faraday rotation and the giant Zeeman splitting of valence and conduction band states, magnetic polarons can also form in DMS. The magnetic polaron is a small region in the crystal in which the magnetic ions and the localized carriers have strongly correlated spins. The spin ordering induces a decrease of the carrier energy, sensed as a Stokes shift between the maximum of the PL line and the maximum of the PLE; the polaron shift is absent if the exciton lifetime τ_{ex} is much smaller than the magnetic polaron lifetime since the polaron formation process is interrupted by exciton recombination. The parameters of the magnetic polarons, such as magnetic polaron energy E_p and the average exchange field created by the localized excitons B_p can be directly determined from the magnetic field dependence of the giant Zeeman splitting of the free-exciton spin levels E_z and the circular polarization of the localized exciton luminescence ρ. The relevant formulae are $E_p = (1/2\pi k_B T)(dE_z/dB)/(d\rho/dB)$, $B_p = (1/\pi k_B T)(dE_z/dB)/(d\rho/dB)^2$. Observations of magnetic polarons can only be made in very restrictive conditions. For example, in $CdTe/Cd_{0.78}Mn_{0.17}Te$ quantum wells with well widths less than 30 Å, magnetic polarons can be observed at temperatures less than 4.2 K, when E_p is comparable with the PL halfwidth. B_p is about 0.6–0.7 T for temperatures less than 10 K (Merkulov et al., 1996). The polaron energy is found to increase and the formation time to decrease with an increase of the Mn concentration.

3.10 Manganites

Doped manganites are an interesting class of materials, which suffer an insulator-metal transition by approaching a critical temperature T_c from the high-temperature side due to crossover from small (Jahn–Teller) polarons to large polarons in the perfectly spin-polarized ferromagnetic state, caused by the enhancement of the one-electron bandwidth W. These materials are also characterized by a 'colossal' magnetoresistance above T_c. The insulator-metal transition is described by the double-exchange theory which includes the transfer integral t of the e_g electrons and the strong on-site exchange interaction between localized t_{2g} spins with $S = 3/2$ and itinerant e_g electrons. When the one-electron bandwidth W of the e_g state becomes smaller than the Hund-rule coupling J, the e_g band splits in two subbands (J-split band) separated by an amount determined by J. e_g carriers in the ferromagnetic ground state are perfectly spin polarized in contrast to conventional itinerant ferromagnetism. The 'colossal' magneto-resistance is explained by an additional mechanism for carrier localization above the transition temperature and a magnetic-field release of the localization, due to strong electron-lattice interaction caused by Jahn–Teller instability of Mn^{3+} ions. In optical spectra this transition is characterized by an increase in intensity of the mid-IR 1.5 eV band with decreasing temperature, and its merger below T_c into a Drude-like band (dotted line in Fig. 3.12).

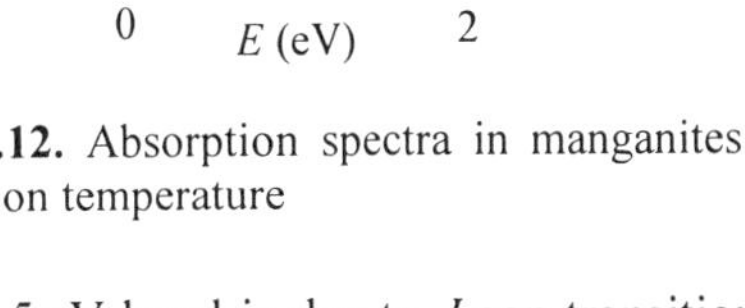

Fig. 3.12. Absorption spectra in manganites above (*solid line*) and below (*dotted line*) transition temperature

The 1.5 eV band is due to J gap transitions between the J-split bands, occupied at high temperature with almost the same number of up- and down-spin electrons. With decreasing temperature, the formation of small polarons makes a pseudogap around the Fermi level. The induced magnetization below T_c modifies the occupation of spin-up and spin-down electrons in both J-split bands, such that at low temperatures all electrons occupy the up-spin state, and the spectral weight transfers to the Drude component. The insulator-metal transition is thus accompanied by a paramagnetic-ferromagnetic transition. The transition temperature in $R_{1-x}Sr_xMnO_3$, with R a rare earth element, is controlled by W, which can be modified by changing the averaged ionic radius $\langle r_A \rangle$ of the perovskite site A. For R = Sm, the ferromagnetic metallic phase disappears below $\langle r_A \rangle = 1.32$ Å (Machida et al., 1998a). The small polaron binding energy E_p can be found by fitting the absorption in the small-polaron band with $\alpha_{sp}(\omega) \approx (1/\hbar\omega)\exp[-(2E_p - \hbar\omega)^2/8E_pE_{vib}]$ where E_{vib} is a characteristic vibrational energy equal to half of the coupled phonon frequency in the low-temperature limit. The peak position and width of the small-polaron absorption band are respectively J and $1.4W$ (Machida et al. 1998b).

Evidence for polaron formation and its interplay with electrical transport and magnetism can also be gathered from the anomalous behavior of Mn and equatorial $O_{xy}^{B_{1g}}$ modes due to local octahedral distortions and coupling to Mn^{3+} e_g electrons in layered manganites $La_{n-nx}Sr_{1+nx}Mn_nO_{3n+1}$ (Romero et al., 1998). In contrast to the pseudocubic compounds $R_{1-x}A_xMnO_3$ with R a trivalent rare earth ion, A a divalent alkaline earth, layered manganites, where $n = 1$ for a single layer, $n = 2$ for double layer and so on, form a tetragonal structure which comprises n $(La_{1-x}Sr_x)MnO_3$ perovskite layers between insulating $(La,Sr)_2O_2$ rocksalt layers. The tetragonal crystal field splits the degenerate Mn e_g orbitals, excluding the Jahn–Teller effect as the mechanism for small polaron formation in paramagnetic-insulator state. Studying the phononic Raman spectra as a function of temperature and magnetic field, it was found that the behavior of the Mn phonons which appear due to local octahedral distortion, is consistent with i) diminishing of local distortion due to a delocalization of Mn^{3+} e_g electrons at $T << T_c$, ii) incoherent character of the local distortion for $T >> T_c$ and iii) the distortions acquire short-range coherence close to the transition temperature due to the proximity to strong localization above T_c. These conclusions are also supported by the behavior of the $O_{xy}^{B_{1g}}$ mode.

Raman spectroscopy can also be used to study the dependence of lattice and charge excitation on doping and temperature. The phonons that appear in the Raman spectrum are an indication of a strong local Jahn–Teller-like lattice distortion in the paramagnetic state, which vanishes gradually below the ferromagnetic transition. The charge excitations are characteristic of the metallic state, their dependence on the doping level and symmetry selection rules being typical for plasmon-like excitations. Raman spectra have revealed a strong coupling between charge, lattice and spin excitations in $La_{1-x}Sr_xMnO_3$, for different doping (different x values) (Björnsson et al., 2000). In manganese perovskites of the general type $R_{1-x}A_xMnO_3$ the B_{1g}-symmetry electronic scattering intensity is drastically reduced near T_c by applying an external magnetic field, as is the magnetoresistivity. This field-dependent scattering rate suggests a field-dependent mobility along the Mn-O bond direction, predicted also in the double-exchange mechanism, in which charge transport involves hopping between Mn^{3+} and neighboring Mn^{4+} e_g states along the Mn-O-Mn bond. The external magnetic field aligns parallel to the magnetic core spin, which is a prerequisite for electron hopping between Mn sites, and so the spin scattering and resistivity is reduced. For example, for $0.2 < x < 0.5$ the paramagnetic insulator-to-ferromagnetic metal transition is accompanied by a decrease of more than 99% in the resistance in the presence of an applied field. In optical spectra, the low-frequency electronic Raman spectrum changes from a diffusive response in the paramagnetic-insulating phase to a flat continuum in the ferromagnetic metallic state (Liu et al., 1998). The effective hopping amplitude in the ferromagnetic metallic state depends on the normalized magnetization $m = M / M_s$ as $t_{\text{eff}}(m) \approx t[(1+m^2)/2]^{1/2}$ where t is the electron kinetic energy. This effective hopping amplitude has a monotonic decrease with temperature followed by a sharp drop at the transition temperature, a similar behavior to the effective electron density per Mn site N_{eff} calculated from the conductivity of the Drude and the mid-IR peak as $N_{\text{eff}} = (2m_e / \pi e^2 N_{\text{Mn}}) \int [\sigma_{\text{Drude}}(\omega) + \sigma_{\text{mid-IR}}(\omega)] d\omega$ with N_{Mn} the number of Mn atoms per unit volume. However, the drop in the effective hopping amplitude is sharper than that of the effective electron density per Mn site, which means that the decrease in N_{eff} with temperature is not entirely explained by $t_{\text{eff}}(m)$; the Jahn–Teller-induced suppression of the double-exchange interaction with a factor $\exp[-E_{\text{JT}} k_B T / (\hbar\omega_{\text{ph}})^2]$ could explain part of it, with ω_{ph} a characteristic phonon frequency. However, to account for the full behavior of N_{eff} one should account for the change in the electron configuration at the ferromagnetic metallic-paramagnetic insulating transition, reflected optically in the shift of the spectral weight from intraband excitations near Fermi level in the ferromagnetic phase to interband excitations in the paramagnetic phase, the latter involving excitation energies of the order of J. From fitting the experimental data, it is found that the effective coupling parameter $\lambda = E_{\text{JT}} / t$ in $La_{0.67}Ca_{0.33}MnO_3$ is approximately 1, which suggests a moderate electron-phonon coupling (Boris et al., 1999).

3.11 Color Centers

Color centers are still a subject of interest, although their main optical properties have been well known for a long time. An early review on the optical properties of color centers is due to Pick (1972). A more recent review is that by Henderson and O'Donnell (1989). Color centers are electrons or holes trapped at defect sites in alkali halide crystals (or more complex ionic solids), the simplest being the F center, which consists of a single electron trapped by the Coulomb potential of an isolated anion vacancy. Since the peak frequency of the optical absorption band varies with the lattice parameter a as $\omega_{\max} \approx d^{-2}$, it can be modeled as an electron in a potential box of diameter a. The F absorption band has an almost Gaussian shape, the oscillator strength being independent of concentration and temperature. The coupling between the electronic center and the lattice can be represented by an effective frequency ω_a, the halfwidth of the F band depending on temperature as $\Gamma^2(T) \approx \coth(\hbar\omega_a / 2k_B T)$. The F center is paramagnetic, with a g value near that of the free electron. Its ground state has an octahedral O_h symmetry with an s-type wavefunction, while the first excited state has a 2p-like wavefunction. Other color centers form from F; for example F^- designates the defect in which two electrons are trapped at one anion vacancy. Crystals containing only F and F^- electronic centers are not in thermal equilibrium at low temperatures, so that F centers combine to form aggregates such as F_2 (composed of two F centers), F_3 (from three F centers), F_4 (from four) and so on. The F_2 center has D_2 symmetry, the F_3 center C_{3v} symmetry, whereas for F_4 two symmetries are possible: a tetrahedral T_d symmetry and a planar combination of four anion vacancies with C_s symmetry. More complex centers are F_A, formed by replacing one nearest-neighbor cation of an F center with an alkali impurity ion (for example Na^+ in KCl) or F_Z when the substitutional ion is divalent (Ca^{2+} in KCl for example). In contrast to these lattice defect centers, V color centers are trapped hole centers, with molecular properties. For example V_K centers are formed when two anions with filled electronic shells are combined by the trapping of a hole to form a molecular ion. The V_H center is a neutral center similar to V_K but with an additional halogen atom in the interstitial site, and annihilates the F center at high temperatures. More complex defects can form by associating lattice defects to V color centers. Additionally, there are substitutional and interstitial hydrogen centers and substitutional OH^- centers. Detailed discussions on the structure and optical properties of these color centers can be found in the review papers mentioned at the beginning of this subsection.

More recent studies have been focused for example, on the bistability of $F_H(OH^-)$ centers in alkali halides. Two bistable configurations, B and R, have been identified in various hosts such as KBr, RbBr, RbI, characterized by partially overlapping electron absorption (blueshifted and redshifted from the F band), and by spectrally well separated stretching mode vibrational absorption lines. At low temperatures there is a reversible conversion between B and R electronic bands, in both directions, with high quantum efficiency. The observed bistability and the measured electronic and vibrational absorption lines at temperatures lower than 10 K suggest a large linear off-center displacement of cation between F and OH^- along the pair axis, which is strongly coupled to the OH^-, OD^- translational/rotational motion. These coupled motions introduce anharmonicity into the total potential, the single- or double-well shape of which decides the possibility of bistability and the preference for B or R configurations in various hosts (Dierolf and Luty, 1996). The relaxation of $F_H(OH^-)$ has also been studied. In KBr for example, it was found that after optical excitation with picosecond pump pulses, different relaxation components have been identified. One of them is a nearly temperature-independent relaxation, with a characteristic decay time of a few ps, due to radiationless electronic transitions during lattice relaxation, which appear mainly when a quantum of stretch vibration is excited. Another component present at all temperatures, slower than 10 ns, is a vibrational relaxation due to the influence of the stretch vibration on the electronic absorption. Finally, the last, temperature-dependent component, with a time constant of about 100 ps at 50 K is due to recovery of the thermal equilibrium between the two $F_H(OH^-)$ configurations, characterized by different electronic absorption bands and different orientations of OH^- with respect to the F centers (Gustin et al., 1996).

The study of F centers is useful for gaining knowledge about the optically excited state of defects with deep electronic levels coupled strongly with phonons in solids. In particular, the superhyperfine spectrum of the relaxed excited state of the F center was obtained from the decrement dips of the quantum efficiency of the F center PL as a function of the magnetic field. Such studies were performed in KCl for magnetic fields below 0.31 T at 4 K, the method being known as optical detection of electron-nuclear double resonance. The data show that the superhyperfine spectrum depends on the angle between the magnetic field and the crystalline axis. The isotropic and anisotropic superhyperfine interaction constants can then be obtained from fitting the angular dependence (Akiyama and Ohkura, 1996).

3.12 Filled-Shell Ions

This category of ions was also studied quite a long time ago, the interest in it fading over recent years. We mention it here for completeness. A review on the optical properties of filled-shell ions can be found in Jacquier (1979). The ground state of filled-shell or closed-shell ions is the completely occupied d^{10} or ns^2 configuration, possible interconfiguration transitions taking place between these states and the excited state configurations d^9s, d^9p or $ns\,np$.

The relative magnitudes of the electron repulsion, crystal-field and spin-orbit splitting of these centers can be determined by calculating the electronic structure with the molecular-orbital method. Due to the Pauli exclusion principle the total wavefunction of a closed-shell many-electron system is antisymmetric and can be written as a single Slater determinant, the states being classified in the Russell–Saunders coupling scheme as $|\,E\;S\;L\;M_S\;M_L\rangle$, with E the average energy of the configuration. The spin-orbit coupling for the electrons outside the closed-shell is considered as a perturbation, the eigenfunctions of these states being linear combinations of Slater determinants, dependent on the specific configuration. When a host ion is substituted by closed-shell ions, several absorption bands appear, besides those of the pure crystal, that with the lowest energy occurring generally in the gap of the crystal. The d orbitals are less sensitive to the ligand field than s orbitals due to the different overlap with the orbitals of surrounding anions. In the crystal the orbital energies are significantly upshifted and their degeneracy is generally lifted, strong mixing appearing between the orbitals of the s impurity center with the wavefunction of valence electrons of the ligand.

By analogy with the atomic model, in the molecular orbital method the electrons are assumed to move in one-electron orbitals (molecular orbitals) of an electrically neutral or charged cluster imbedded in the field of the rest of the crystal. The cluster includes the central doping ion and the nearest-neighbor ligands, the electronic interactions between electrons of different orbitals of various neighbors in this pseudomolecule being also accounted for. The total wavefunction of the cluster, as for an atom, is expressed as a single antisymmetrized product of molecular orbitals (MO), grouped into core and valence sets. The core MOs include the orbitals that do not participate in the bond and are approximately equal to the inner-shell atomic orbitals, while the best valence MO, ψ_k, is taken as a linear combination of valence atomic orbitals $\psi_k = \sum_{s,i} c_{ksi}\phi_{si}$ with ϕ_{si} the atomic orbital of atom s. The MOs are completely determined by solving the secular equation $\det|F_{si,tj} - S_{si,tj}E_k| = 0$ where $F_{si,tj} = \langle\phi_{si}|F|\phi_{tj}\rangle$, $S_{si,tj} = \langle\phi_{si}|\phi_{tj}\rangle$ with F the Hartree–Fock Hamiltonian of the cluster. Several approximations are used to solve this equation.

3.13 Optical Properties of Donors and Acceptors in Semiconductors

Unlike the impurities we have encountered up to now in insulating crystals, the impurities in semiconductor materials have the possibility to participate in the conduction properties of the host materials, upon thermal activation. The properties of semiconductor materials are drastically modified by the introduction of donor or acceptor states, whereas insulating crystals modify the properties of the impurities relative to their state in free space.

In semiconductors, the absorption on donor or acceptor levels can take place between different dopant levels or between dopant levels and the conduction and valence bands (see Fig. 3.13)

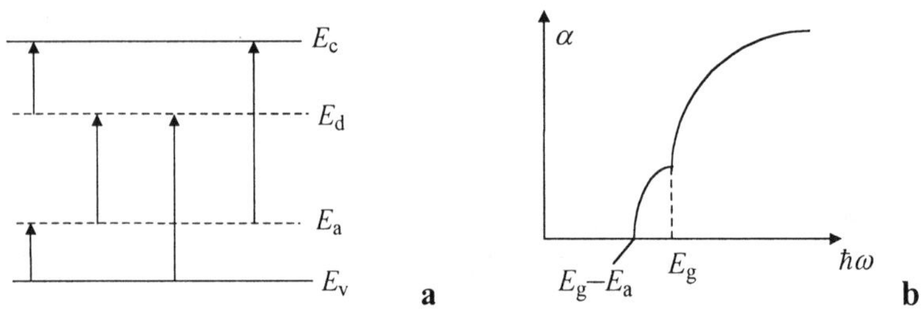

Fig. 3.13. (**a**) Possible absorption transitions in doped semiconductors and (**b**) the absorption spectrum in direct semiconductors doped with acceptors

When absorption takes place on the acceptor energy level, for example, the spectrum has a similar threshold as in the fundamental absorption case, but is smaller in intensity and shifted towards smaller energies. Figure 3.13 shows the total absorption spectrum, the shoulder at lower energies representing the contribution of acceptors.

The absorption on donors or acceptors can be observed for moderate dopant concentrations in doped semiconductors and for small concentrations of dopants in compensated semiconductors. The absorption coefficient is proportional to the concentration of donors or acceptors, and is much smaller than that for the fundamental absorption due to the smaller density of states. The absorption on dopant levels close to the conduction or valence bands can be observed in the far IR ($\lambda < 25\ \mu m$) and at low temperatures, to avoid the ionization of the impurities.

Besides the fundamental level, donors or acceptors can also have excited levels situated in the forbidden band, which can also take place in absorption. The spectrum in this case looks like that in Fig. 3.14.

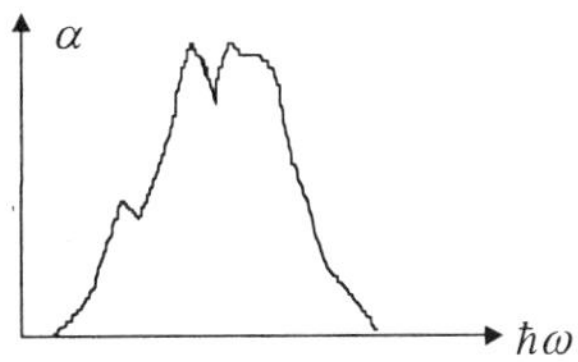

Fig. 3.14. Absorption spectrum of donors or acceptors when their excited levels are situated in the energy gap

The absorption begins to rise when the photon energy becomes equal to the first excited state $n = 1$ of the dopant, the other maxima in the absorption spectrum corresponding to $n = 2,3....$ The subsequent decrease of α with $\hbar\omega$ is due to the decrease in the density of states with the departure from the bottom of the conduction band.

For high concentrations of randomly distributed donors or acceptors the discrete energy levels of impurities become broader and can eventually merge with either conduction or valence bands. As a result, band tails in the density of states appear (see Fig. 3.15; the density of states for an undoped semiconductor is represented by a dotted line)

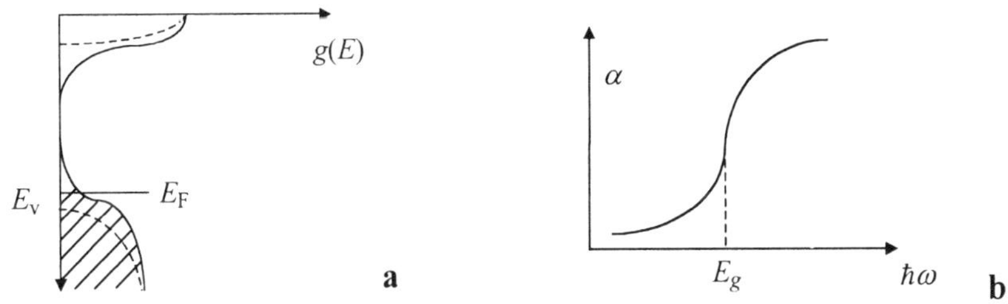

Fig. 3.15. (**a**) Density of states for high concentrations of randomly distributed donors or acceptors in a semiconductor and (**b**) the corresponding absorption spectrum

The absorption spectrum in this case has an exponential decay, instead of a sharp decrease, at the fundamental bandgap energy, described by the Urbach formula

$$\mathrm{d}(\ln\alpha)/\mathrm{d}(\hbar\omega) \approx 1/k_{\mathrm{B}}T. \tag{3.30}$$

The properties of different donor and acceptor states can be determined by optical measurements. For example, the binding energies of shallow donors in GaN can be determined by Zeeman spectroscopy. By measuring at 4 K the rate splitting with the magnetic field of the $1\mathrm{s} \rightarrow 2\mathrm{p}^{\pm}$ transitions, given by $\Delta E = hB/2\pi m_{\mathrm{eff}}$, one can measure the effective mass $m_{\mathrm{eff}} = 0.22 m_0$. The effective-mass theory predicts that the ground-state binding energy $E_{\mathrm{D}} = 13.606\,(m_{\mathrm{eff}}/m_0)/\varepsilon^2$ of the observed donor is 4/3 times the energy of $1\mathrm{s} \rightarrow 2\mathrm{p}^{\pm}$ transitions. So, from the peak position and the magnetic field splitting it is possible to determine both E_{D} and the static dielectric constant ε (Moore et al., 1997).

Lyman-type transitions are observed for some impurities, for example for Be acceptors in GaAs. At 1.9 K, a series of sharp lines at 134.42, 166.76 and 182.3 cm^{-1} are identified as the G, D and C lines, with their integrated intensities in the ratio 11.5:100:57.6 (Lewis et al., 1996). The electronic nature of these lines is confirmed by their behavior with temperature: they sharpen and grow in intensity as the temperature decreases. The ratios between the lines remain constant, although the PL intensities decrease with temperature due to the thermal ionization of acceptor sites.

The Luttinger Hamiltonian parameters for the valence band can be obtained by studying the spin-flip Raman scattering of holes bound to acceptors, as has been done, for example, in p-type nitrogen-doped zinc selenide (Orange et al., 1997). Due to the strain, the ground state of acceptor-bound hole is split into two (in light hole (LH) and HH bands), the splitting being measured by the resonant excitation of the PL at the energy of the acceptor-bound exciton emission $\mathrm{A}^0\mathrm{X}$. If the excitation corresponds to a transition between the lower acceptor level to the excitonic state, PL can occur at an energy corresponding to a transition from the excitonic state to the upper acceptor level, the difference between the excitation and emission energies giving the strain splitting. The PL signal is similar to a Raman signal in the sense that it has a constant energy shift relative to the excitation. To confirm the fact that these lines are related to the split of the valence band, magneto-optic spectroscopy of the spin-flip Raman scattering has been performed for different orientations of the magnetic field with respect to the normal to the sample, the latter being parallel to the incident light and the direction of strain. In the Faraday geometry two lines are observed, corresponding to the transitions $m_J = -3/2$ to $-1/2$, and $-3/2$ to $1/2$, respectively, whose positions depend on the orientation of the magnetic field. This anisotropy is related to the valence band states, the conduction band states being assumed isotropic. From the energy shifts of these lines, the parameters of the Luttinger Hamiltonian can be determined from fitting.

The Luttinger parameters for ZnSe doped with Li can also be determined by studying the Fano resonance between LO-phonon-coupled excited states of Li and the valence band of ZnSe. In IR absorption for the crystal-field split $2P_{5/2}$ state of Li, two twin peaks were observed, the Luttinger parameters obtained from the position of these peaks and the cyclotron mass in the [111] direction being estimated as $\gamma_1 = 6.44$, $\gamma_2 = 2.58$, $\gamma_3 = 2.74$. The lineshape of the Fano resonance is modeled as $F(E,q) = (q+E)^2/(1+E^2)$, with $q = 1.07$ the distortion parameter, and $E = [\hbar\omega - (E_{\rm bh} + \hbar\omega_{\rm opt} + f(\hbar\omega))]/(\Gamma/2)$, where $E_{\rm bh}$ is the energy of the bound hole, $\hbar\omega$ that of the incident photons, $\hbar\omega_{\rm opt}$ of the optical phonons, $\Gamma = 16.8\ {\rm cm}^{-1}$ the spectral width of the coupled state, and $f(\hbar\omega) = 2.8\ {\rm cm}^{-1}$ the strength of the hole-LO-phonon coupling (Nakata et al., 1999).

Bound excitons in Be-doped Si are assumed to form in two stages: the electron binds first to the Be isoelectronic complex radiative centers, and the hole binds then to the electron to form the bound exciton. (An impurity replacing a host atom with the same valence orbitals is called isoelectronic.) The binding energy of the bound exciton relative to the two free particles can be determined by measuring the PL dependence on an applied hydrostatic pressure P. The binding energy is given by $E_{\rm ex} = E_{\rm ig} - ({\rm d}E_{\rm ig}/{\rm d}P)P - E_{\rm PL}$ where $E_{\rm ig}$ is the indirect bandgap energy of Si and $E_{\rm PL}$ the zero-phonon PL peak in the presence of pressure. The experimental results, which show a change in the exciton binding energy with pressure and the disappearance of the PL above 60 kbar are most accurately explained with models of the electron-binding Hamiltonian which are not based on the Coulomb interaction, but on the creation of a potential well which binds the electron to the Be isoelectronic complex (Kim et al., 1996).

The concentration of interstitial vacancy and antisite pairs in a host can be determined from the downshift of the LO-phonon Raman peak in samples with defects. These defects are introduced in semi-insulating GaAs, for example, by neutron-transmutation doping, and are accompanied by a shift of the TO phonon in irradiated samples, much smaller, however, than the peak shift of the LO phonon. The corresponding frequency shift for vacancies is $\omega_{\rm LO}\Delta\omega_{\rm LO} = -12\Omega^2 N_{\rm V}/N$ where $N_{\rm V}$ is the number of vacancies, N the number of primitive unit cells, and $\Omega = 4\pi n_{\rm p}Q^2/\varepsilon_\infty m_{\rm r}$ is the ionic plasma frequency, with $n_{\rm p}$ the density of Ga-As pairs, $m_{\rm r}$ their reduced mass and Q the effective charge, which includes the local charge and a contribution from the nonlocal effective charge. The effect of the $N_{\rm a}$ antisite pair defects on the phonon frequency is $2\omega_{\rm LO}\Delta\omega_{\rm LO} = -4.68\Omega^2 N_{\rm a}/N$. Knowing the value of Q, the density of vacancy and antisite pairs has been determined experimentally as 2.2% and 1.4%, respectively (Kuriyama et al., 1996).

The temperature dependence of PL intensities for impurities has several forms. For example, some Cd-related defects, including several $\rm Cd_A$, $\rm Cd_B$ defects in Si, show a temperature dependence of the zero-phonon line of the form $I(T) = I(0)/[1 + GT^{3/2}\exp(-\Delta E/k_{\rm B}T)]$, where G is the band density of states relative to the defect density of states, and ΔE, which is the quantity derivable from this dependence, is the thermal binding energy (McGlynn et al., 1996). These defect lines show no Zeeman splitting or shift, which suggests that they are due to recombination of tightly bound holes to loosely bound electrons at isoelectronic centers. This behavior is typical for many defects in Si; if a is the electron-phonon coupling, the Huang–Rhys factor for sidebands is $S = a^2/(2m\hbar\omega^3)$.

In other materials, such as for example in GaN, the defects are characterized by sharp lines in the Raman spectrum between 95–250 cm^{-1}, which decrease exponentially in intensity with temperature and are excitable only in the range 2–2.5 eV. The activation energies of the lines, obtained from fitting the expression $I(T) = A/[1 + C\exp(-E_{\text{act}}/k_{\text{B}}T)]$, with A, C constants, are 10–60 meV. This temperature dependence suggests that the lines are not vibrational Raman lines, since they should in this case increase with temperature, but correspond to defect lines in which the involved electron states have a thermal occupation. Moreover, since they are excitable only in a certain range, the defects must be vibronicaly excited with a resonance process having this temperature dependence (Siegle et al. 1998).

Not only donors or acceptors can be found in semiconductors, but also donor-acceptor pairs can be created on illumination. An electron trapped at a donor and a hole trapped at an acceptor tend to recombine if the temperature is not high enough to eject them again in the conduction or valence bands. If the transition energy $E_{\text{g}} - (E_{\text{A}} + E_{\text{D}})$ is far larger than the energy of acoustic phonons, multiphonon processes become highly improbable, and radiative recombination dominates. The real transition energy for a pair at distance r_{AD} differs from $\hbar\omega_\infty = E_g - (E_{\text{A}} + E_{\text{D}})$ in the first approximation through a term that accounts for the polarization effects of trapped electrons or holes $\hbar\omega(r_{\text{AD}}) = \hbar\omega_\infty + e^2/Kr_{\text{AD}}$. For donor-acceptor (D-A) pairs with weak coupling to phonons, few phonon replicas are observed in addition to the zero-phonon spectrum, whereas for strong phonon coupling, each D-A with a well-defined r_{AD} gives a large bell-shaped spectrum, the resulting emission being a superposition of all these spectra. The probability of emitting $m = 0,1,2..$ phonons assisting the D-A transition is given by a Poisson distribution. When the coupling with the lattice is linear, for large values of the Huang–Rhys coupling constant S or at sufficiently high temperatures, the spectrum is nearly Gaussian

$$I(\hbar\omega, r_{\text{AD}}) = (1/\sigma\sqrt{2\pi})\exp[-(\hbar\omega - \hbar\omega_0)^2/2\sigma^2], \tag{3.31}$$

the constant $\hbar\omega_0$ depending on r_{AD}. The wide bell-shaped spectrum observed during photoexcitation is then $I(\hbar\omega) = \int_0^\infty N(r_{\text{AD}})I(\hbar\omega, r_{\text{AD}})\text{d}r_{\text{AD}}$, where $N(r_{\text{AD}})$ is the number of emitting centers in the shell $\text{d}r_{\text{AD}}$, $N(r_{\text{AD}})\text{d}r_{\text{AD}} = \text{const.} \times \exp[-4\pi r_{\text{AD}}^3 N/3]4\pi r_{\text{AD}}^2 \text{d}r_{\text{AD}}$.

For example, in nitrogen-doped ZnSe, neutral shallow or deep donor-acceptor pairs with the electron and hole in the i, j excited states, respectively, are created on illumination with pump photons of energy $E_{\text{exc}} = E_{\text{g}} - E(\text{A}_j) - E(\text{D}_i) + e^2/\varepsilon R_{ij} + J_{ij}(R_{ij})$ where R_{ij} is the pair separation and J_{ij} are corrections to the Coulomb effect due to the overlap of the excited donor and acceptor wavefunctions. After donor-acceptor pair formation, the electron and hole thermalize into the ground state of the impurity before recombination, so that the selective PL spectrum shows a series of sharp resonance lines for $E_{\text{lum}} = E_g - E(A_{1S3/2}) - E(D_{1S}) + e^2/\varepsilon R_{ij} + J_{11}(R_{ij})$, for each (i, j) excited level. Since the pair interacts with phonons during thermalization and even during recombination, the R dependence is arbitrary, the process involving both local phonon modes in the vicinity of the impurity and lattice modes, such that $E_{\text{exc}} - E_{\text{lum}} = \hbar\omega_{\text{ph}}$ (Morhain et al., 1996).

In semiconductors codoped with both donor and acceptor impurities, such as ZnSe codoped with N and Cl, the zero-phonon emission line appears at $E(\hbar\omega) \cong E_g - (E_D + E_A) - (e^2/2\pi\varepsilon)(N_t^{2/3}/N^{1/3})$ where N_t is the concentration of charged impurities and $N = |N_A - N_D|$ is the density of uncompensated carriers. The donor-acceptor pair recombination band is broadened and shifted for increasing Cl concentration, due to charged ionized impurities that cause potential fluctuations in crystals (Behringer et al., 1998).

More complex defects can exist in semiconductor materials, such as for example the presence of H_2^* dimers in Ge. This defect was studied by Budde et al. (1996), and is represented in Fig. 3.16.

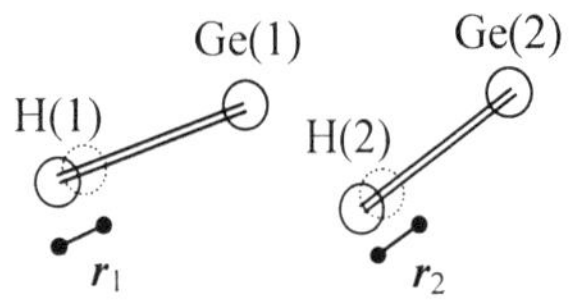

Fig. 3.16. Complex defects in semiconductor materials: H_2^* dimers in Ge

H_2^* dimers are present in optical spectra through Ge-H stretch or bend modes associated to trigonal site symmetry, as results from their behavior when a uniaxial stress is applied (they split into two for [111] stress, but do not split for [100] stress). The fact that these are local vibrational modes of H_2 coupled to Ge is confirmed by the downward shift of some lines associated with these defects by a factor of $\sqrt{2}$ when the proton is replaced by a deuteron, according to the theoretical relation between the Ge-H and Ge-D stretching frequencies: $\omega_H/\omega_D = \sqrt{m_{r,D}/m_{r,H}}$, where $1/m_{r,H,D} = 1/m_{H,D} + 1/m_{Ge}$ are the reduced masses. The vibration frequencies of the dimmer defect are calculated by diagonalizing the harmonic Hamiltonian of the diatomic molecule, or, for a better approximation, by local-density-functional cluster theory. The experimental data can give us, however, the ratio of the effective charges associated to the two pairs of the H_2^* dimer defect: $e_1/e_2 \cong I_1/I_2$, where I_i are the integrated absorption of the modes. The resulting ratio is 1.1 for Ge.

In far-IR, the photoexcited electrons and holes are captured by the ionized impurities, a process followed by a subsequent donor-acceptor recombination, depending on the distance between donors and acceptors. After recombination, the ionized impurities produce an inhomogeneous electric field which Stark shifts the energy levels of neutral impurities, and broadens their linewidths. The broadening of the impurity absorption lines is caused by a series of other mechanisms, besides the Stark effect, such as for example, the phonon lifetime broadening due to the interaction of bound carriers with phonons, strain-induced broadening due to dislocations, impurities and precipitates, and concentration broadening due to the overlap of the extended hydrogenic impurity wavefunctions. To isolate the effect of electric-field broadening, experiments should be made for a constant net concentration of impurities. The effect of the electric field on the external potential $V(\boldsymbol{r})$ of an impurity expressed in spherical coordinates, is found by expanding $V(\boldsymbol{r})$ in a Taylor series. The $l=0$ term produces an r-dependent shift of the energy levels, while the Stark shift corresponds to the $l=1$ term. In weak electric fields E, the shifted energy levels of an impurity with principal quantum number n, and magnetic quantum number m, are

$$E=-e^2/(2an^2)+(3/2)n(n_1-n_2)eEa-a^3[17n^2-3(n_1-n_2)^2-9m^2+19]E^2n^4/16, \tag{3.32}$$

where a is the effective Bohr radius and n_1, n_2 are non-negative integers such that $n=n_1+n_2+|m|+1$. The term linear in E – the linear Stark term – produces a linewidth increase proportional to $(N_i/N_0)^{2/3}$ where N_i, N_0 are the concentrations of ionized and neutral impurities, respectively, whereas the quadratic Stark term produces a linewidth increase which scales as $(N_i/N_0)^{4/3}$. Additionally, there is also a contribution to the linewidth broadening proportional to N_i/N_0 due to the $l=2$ term in the Taylor expansion of the potential, called quadrupole broadening, caused by the interaction between the electric quadrupole moment of the neutral impurity and the field gradient induced by the ionized impurities. The last contribution to linewidth broadening dominates at small N_i, the quadratic Stark field predominates at high N_i, whereas the linear Stark effect is usually neglected (Itoh et al., 1996). In highly compensated p-type semiconductors, such as Ge, the study of the broadening of the absorption line of Ga acceptors as a function of excitation light intensity with above-bandgap energy revealed a $N_i^{4/3}$ dependence, which corresponds to a second-order Stark effect (Harada et al., 1996). In neutron-transmutation doped Ge, on the other hand, the linewidth was observed to broaden proportional to N_i, and not to N_i/N_0. This behavior was explained by the fact that the ionized impurities are correlated at low temperatures, and not randomly distributed. The electrons and holes redistribute among the randomly distributed donors and acceptors such that the resulting state corresponds to an energy minimum. As a result, the average strength of the electric field is reduced, the ionized impurities being correlated at energies much smaller than the Coulomb interaction between majority impurities, i.e. for $k_BT \ll e^2N_{mj}^{1/3}/\varepsilon$, and randomly distributed for $k_BT \gg e^2N_{mj}^{1/3}/\varepsilon$, where N_{mj} is the concentration of majority impurities (Itoh et al., 1996).

The magnetic circular dichroism of the optical absorption, which is the differential absorption of left- and right-circularly polarized light propagating along an external magnetic field, can be used to identify different defects in GaAs. Different arsenic-antisite-related defects, such as As-antisite, As-antisite-related double donor, As-antisite defect with antistructure pair, and so on, produce signals with different forms in magnetic circular dichroism of the optical absorption (Koschnick et al., 1998). These can be easily observed and identified due to the high abundance of magnetic isotopes of both lattice atoms in GaAs.

The study of irradiation-induced PL in 4H and 6H polytypes of SiC, and the optically detected magnetic resonance (ODMR) signals have demonstrated the formation of primary defects such as vacancies. ODMR spectra measure the fractional change in the PL induced by a magnetic resonance in the excited state of a local defect, as a function of the magnetic field. The different charge states of the vacancy have a strong lattice relaxation (Jahn–Teller effect) in Si. The PL consisted of a number of sharp no-phonon lines equal to the number of inequivalent sites in the two SiC polytypes, which are each split in two in ODMR spectra. The splitting is dependent on the angle θ between the magnetic field and the c axis of the crystal, characteristic for a spin-triplet state (S = 1) with axial symmetry, and which can be fitted by $g\mu_B B = \hbar\omega \pm (D/2)[3(\cos\theta)^2 - 1] + \mathrm{A}M_I$ with g the gyromagnetic ratio, μ_B the Bohr magneton, D the zero-field splitting of the magnetic sublevels and A the isotropic hyperfine interaction. g and D can be identified from experimental data. If the ODMR signal is obtained for resonant excitation energies identical to the no-phonon lines in the PL spectrum, it is possible to resolve the hyperfine structure of ODMR signals, each line having a symmetric pair of inner and outer satellites split at 0.3 and 0.6 mT (A and $2A$) respectively, which overlap partially with the central line, and which have intensities about 27% and 4%, respectively, of the central line. The weakness of the satellites is explained by the low natural abundance of nuclear-spin active isotopes of both Si and C. The fact that these lines are related to primary irradiation defects is confirmed by the increase of the PL and ODMR signals with the irradiation dose and by the insensitivity of the signals on the impurity content; they were assigned to internal transitions from an excited state of the neutral Si vacancy at the inequivalent sites (Sörman et al., 2000).

References

Akimoto, R., K. Ando, F. Sasaki, S. Kobayashi, and T. Tani (1998) Phys. Rev. B **57**, 7208.

Akiyama, N. and H. Ohkura (1996) Phys. Rev. B **53**, 10632.

Arima, A., H. Horie, and Y. Tanabe (1954) Prog. Theor. Phys. **11**, 143.

Bak, J., C.L. Mak, R. Sooryakumar, U.D. Venkateswaran, and B.T. Jonker (1996) Phys. Rev. B **54**, 5545.

Balda, R., J. Fernández, A. de Pablos, J.M. Fdez-Navarro, and M.A. Arriandiaga (1996a) Phys. Rev. B **53**, 5181.

Balda, R., J. Fernández, J.L. Adam, and M.A. Arriandiaga (1996b) Phys. Rev. B **54**, 12076.

Barker Jr., A.S. and A.J. Sievers (1975) Rev. Mod. Phys. **47**, Suppl. No. 2.

Beck, W., and D. Ricard, and C. Flytzanis (1998) Phys. Rev. B **58**, 11996.

Behringer, M., P. Bäume, J. Gutowski, and D. Hommel (1998) Phys. Rev. B **57**, 12869.

Björnsson, P., M. Rübhausen, J. Bäckström, M. Käll, S. Eriksson, J. Eriksen, and L. Börjesson (2000) Phys. Rev. B **61**, 1193.

Boris, A.V., N.N. Kovaleva, A.V. Bazhenov, P.J.M. van Bentum, Th. Rasing, S.W. Cheong, A.V. Samoilov, and N.C. Yeh (1999) Phys. Rev. B **59**, 697.

Budde, M., B.B. Nielsen, R. Jones, J. Goss, and S. Öberg (1996) Phys. Rev. B **54**, 5485.

Callaway, J. (1991) *Quantum Theory of the Solid State*, Academic Press, Boston, 2nd edition.

Catunda, T., L.A.O. Nunes, A. Florez, Y. Messaddeq, and M.A. Aegerter (1996) Phys. Rev. B **53**, 6065.

DiBartolo, B. (1979) in *Luminescence of Inorganic Solids*, ed. D. DiBartolo, Plenum Press, New York, p.15.

Dierolf, V. and F. Luty (1996) Phys. Rev. B **54**, 6952.

Downer, M.C. (1989) in *Laser Spectroscopy of Solids II*, ed. W.M. Yen, vol.65 in Topics in Applied Physics, Springer, Berlin, Heidelberg, p.29.

Ellens, A., H. Andres, H.L.H. ter Heerdt, R.T. Wegh, A. Meijerink, and G. Blasse (1997) Phys. Rev. B **55**, 180.

Farah, W., D. Scalbert, and M. Navrocki (1996) Phys. Rev. B **53**, 10461.

Farah, W., D. Scalbert, M. Navrocki, J.A. Gaj, E. Janik, G. Karczewski, and T. Wojtowicz (1998) Phys. Rev. B **57**, 8770.

Feofilov, S.P., A.A. Kaplyanskii, R.I. Zakharchenya, Y. Sun, K.W. Jang, and R.S. Meltzer, (1996) Phys. Rev. B **54**, 3690.

Fuhs, W., I, Ulber, G, Weiser, M.S. Bresler, O.B. Gusev, A.N. Kuznetsov, V.K. Kudayarova, E.I. Terukov, and I.N. Yassievich (1997) Phys. Rev. B **56**, 9545.

A.J. García and J. Fernandez (1997) Phys. Rev. B **56**, 579.

F. Geist, H. Pascher, N. Frank, and G. Bauer (1996) Phys. Rev. B **53**, 3820.

Graf, F.R. A. Renn, U.P. Wild, and M. Mitsunaga (1997) Phys. Rev. B **55**, 11225.

Gregorkiewicz, T., D.T.X. Thao, J.M. Langer, H.H.P.Th. Bekman, M.S. Bresler, J. Michel, and L.C. Kimerling (2000) Phys. Rev. B **61**, 5369.

Gustin, E., M. Leblans, A. Bouwen, and D. Schoemaker (1996) Phys. Rev. B **54**, 6977.

Harada, Y., K. Fujii, T. Ohyama, K.M. Itoh, and E.E. Haller (1996) Phys. Rev. B **53**, 16272.

Henderson, B. and K.P. O'Donnell (1989) in *Laser Spectroscopy of Solids II*, ed. W.M. Yen, vol.65 in Topics in Applied Physics, Springer, Berlin, Heidelberg, p.123.

Hoerstel, W., W. Kraak, W.T. Masselink, Y.I. Mazur, G.G. Tarasov, E.V. Kuzmenko, and J.W. Tomm (1998) Phys. Rev. B **58**, 4531.

Holstein, T., S.K. Lyo, and R. Orbach (1981) in *Laser Spectroscopy of Solids*, eds. W.M. Yen and P.M. Selzer, vol.49 in Topics in Applied Physics, Springer, Berlin, Heidelberg, p.39.

Huber, D.L., D.S. Hamilton, and B.B. Barnett (1977) Phys. Rev. B **16**, 4642.

Imbusch, G.F. (1979a) in *Luminescence of Inorganic Solids*, ed. B. DiBartolo, Plenum Press, N.Y., p.135.

Imbusch, G.F. (1979b) in *Luminescence of Inorganic Solids*, ed. B. DiBartolo, Plenum Press, N.Y., p.155.

Imbusch, G.F. and R. Kopelman (1981) in *Laser Spectroscopy of Solids*, eds. W.M. Yen and P.M. Selzer, vol.49 in Topics in Applied Physics, Springer, Berlin, Heidelberg, p.1.

Itoh, K.M., J. Muto, W. Walukiewicz, J.W. Beeman, E.E. Haller, H. Kim, A.J. Mayur, M.D. Sciacca, A.K. Ramdas, R. Buczko, J.W. Farmer, and V.I. Ozhogin (1996) Phys. Rev. B **53**, 7797.

Jacquier, B. (1979) in *Luminescence of Inorganic Solids*, ed. B. DiBartolo, Plenum Press, New York, p.235.

Jandl, S., P. Dufour, T. Strach, T. Ruf, M. Cardona, V. Nekvasil, C. Chen, and B.M. Wanklyn (1995) Phys. Rev. B **52**, 15558.
Judd, B.R. (1962) Phys. Rev. **127**, 750.
Kim, S., I.P. Herman, K.L. Moore, D.G. Hall, J. Bevk, (1996) Phys. Rev. B **53**, 4434, 1996.
Koschnick, F.K., K.H. Wietzke, and J.M. Spaeth (1998) Phys. Rev. B **58**, 7707.
Kuriyama, K., S. Satoh, and M. Okada (1996) Phys. Rev. B **54**, 13413.
Lewis, R.A., T.S. Cheng, M. Henini, and J.M. Chamberlain (1996) Phys. Rev. B **53**, 12829.
Liu, G.K., R. Cao, and J.V. Beitz (1996) Phys. Rev. B **53**, 2385.
Liu, H.L., S. Yoon, S.L. Cooper, S.W. Cheong, P.D. Han, and D.A. Payne (1998) Phys. Rev. B **58**, 10115.
Lyo, S.K. (1971) Phys. Rev. B **3**, 331.
Mac, W., M. Herbich, N.T Khoi, A. Twardowski, Y. Shapira, and M. Demianiuk (1996) Phys. Rev. B **53**, 9532.
Macfarlane, R.M., R.S. Meltzer, and B.Z. Malkin (1998) Phys. Rev. B **58**, 5692.
Machida, A., Y. Moritomo, and A. Nakamura (1998a) Phys. Rev. B **58**, 12540.
Machida, A., Y. Moritomo, and A. Nakamura (1998b) Phys. Rev. B **58**, 4281.
Martín, I.R. V.D. Rodríguez, V.R. Rodríguez-Mendoza, and V. Lavín (1998) Phys. Rev. B **57**, 3396.
Matsuda, Y. and N. Kuroda (1996) Phys. Rev. B **53**, 4471.
McGlynn, E., M.O. Henry, K.G. McGuigan, M.C. doCarmo (1996) Phys. Rev. B **54**, 14494.
Meltzer, R.S. and K.S. Hong (2000) Phys. Rev. B **61**, 3396.
Mendonca, C.R., B.J. Costa, Y. Messaddeq, and S.C. Zilio (1997) Phys. Rev. B **56**, 2483.
Merkulov, I.A. G.R. Pozina, D. Conquillat, N. Paganotto, J. Siviniant, P.J. Lascaray, and J. Cibet (1996) Phys. Rev. B **54**, 5727.
Moore, W.J., J.A. Freitas Jr., and R.J. Molnar (1997) Phys. Rev. B **56**, 12073.
Morgan, G.P. and W.M. Yen (1989) in *Laser Spectroscopy of Solids II*, ed. W.M. Yen, vol.65 in Topics in Applied Physics, Springer, Berlin, Heidelberg, p.77.
Morhain, C., E. Tournié, G. Neu, C. Ongaretto, J.P. Faurie (1996) Phys. Rev. B **54**, 4714.
Nakata, H., K. Yamada, and T. Ohyama (1999) Phys. Rev. B **60**, 13269,
Nipko, J., M. Grimsditch, C.K. Loong, S. Kern, M.M. Abraham, and L.A. Boatner (1996) Phys. Rev. B **53**, 2286.
Ofelt, G.S. (1962) J. Chem. Phys. **37**, 511.
Orange, C., B. Schlichtherle, D. Wolverson, J.J. Davies, T. Ruf, K.I. Ogata, and S. Fujita (1997) Phys. Rev. B **55**, 1607.
Pappalardo, R.G. (1979) in *Luminescence of Inorganic Solids*, ed. B. DiBartolo, Plenum Press, New York, p.175.
Park, Y., H. Kim, I. Hwang, J.E. Kim, H.Y. Park, M.S. Jin, S.K. Oh, and W.T. Kim (1996) Phys. Rev. B **53**, 15604.
Parrot, R. (1979) in *Luminescence of Inorganic Solids*, ed. B. di Bartolo, Plenum Press, New York, p. 393.
Pick, H. (1972) in *Optical Properties of Solids*, ed. F. Abelès, North-Holland, Amsterdam, p.653.
Przybylinska, H., W. Jantsch, Yu. Suprun-Belevitch, M. Stepikhova, L. Palmetshofer, G. Hendorfer, A. Kozanecki, R.J. Wilson, and B.J. Sealy (1996) Phys. Rev. B **54**, 2532.
Pucker, G., K. Gatterer, H.P. Fritzer, M. Bettinelli, and M. Ferrari (1996) Phys. Rev. B **53**, 6225.
Racah, G. (1942) Phys. Rev. **62**, 438.
Racah, G. (1943) Phys. Rev. **63**, 367.
Reynolds, D.C., D.C. Look, B. Jogai, C.W. Litton, T.C. Collins, W. Harsh, and G. Cantwell (1998) Phys. Rev. B **57**, 12151.
Riley, M.J., E.R. Krausz, N.B. Manson, and B. Henderson (1999) Phys. Rev. B **59**, 1850.

Romero, D.B., V.B. Podobedov, A. Weber, J.P. Rice, J.F. Mitchell, R.P. Sharma, and H.D. Drew (1998) Phys. Rev. B **58**, 14737.
Saisudha, M.B. and J. Ramakrishna (1996) Phys. Rev. B **53**, 6186.
Sardar, D.K. and S.C. Stubblefield (1999) Phys. Rev. B **60**, 14724.
Savoini, B., J.E. Muñoz Santiuste, and R. Gonzãlez (1997) Phys. Rev. B **56**, 5856.
Semenov, Yu.G., V.G. Abramishvili, A.V. Komarov, and S.M. Ryabchenko (1997) Phys. Rev. B **56**, 1868.
Siegle, H., A. Kaschner, A. Hoffman, and I. Broser (1998) Phys. Rev. B **58**, 13619.
Sörman, E., N.T. Son, W.M. Chen, O. Kordina, C. Hallin, E. Janzén (2000) Phys. Rev. B **61**, 2613.
Strauss, E., W.J. Miniscalco, J. Hegarty, and W.M. Yen (1981) J Phys. C**14**, 2229.
Twardowski, A., D. Heiman, M.T. Liu, Y. Shapira, M. Demianiuk (1996) Phys. Rev. B **53**, 10728.
van Vleck, J.H. (1937) J. Phys. Chem. **41**, 67.
Weber, M.J. (1981) in *Laser Spectroscopy of Solids*, eds. W.M. Yen and P.M. Selzer, vol.49 in Topics in Applied Physics, Springer, Berlin, Heidelberg, p.189.
Weston, S.J., M. O'Neill, J.E. Nicholls, J. Miao, W.E. Hagston, and T. Stirner (1998) Phys. Rev. B **58**, 7040.
Yasuhira, T., K. Uchida, Y.H. Matsuda, and N. Miura (2000) Phys. Rev. B **61**, 4685.
Yugami, H., R. Yagi, S. Matsuo, and M. Ishigame (1996) Phys. Rev. B **53**, 8283,
Zhang, Z., K.T.V. Grattan, and A.W. Palmer (1993) Phys. Rev. B **48**, 7772.

4. Bulk Materials

Optical spectroscopy of bulk materials has a long tradition in identifying and studying the electronic and vibrational excitations. The focus now is either on the study of quite exotic materials, or on finding suitable and relatively simple analytical formula for the description of bulk materials. Of importance is also the investigation of anharmonic properties, isotope effects, and in compounds the dependence of several parameters on the concentration, temperature or pressure. In this chapter we deal with bulk dielectric crystals, semiconductors and metals. There is no basic difference between the optical properties of bulk dielectrics and semiconductors; the only difference between them coming from the value of the energy gap. The excitonic and band-to-band transitions in dielectric crystals and semiconductors are thus excited with electromagnetic radiation in different energy ranges. Phase transitions in bulk materials will also be briefly presented at the end of the chapter. The most studied materials from the point of view of optical properties are the semiconductors. Optical properties of bulk semiconductors as well as the optical methods used to study their structure, excitations and interactions is a well-documented subject, reviewed in several books (see for example Basu (1997) and the references therein, and Klingshirn (1997) for the optical spectroscopy of bulk semiconductors). The phonon spectra of bulk semiconductors and other bulk materials, mainly studied using Raman scattering, have been reviewed, for example, in the eight volumes edited by Cardona and dedicated to light scattering in solids. The series started in 1975 (see Cardona (1975)) and still continues (see Cardona and Güntherodt (1999 and 2000)). Our intention in this chapter is not to repeat or even summarize the work in this field, but only to point to some parameters that can be determined through optical measurements, and to discuss some unusual properties of bulk materials evidenced in optical experiments.

4.1 Electronic Structure

In Chap. 2 we have already discussed how to determine the type of band structure (direct or indirect) in bulk materials and the symmetry of states in the conduction and valence bands from measurements of the absorption coefficient. In this section we discuss in more detail the determination of some basic parameters of the electronic structure such as the energy gap, effective mass and carrier concentrations. The bandgap is an essential parameter of bulk semiconductors; its value, as well as the values of the crystal-field splitting and spin-orbit splitting can be optically determined from absorption curves. Such measurements have been performed, for example, in bulk-like wurtzite CdTe nanocrystals large enough to suppress confinement effects, hosted in a sodium borosilicate matrix (Lefebvre et al., 1996). The absorption curve at 2 K showed two sharp peaks corresponding to the fundamental free-exciton gap $E_g^A = 1653$ meV, and B excitons, the splitting between peaks being $E_g^A - E_g^B = 46.5$ meV. The crystal-field splitting Δ_{cr} obtained from $E_g^A - E_g^B = (\Delta_{cr} + \Delta_{so})/2 - [(\Delta_{cr} + \Delta_{so})^2 - 8\Delta_{so}\Delta_{cr}/3]^{1/2}/2$ is found to be $71.6\,\mathrm{meV}$ for a value of the spin-orbit splitting of $\Delta_{so} = 927\,\mathrm{meV}$.

The absorption curve of GaN near the band edge (see Fig. 4.1) shows a sharp peak due to the free exciton at 353.55 nm and a shoulder at 354.67 nm due to bound excitons. The temperature dependence of the fundamental bandgap obtained from this absorption curve with a Varshni-like expression

$$E_g(T) = E_g(0) - aT^2/(T + b), \tag{4.1}$$

where a and b are fitting parameters, can be used to determine the parameter $b = 737.9\,\mathrm{K}$ (Manasreh, 1996).

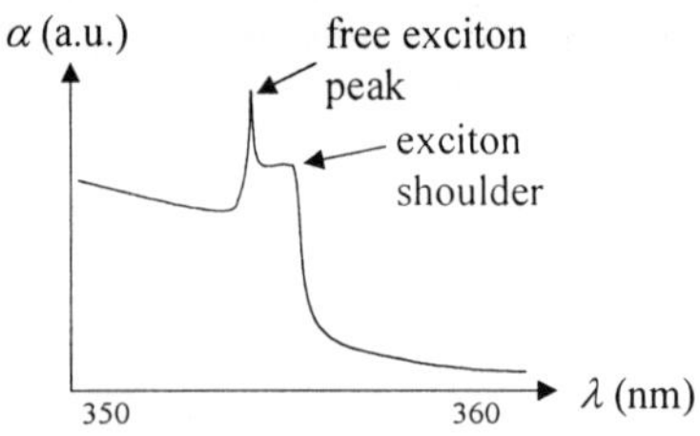

Fig. 4.1. Absorption spectrum in GaN near the band edge

In some materials the energy bandgap is anisotropic. This anisotropy, encountered especially in layered-type materials, can be identified in polarization-dependent absorption measurements. An example of anisotropic materials are the diamagnetic semiconductors ReS_2 and $ReSe_2$, which belong to the layered-type transition-metal dichalcogenides, and crystallize in a lattice with strong covalent bonds within a layer and weak interaction (of van der Waals type) between layers (Ho et al., 1998a). Absorption measurements between 25–500 K in these triclinic symmetry materials showed a shift in the transmittance spectra of $\boldsymbol{E} \parallel \boldsymbol{b}$ towards lower energies compared to those obtained for $\boldsymbol{E} \perp \boldsymbol{b}$. The absorption coefficient calculated from transmission spectra was found to depend on frequency as $\alpha \cong (\hbar\omega - E_g)^n$ with $n = 2$, suggesting that indirect allowed transitions are dominant. The bandgaps for parallel and perpendicular polarized light with respect to the b axis are 1.35 eV and 1.38 eV respectively for ReS_2 and 1.17 and 1.2 eV for $ReSe_2$. The bandgap energies for $\boldsymbol{E} \parallel \boldsymbol{b}$ are smaller than for $\boldsymbol{E} \perp \boldsymbol{b}$ in both materials. The temperature of the gap can be fitted either with a Varshni-like expression or with a Bose–Einstein dependency

$$E_{gi}(T) = E_{i\mathrm{B}} - a_{i\mathrm{B}}\{1 + 2/[\exp(\theta_{i\mathrm{B}} / k_{\mathrm{B}}T) - 1]\}, \tag{4.2}$$

where $i = \parallel, \perp$, $a_{i\mathrm{B}}$ is the strength of the exciton-phonon interaction and $\theta_{i\mathrm{B}}$ the average phonon temperature. At high temperatures the Varshni and Bose–Einstein relations are related through $a_i = 2a_{i\mathrm{B}} / \theta_{i\mathrm{B}}$. Another example of an optical anisotropic material is the HgI_2 tetragonal crystal, where different dielectric response functions were found for the electric vector of the incident light parallel or perpendicular to the c axis (Yao et al., 1997). In layered media, such as for example in $La_{2-x}Sr_xCuO_4$, a high-temperature cuprate superconductor material, the optical response is also anisotropic. Ellipsometry measurements in this material, for different x values and different polarization directions, have revealed that the phononic contribution to the optical conductivity is higher than the electronic contribution in IR (Henn et al., 1997a).

From the Varshni or Bose–Einstein temperature dependence of the energy gap or other critical points, it is possible to monitor the temperature in direct semiconductors during in situ molecular beam epitaxy (MBE) growth. In this case the bandgap is determined from PL measurements, photoreflectance or ellipsometry. The temperature of the sample can also be found by measuring the temperature-dependent phonon frequencies in Raman scattering (Herman, 1995).

Stronger-than-normal temperature dependencies for the real and imaginary parts of the dielectric constant are found in materials with hopping motion of localized charge carriers. An example of such a material is boron carbide, in which the real and imaginary parts of the dielectric constant show a strong increase with increasing temperature, and with decreasing frequency (Samara et al., 1993). The real part of the dielectric function, for constant frequency and temperature, shows a non-monotonic dependence on the C content, with a maximum at about 13 at.% C. As in covalent semiconductors, the electronic contribution to $\varepsilon_{1\mathrm{e}}$ is larger than the lattice contribution, and the low-frequency dielectric constant is dominated by the dipolar contribution associated with the hopping motion of localized carriers. These carriers are supposedly bipolaronic holes in $B_{11}C$ icosahedra.

Modulated reflectance or transmittance measurements are used to determine the critical point energies. In strained $Si_{1-x}Ge$ layers grown on Si for example, the modulated reflectance spectra is given by $\Delta R / R = \mathrm{Re}[A\exp(i\theta)(E - E_{\mathrm{g}} + i\Gamma)^{-n}]$ where A is the signal amplitude, θ its phase factor, Γ is the line broadening, E_{g} the energy of the critical point in the combined density of states, E the photon energies and n = 2.5–4 depending on the type of the critical point (Ebner et al., 1998). The transitions corresponding to E_1, $E_1 + \Delta$, E_0', E_0, $E_0 + \Delta_0$, $E_2(\mathrm{X})$ and E_1' critical points are easily determined from the sharp peaks in the differential reflectance spectra, and their energy dependence on the concentration x can thus be calculated. The energy dependence on concentration is caused by the strain modification of the electronic structure for different concentrations. The bandgap energy E_g is also determined and its temperature behavior can again be fitted with either of the two formulas indicated above.

There are still materials for which the electronic band structure is not well known, for example HgSe, optical measurements being invaluable as a means to clarify the situation. Recent magneto-optical studies at liquid He temperatures have shown that HgSe is not a semiconductor, but a semimetal with inverted-type electronic band structure (von Truchseß et al., 2000). The decisive evidence has been the observation in the magneto-optical transmission spectra of an asymmetric resonance line shape for interband transitions with high-frequency tail, between Landau levels. This can only happen for nonparabolic systems where the initial and final Landau levels have opposite curvatures, the interband separation increasing with increasing the wavevector component parallel to the direction of the applied $\boldsymbol{B}$ field.

When a magnetic field is applied to GaN the double degeneracy of the valence subbands A, B, C is lifted, they become mixed and new eigenstates and selection rules are established. The doubly degenerate s-type conduction band S is also split. The bands are characterized by the c-axis component of the total angular momentum $J_z = \pm 1/2$ for S, B, C and $J_z = \pm 3/2$ for A. The crystal field Δ_1, spin-orbit parameters Δ_2 and Δ_3, as well as the energy and oscillator strengths of HH exciton transition X_A, and LH exciton transitions $X_{B,C}$ were determined in GaN from magnetoreflectance experiments in the Faraday configuration (Campo et al., 1997). When there is no magnetic field the oscillator strengths are $\alpha_B = a^2 \alpha_A$, $\alpha_C = (1-a^2)\alpha_A$, $\alpha_A = \alpha_B + \alpha_C$, where $a = a_+ = a_- = 1/[x(1/x^2+1)^{1/2}]$ with $x = \{-(\Delta_1 - \Delta_2) + [(\Delta_1 - \Delta_2)^2 + 8\Delta_3^2]^{1/2}\}/(2^{3/2}\Delta_3)$. In the presence of the magnetic field, the oscillator strengths become dependent on the circular polarization and there is a splitting ΔE_S for the conduction band and $\Delta E_{A,B,C}$ for valence subbands where $\Delta E_i = E_i(+J) - E_i(-J)$ with $i = \mathrm{A,B,C}$. Only $\delta E_A = \Delta E_A - \Delta E_S$, $\delta E_B = \Delta E_B + \Delta E_S$ and $\delta E_C = \Delta E_A + \Delta E_C$ can be measured in the Faraday configuration due to the selection rules. Then, the theoretical values of the reflectance and polarization rate are fitted with their experimental counterparts to obtain $\Delta_1 = 1.3$ meV, $\Delta_2 = 6.6$ meV, $\Delta_3 = 5.3$ meV and $a^2 = \alpha_B/\alpha_A = 0.68$ for GaN. From the dependence of the differential reflectance on the magnetic field $\Delta R(B)$, the splitting $\delta E(B) = g_{\mathrm{eff}} \mu_B B \delta J_z$ with $\delta J_z = \pm 1$ for the three excitons are extracted, and the corresponding effective g factors are determined. The g factor for the S band is obtained from PL experiments in a magnetic field perpendicular on the c axis.

Magneto-optical methods, in particular the optical detection of the cyclotron resonance, can be used to determine the electron effective mass tensor. The general expression of the cyclotron effective mass is:

$$m_{\mathrm{eff}} = [(m_1 m_2 m_3)/(m_1 H_1^2 + m_2 H_2^2 + m_3 H_3^2)]^{1/2}, \tag{4.3}$$

where $H_{1,2,3}$ are director cosines of the magnetic field direction with the principal axes of the mass tensors $m_3 = m(\mathrm{L})$, $m_2 = m(\mathrm{K})$, $m_1 = m(\Gamma)$. The effective mass tensor of 4H SiC, for example, are $m(\mathrm{L}) = 0.33 \pm 0.01 m_0$, $m(\Gamma) = 0.58 \pm 0.01 m_0$ and $m(\mathrm{K}) = 0.31 \pm 0.01 m_0$ (Volm et al., 1996).

IR cyclotron resonance experiments were performed in a transmission configuration for GaSb at both Γ and L points in magnetic fields up to 500 T (Arimoto et al., 1998). At temperatures higher than 100 K the resonances from Γ and L points are separated, the latter appearing for $B \geq 60$ T. The longitudinal and transverse effective masses at L point, $m_l = 1.4 m_0$ and $m_t = 0.085 m_0$, respectively, are determined from the effective mass dependence on the photon energy for magnetic fields parallel to $\langle 110 \rangle$ and $\langle 111 \rangle$ directions. The magnetic field induces a $\Gamma - \mathrm{L}$ crossover at 125 T for $\boldsymbol{B} \parallel \langle 100 \rangle$ and nonparabolicity is observed when $\mathrm{L} > \Gamma$. At 200 K the difference between the conduction band minima at L and Γ points is 86 meV.

The free-carrier concentration in as-grown n-type GaN can be determined from IR reflection or Raman scattering. The dielectric function in an oscillator model can be written as (Wetzel et al., 1996):

$$\varepsilon(\nu) = \varepsilon_\infty [1 + (\bar{\nu}_\mathrm{L}^2 - \bar{\nu}_\mathrm{T}^2)/(\bar{\nu}_\mathrm{T}^2 - \bar{\nu}^2 + i\bar{\nu}\Gamma) - \bar{\nu}_\mathrm{p}^2 / \bar{\nu}(\bar{\nu} - i\gamma)], \tag{4.4}$$

where $\bar{\nu} = \omega / 2\pi c$ denote frequencies expressed in [m^{-1}], $\bar{\nu}_\mathrm{p}$, $\bar{\nu}_\mathrm{L}$ and $\bar{\nu}_\mathrm{T}$ being the quantities corresponding to the plasma frequency, longitudinal and transverse optical phonon frequencies. The plasma frequency $\omega_\mathrm{p}^2 = Ne^2 / [\varepsilon_0 \varepsilon(0) m_\mathrm{eff}]$ is directly proportional to the concentration of free carriers N. The minimum of the IR reflection coefficient

$$R(\nu) = [1 - \varepsilon^{1/2}(\nu)] / [1 + \varepsilon^{1/2}(\nu)], \tag{4.5}$$

as well as its position $\bar{\nu}_\mathrm{min}$, are very sensitive to N. The relation between N and $\bar{\nu}_\mathrm{min}$ is expressed by three parameters $N = N_0 (\bar{\nu}_\mathrm{min} - \bar{\nu}_0)^r$, which take different values for different IR active modes. For example, for E_1 modes $N_0 = 5.3 \times 10^{16}$ cm^{-3}, $\bar{\nu}_0 = 774$ cm^{-1}, $r = 0.876$, while for the A_1 phonon that corresponds to the phonon-plasmon peak in Raman spectra, $N_0 = 1.1 \times 10^{17}$ cm^{-3}, $\bar{\nu}_0 = 736$ cm^{-1}, $r = 0.764$. The Raman scattering measurements were performed at 27 GPa because at lower pressures the LO mode is strongly distorted by the plasmon mode, the coupled mode being overdamped.

Time-resolved reflectance measurements can be used to determine the effective electron density in Γ valleys in n-type GaAs, for example (van Dantzig and Planken, 1999). In a sample with high carrier density (2×10^{18} cm^{-3}), the excitation with high-intensity far-IR laser pulses, when no additional carriers are created, induces a shift of the plasma frequency to lower values, the magnitude of the shift increasing with the pump intensity. The pump-induced decay at 300 K occurs in a time interval of the order of 3 ps, following the change in the effective carrier density due to electron scattering between Γ and L valleys. As follows from (4.2), the reflectance minimum is close to the plasma frequency $\omega_p = e[N_{eff}/(\varepsilon_\infty m_{eff})]^{1/2}$, the reflectance rising sharply above ω_p and becoming almost flat at longer wavelengths (see Fig. 4.2 for an unperturbed sample).

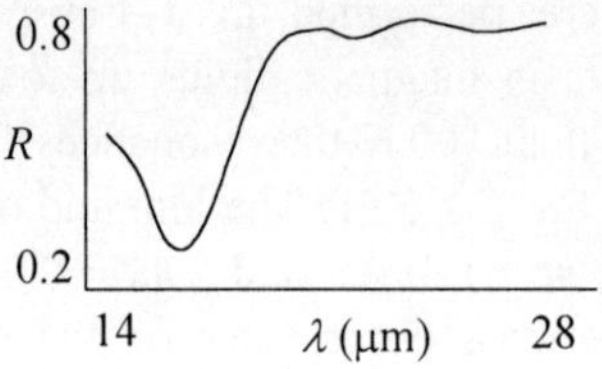

Fig. 4.2. Typical reflectance spectrum of GaAs at room temperature

The differential reflectivity ΔR is positive for excitation frequencies larger than ω_p and negative for smaller excitation frequencies. The effective number of carriers in Γ valleys, which contribute to the plasma frequency, is determined by fitting the experimental change in ΔR at different probe delays. N_{eff} decreases on pulse heating due to electron scattering in L valleys, the backscattering from L into Γ valleys taking place in picoseconds.

As an example of the effort to find simple formulas for the electronic contribution to optical properties we mention here the analytical formulae for the refractive index of alkali halides proposed by Johannsen (1997). Supposing that the bottom of the conduction band of these materials has s symmetry with minima at the Γ, X points in the Brillouin zone, and that the valence band is constituted from p electrons, the imaginary part of the electronic contribution to the dielectric constant can be approximated by $\varepsilon_{im}^{el}(E) = CP^2\rho/E^2$ for energies above the onset of the optical absorption E_0, and zero below it. Here P is the average transition matrix and ρ the joint density of states, assumed constant. The electronic contribution $\Delta\varepsilon_{re}^{el}(E) = -(CP^2\rho/\pi E^2)\ln|1 - E^2/E_0^2|$ to the refractive index $n(E) = [1 + \Delta\varepsilon_{re}^{el}(E) + \Delta\varepsilon_{re}^{TO}(E)]^{1/2}$ is obtained from the imaginary part of the corresponding dielectric constants using the Kramers–Krönig relations. The optical phonon contribution in the low-frequency region is given by $\Delta\varepsilon_{re}^{TO}(E) = (\varepsilon_0 - n_\infty^2)E_{TO}^2/(E_{TO}^2 - E^2)$ with $n_\infty^2 = 1 + CP^2\rho/(\pi E_0^2)$. These simple formulae are in good agreement with experimental data for alkali halides. The effect of pressure on the refractive index in these materials is investigated in Johannsen et al. (1997).

4.2 Excitons in Bulk Semiconductors

Although in principle excitons can exist in any dielectric material, most studies refer to excitons in semiconductors, where they are easier to excite optically. The main properties of excitons are their associated oscillator strengths, effective masses and binding energies. The strength and binding energy of excitons can be most simply determined from the absorption spectrum near the fundamental absorption edge, which includes the contributions of both discrete and continuum excitons. In $CuGaS_2$ for example, a ternary chalcopyrite semiconductor with a direct bandgap of 2.5 eV at room temperature, the absorption spectrum near the fundamental edge showed a series of sharp lines E_n inside the bandgap fitted by $E_n = E_g - R_{ex}/n^2$ where $n = 1,2,3\ldots$, R_{ex} is the effective Rydberg of excitons and E_g the energy gap (Bellabarba et al., 1996). These lines are assigned to the quantum states of electron-hole pair relative motion. For large n values, as the fundamental absorption edge is approached, the discrete lines overlap forming a continuum, the total absorption being in this case $\alpha(\hbar\omega) = A\{2R_{ex}[\delta(\hbar\omega - E_n)/n^3] + \Theta(\hbar\omega - E_g)/[1 - \exp[2\pi R_{ex}/(\hbar\omega - E_g)]^{1/2}]\}$ with $\Theta(\hbar\nu - E_g)$ the step function. If the exciton-phonon coupling constant $g = Vm_r E_d^2/\hbar^3 Mv$ with M, V the mass and volume of the unit cell, v the sound velocity and E_d the deformation potential, is very small compared to unity, the absorption lineshape is obtained by convolving the above expression with a Lorentzian $\Gamma\pi/[(\hbar\omega)^2 + \Gamma^2]$ (with a Gaussian, in the opposite case). The resulting the absorption lineshape is then a superposition of Lorentzian-broadened lines with halfwidth half-maximum (HWHM) $\Gamma_n = \Gamma_c - (\Gamma_c - \Gamma_1)/n^2$, where Γ_c is the HWHM of the continuum exciton and Γ_1 that of the discrete exciton. Fitting this formula with the experimental data, the following parameters were obtained: $g = 0.02$, $R_{ex} = 29$ meV, $\Gamma_1 = 2.2$ meV and the exciton relative mass $m_r = 0.12\,m_0$. The oscillator strength calculated with these values is 0.02. More information can still be obtained from the absorption curve in this material (see Fig. 4.3.)

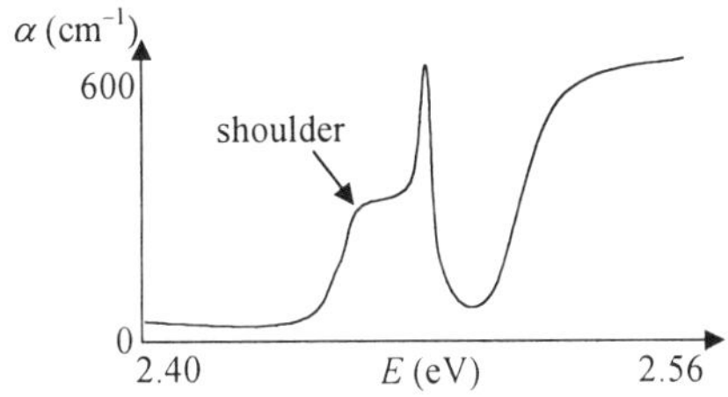

Fig. 4.3. Absorption spectrum of $CuGaS_2$ at 5K

The shoulder in the absorption curve at energies below exciton absorption is due to transitions between shallow acceptor or donor levels and conduction and valence bands. These transitions have a threshold smaller than E_g with an amount equal to the binding energy of the impurity. In particular, the contribution to the absorption from transitions between acceptor-states and the conduction band is given by

$$\alpha(\hbar\omega) = B[(\hbar\omega + E_A - E_g)^{1/2}(m_e/m_0)^{3/2}/\hbar\omega][1 + m_e(\hbar\omega + E_A - E_g)/m_h E_A]^4, \quad (4.6)$$

where B is a proportionality constant depending on the interband transition matrix element and on the number of unfilled acceptor states. By fitting this expression with experimental data, the binding energy of acceptors, due to Cu vacancies, was found to be $E_A = 0.15\,\text{eV}$.

In anisotropic crystals, such as ZrS_3 and $ZrSe_3$, the exciton absorption bands are also anisotropic. For these materials the optical spectrum near the band edge is dominated at low temperatures (2 K) by exciton absorption bands that can be observed only for light polarized along the chain direction, $\boldsymbol{E} \parallel b$. The absorption spectra and the high dichroism of these materials is due to the difference in oscillator strengths for $\boldsymbol{E} \parallel b$ and $\boldsymbol{E} \perp b$, caused by the wavefunction anisotropy at the band edges, due to paired anions with p orbitals (Kurita et al., 1993).

The exciton binding energy at either zero-field or high magnetic fields can be determined by measuring the absolute energy of the magnetoexciton absorption peak. The magnetoexciton binding energy at high fields can be calculated with the adiabatic method and can be approximated by (Tang et al., 1997) $E_b(n) \cong 1.6R[\gamma/(2n+1)]^{1/3}$ where R is the zero-field binding energy, n is the conduction band Landau level index and $\gamma = \hbar^3 B(4\pi\varepsilon_0\varepsilon_r)^2 / m_r^2 e^3$ with m_r the exciton reduced mass. From spectral transmittance curves at different magnetic fields the exciton reduced mass can be extracted from the magnetoexciton energy shift. This method was applied for InAs by Tang et al. (1997).

The temperature dependence of free exciton emission in GaN, for example, is described by $I_{PL}(T) = I(T=0)/[1 + C\exp(-E_{FE}/k_B T)]$, where C is the ratio between nonradiative and radiative transition rates, and E_{FE} was fitted to 26.7 meV (Kovalev et al., 1996). Free excitons show a temperature broadening of the high-energy part of the PL line and a sharp cut of the low-energy part due to the low density of exciton states near the polaron bottleneck and an inefficient kinetic energy relaxation by long-wavelength acoustic phonons in the low-energy side of the line. The high-energy part reflects the thermal distribution of kinetic energies of excitons and their interaction with acoustic phonons and impurities imposed by the $\boldsymbol{k}$ conservation rule. LO sidebands of the free exciton line in PL can be used to extract the kinetic energy distribution of the exciton gas, since this sideband is due to the transitions in which excitons with any wavevector can recombine due to the breakdown of the momentum conservation rule. Fitting the PL lineshape of the 2 LO phonon replica measured as a function of temperature with $I_{PL}(E_{ex}) = E_{ex}^{1/2}[\exp(-E_{KE}/k_B T)]$, where E_{KE} is the excitonic kinetic energy and $E_{ex}^{1/2}$ corresponds to the density of states in the parabolic exciton band in direct bandgap semiconductors, it was found that free excitons are in thermal equilibrium with the lattice. Additional PLE measurements were performed to distinguish between the two mechanisms of free exciton formation: binding of free electron-hole photogenerated pairs or phonon-assisted exciton formation. The second mechanism would be distinguished by a series of LO-phonon cascades in the exciton band, which was observed experimentally.

The temperature dependence of energies and broadening parameters of band-edge excitons in $ReS_{2-x}Se_x$ single crystals was determined by Ho et al. (1998b) in a temperature range 25–300 K using piezoreflectance. The piezoreflectance line shape is fitted with

$$\Delta R / R = \mathrm{Re}[\Sigma_{i=1}^{2} A_i^{ex} \exp(i\phi_i^{ex})(E - E_i^{ex} + i\Gamma_i^{ex})^{-2}], \quad (4.7)$$

where A_i^{ex} and ϕ_i^{ex} are the amplitudes and phases of the lineshape, and E_i^{ex} and Γ_i^{ex} are the energies and broadening parameters of interband excitonic transitions. The dependence of E_i^{ex} on the Se composition x is fitted with a parabolic function $E_i^{ex}(x) = E_i^{ex}(0) + b_i x + c_i x^2$, the temperature dependence of the same parameter being described by either the Varshni empirical formula or by the Bose–Einstein formula. The broadening parameter depends on temperature as $\Gamma(T) = \Gamma_0 + \Gamma_{LO} /[\exp(\theta_{LO} / T) - 1]$ where θ_{LO} is the LO phonon temperature. In the presence of exciton-lattice interaction additional terms must be added in the fitting formula for both transition energy and broadening parameter.

A shift of the absorption edge for the Wannier exciton peak, accompanied by quenching and broadening, was observed in thin epitaxial GaN in the presence of an electric field (Franz–Keldysh effect). A similar effect is observed in other materials for Frenkel excitons. Although several theories predicted an exponential dependence of the absorption on energy, the experiment showed that, at least in GaN, the absorption scales as f^{-s} with $s = 0.65$ at 300 K, where $f = eFa_{ex} / R_{ex}$, with F the electric field, R_{ex} the exciton Rydberg and a_{ex} the exciton radius (Binet et al., 1996). For $f > 0.05$ the exciton peak shifts linearly with F.

The radiative recombination lifetime T_1 and the dephasing time $T_2 = 2T_1$ of bound excitons in GaAs can be determined from the spectral hole burning method. The excitons bound to an impurity in a bulk semiconductor are modeled as zero-dimensional (0D) electron-hole systems since the electron-hole pairs are confined in small volumes by the Coulomb attraction. In high-purity semiconductors where the distance between impurities is much larger than the Bohr radius, each bound exciton is equivalent to an isolated two-level system, the optical properties of the bound exciton being similar to those of a quantum dot (a 0D quantum confined system), although in the latter case the electrons and holes are confined by the potential barrier around the dot. In 0D systems it is expected that (i) the oscillator strength is greatly enhanced due to exciton localization, and (ii) the optical-dipole dephasing time T_2 is drastically enhanced due to the reduced optical-dipole scattering rate in 0D systems. The homogeneous linewidth of bound excitons in GaAs was determined in a hole burning experiment (Oohashi et al., 1996), T_2 being derived from the width of the hole γ_{hole} burned in the absorption spectrum as $T_2 = (\pi\gamma_{hole} / 2)^{-1}$. T_2 was found to be of about 300 ps, an order of magnitude greater than T_2 of 2D excitons in a quantum well, which confirms the 0D behavior of bound excitons. The localization energy of bound excitons, E_{loc} can be determined if the oscillator strength of free and bound excitons are known. It can be extracted from $f_{\mathrm{bound}} = 8\pi f_{\mathrm{free}}[\hbar^2/2(m_e + m_h)E_{\mathrm{loc}}]^{3/2}$, the corresponding value in the high-purity GaAs sample being $E_{\mathrm{loc}} = 0.9\,\mathrm{meV}$ for $f_{\mathrm{bound}} = 5.8$.

4.3 Anharmonic Effects

Anharmonic phonon-phonon interactions include three-phonon up-conversion and down-conversion processes. These have been studied for example in BeO (Morell et al., 1996), which is a metal oxide ceramic with a wurtzite structure where Be atoms are tetrahedrally coordinated to four O atoms and vice-versa, all atoms occupying $C_{3\mathrm{v}}$ sites. The anharmonicity of the potential, observed in first-order Raman scattering, has as effects the broadening (decrease in lifetime) and frequency downshift of the normal vibrational modes of the lattice, as the temperature increases. The up-conversion contribution vanishes as $T \to 0$, whereas the down-conversion processes have a finite value even at very low temperatures due to the spontaneous decay of phonon in two lower-energy phonons, the total momentum being conserved in the process. The contribution of three-phonon processes to the frequency shift and the linewidth are expressed as

$$\Delta\omega = A[1 + \textstyle\sum_{j=1}^{2} 1/(\exp x_j - 1)], \quad \Delta\Gamma = B[1 + \textstyle\sum_{j=1}^{2} 1/(\exp x_j - 1)], \tag{4.8}$$

where $\sum_{j=1}^{2} x_j = \hbar\omega_0 / k_B T$ with ω_0 the harmonic frequency. The next higher-order contribution to anharmonicity comes from the quartic term, and is caused by elastic collisions of two phonons from the thermal bath. The quartic contribution leads to a frequency shift of the phonon modes, with no effect on their lifetimes. Besides anharmonicity, volume expansion contributes also to a shift of phonon frequencies, without influencing, again, their linewidth. The frequency shift due to thermal expansion is

$$\Delta\omega_{\mathrm{th}}(T) = \omega_0[\exp(-\gamma \textstyle\int_0^T \alpha_V \mathrm{d}T) - 1], \tag{4.9}$$

with γ the mode Grüneisen parameter, and α_V the coefficient of volume expansion. The frequency shift is sometimes expressed also in terms of the linear thermal expansion coefficient $\alpha \cong \alpha_V / 3$. The frequencies of the phonons created in up- and down-conversion processes, as well as their oscillator strengths are fitting parameters for experimental data. Thermal expansion is dominant in GaAs, while anharmonicity is the main source of frequency shifting and broadening in AlAs Raman modes (Jiménez et al., 1998). The temperature sensitivity of Raman scattering makes this method suitable as a probe for the local temperature.

More quantitative models for the anharmonic lattice potential can be imagined. For example, the temperature broadening and the asymmetry of Raman lines can be explained with an asymmetric double-well potential as in Fig. 4.4, although anharmonicity is a nonlocal phenomena, associated to continua, whereas the double-well potential is localized. In this model the Raman spectrum is taken as a superposition of Raman lines due to different transitions between higher levels in the asymmetric potential for ionic motion, and the decay of optical modes in two counterpropagating acoustic phonons is also accounted for. At low temperatures only the $v=0 \to 1$ transition is assumed to occur and the optical phonon of frequency ω decays in two acoustic phonons with opposite wavevectors and frequencies $\omega/2$, the temperature dependence of the spontaneous Raman frequency and linewidth being

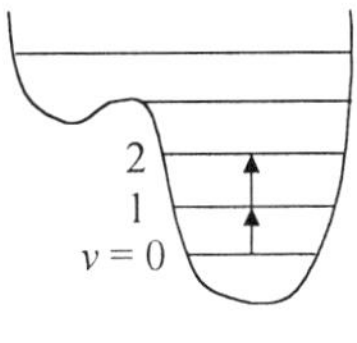

Fig. 4.4. Asymmetric double-well potential model describing anharmonicity

$$\omega(T) = \omega_0 - \Delta[2\bar{n}(\omega_0/2, T) + 1], \qquad \Gamma(\omega, T) = \Gamma_0[2\bar{n}(\omega/2, T) + 1], \tag{4.10}$$

with $\bar{n}(\omega,T) = 1/[\exp(\hbar\omega/k_B T) - 1]$. The parameters ω_0, Δ, Γ_0, which characterize the shape of the anharmonic potential, are obtained by fitting experimental data. At higher temperatures, above 200 K, the transitions $v=1 \to 2$, $2 \to 3$ contribute to the broadening and asymmetry of Raman lines, an additional term due to scattering on thermal photons being added to the linewidth (Schwarz and Maier, 1997).

Multiphonon Raman scattering can be caused by Fröhlich interaction. Fröhlich activation explains the appearance of some Raman modes, which are present only for light polarization parallel to the chain structure in $SrCuO_2$ and $Sr_{0.5}Ca_{0.5}CuO_2$. These modes are characterized by an anomalous increase in intensity at cooling, explained by the following expression for two-LO-phonon resonant Raman scattering: $\mathrm{d}S/\mathrm{d}\Omega \approx \int_0^\infty dq\,|F(q)|^2/q^2$ where $F(q)$ is the sum of functions related to the different types of intermediate states. The magnitude of the phonon wavevector q is defined in real crystals only for values between $1/l$ with l the mean free phonon path and $1/a$ where a is the lattice parameter. The increase of line intensity with decreasing temperature is due to the $1/q^2$ term, which increases rapidly due to the increase of l (Abrashev et al., 1997).

The frequency and linewidths of phonon modes can vary throughout a family of materials, due to anharmonic effects. For example, in the $RTiO_3$ family of Mott–Hubbard insulators (R = La, Ce, Pr, Nd, Sm, Gd) the increase of the orthorhombic distortion from $LaTiO_3$ to $GdTiO_3$, induces changes in the electronic structure, a significant softening and broadening of the phonon modes accompanied by an increased asymmetry and a decrease of the scattering rates from $GdTiO_3$ to $LaTiO_3$. The electronic background undergoes redistribution of the spectral shape since the number of free carriers increases systematically from Gd to La due to the increase of R^{3+} vacancies (Reedyk et al., 1997).

4.4 Isotope Effects

Isotopic effects in dielectric crystals include changes in the optical spectrum due to isotope replacement. Isotopic studies are useful for the investigation of average atomic mass influence on the dynamical properties of lattice, including the frequency, linewidths and lifetimes of phonons, on the isotopic-disorder effects and on the electronic structure. In pure semiconductors the phonon frequency $\omega(q) \approx m^{-1/2}$, with an additional q-dependency being expected in compounds. In compound materials the effects depend on which constituent atom is substituted by isotopes. This property is used for identification of some TO phonons. Moreover, polar and nonpolar optical modes can have different isotope shifts. In CdS for example, the frequency of polar phonons scale with the reduced mass, whereas the frequency of the Raman active nonpolar modes scales with the mass of either Cd or S atoms. In particular, the frequency shift of the free and bound excitons was explained by supposing a dependence of the energy gap on temperature and isotope mass as

$$E(M,T) = E_0 - (A_{\mathrm{Cd}} / \omega_{\mathrm{Cd}} M_{\mathrm{Cd}})[\bar{n}(\omega_{\mathrm{Cd}},T) + 1/2] - (A_{\mathrm{S}} / \omega_{\mathrm{S}} M_{\mathrm{S}})[\bar{n}(\omega_{\mathrm{S}},T) + 1/2], \tag{4.11}$$

with $\bar{n}(\omega,T)$ the Bose–Einstein occupation number. The parameters A_{Cd} and A_{S} are determined from fitting experimental data (Zhang et al., 1998). Similar results hold for CuCl, where, due to the large mass difference between the constituents, acoustic mode eigenvectors are dominated by Cu displacements, and optical modes by Cl displacements. In this material, since $\omega_{\mathrm{Cu,Cl}} \approx 1/m_{\mathrm{Cu,Cl}}^{1/2}$, the isotope shift of the bandgap is $\partial E(T=0,m)/\partial m_{\mathrm{Cu,Cl}} = -A_{\mathrm{Cu,Cl}} m_{\mathrm{Cu,Cl}}^{2} / 4\omega_{\mathrm{Cu,Cl}}$, the respective values being –76 μeV/amu for Cu and +364 μeV/amu for Cl isotopes (Göbel et al., 1998). In general, the isotope shift of the nth phonon in compounds is given by (Zhang et al., 1997)

$$\partial \omega_n / \partial m_i = -(\omega_n / 2m_i) \Sigma_\alpha | u_{i,\alpha}(n) |^2, \tag{4.12}$$

where m_i is the mass of atom i, and $u_{i,\alpha}(n)$ with $\alpha = x, y, z$ is the ith component of the orthonormal eigenvector of mode n along the direction α. The phonon eigenvectors can be determined by measuring the phonon shifts. This law is valid, for example, in $YBa_2Cu_3O_7$ where the ions of several sublattices participate in the mixed vibration (Henn et al., 1997b). In GaN for example, it was found that the frequencies of polar optical modes scale with $1/m_{\mathrm{r}}^{1/2}$ where $m_{\mathrm{r}}^{-1} = m_{\mathrm{Ga}}^{-1} + m_{\mathrm{N}}^{-1}$ is the reduced mass, whereas the frequency shifts of the two nonpolar modes differ from that calculated in terms of the reduced mass. The difference is caused by the coupling of nonpolar normal modes of Ga and N, coupling which can be described in the harmonic approximation using three force constants. The force constant responsible for coupling, and those which characterize the pure vibrations of Ga and N, as well as the Ga and N contributions to the coupled modes, are determined by fitting. For high-frequency phonons the disorder induced by the isotopes of N produces a phonon shift

$$\Delta\omega = (\omega^3 / 12)\sum_i x_i (1 - m_i / \overline{m})^2 \int_0^\infty \mathrm{d}\omega' N_\mathrm{d}(\omega') / (\omega^2 - \omega'^2), \tag{4.13}$$

with the normalization $\int_0^\infty \mathrm{d}\omega' N_\mathrm{d}(\omega') = 12$, which expresses the fact that there are 12 phonon states in the unit cell. Here x_i is the concentration of the ith N isotope, and $\overline{m}$ the average N mass (Zhang et al., 1997). Strong isotope-disorder-induced broadening of the TO(Γ) phonon is also observed in ZnSe due to the large one-phonon density of states in the corresponding energy range, no systematic change in LO(Γ) phonons being found. The energy of TO(Γ) phonon deviates from the $1/m_\mathrm{r}^{1/2}$ law, and the linewidth increases, due to the disorder-induced self-energy from scattering of zone-center phonons into others with similar frequency but different wavevectors. In ZnSe the energy gap increases by 214 μeV/amu when the Zn mass is increased, and by 216 μeV/amu when the Se mass increases. The isotope mass influences also the electron-phonon interaction: although the electronic properties are not significantly modified, the lattice constant is dependent on the isotope mass as an integral over all occupied phonon states with the Grüneisen parameter as a weighting factor (Göbel et al., 1999).

Substitution of the light isotope by a heavy one increases both the binding energy E_b and the energy of polariton longitudinal-transverse-splitting, as a result of renormalization of Wannier–Mott excitons. This phenomenon was observed by studying the reflection and intrinsic PL spectra of mixed LiH_xD_{1-x} at low temperatures (Plekhanov, 1996). The binding energy, determined from the peak energy of the two excitons present in the spectra E_1 and E_2 as $E_\mathrm{b} = m_\mathrm{r}e^4 / 2\hbar^2\varepsilon^2 = 4(E_2 - E_1)/3$, shows a nonlinear dependence on x due to fluctuation broadening of conduction and valence bands, connected with isotope disorder. This nonlinear dependence can be fitted with a second-order polynomial.

4.5 Effects Due to Alloying

For materials with mixed layers an effective dielectric function ε can be defined for the system. Supposing that the material consists of constituents A, B with corresponding dielectric functions ε_A, ε_B in the effective-mass approximation, the effective dielectric function is given by $f_\mathrm{A}(\varepsilon_\mathrm{A} - \varepsilon)/(\varepsilon_\mathrm{A} + 2\varepsilon)$ $+ f_\mathrm{B}(\varepsilon_\mathrm{B} - \varepsilon)/(\varepsilon_\mathrm{B} + 2\varepsilon) = 0$, with f_A, f_B the relative volume fractions of A, B constituents, respectively.

The frequencies of optical phonons depend on the concentration of different constituents in an alloy. For example, in $Ti_{1-x}Sn_xO_2$ the x dependence is caused mainly by octahedral distortion, the different thermal expansivities and anisotropic compressibilities of the two materials playing a crucial role. The frequency and linewidth of A_{1g} and E_g Raman modes of TiO_2 do not, however, depend linearly on x, although both TiO_2 and SnO_2 have a rutile structure with tetragonal symmetry and two molecular units per primitive cell (Hirata et al., 1996). The concentration dependence of optical modes was also studied in $YBa_{2-x}La_xCu_3O_7$ ceramics (Wegerer et al., 1996). The focus was on the coupled O(2)-O(3)-out-of-phase and O(2)-O(3)-in-phase modes. The study has showed that La^{3+} ions occupy predominantly Ba sites for $x < 0.5$, and act as an isovalent substitution for Y^{3+} when $x > 0.72$. The sites predominantly occupied by La ions are determined optically by observing the shift of the O(2)-O(3)-out-of-phase mode. This mode would shift to lower frequencies if La replaced Y ions, and remains almost constant when La substitutes Ba ions. With increasing x, a transition from orthorhombic to tetragonal symmetry takes place for $x = 0.4$. The O(2)-O(3)-in-phase and out-of-phase modes do not mix in the tetragonal crystal with higher D_{4h} symmetry, but do mix in the orthorhombic phase with lower symmetry D_{2h}. With increasing decoupling between these two modes, as a result of phase transition, the frequency of the out-of-phase mode shifts toward higher values. From the coupled-mode Hamiltonian treatment, it follows that the frequency difference between the in-phase and out-of-phase modes is $\Omega = 2[k_T^2 + (k_\alpha + k_\beta)^2]^{1/2}$ where $2k_T$ is the energy separation of the decoupled modes in the tetragonal phase, k_α the coupling coefficient due to the presence of CuO chains or chain fragments in the orthorhombic crystal, which exist even if the a, b axes lengths are equal, and k_β is the coupling due to orthorhombic distortion of the unit cell, present when $a \neq b$. Fitting with experimental data, for $k_T = 24.7\ \text{cm}^{-1}$ calculated from linear muffin-tin orbital model, gives $k_\alpha = -52.9\ \text{cm}^{-1}$ and $k_\beta = 10\ \text{cm}^{-1}$ (Wegerer et al., 1996).

4.6 Deformation Potentials

When an impurity is introduced into a perfect lattice or a stress/strain is applied, the lattice becomes deformed (see also Sect. 3.4). Here we deal with the effects of applied stress. The hydrostatic, tetragonal and trigonal deformation potential can also be obtained from optical measurements, more exactly from two-photon excitation spectroscopy, which has a linewidth about one order of magnitude smaller than linear optical spectroscopy. In particular, such a study has been performed on ZnSe by Fröhlich et al. (1995) for different configurations specified by the direction of the wavevector $\boldsymbol{k}$ and of the applied force $\boldsymbol{F}$ with respect to the crystalline axes. Four configurations have been used: A for which $\boldsymbol{F} \parallel [001]$, $\boldsymbol{k} \parallel [1\bar{1}0]$, B when $\boldsymbol{F} \parallel [111]$, $\boldsymbol{k} \parallel [1\bar{1}0]$, C with $\boldsymbol{F} \parallel [11\bar{2}]$, $\boldsymbol{k} \parallel [1\bar{1}0]$ and D for which $\boldsymbol{F} \parallel [11\bar{2}]$, $\boldsymbol{k} \parallel [111]$. The ground exciton state of ZnSe is split into orthoexciton and paraexciton subbands. The interaction Hamiltonian in the presence of strain is

$$H = H_{\text{exch}} + H_{\text{l}}, \tag{4.14}$$

where

$$H_{\text{exch}} = (1/4 - \boldsymbol{\sigma}_{\text{e}} \cdot \boldsymbol{\sigma}_{\text{h}})\Delta_{\text{exch}}, \tag{4.15}$$

is the exchange Hamiltonian that characterizes the interaction between electrons and holes, with spin operators $\boldsymbol{\sigma}_{\text{e}}$, $\boldsymbol{\sigma}_{\text{h}}$, respectively, and

$$H_{\text{l}} = a\text{Tr}(e) - b[(l_{\text{h},x}^2 - l_{\text{h}}^2/3)e_{xx} + \text{c.p.}] - d[(l_{\text{h},x}l_{\text{h},y} + l_{\text{h},y}l_{\text{h},x})e_{xy} + \text{c.p.}], \tag{4.16}$$

describes the effects of uniaxial stress. Here a, b, d are the hydrostatic, tetragonal and trigonal deformation potentials referring, respectively, to the difference of hydrostatic shift between conduction and valence bands, and (the last two) to the stress-induced splitting of the valence band, e_{ij} are the components of the strain tensor and c.p. denotes cyclic permutation with respect to x, y, z. In the absence of an applied stress only dipole-allowed orthoexciton transitions are possible, paraexcitons being excited in the presence of stress due to state mixing. The second term in H_{l} vanishes in the A configuration, and the third term equals zero for the B configuration. Thus, by applying an external strain, the two remaining adjustable parameters in H_{l} can be determined. The exchange energy $\Delta_{\text{exch}} = 0.33$ meV is determined from the difference between the orthoexciton and paraexciton transition energies, the latter being obtained by extrapolating to zero the data for $P \neq 0$. The elastic constants S_{ij} can be obtained by several measurements in the D configuration, for which since $\boldsymbol{k}$ is not parallel to any principal axis of the dielectric tensor, the dispersion relation for the polariton involving longitudinal-transverse mixed states is:

$$c^2k^2/\omega^2 = \varepsilon_{11}\varepsilon_{33}/(\varepsilon_{11}\sin^2\varphi + \varepsilon_{33}\cos^2\varphi), \tag{4.17}$$

with φ the angle between $\boldsymbol{k}$ and $\hat{e}_3$, calculated by diagonalizing the Hamiltonian for the stress along $[11\bar{2}]$. This angle is related to the elastic constants by $\tan\varphi = 2/[z + (z^2 + 4)^{1/2}]$, $z = (1/\sqrt{2})[2(S_{11} - S_{12})b + 5S_{44}d]/[2(S_{11} - S_{12})b - S_{44}d]$. Moreover, the dispersion relations in the absence of an applied stress give the nonlinear resonances on the upper and lower polariton branches for orthoexcitons.

The piezo-optic coefficients, which characterize the change in the dielectric constant when a stress is applied, can also be optically determined. There are three independent piezo-optical coefficients P_{11}, P_{12} and P_{44} for cubic crystals of the O, O_{h} and T_{d} classes, and only two (since $P_{11} - P_{12} = P_{44}$) for polycrystalline and amorphous materials. The piezo-optical coefficients for InP (T_{d} class) have been determined from spectroscopic ellipsometry (for above gap excitations) or from transmission measurements (for excitations below the fundamental gap) of the dielectric constant (Rönnow et al., 1998). P_{11} and P_{12} are determined, respectively, from the dielectric tensors $\varepsilon^{\|}$ and $\varepsilon^{\perp}$, parallel and perpendicular to the stress applied along the [001] direction. P_{44} has been calculated from $\varepsilon^{\|} - \varepsilon^{\perp}$ for an applied stress along [111]. The piezo-optical coefficients are determined from the shifts of the critical point energies under an applied stress. For example, the contribution of the 2D critical point E_1 at the dielectric constant is $\varepsilon(E) = -A(E_1^2/E^2)\exp(\mathrm{i}\phi)\ln[1-(E-\mathrm{i}\Gamma_\mathrm{A})/E_1]$, while the contribution of $E_1 + \Delta_1$ is $\varepsilon(E) = -B[(E_1+\Delta_1)^2/E^2]\exp(\mathrm{i}\phi)\ln[1-(E-\mathrm{i}\Gamma_\mathrm{B})/(E_1+\Delta_1)]$. The energies, strengths (A, B) and lifetime broadenings are determined from fitting. The formulae that relate the shifts of the critical point energies to the applied stress for different stress directions can be found in Rönnow et al. (1998). The piezo-optical coefficients (and the deformation potentials) can be determined from either the shift in the critical-point energies or from the change in the oscillator strengths with stress.

4.7 Optical Determination of Magnetic Properties

From the resonant two-magnon Raman scattering in cuprate antiferromagnetic insulators as the single-layer $Sr_2CuO_2Cl_2$ and the bilayer $YBa_2Cu_3O_{6+\delta}$, one can determine the value of the superexchange constant J (Blumberg et al., 1996). The antiferromagnetic interaction between spins $\boldsymbol{S} = 1/2$ localized on Cu atoms in the CuO_2 planes is described by a Heisenberg Hamiltonian on a 2D square lattice $H = J\sum_{(i,j)}(\boldsymbol{S}_i \cdot \boldsymbol{S}_j - 1/4)$ where the sum is performed over nearest-neighbor Cu pairs. The two-magnon scattering is a short-wavelength magnetic excitation in the antiferromagnetic lattice, involving a photon-stimulated virtual charge-transfer excitation that exchanges two spins, with the superexchange constant J. The coupling between light and spin degrees of freedom is described by the London–Fleury Hamiltonian (Fleury and London, 1968)

$$H = \alpha\sum_{(i,j)}(\hat{e}_i \cdot \boldsymbol{R}_{ij})(\hat{e}_f \cdot \boldsymbol{R}_{ij})\boldsymbol{S}_i \cdot \boldsymbol{S}_j, \tag{4.18}$$

where α is the coupling constant, $\boldsymbol{R}_{ij}$ is the vector which connects two nearest-neighbor sites i, j, and $\hat{e}_i$, $\hat{e}_f$ are the polarization vectors of the incoming and outgoing photons, respectively. The sum is taken over nearest-neighbor Cu pairs. Including the final-state interaction in this Hamiltonian, the two-magnon Raman frequency shift is estimated to be near (2.7–2.8)×J in the B_{1g} scattering geometry for a single-layer 2D lattice with D_{4h} symmetry. The value of J is then obtained from the position of this peak. The scattering intensity of the two-magnon line has a strong non-monotonic dependence on the excitation energy: it is very weak for energies near the charge transfer gap and increases substantially above it. Triple-magnetic resonances must also be considered to explain all the features in the resonant Raman spectra.

Two-magnon scattering is more intense than one-magnon scattering in all antiferromagnetic isolating phases of Cu-O based high-temperature superconductors. In materials with isolated CuO_4 square planar units of Cu^{2+} ions stacked on top of each other in a staggered manner along the c axis, as in tetragonal Bi_2CuO_4, the anisotropic exchange antiferromagnetic Hamiltonian is

$$H = -2\sum_{(i,j)\text{pair}} (J_{ij}\boldsymbol{S}_i \cdot \boldsymbol{S}_j + D_{ij}S_i^z S_j^z), \tag{4.19}$$

where the sum is performed over two sublattices, and J_{ij}, D_{ij} are the isotropic and anisotropic parts of the exchange interaction between $\boldsymbol{S}_i$, $\boldsymbol{S}_j$, respectively. The spin-wave dispersion relation obtained by diagonalizing this Hamiltonian is $\hbar\omega(\boldsymbol{k}) = 2S\{[\mu + J_{11}(\boldsymbol{k})]^2 - J_{12}^2(\boldsymbol{k})\}^{1/2}$ with $\mu = J_{12}(0) + D_{12}(0) - J_{11}(0) - D_{11}(0)$. The two-magnon cross-section in the presence of magnon-magnon interaction characterized by strength b is $I(\omega) \approx \text{Im}[G_0(\omega)/(1 + bG_0(\omega))]$ where $G_0 = (1/N)\sum_k \phi(\boldsymbol{k})/[\omega^2 - 4\omega^2(\boldsymbol{k})]$ with $\phi(\boldsymbol{k})$ the polarization-dependent weighting function. This function, as well as b, takes different forms for different symmetries, so that by measuring the two-magnon light scattering, one can extract information about the symmetry of the system. Analytical formulae for $\phi(\boldsymbol{k})$ for different symmetries can be found in Konstantinović et al. (1996).

In antiferromagnetic two-layer systems, two-magnon Raman scattering can be used to determine the interplanar exchange constant J_2. The value of this constant obtained from the one-gap Hubbard model in the London–Fleury approximation is not in agreement with experimental values. It can be obtained directly from experimental data by measuring the Raman scattering in A_{1g} and B_{1g} channels. Neglecting the final-state magnon-magnon interaction, the Raman intensities in these channels are

$$I_{A1g}(\Omega) \approx \sum_q (J_2/4J_1)^2[2J_1(1-\nu_q)/\Omega_1(\boldsymbol{q})]^2\delta[\hbar\omega_i - \hbar\omega_f - 2\Omega_1(\boldsymbol{q})], \tag{4.20}$$

$$I_{B1g}(\Omega) \approx (1 + J_2/4J_1)^2\sum_q [2J_1\nu_q/\Omega_1(\boldsymbol{q})]^2\delta[\hbar\omega_i - \hbar\omega_f - 2\Omega_1(\boldsymbol{q})], \tag{4.21}$$

where J_1 is the intralayer exchange coupling, $\Omega = \omega_i - \omega_f$, $\nu_q = (\cos q_x - \cos q_y)/2$, and $\Omega_1(\boldsymbol{q}) = 2J_1[(1-\nu_q)(1+\nu_q + J_2/2J_1)]^{1/2}$ is the dispersion of the spin-wave excitation spectrum in the two-layer antiferromagnetic system, with $\boldsymbol{q}$ in the first Brillouin zone. The magnon energy is gapless at $\boldsymbol{q} = 0$ and has a gap $\Omega_1(\boldsymbol{Q}) = 2\sqrt{J_1J_2}$ at $\boldsymbol{Q} = (\pi,\pi)$. If $\Omega < 2\Omega_1(\boldsymbol{Q})$ only magnons at the center of the Brillouin zone are excited, the intensities in both A_{1g} and B_{1g} channels scaling as Ω^3. For $\Omega > 2\Omega_1(\boldsymbol{Q})$ magnons with $\boldsymbol{q} \cong \boldsymbol{Q}$ can also be excited, the A_{1g} scattering intensity increasing rapidly, whereas B_{1g} still scales with Ω^3. J_2 can be estimated from $\Delta I = \{I_{A1g}[2\Omega_1(\boldsymbol{Q}) + \delta\omega] - I_{A1g}[2\Omega_1(\boldsymbol{Q}) - \delta\omega]\} \times \{I_{A1g}[2\Omega_1(\boldsymbol{Q}) - \delta\omega]\}^{-1} \cong 256J_1^4/\Omega_1^4(\boldsymbol{Q})$. Similar results are obtained when the interaction between final states is considered, the respective renormalized exchange coupling constants being obtained, in this case, as $J_1 \to J_1(1 + r/2S)$, $J_2 \to J_2(1 + r'/2S)$, where $2S$ is the number of orbitals at a given site, $r = 1 - (1/N)\sum_q (1-\nu_q)[4J_1(1+\nu_q) + J_2]/2\Omega_1(\boldsymbol{q})$, $r' = 1 - (1/N)\sum_q 4J_1(1-\nu_q) \times [2\Omega_1(\boldsymbol{q})]^{-1}$ (Morr et al., 1996).

Non-optical parameters, such as magnetic specific heat and the inverse of magnetic correlation length, can be obtained from the temperature dependence of the low-frequency Raman spectrum in (*c*,*c*) polarization, above the spin-Peierls transition. This quasi-elastic scattering appears near the incident laser light as a tail with strong temperature dependence and Lorentzian profile, described by

$$I(\omega)=\frac{\gamma\omega}{1-\exp(-\hbar\omega/k_{\mathrm{B}}T)}\frac{C_{\mathrm{m}}TD_{\mathrm{T}}q^2}{\omega^2+(D_{\mathrm{T}}q^2)^2}, \tag{4.22}$$

where D_{T} is the thermal diffusion constant, C_{m} the magnetic specific heat, and $q \cong q_0 \sin(\theta/2)$, with q_0 the wavevector of incident light and θ the scattering angle. Both C_{m} and D_{T} can be obtained from the quasi-elastic scattering caused by energy-density fluctuations in the spin system. The latter parameter vanishes at T_{sp} and is proportional to the inverse of the magnetic correlation length ξ. The temperature dependence of ξ is $1/\xi \approx (T-T_{\mathrm{sp}})^{1/2}$ (Kuroe et al., 1997).

The critical temperature of the paramagnetic EuS phase in the macroscopic ferrimagnet $Co_{1-x}(EuS)_x$ can also be determined by optical means (Fumagalli et al., 1998). Studying through MOKE the temperature dependence of the magnetic-exchange behavior between the EuS and Co phases, it was found that the Curie temperature of the EuS phase has a strong enhancement for $x \geq 0.2$, reaching 160 K for the concentration value $x = 0.4$, for which percolation from metallic to semiconductor behavior occurs. The antiferromagnetic coupling between the two macroscopic phases across the phase boundaries is crucial for this transition. The polar Kerr spectrum is measured at two energies (2.15 eV and 2.8 eV) for which the contribution of the Co phase is almost the same but that of EuS has opposite signs. The EuS contribution is separated by differentiating the two curves, and the polar Kerr rotation as a function of the magnetic field B is fitted with $\theta_{\mathrm{K}}^{\mathrm{para}}/B = \chi NJ(J+1)g^2\mu_{\mathrm{B}}^2/[3k_{\mathrm{B}}(T-\Theta_{\mathrm{c}})]$, where $\chi = \theta_{\mathrm{K}}(M)/M$, $J = 7/2$ and $g = 2$, to extract Θ_{c}. Then T_{c} is calculated from the plot of the polar Kerr rotation as a function of temperature for a photon energy 2.8 eV (where the EuS contribution is dominant). The EuS contribution is separated after subtracting the Co contribution to θ_{K}, calculated by means of additional measurements. At temperatures higher than T_{c} the EuS phase does not make any contribution to the polar Kerr rotation.

The magnetic moment of bulk semiconductors, or the Landé *g* factor that is modified by spin-orbit interactions, can be determined from observations of quantum beats in time-resolved PL. In the experiment the circularly polarized excitation propagates along the *y* direction, the magnetic field is applied along *z* and the detector is placed along $-y$, in a backward configuration (Oestreich, 1996). The magnetic field perpendicular to the excitation produces Larmour precessions of electron spins, the PL for σ^+ and σ^- polarizations showing oscillations in intensity with a phase difference of π and a frequency $\omega_{\mathrm{L}} = g_{\mathrm{L}}\mu_{\mathrm{B}}B/\hbar$. The oscillations are associated with quantum beats between

Zeeman split electron levels ($\Delta E = \hbar\omega_L$) with the spin quantization axis parallel to z. The g factor can be extracted directly from the frequency of quantum beats observed in a time-integrated PL measurement at different excitation energies corresponding to LH or HH excitons. Moreover, the temperature dependence of the g factor, determined from the temperature variation of the time-resolved PL oscillations, can be fitted with $g_L(E,T) = g_c(6.3 + 0.037T[\mathrm{K}])E[\mathrm{eV}]$, where g_c is the g factor at the conduction band minimum and E is the electron excess energy.

The g value of excitons, photoexcited electrons, free holes and holes bound to acceptors can also be determined from spin-flip Raman scattering experiments. An excellent review on this subject is that due to Scott (1980). Through linewidths, spin-flip scattering distinguishes also between free and bound electrons, electrons and holes, effective-mass and non-effective-mass acceptors (for non-effective-mass acceptors, g is different from zero for magnetic fields perpendicular to the c axis). Hole spin-flip scattering is weaker than electron spin-flip by a factor of $[\mu_B gH/(E_g - \hbar\omega)^2]^2$, and thus harder to detect except very close to resonance. For common III-V or II-VI semiconductors, the electron gyromagnetic ratio is often negative, so that spin-down conduction band levels are higher in energy than spin-up levels with the same orbital quantum numbers.

4.8 Rotational Spectra of Solids

Although, in general, only vibrational and electronic spectra are observed in solid-state materials, there is a noticeable exception: solid hydrogen. In this material four pure rotational Raman scattering lines have been observed at room temperature and high pressures, corresponding to the $v = 0$ fundamental vibrational level. The energies of the rotational levels characterized by the quantum number J are $E(J) = B_0J(J+1) - D_0J^2(J+1)^2 + H_0J^3(J+1)^3 + \ldots$ where the last two terms represent corrections to the rigid rotor model. J is a good quantum number even in the condensed phase of H_2, which is a hexagonal-centered phase solid. From the pressure dependence of the molecular rotational lines it is possible to infer the density dependence of the hydrogen internuclear distance, since B_0 is related to the internuclear distance averaged for $v = 0$ through $B_0 = (h/8\pi^2 cm_r)\langle 1/r^2\rangle_0$, where m_r is the reduced mass for the nuclear motion. The frequencies of the rotational lines depend on density as $\omega(\rho) = \omega(0) + a\rho^b$, the internuclear distance being 0.745 Å in the liquid phase and 0.735 Å in the solid phase of hydrogen (Ulivi et al., 1998). A slight dependence on pressure (density) of the internuclear distance has been observed: in the fluid phase it decreases up to freezing transition with a negative curvature, while in the solid phase the density decrease is less significant, the curvature of the density dependence of the internuclear distance becoming slightly positive. In solid hydrogen the presence of anisotropic components of the intermolecular potential induces jumps of rotational excitations between neighboring molecules, uniformly distributed over the lattice. These nonlocal excitations, called rotons, can even interact with vibrons (quanta of vibrational motions), generating bound and free scattering states (Moraldi et al., 1998).

4.9 Many-Body Effects

In degenerate n-type bulk InAs, an enhancement of the absorption at the optical bandgap was observed even under degenerate doping conditions where the Burstein–Moss shift is present. This enhancement, determined from the Fourier transform of PL and PLE, changes its spectral position in the same way as the Fermi energy E_F with increasing doping concentration, and is called Fermi-edge singularity (FES) (Fuchs et al., 1993). As shown by Mahan (1990) it is caused by the many-body correlation of minority carriers (holes) with the degenerate electron gas, which increases the electron-hole overlap. The correlation is more pronounced at the Fermi edge due to the suppression of electron-electron scattering as a result of the exclusion principle, and leads to a logarithmic singularity of the interband absorption at E_F even for a high-density electron gas. The FES, although observable in degenerate bulk semiconductors, is more pronounced in low-dimensional structures when the doping region is separated from the high-density electron gas. In bulk semiconductors, a broadening of FES over the energy range $(m_e / m_h)\delta E_F$ (where δE_F is E_F measured from the bottom of conduction band) is observed, due to the finite hole mass. If this energy range is sufficiently small (Mahan, 1990), FES can be observed in absorption spectra, the absorption edge in this case being $\alpha(\omega) = \alpha_0(\omega)[\xi_0 /(\omega - \omega_c)]^{2\Delta}$ where $\alpha_0(\omega) \approx (\hbar\omega - E_g)^{1/2}$ is the absorption coefficient in the absence of FES with E_g the fundamental bandgap in the one-particle picture. ξ_0 is the cutoff frequency and $\omega_c = E_g + \delta E_F + \Sigma_x$ the threshold frequency, where $\Sigma_x = -3e^2 k_F / 4\pi$ is the exchange contribution of the electron gas that decreases the bandgap. The coupling parameter due to the attractive Coulomb potential with Fermi–Thomas screening is $\Delta \cong R_{se} /[12\ln(1 + 6 / R_{se})]$ where $R_{se} = (3/4\pi n)^{1/3} e^2 m_{eff} /(4\pi\varepsilon_0\varepsilon\hbar^2)$ is the ratio of the radius of the volume that contains one unit charge of the Fermi sea to the excitonic Bohr radius. The lifetime broadening can be included in this picture by convoluting $\alpha(\omega)$ with a Lorentzian distribution, the halfwidth of which is the fitting parameter.

FES can be easily visualized in time-resolved pump-probe transmission experiments. The differential transmission spectrum $\Delta T / T_0 = -\Delta\alpha d$ shows a peak (a spectral hole in the differential absorption), which almost coincides with the energy of the laser excitation. In the presence of FES this peak is redshifted and asymmetric, the differential transmission showing a negative peak (an increase of absorption) at higher energies (see Fig. 4.5). These characteristics attributed to FES where observed in GaAs at 20 K (Shah, 1999).

In some bulk semiconductors like GaAs there is an additional repulsive interaction between electrons and holes, due to electron-electron exchange interactions in the many-body system. In two-particle systems this repulsive interaction is called electron-hole exchange interaction and is described in the isotropic case by the Hamiltonian

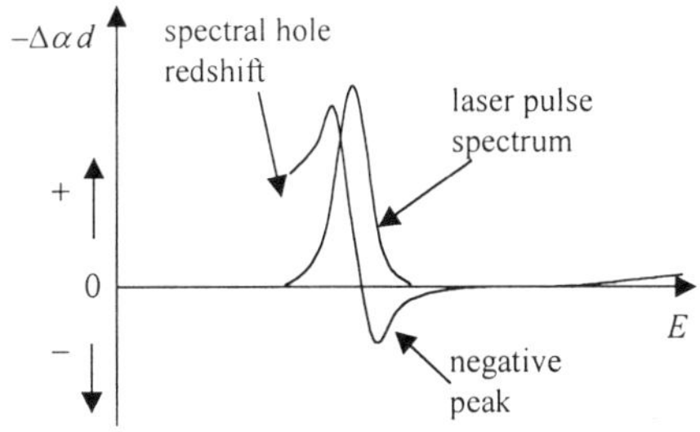

Fig. 4.5. FES signature in pump-probe transmission

$$H_{\text{exch}} = (1/8)\Delta_{\text{exch}}(3 - 4\boldsymbol{\sigma}\cdot\boldsymbol{J}) + \Delta_{\text{LT}}\delta_{J1}(1 - M_J^2), \tag{4.23}$$

where $\boldsymbol{\sigma}$ is the spin vector of the electron, $\boldsymbol{J}$ the angular momentum of the hole and Δ_{LT} the longitudinal-transversal splitting. The electron-hole exchange interaction splits the ground exciton state into a spin-allowed singlet-triplet mixed state with $J = 1$ called an othoexciton and a spin-forbidden pure triplet state with $J = 2$ called a paraexciton. The energy difference between these states is Δ_{exch}. Only orthoexcitons can interact with photons and form polaritons. In one-photon absorption transitions to lower 1s-exciton states are electric dipole forbidden, whereas in nonlinear spectroscopy transitions to both ortho- and paraexcitons can be induced due to the different dipole transitions (electric and magnetic) induced by illuminating the sample with strong laser light of energy $E_g/2$. The orthoexciton absorption peak ($1S_O$) is caused by the reabsorption of part of the second harmonic radiation produced inside GaAs. This process is a one-photon dipole allowed process. The absorption peak due to paraexcitons ($1S_P$), situated at a slightly lower energy, is a two-photon absorption process involving an electric dipole transition to an intermediate spin-allowed state of odd parity followed by a magnetic dipole transition to a spin-forbidden odd parity final state. It is not a common two-photon absorption process since it does not involve two consecutive electric dipole transitions to even parity states. $\Delta_{\text{exch}} = 0.38\,\text{meV}$ is the difference energy between the two peaks corresponding to $1S_P$ and $1S_O$ (Michaelis et al., 1996). Additional two-photon magnetoabsorption experiments in the Faraday configuration confirm the assignment of the peaks. At high magnetic fields the field-induced mixing of states produces new transitions corresponding to paraexcitons with $M_J = \pm 2$.

The two-photon resonant absorption in cuprous oxide (CuO_2) keeps the temperature of orthoexcitons quite low. In this way, the orthoexcitonic density becomes sufficiently high to allow the study of a nearly ideal Bose gas (Goto et al., 1997). The orthoexcitonic system favors Bose–Einstein condensation (BEC) due to the small exciton mass. BEC cannot be observed at one-photon excitation since the temperature (density) is higher (smaller) than the critical temperature (density) for condensation. Time-resolved PL spectra of the orthoexcitonic system suggest that it consists of two parts. One part follows the Bose–Einstein statistics with a chemical potential $\mu = 0$, the other part being located at a spectral position corresponding to zero kinetic energy. The temperature and chemical potential μ of the orthoexciton gas are extracted by fitting the time-resolved phonon-assisted PL to the energy distribution of an ideal Bose gas $\overline{n}(\varepsilon) = (g/2\pi^2)(2m/\hbar^2)^{3/2}(\varepsilon - E_0)^{1/2}/\{\exp[(\varepsilon-\mu)/k_BT]-1\}$ where g is the ground-state multiplicity. Experimental results show that for time delays less than 30 ns and T = 8–10 K the chemical potential is zero (BEC occurs) while for greater delays $\mu < 0$. The density of the orthoexciton gas in the latter case is $n = \int_{E_0}^{\infty} \overline{n}(\varepsilon)\mathrm{d}\varepsilon$, the critical condensation temperature T_0 being extracted from $n = 2.612g(2\pi m k_B T_0)^{3/2}/\hbar^3$. The density of the Bose–Einstein condensate part, n_0, at temperatures smaller than the critical temperature is given by $n_0/n = 1-(T/T_0)^{3/2}$.

4.10 Coupled Excitations

One of the coupled excitations is the phonon polariton. The dispersion of the two lowest E-symmetry phonon polaritons in the pure ferroelectric crystal $PbTiO_3$ has been obtained in the range $q \leq 3000\ \mathrm{cm}^{-1}$ using FWM in the forward arrangement (Loulergue and Etchepare, 1995). The forward FWM enables the excitation of phonons with small wavevectors, especially the polar modes. The FWM signal is proportional to the square of the induced polarization change

$$S(t) \approx [A\exp(-t/\tau) + \textstyle\sum_{i=1,2} A_i \exp(-t/\tau_i)\sin(\omega_i t + \varphi_i)]^2, \tag{4.24}$$

representing the uncoupled contributions from the relaxation and from two damped harmonic oscillators, whose phase factors vanish in the nonresonant regime. The pure relaxation process characterizes the order-disorder contribution. Assuming the material response function described by two harmonic oscillators with frequencies ω_1 and ω_2, the Fourier transform of the signal shows peaks at $2\omega_1$, $2\omega_2$ and $\omega_1 \pm \omega_2$. Additional peaks at ω_1 and ω_2 appear if there is a strong relaxation process with a characteristic time of the same order as the oscillator damping. By plotting the Fourier transform of the spectrum as a function of the angle between the pump beams (the wavevectors of excited phonons), the values $\omega_1(q)$, $\omega_2(q)$ are identified from the peaks of the Fourier transform. Since no peaks are found at ω_1 and ω_2, no pure relaxation process occurs in the material. The results also show that the phases of the two lowest polariton branches change with q, and that their dampings are almost constant over the q range. The obtained dispersion curves are shown in Fig. 4.6.

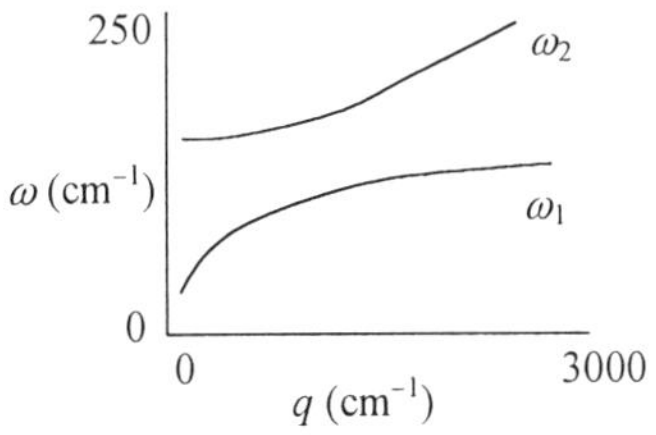

Fig. 4.6 Typical phonon polariton dispersion curve in $PbTiO_3$

Information about the damping mechanism of phonon polaritons can be obtained from the linewidth of the spontaneous Raman scattering, or from the linewidth of the stimulated Raman gain. The first method has low resolution at low frequencies but a better signal-to-noise ratio than the second method. A typical dependence of the linewidth on the phonon wavevector is given in Fig. 4.7.

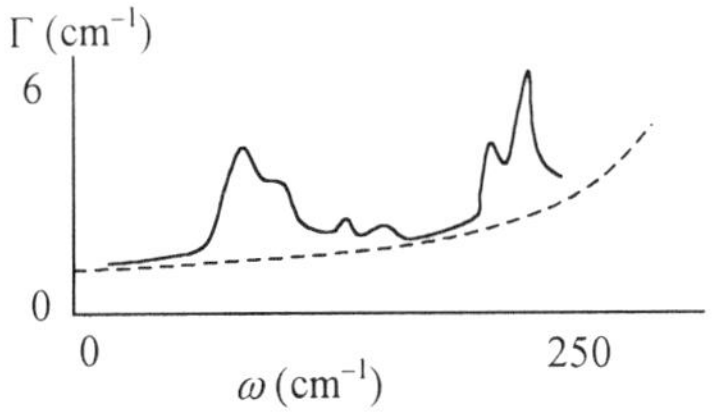

Fig. 4.7. Dependence of the linewidth of the A_1 mode of $LiNbO_3$ on the phonon wavevector

Such a curve has been obtained for the A_1 mode of $LiNbO_3$ (Schwarz and Meier, 1998). The dotted line represents the contribution of the decay into two acoustic phonons, described by $\Gamma_d(\omega,T)=\Gamma_0(\omega/\omega_0)^3[2\bar{n}(\omega/2,T)+1]$. To account for the experimental curve, other scattering mechanisms must be considered: elastic scattering of the polariton at imperfections, which contribute with $\Gamma_{sc}(\omega)$ $=Cq^2\omega^2/v_{gr}(\omega)$ where C depends on the unit cell volume and on the difference between the mass of the defect and the average mass in the crystal. Coupling to low-frequency modes can also contribute to the linewidth dependence on ω. Each low-frequency mode that is coupled to the polariton by a coupling constant k_i, can be treated as a damped oscillator with resonant frequency ω_i, mass m_i, damping constant Γ_i. Coupling to these low-frequency modes due to internal strains or defects produces peaks in the linewitdh, the parameters of the low-frequency modes being obtained from fitting. Polariton damping due to anharmonic decay of the phonon part of the polariton in two acoustic phonons and scattering at crystal defects are dominant at frequencies below the optical-phonon frequency, whereas coupling to low-frequency excitations dominates at higher frequencies.

The PL decay behavior of exciton polaritons, a coupled mode between excitons and photons, was studied in PbI_2 (Makino et al., 1998). It was assumed that the decay of the zero-phonon band is due to the exciton polariton diffusion inside the crystal (from the surface), the PL decay being modeled by $I(t) = I_0(Dt)^{-d/2}\exp(-t/\tau_0) + I_1\exp(-t/\tau_1)$, where D is the diffusion coefficient, d the dimension in which the diffusion takes place and τ_0 the decay time of the zero-phonon band. This decay time changes from about 50 ps near the energy of the longitudinal exciton, to 140 ps a little below the energy of the transverse exciton. Therefore, it is associated to intraband relaxation of polaritons towards the bottleneck region. The exciton polariton lifetime is reflected in the decay of the 2 LO sideband, in about 1–4 ns.

In materials where different types of excitons can form, as in CdS, quantum beats of exciton polaritons are observed (Weber et al., 1997). When A- and B-excitons are simultaneously excited, beats are observed with femtosecond transient FWM, with two periods: 275 fs and 4 ps. The corresponding energy differences are obtained from the period T as $\Delta E = h/T$. The FWM signal looks like that in Fig. 4.8.

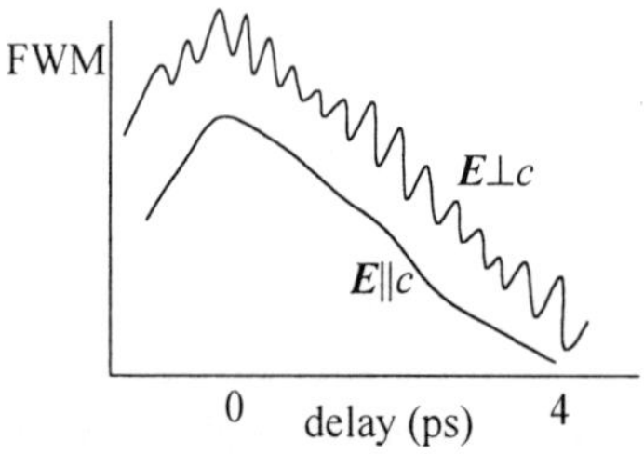

Fig. 4.8. FWM signal in CdS for two polarizations

The shorter period quantum beat is due to quantum interferences between A, B excitons, while the longer period beat is between the lower and intermediate B-polariton branches caused by the $\boldsymbol{k}$-linear term in the dispersion relation for $\boldsymbol{k} \perp c$. This term gives rise to an intermediate polariton branch, besides the upper and lower branches, which does not appear for $\boldsymbol{E} \parallel c$ when only the B exciton is dipole allowed. The $\boldsymbol{k}$-linear term in the valence subband dispersion is difficult to detect in linear optical spectroscopy due to the small energy split, which is obscured by any inhomogeneity. It is, however, possible to observe it in FWM, which detects the homogeneous linewidth even in the presence of inhomogeneous broadening.

In direct gap semiconductors such as GaN there is a strong coupling of free excitons with photons of the same energy, and the valence band splits by crystal field and spin-orbit effects into one subband of Γ_9 and two lower subbands of Γ_7 symmetry, corresponding to A, B, C excitons, as in CdS. The crystal-field and spin-orbit splitting parameters are, however, smaller than in CdS, so that the coupling of the electromagnetic field with the three excitons cannot be separated into three independent contributions. In this case the crystal-field and spin-orbit splitting parameters (as well as the energies of the transverse excitons) are determined by fitting the reflection curve with a dielectric function obtained as a solution of three coupled equations corresponding to the three excitons (Stepniewski et al., 1997).

Coupled plasmon-LO phonons are observed in both semiconducting and semi-insulating InP (Cuscó et al., 1998) at quite low laser powers. The photo-excited plasma density varies almost linearly with the incident laser power, being, however, about one order of magnitude lower in semi-insulating materials due to efficient recombination of photoexcited carriers at impurities. The coupled plasmon-LO phonon mode is distinguished from the LO phonon in that it shifts to higher energies and broadens with increasing laser power, whereas the LO phonon has an almost constant frequency and linewidth.

Coupling between phonon or exciton modes and the modes of the electromagnetic radiation is facilitated by the introduction of the bulk sample in a microcavity. If the microcavity mode is in resonance with, say, the fundamental excitonic transition of the bulk sample, the optical properties of the bulk change dramatically due to the coherent interaction between exciton and photon modes. In the weak coupling regime the interaction affects only the efficiency of the radiative decay, whereas in the strong coupling regime cavity polaron modes with anticrossing behavior appear in the radiation spectrum. In time-resolved emission, Rabi oscillations occur between the two modes, corresponding to coherent energy transfer between exciton and photon modes. Studying the light emission from a $\lambda/2$ bulk GaAs microcavity, it was found that the coherent signal from elastically scattered light dominates the emission spectrum, and that scattering originates mainly from the lower polariton branch (Gurioli et al., 1999).

4.11 Relaxation Phenomena in Bulk Semiconductors

The study of relaxation phenomena in bulk semiconductors following excitation with ultrashort laser pulses is a flourishing area of research. For a comprehensive review of the subject see Shah (1999). A large variety of nonequilibrium, nonlinear or transport phenomena can be investigated using time-resolved spectroscopy, and new phenomena such as many-body effects, coherent excitations and dephasing can be experimentally studied and interpreted. Besides free carriers and phonon dynamics, many studies focus on excitons because they are neutral and thus have a longer dephasing time than charged carriers.

4.11.1 Exciton Relaxation

The dynamics of resonantly excited excitons can be analyzed in time-resolved pump-probe reflectance experiments. The band-edge optical response in all direct widegap semiconductors is dominated by excitonic effects. Bleaching of exciton resonance is expected to occur at high excitation intensities, due to both phase-space filling and screening caused by Coulomb interaction between electrons and holes. An exciton gain was, however, experimentally evidenced at high densities in GaN, a material in which the exciton binding energy is about 20 meV (Hess et al., 1998). A time-resolved pump-probe experiment was performed in order to elucidate the mechanisms that induce such exciton dynamics. Exciting resonantly the excitons with 250 fs pulses, a photoexcited exciton population with $\boldsymbol{q} \cong 0$ is created, which bleaches the sharp resonances of A, B free excitons in reflectance. The dynamics of $\boldsymbol{q} \cong 0$ excitons is measured by changes in the reflectance spectrum. The results suggest that at 4 K the dominant excitonic relaxation process is trapping at defects and impurities, with a time constant of about 16 ps, a much longer characteristic time of 375 ps being observed at temperatures higher than 60 K, associated with the emergence of intrinsic radiative recombination. Free excitons are first localized, and then recombine. At temperatures higher than 60 K the free-exciton population is practically in thermal equilibrium with the neutral donor bound excitons, which are thermally ionized and feed the free A and B exciton population. The radiative recombination lifetime given by $\tau_{\mathrm{rad}} = 0.5(5\pi/3)^{3/5}(h^2 c^7 M^2 \alpha L^3 / C^4 \omega_{\mathrm{A}})/\omega_{\mathrm{A}}$ where M is the mass/unit volume of unit cell, α the polarizability, L the thickness of the excited layer, C the deformation potential and $\hbar\omega_{\mathrm{A}}$ the energy of exciton A, for example, has a value of about 300 ps, comparable to the experimentally found value of 375 ps.

The time constant for free-exciton formation in CdS at 8 K can be determined by measuring the ps time-resolved PL at a LO-phonon-assisted Stokes sideband of the free exciton (Prabhu et al., 1996). In contrast to the zero-phonon free-exciton peak, which characterizes the behavior of $\boldsymbol{k} = 0$ excitons, the LO-phonon-assisted Stokes sidebands reflect the time evolution of the total exciton population. The time constant for free-exciton formation, τ_1, was obtained by fitting the spectrally integrated up-conversion PL at the A_1-2LO Stokes sideband of the free exciton with $A[\exp(-t/\tau_2) - \exp(-t/\tau_1)]$, where $\tau_1 =$ 7–10 ps, and $\tau_2 \cong$ 1000 ps is the decay time of exciton population. τ_1 was found to be independent of photon energy, which suggests a rapid carrier thermalization and energy relaxation in the continuum before the formation of excitons. In this case, excitons form from electrons and holes with small wavenumbers, near the band-edge, the PL lifetime being related to the exciton formation time. The other alternative mechanism, of hot-exciton formation, followed by cascade emission of LO phonons was discredited by experiment.

The same experimental method was also applied to the study of cooling dynamics of excitons in GaN at 20 K (Hägele et al., 1999). At about 10 ps after laser-pulse excitation, the PL spectrum showed a bound exciton peak (D^0X) at 3.478 eV and the free exciton peaks A and B at 3.49 eV and 3.5 eV, respectively, as well as an oscillatory feature at 3.54 eV attributed to the first excited state (n = 2) of the A exciton. At longer delays, for example at 200 ps after laser excitation, the peaks corresponding to B and A (n = 2) vanish due to carrier cooling. The dropping of exciton population at high energies indicates that the exciton temperature is approximately equal to the lattice temperature. Time-resolved PL experiments on LO replicas of the free exciton peak can determine quantitatively the transient exciton temperature. The lineshape of the mth LO-phonon replica is $I_m^{\mathrm{LO}}(E) \cong E^{1/2} \exp(-E / k_{\mathrm{B}} T) W_m(E)$ where E is the kinetic energy of the exciton and $W_m(E)$ is the probability for an electron-hole pair to emit one photon and m LO phonons. For the first two phonon replicas in polar semiconductors $W_1(E) \propto E$ and $W_2(E)$ is constant, respectively, and the total number of excitons is proportional to the spectral integral over the second replica. The formation and lifetime of excitons is therefore given by the temporal evolution of the 2-LO replica. After 20 ps the shape of both 1-LO and 2-LO replicas show a thermalized exciton distribution in agreement with the above formula, the transient exciton temperature being obtained from fitting the high-energy Boltzman tail of the 1-LO phonon replica. The exciton formation time was found to be faster than 10 fs, the biexponential lifetime of the free A exciton being a signature of the existence of two decay mechanisms: capture into a bound exciton (in 50 ps) and a longer decay in 240 ps. The dependence of the exciton temperature T on time can be analy-tically found from the energy loss rate equation of excitons in a bath of acoustic phonons at lattice temperature T_{L}. The result is $T(t) = T_{\mathrm{L}}[\exp(T_{\mathrm{L}}^{1/2}\beta(t)) + 1]^2 / [\exp(T_{\mathrm{L}}^{1/2}\beta(t)) - 1]^2$ where $\beta(t) = \alpha t + T_{\mathrm{L}}^{-1/2} \ln[(T_0^{1/2} + T_{\mathrm{L}}^{1/2})/(T_0^{1/2} - T_{\mathrm{L}}^{1/2})]$ with $\alpha \approx E_{\mathrm{d}}^2 m_{\mathrm{X}}^{5/2} / \rho$ a constant depending on the deformation potential E_{d}, density mass of the crystal ρ and exciton mass m_{X}, and $T(0) = T_0$ is the initial exciton temperature. Fitting the experimental transient exciton temperature with this formula it is possible to determine the effecttive acoustic deformation potential E_{d}, which is 13 eV for GaN where $m_{\mathrm{X}} = 0.46\, m_0$. The exciton temperature reaches the lattice temperature within 150 ps.

Exciton dynamics can be studied also using quantum beats. Quantum beats were observed in the excitonic system of many bulk semiconductors, since the discrete exciton levels are closely spaced (see Shah (1999) and the references therein). For example, in GaAs a coherent polarization between HH and LH continuum states is produced in a pump-probe experiment with 20 fs pulses having an excitation energy near the band edge. This coherent polarization has a strong contribution to the nonlinear optical response, time-resolved transmission experiments being used for the study of nonequilibrium dynamics of elementary excitations. The HH-LH splitting can be extracted from the beat frequency. If the crystal is slightly strained, the different additional contributions to the excitonic HH-LH quantum beats can be identified and separated via polarization selection rules (Joschko et al., 1998). Quantum beats were also observed in time-resolved FWM experiments between A and B excitons in GaN (Zimmermann et al., 1997). The beating period of 0.52 ps indicates a splitting between A and B excitons of 7.98 meV. Moreover, a biexcitonic binding energy of 5.7 meV is estimated from the power spectrum of the FWM. The biexciton, located on the lower-energy side of the A exciton, is only observed if the two pulses have perpendicular linear polarizations. The beating between excitons and biexcitons is observed only in time-resolved FWM, and not in FWM transients due to the homogeneous broadening of the resonances in GaN.

Quantum beats with LO phonon frequency in GaAs resonantly excited at the excitonic energy were observed in FWM experiments, caused by the interference between different states with different numbers of virtual phonons. These phonons form when electrons close to the band edge do not have enough energy to form a polaronic state by emitting real LO phonons. The optical response of the sample was coherently controlled through the phase coherence of the excitation produced by a pair of phase-locked pulses that induced interfering polarizations in the sample (Castella and Ziemmermann, 1999).

4.11.2 Carrier Relaxation

To study the nonequilibrium carrier dynamics in semiconductors it is necessary to excite the sample above the bandgap. The photogenerated nonequilibrium electron-hole pairs relax towards the band edge by several scattering processes. The carrier relaxation can be studied by intraband far-IR (FIR) spectroscopy, in particular by two-color picosecond time-resolved cyclotron resonance (CR). This method was used for InSb, where the near-IR (NIR) pulses created the nonequilibrium carrier distribution, and CR in FIR absorption recorded their dynamics (Kono et al., 1999), in particular the time evolution of effective mass, carrier density and scattering times of nonequilibrium carriers in the presence of a magnetic field. The effective mass is computed from the CR peak position, and the scattering time from its width. Experimental results have shown that the electron effective mass increases as a function of the time delay between NIR and FIR pulses, whereas the scattering time decreases. This behavior is due to carrier-carrier interaction leading to bandgap renormalization; a bandgap decrease for high-density photogenerated carriers, and a corresponding decrease of their effective mass are observed. At longer time delays, when the carrier density approaches its equilibrium value, the effective mass must therefore increase.

Time-resolved PL spectra under high excitation were used to extract hot-carrier effects from the optical properties of GaN (Binet et al., 1999). In particular, Mott transitions were observed as transitions from the free or bound exciton recombinations at low excitations to free-electron/free-hole recombinations in electron-hole plasma at high-excitation densities. Above the Mott transition the electron-hole plasma contains hot carriers. At low excitation densities and 30 K the PL spectrum has a main peak at 3.48 eV due to excitons bound to neutral donors (D^0X). For high pulse widths, during which the steady state is reached, the high-energy part of the PL spectrum (from 3.5 to 3.6 eV) becomes exponentially dependent on energy at increasing excitation density. At the same time the amplitude of the high-energy tail increases exponentially for carrier densities higher than 3×10^{18} cm^{-3}, due to hot-carrier effects. On the other hand the low-energy side of the spectrum broadens at increasing power densities, and several, almost periodic structures identified as phonon replicas of the hot plasma appear below the main peak (see Fig. 4.9). The main PL peak (D^0X) is redshifted due to bandgap renormalization effects and can be used to characterize the Mott transition. If the photogenerated electrons and holes have energies well above the respective quasi-Fermi energies, and their distributions are Maxwellian, $I_{PL}(\hbar\omega) \propto \exp[-(\hbar\omega - E_g)/k_B T_e]$ with T_e the common temperature of electrons and holes.

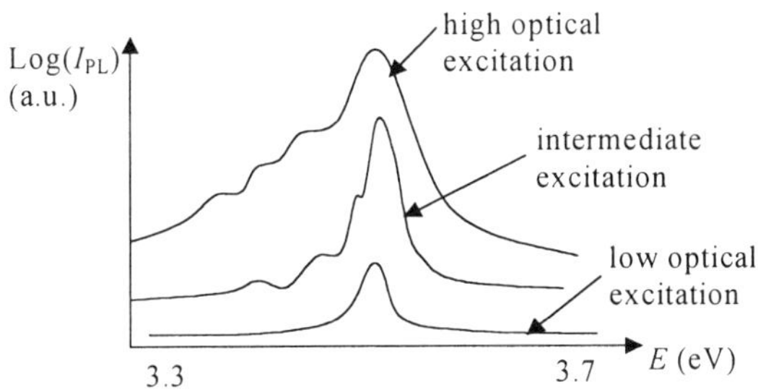

Fig. 4.9. PL spectrum in GaN under low and high optical excitation

T_e can be calculated from the high-energy tail of the PL spectra represented in a logarithmic scale at various optical excitations. For example, at 1.7 MW/cm^2 the carrier temperature is T_e =150 K, well above the bath temperature of 30 K, indicating a hot-carrier population. On illumination of GaN with photon energies larger than the bandgap E_g = 3.5 eV, for example at E = 4.7 eV, the excess energy is mainly received by electrons, which have a much smaller mass than the holes. Two relaxation paths for the photoexcited electrons located in the high-energy tail of the distribution function are possible: interaction with the lattice or direct collisions with the electron gas formed from the electrons located at the bottom of the conduction band. The two relaxation mechanisms have equal contributions for a critical carrier density N_{cr}, found in this case equal to 2.4×10^{19} cm^{-3}; for lower carrier densities the relaxation primarily occurs through LO-phonon emission, while the electron gas heating is mainly due to electron-electron collisions for densities larger than 10^{18} cm^{-3}. For sufficiently low lattice temperatures, the electron temperature follows an Arrhenius law with an activation energy of 92 meV, the power dissipated by the electron gas per photoexcited electron being $P(T_E) = (E_{LO}/\tau)\exp(-E_{LO}/k_B T_e)$, where E_{LO} is the LO-phonon energy. The experimental value for the characteristic time τ is 9 fs. The phonon replicas at the low-energy part of the spectrum are due to breaking of the momentum conservation law, the first phonon replica occurring for example at $E_g - E_{LO} + 2k_B T_e$, whereas the peak energy associated to interband transitions occurs at $E_g + k_B T_e/2$. The total redshift $\Delta E = -\alpha_R N^{1/3} + k_B T_e/2$ of the main PL peak for carrier densities larger than 1.8×10^{18} cm^{-3} is due to both bandgap renormalization, represented by the first term with a renormalization coefficient α_R = 2.1×10^{-8} eV×cm, and a thermal blueshift of $k_B T_e/2$. For carrier densities lower than 1.8×10^{18} cm^{-3}, the PL peak position is practically unchanged due to the compensation of the redshift caused by bandgap renormalization with the blueshift caused by the screening of the Coulomb potential. The exciton binding energy decreases due to this screening as $-\alpha_R N^{1/3}$ and vanishes at the Mott transition between free excitons and free electron-hole pairs, which occurs at a carrier density of $1.8\times10^{18}\,cm^{-3}$.

The nonequilibrium electron behavior in intrinsic InAs was studied with time-resolved up-conversion of hot PL (Nansei et al., 1999). The up-conversion PL gives the product of electron and hole distributions at a given transition energy. The temporal evolution of the PL has two components: one in the ps range due to electron cooling and the other in the fs range due to the partly redistributed nonequilibrium electrons in thermalization. The holes from HH, LH and split-off bands are photoexcited to the conduction band, where they relax faster than electrons due to their more frequent scattering with other carriers, and more efficient interaction with phonons. The temporal behavior of the hot PL, caused by the recombination of relaxed HH and hot electrons reflects the temporal behavior of the carrier temperature and of the chemical potential, which is a function of carrier temperature and Fermi energy. The time evolution of these parameters can be estimated from experimental data of the decay of both Stokes and anti-Stokes sides of the PL.

Similar up-conversion PL studies on hot-carrier cooling and exciton formation have been performed on GaSe (Nüsse et al., 1997). In contrast to polar semiconductors with high structural symmetry, where the carrier-phonon coupling is dominated by the long-range Fröhlich interaction involving LO phonons, in GaSe with low site symmetry, short-range deformation potential coupling is predominant. When the carrier energy decreases during cooling at values below the exciton binding energy, the attractive Coulomb force between electrons and holes can create excitons, which are incoherently occupied by thermalized electron-hole pairs. Time-resolved PL experiments at 46 K show that carriers relax initially via a Fröhlich-type interaction with LO phonons, the subsequent slower cooling regime being dominated by deformation potential interaction with nonpolar LO phonons. The two respective decay times of the cooling temperature $T(t)$ are $t_1 = 0.1$ ps and $t_2 = 5.7$ ps. The carrier temperature T_c is found by supposing that the PL for excitation energies above the bandgap is due to recombination of free electron-hole pairs described by Boltzmann statistics. Fitting the experimental T_c with the analytical solutions of the cooling rates due to Fröhlich interaction and the deformation potential, it is possible to extract the value of the deformation potential as $\Delta = 5.5$ eV/Å. The carrier relaxation is accompanied by a shift of the PL peak towards the position of the fundamental free exciton, since a buildup of excitonic state population during cooling takes place for time intervals larger than 1 ps. The exciton formation time is inversely proportional to the carrier density.

Exciting a CdS crystal at room temperature with 2 eV pulses, a nonequilibrium carrier population is formed due to strong two-photon absorption, which relax by emitting a strong PL in the green range of the visible spectrum. Studying the two-photon excited PL in this material, it was possible to identify the source of PL as the shallow localized states inside the bandgap associated to the valence band A (Lami and Hirliman, 1999). The PL from this direct band semiconductor is weaker for excitation light polarized parallel to the *c* axis of the crystal, since in this geometry transitions between the highest valence band A and the conduction band are forbidden. The relaxation of the nonequilibrium carrier population occurs mainly by interactions with LO phonons, the temperatures and chemical potentials of thermalized electrons and holes being estimated assuming a Fermi–Dirac distribution of carriers. The corresponding values are $T_e = 14\,000$ K, $\mu_e = -5.5$ eV and $T_h = 3500$ K, $\mu_h = -2$ eV. Degenerate pump-probe experiments allow the estimation of the variation of integrated probe intensity (carrier population) with time (see Fig. 4.10).

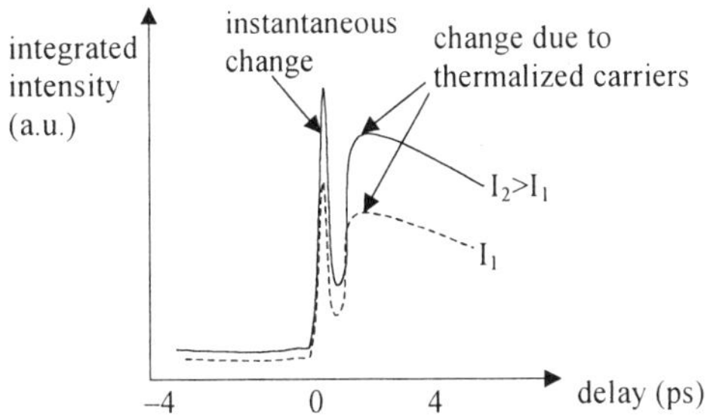

Fig. 4.10. Time dependence of integrated probe intensity in CdS for different optical intensities

The instantaneous change of the integrated intensity is due to the creation of a carrier population with a significant excess of kinetic energy, the thermalized carrier contribution being responsible for the signal rise that follows the instantaneous response. The subsequent slow decay is due to carrier relaxation. Fitting the data with a set of coupled equations between the thermalized and nonequilibrium carrier populations, it is found that the thermalization time of 350 fs is independent of the optical excitation, and the radiative relaxation time varies from 100 ps at lower excitation intensities to 40 ps for higher excitations. The excess of the kinetic energy of electrons, of 1.34 eV, needs an emission of about 35 LO phonons with 38 meV energy for complete dissipation. The electron-phonon interaction time, calculated from the thermalization time, is about 10 fs, five times shorter than in GaAs.

Transient sub-ps Raman spectroscopy was used to study the electron velocity overshot distribution and the nonequilibrium phonon population in a GaAs-based p-i-n nanostructure on which a high electric field was applied. Measurements were made for different electric field intensities and different electron densities (Grann et al., 1996a). Usually, highly energetic electrons thermalize with lattice via emission of LO phonons, the relaxation mechanism and the velocity distribution of carriers being determined via Raman or PL spectroscopy (see references in Grann et al. (1996a)). The electron overshot regime in undoped GaAs occurs within the first picosecond after photoexcitation, when the dominant cooling process for electrons is intervalley scattering, and the effective temperature of nonequilibrium phonons temporarily overshoots that of hot electrons. In this short time interval the photoexcited electrons in the polar GaAs semiconductor have a nonequilibrium distribution, which differs from either Maxwell–Boltzmann or Fermi–Dirac distributions. An externally applied high electric field enhances the nonequilibirum nature of electron distribution. In the electron velocity overshot regime, electrons travel with average velocities higher than the steady-state values. To study this regime, scattered light from single-particle excitations associated with spin-density fluctuations, which predominantly probe electron transport in Γ valleys, were monitored. The energy conservation relation for light scattering from electrons requires that $\hbar\omega = \hbar\boldsymbol{q}\cdot\boldsymbol{V} + \hbar^2 q^2 / 2m_{\text{eff}}$, where $\boldsymbol{V}$ is the electron velocity, $\hbar\omega$ the energy shift of scattered photons, and $\boldsymbol{q} = \boldsymbol{k}_i - \boldsymbol{k}_f$ the transmitted wavevector in the scattering process. Detailed calculations showed that the Raman scattering cross-section is directly proportional to the number of electrons with the velocity component V_q in the $\boldsymbol{q}$ direction and inversely proportional to the effective electron mass. If $T_{\text{e}} \cong 600\,\text{K}$ is the electron temperature, the Kane model predicts that the effective electron mass in the scattering process is $m_{\text{eff}} \cong [E_{\text{g}} m_{\text{c,eff}} (1 + 2k_{\text{B}}T_{\text{e}} / E_{\text{g}})]/(E_{\text{g}} - m_{\text{c,eff}} V_q^2)$ with $m_{\text{c,eff}}$ the effective mass at the Γ point. So, the electron velocity distribution as a function of V_q can be directly determined from Raman scattering, and the drift velocity can then be calculated by performing a weighted average over the electron-velocity distribution. The existence of the velocity-overshot regime was confirmed by the reduction in electron drift velocity for 3 ps data compared to 0.6 ps data. A typical electron velocity distribution looks like that in Fig. 4.11.

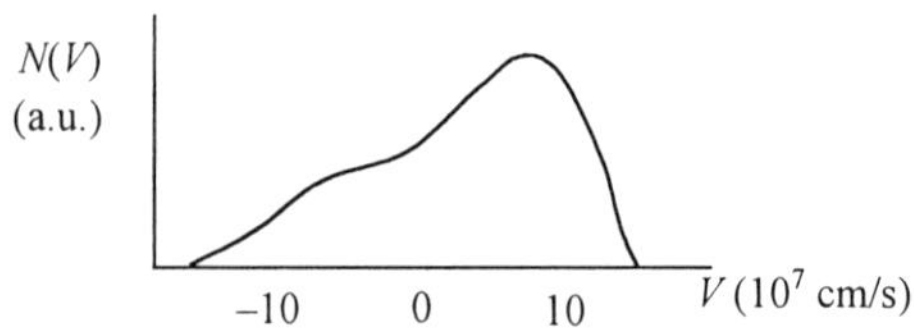

Fig. 4.11. Electron velocity distribution in the overshot regime, as determined from transient Raman spectroscopy

Raman spectroscopy can also be used to calculate the nonequilibrium LO phonon population from $n(\omega_{\text{LO}}) = 1/(I_{\text{S}} / I_{\text{AS}} - 1)$ where I_{S}, I_{AS} are the intensities of Raman Stokes and anti-Stokes LO phonon lines. The nonequilibrium phonon population was found to increase with either photoexcited electron density or electric field (electron acceleration) except for low electron densities when the LO phonon population decreases with the electric field. In this case, the phonon population increase with electron acceleration is overshadowed by the decrease of electron concentration due to the fact that electrons leave the scattering volume or suffer intervalley scattering.

4.11.3 Phonon Relaxation

When the density of photoexcited carriers is high, carrier-carrier interaction dominates and carriers thermalize to hot plasma, which cools to the lattice temperature by emitting optical phonons. The photoexcited carriers relax by consecutive emission of optical phonons until their energy is lowered below the threshold for optical phonon emission. This relaxation path, called phonon-cascade, generates a large population of nonequilibrium optical phonons.

A large population of optical phonons can also be produced using coherent mixing techniques such as infrared time-resolved coherent anti-Stokes Raman scattering (CARS) (Ganikhanov and Vallé, 1997) that offers the possibility to investigate the relaxation of coherent TO phonons. In this method coherent Raman excitation of the phonon population at frequency $\omega_i - \omega_f \cong \omega_{\mathrm{TO}}$ and with wavenumber $\boldsymbol{k}_i - \boldsymbol{k}_f$ is done in the transparent region of the bulk material using two synchronized pulses ω_i and ω_f. Coherent anti-Stokes Raman scattering at $\omega_{\mathrm{p}} + \omega_{\mathrm{ex}}$ of a delayed coherent probe of frequency ω_{p} monitors the temporal evolution of excitation coherence. The frequency of transverse optical phonons is 270 cm^{-1} in GaAs and 370 cm^{-1} in InP. The CARS signal for InP is shown in Fig. 4.12; the dephasing time can be extracted from the linear decay of the logarithm of CARS signal.

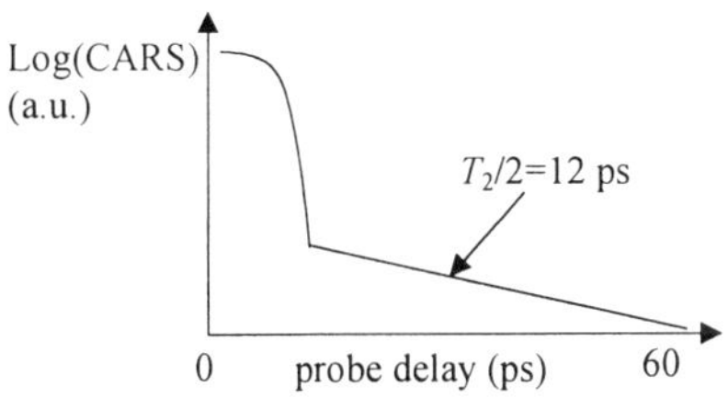

Fig. 4.12. CARS signal at 82 K in InP

The phonon coherence can be lost due to anharmonic interaction with other phonon modes, the most probable relaxation channels being due to the following three particle interactions: (i) up-conversion processes where the coherent phonon is scattered by a thermal phonon $\omega_i^+(\boldsymbol{q})$ into a phonon of higher energy $\omega_j^+(-\boldsymbol{q})$, and (ii) down-conversion processes where the initial phonon splits into two lower-energy phonons $\omega_i^-(\boldsymbol{q})$ and $\omega_j^-(\boldsymbol{q})$. In both cases the initial coherent phonon is destroyed and the lifetime and dephasing times are related by $T_1 = T_2/2$. Overtones correspond to $i = j$ and combination channels are associated with $i \neq j$.

The CARS signal dependence on temperature allows the identification of the two-phonon relaxation paths, since the relaxation rates in both cases are connected to the occupation number of the final phonons and thus to their final energies. Due to the symmetry of the phonon dispersion curve, the only possible down-conversion in InP is the TO phonon decay into two LA phonons ($\omega_{\mathrm{LA}} = \omega_{\mathrm{TO}}/2$)

of opposite wavevectors near the X and L critical points. The temperature dependence of the decay rate is $\Gamma^{\text{ov}}(T) = \gamma_{\text{a}}^{\text{ov}}[1 + 2\bar{n}(\omega_{\text{TO}}/2, T)]$ where $\gamma_{\text{a}}^{\text{ov}}$ is the coupling anharmonic constant and $\bar{n}(\omega, T)$ the phonon occupation number. This overtone channel does not account fully for the experimental results, since the theoretical dephasing time $T_2 = 2/\Gamma^{\text{ov}}$ depends too slowly on temperature compared to measurements. An additional relaxation channel involving states with small wave-vectors must therefore be considered, namely the up-conversion of the TO phonon into a LO phonon with the absorption of a low-energy LA or TA phonon with $\omega_{\text{A}} = 45\ \text{cm}^{-1}$. The total decay rate is then $\Gamma_{\text{TO}}^{\text{InP}}(T) = \gamma_{\text{a}}^{\text{ov}}[1 + 2\bar{n}(\omega_{\text{TO}}/2, T)] + \gamma_{\text{a}}^{\text{up}}[\bar{n}(\omega_{\text{A}}, T) - \bar{n}(\omega_{\text{LO}}, T)]$, in agreement with the experiment. In GaAs, the TO phonon decays into TA and LA phonons.

A similar study in n-type GaAs (Valleé et al., 1997), where time-resolved CARS was measured at different temperatures (10–300 K) as a function of the electron density injected by doping, revealed that for densities greater than $10^{16}\ \text{cm}^{-3}$ the hybrid mode dephasing time (T_2^+) is much smaller than the dephasing time of the bare LO phonon (T_2^0) due to the coherent decay of the LO-plasmon mixed mode in the weak coupling regime. Two hybrid excitations that mediate the carrier-lattice energy exchange in n-GaAs are caused by the long-range electrostatic coupling of LO phonons with the charge-density wave of the injected plasma. The degree of mixing of the hybrid modes with frequencies $\omega^{\pm}$ depends on the frequency of the bare excitations, i.e. on the nature and density of the plasma. For low plasma densities $1/T_2^+ = 1/T_2^0 + \omega_{\text{p}}^2\Omega_{\text{p}}^2/(2\omega_{\text{LO}}^4\langle\tau\rangle_\infty)$, where $\Omega_{\text{p}}^2 = \omega_{\text{LO}}^2 - \omega_0^2$ with ω_0 the TO phonon frequency and $\langle\tau\rangle_\infty$ is the average of the velocity-dependent electron momentum relaxation time. From CARS decay experiments for probe time delays larger than the pulse duration, and under the resonant condition $\omega^+ \cong \omega_i - \omega_f$, all parameters ω^+, T_2^+ and $\langle\tau\rangle_\infty$ can be determined for different electron densities.

The decay of Raman scattering can be used to study the generation of nonequilibrium LO phonons after excitation with subpicosecond laser pulses in InP and InAs, as well as their different relaxation mechanisms (Grann et al., 1996b). The thermalization of nonequilibrium carriers produces details about the band structure and the electron-phonon interaction. Experiments have shown that in InP the Raman decay is due to LO phonon lifetime while in InAs it is determined by the time required for electrons to return to the Γ valley from the L valleys of the conduction band. In other materials, such as GaAs at 77 K, the slow decay time suggests a buildup of nonequilibrium phonons and the appearance of a phonon 'bottleneck', in which phonons and carriers reach a common energy, the reabsorption of phonons slowing the carrier cooling process.

4.12 Optical Properties of Metals

Metals are similar to highly doped bulk semiconductors, and are characterized by a very high density of conduction electrons, orders of magnitude higher than in bulk semiconductors. As in degenerate semiconductors, the Fermi level in metals is located inside the conduction band, and the electrons can be classified as free or bound, their contributions to the dielectric constant being additive. Two types of transitions are encountered in metals: thresholdless intraband transitions due to free electrons, which do not conserve the momentum, and momentum-conserving interband transitions. Intraband transitions dominate the low-frequency spectrum, whereas interband transitions govern the optical properties of simple metals like noble metals and their alloys at higher phonon energies. The distinctive feature of the optical response of metals is their very high reflectance at low frequencies (visible and IR regions), which implies that most optical studies are done by monitoring the reflectance, usually at non-normal incidence, transmission studies being only possible for very thin metal films. Bulk metals are mainly used when the anisotropy of optical properties is under scrutiny. The electromagnetic field penetrates only short distances into the metal, to a distance known as the skin depth, and for this reason the surface quality must be very high.

The optical properties of metals have been well known for a long time, few studies being published in recent years. One of the most complete reviews on the subject is that of Abelès (1972), which will guide us in the brief presentation that follows.

If the relaxation time τ is such that the mean free path of electrons $l = v\tau$ is much smaller than the skin depth, the dielectric constant due to intraband transitions is given by the Drude equations derived in Sect. 2.3.2, in which $\delta = 1/\tau$, N_0 denotes the number of conduction electrons/unit volume and the electron mass must be replaced by the optical mass m_{opt}. For an isotropic metal the optical mass is defined as

$$N_0 / m_{\text{opt}} = \int v \mathrm{d}S_{\text{F}} /(12\pi^3\hbar), \tag{4.25}$$

the integral being performed over the Fermi surface S_{F}.

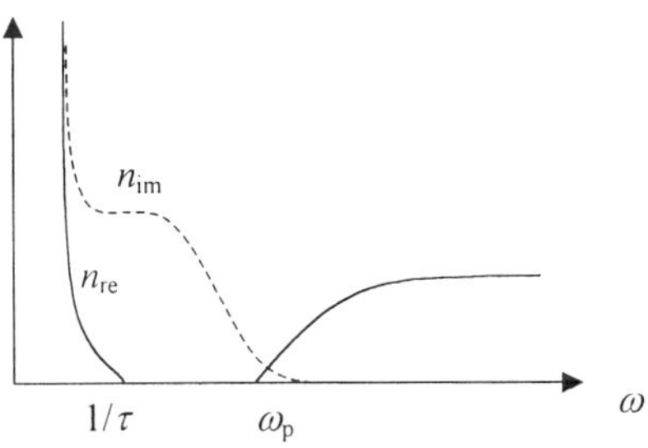

Fig. 4.13. Typical behavior of the real and imaginary parts of the refractive index in metals

As discussed also in Sect. 2.3.2, the behavior of the dielectric constant with frequency depends on the frequency region. For high frequencies $\omega > \omega_{\text{p}}$, where ω_{p} is the plasma frequency of the metal, $n_{\text{re}} \cong [1-(\omega_{\text{p}}/\omega)^2]^{1/2} \cong 1$ and

$n_{\text{im}} \cong \omega_{\text{p}} /(2\omega^2\tau) \cong 0$, the reflectance being very low. This is the transparent region. The reflecting region occurs when $1/\tau < \omega < \omega_{\text{p}}$. In this case the real part of the dielectric constant is negative, and $n_{\text{re}} \cong \omega_{\text{p}} /(2\omega^2\tau)$ is smaller than $n_{\text{im}} \cong \omega_{\text{p}} / \omega > 1$. For low frequencies $0 < \omega\tau < 1$ the real and imaginary parts of the refractive index are linked through the Hagen–Rubens relation $n_{\text{re}}^2 \cong n_{\text{im}}^2 \cong \omega_{\text{p}}^2\tau / 2\omega$ and strong light absorption takes place. The behavior of the dielectric constant in the three regions is schematically displayed in Fig. 4.13. The skin depths for the absorption, reflecting and transparent regions are given respectively by $\delta \cong c(2/\omega\tau)^{1/2} / \omega_{\text{p}}$, $\delta = c/\omega_{\text{p}}$ and $\delta \cong 2c\omega\tau / \omega_{\text{p}}$.

The optical mass introduced through the Drude equation for metals is different from the thermal mass of the conduction electrons m_{th} determined from specific heat measurements. Their ratio is $m_{\text{th}} / m_{\text{opt}} = (S_{\text{F}} / S_{\text{F}}^0)^2 \langle v_{\text{F}} \rangle \langle 1/v_{\text{F}} \rangle$, where S_{F}^0 is the area of a sphere with the same volume as that surrounded by the Fermi surface, $\langle v_{\text{F}} \rangle = (1/S_{\text{F}})\int v\text{d}S_{\text{F}}$ and $\langle 1/v_{\text{F}} \rangle = (1/S_{\text{F}})\int \text{d}S_{\text{F}} / v$. $m_{\text{th}} > m_{\text{opt}}$ if there is no contact between the Fermi surface and the limits of the Brillouin zone, the reverse being true in the case of contact. The difference between the thermal and optical mass is an indication of the anisotropy of the Fermi surface. In noble metals, for example, $m_{\text{th}} / m_{\text{opt}} \cong 1.18$.

If the relaxation time τ varies across the Fermi surface, The Drude–Lorentz expression for the dielectric constant must be replaced by:

$$\varepsilon = 1 - (e^2 / 3\pi^2 h^2 \omega)\int v\text{d}S_{\text{F}} /(\omega + \text{i}/\tau). \tag{4.26}$$

Defining the optical resistivity as $\rho_{\text{opt}} = 4\pi\varepsilon_{\text{im}} /(1-\varepsilon_{\text{re}})\omega$ where $\varepsilon = \varepsilon_{\text{re}} + \text{i}\varepsilon_{\text{im}}$, a measure of the joint anisotropy of the velocity and relaxation time of conduction electrons on the Fermi surface is given by the ratio of the optical to the electrical resistivity $\rho_{\text{opt}} / \rho_{\text{el}} = (\int v\tau\text{d}S_{\text{F}})(\int v\text{d}S_{\text{F}} / \tau)/(\int v\text{d}S_{\text{F}})^2 = \langle\tau\rangle\langle 1/\tau\rangle \geq 1$. $\rho_{\text{opt}} = \rho_{\text{el}}$ only for an isotropic τ. $\rho_{\text{opt}} / \rho_{\text{el}}$ is about 1.5 for noble metals. The relaxation time determined by optical means is smaller than the value obtained from electrical resistivity measurements, due to electron-electron interactions in metals, and is also temperature dependent, especially in the low-temperature range.

When the relation $l << \delta$ is no longer valid, the electron free path and the penetration depth having the same order of magnitude, the interaction of electrons with the metal surface is taken into account by the introduction of an effective dielectric constant $(1/\varepsilon_{\text{eff}})(1-\gamma) = 1/\varepsilon$ via the reflection coefficient r, ε being given by the Drude equation and $\gamma = v(1-r)(n_{\text{re}} + \text{i}n_{\text{im}})/c[1+\text{i}/(\omega\tau)]$. This weakly anomalous skin effect becomes important at low temperatures. The inequality $l >> \delta$ corresponds to the extremely anomalous skin effect, in which the electrons are located on surfaces of constant phase during the light propagation through the metal. This case is encountered in the far-IR or microwave regions where the electrons travel almost normal to the wavevector of light, producing a current analog to the dc transport. This phenomenon allows the experimental mapping of the Fermi surface in metals.

The geometry of the Fermi surface in diamagnetic or paramagnetic metals can be determined using the de Haas–van Alphen (dH–vA) effect (Hjelm and Calais, 1993), which consists in the observation of oscillations, periodic in the reciprocal field due to subsequent passing of Landau tubes through the Fermi surface. The oscillation frequency is proportional to the external cross-sectional area of the Fermi surface perpendicular to the magnetic field. In applied magnetic fields the Fermi surface splits into two almost identical areas A_1 and A_2 which modulate the amplitude of dH–vA oscillations with a cosine function of argument $\pi R = \pi\hbar(A_1 - A_2)/2\pi eB$. The anisotropy of the Fermi surface can be deduced by estimating R (up to an unknown integer) from the measured dH–vA amplitudes. An absolute value of R can be determined from the bulk susceptibility. The Zeeman splitting of the cyclotron orbits on the Fermi surface sheet is isotropic in alkali metals and higher than for free electrons, whereas in noble metals it is anisotropic and only slightly enhanced compared to free electrons. A strong anisotropy of the Zeeman splitting occurs for transition metals.

Several improvements to the simple Drude model described above have been made. One of them, valid for photon energies below the interband absorption edge, includes the contribution of core polarisability ε_∞. This model in which $\varepsilon_{\mathrm{re}} = 1 + \varepsilon_\infty - \omega_{\mathrm{p}}^2\tau^2/(1+\omega^2\tau^2)$, $\varepsilon_{\mathrm{im}} = \omega_{\mathrm{p}}^2\tau/\omega(1+\omega^2\tau^2)$ and $\tau^{-1} = \tau_0^{-1} + \beta(\hbar\omega)^2$ describes the intraband transitions even in polycrystalline metals (Boyen et al., 1997). The eventual deviations from this formula in polycrystalline metals are due to grain-boundary scattering and surface scattering. Another cause of deviation from the simple Drude expression, evidenced in polycrystalline $AuIn_2$ films is the non-free-like density of states that are quite different from the standard band shapes in simple ordered metals. As a result, a peak in the imaginary part, and a dip in the real part of the dielectric constant appear in the polycrystalline metal, in contrast to the smooth behavior of the Drude expression. The frequency dependence of the dielectric constant is also smoother at low frequencies. The contribution of interband transitions ε_{b} to the Drude expression can be accounted for by writing the dielectric function as $\varepsilon = \varepsilon_{\mathrm{b}} - (\omega_{\mathrm{p}}/\omega)^2/[1 + \mathrm{i}/(\omega\tau)]$ (Brouers, 1998).

Since the optical properties of metals are usually experimentally determined from thin films, the influences of interferences and anomalous skin effects must be accounted for (Abelès, 1972). The transmittance of a metallic film with thickness d comparable with the skin depth and bounded by a substrate with a refractive index n_{s} is given by $T = 4n_{\mathrm{s}}(\omega/\omega_{\mathrm{p}})^2\sinh^2(d/\delta)$, from which the plasma frequency can be determined from the slope of T. Although T is independent of the relaxation time, the absorption $A = 1 - R - T$ depends on τ as $An_{\mathrm{s}}/T = (\omega_{\mathrm{p}}\tau/2\omega^2)[\chi + \sinh\chi\cos\chi + (3v/8c)(1-p)(1+\cosh^2\chi)\omega_{\mathrm{p}}\tau]$ with $\chi = d/\delta$. The first two terms in this expression are due to interference effects while the third is caused by the joint interference and anomalous skin effects. The relaxation time τ can be determined from the slope of A as a function of λ^2, as shown in Fig. 4.14. The relaxation time in a gold film with $d = 180$ Å and $d/\delta = 0.75$ for example is about 10 fs.

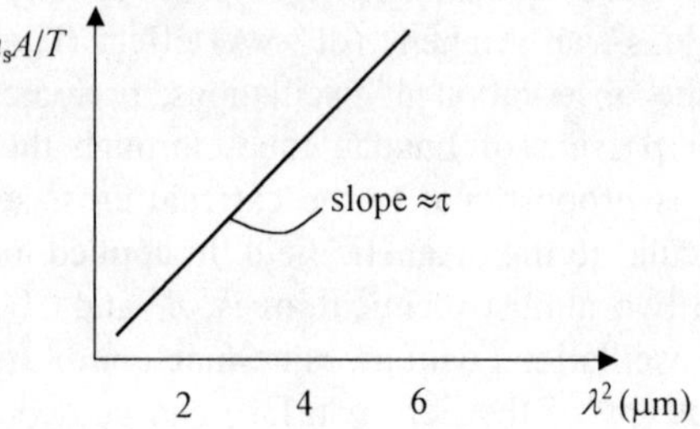

Fig. 4.14. Absorption dependence on wavelength in a thin gold film

The theory of interband transitions in semiconductors can be used also for metals (Abelès, 1972). The contribution of this type of transitions for different types of metals can be summarized as follows:

(i) Alkali metals (Na, K, Rb, Cs) have a nearly free-electron behavior and their conductivity is given by $\varepsilon_{\rm im}(\omega) = 4me^2 |V_{110}|^2 (\omega_1 - \omega)(\omega - \omega_0)/(\hbar\omega)^4 G_{110}$ where G_{110} is the reciprocal lattice vector, V_{110} the pseudopotential Fourier coefficient, $\hbar\omega_0 = (\hbar^2/2m)(G_{110} - 2k_{\rm F})G_{110}$ and $\hbar\omega_1 = (\hbar^2/2m)(G_{110} + 2k_{\rm F})G_{110}$ with $k_{\rm F}$ the Fermi wavevector of free electrons. $|V_{110}|$, proportional to the square of the intensity of the absorption band, can be estimated from measurements of the optical conductivity.

(ii) Polyvalent metals (Pb, In, Al, Zn, Mg) present only critical points of the second kind, i.e. the joint density of states behaves like $\omega^2(\omega^2 - \omega_{\rm G}^2)$ for $\omega > \omega_{\rm G}$ and is zero for $\omega < \omega_{\rm G}$, where $\hbar\omega_{\rm G} = 2|V_{\rm G}|$ with $V_{\rm G}$ a Fourier coefficient of the potential. The absorption spectrum thus has a sharp edge near $2|V_{\rm G}|$, from which $|V_{\rm G}|$ is optically determined.

(iii) Noble metals (Cu, Au, Ag) have two strongly hybridized groups of bands: an s-p group, analogous to the nearly-free-electron band in polyvalent metals, and a d-electron group at about 2–4 eV below the Fermi level, which can be excited in the visible and the near UV. The absorption edge is determined by the $d \rightarrow$ Fermi level transition.

(iv) Transition metals have a more complicated electronic band structure than normal metals and are divided into three categories: ferromagnetic (Ni), antiferromagnetic (Cr) and rare-earth (Eu, Yt). A more detailed discussion of these categories can be found in Abelès (1972).

The nonequilibrium electron dynamics at metal surfaces can be investigated by double-pump/reflectivity probe measurements with ultrashort laser pulses (see Fig. 4.15). Two delayed pump pulses are needed for obtaining information on nonequilibrium electron dynamics from the surface reflectivity, as a function of the delay between the pump pairs. The ultrashort pump pulses heat the electrons at the metal surface, which thermalize subsequently in a very short time to a Fermi–Dirac distribution by electron-electron scattering. Peak electronic temperatures of thousands of K above the equilibrium temperature can be obtained due to the small heat capacity of electrons. The transfer of heat from the electronic system to the lattice via electron-phonon coupling is accompanied by the ballistic transport of nonthermalized electrons and by the diffuse transport of thermalized electrons into the bulk. The competition between electron-phonon coupling and hot-electron transport can be studied by two-pump excitation. When the two pump pulses coincide in time, a pronounced dip in ΔR is measured, directly proportional to the change in surface temperature.

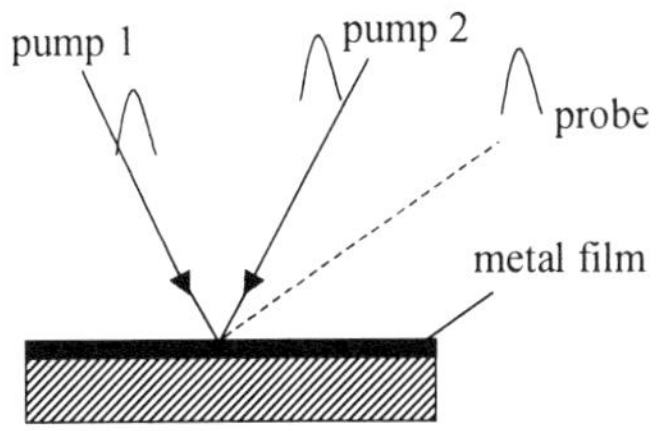

Fig. 4.15. Two-pump probe set-up for electron-phonon coupling determination

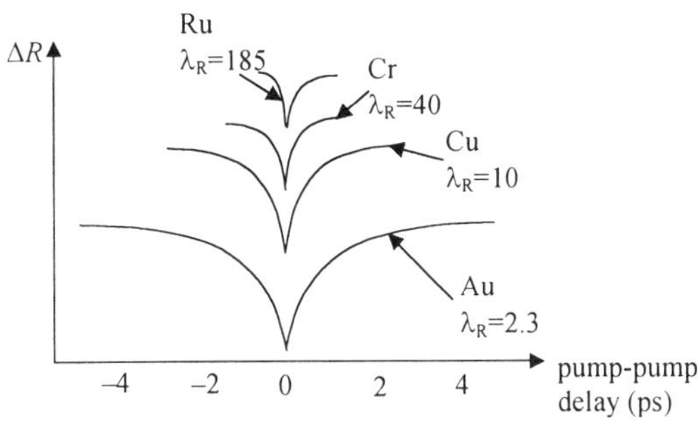

Fig. 4.16. Reflection dip and the corresponding electron-phonon coupling coefficient in different metals

In Fig. 4.16 the electron-phonon constant λ_R for different metals must be multiplied with 10^{16} [$Wm^{-3}K^{-1}$] (Bonn et al., 2000). The dip in reflection is narrower with increasing electron-phonon strength and vanishes if the two pumps do not coincide. This extreme sensitivity of the reflection dip to the electron-phonon constant is the basis of an accurate method to determine λ_R. Experimental results can be reproduced by a set of coupled equations for the temperatures of the electron and phonon subsystems. The electron-phonon coupling tends to localize the heat at the surface, while electron diffusion transports it from the surface into the bulk. The resulting surface temperature is smaller for shorter pump-pump delays since the heat penetrates deeper into the film because the electron thermal conductivity is larger due to the higher transient electronic temperature.

In metal-dielectric granular films, a significant enhancement of Raman scattering in the visible and the IR due to the highly spatial correlated giant field fluctuations of the local field $\boldsymbol{E}(\boldsymbol{r})$ has been observed. The Raman enhancement can reach 10^6, the scattered local fields at the Stokes frequency being concentrated in sharp, well-separated peaks (Brouers et al., 1998). The Raman enhancement due to the presence of metal grains on dielectric substrate is (Brouers et al., 1997) $A = \langle |\varepsilon(\boldsymbol{r})|^2 |E(\boldsymbol{r})|^4 \rangle / (\varepsilon_d^2 |E_0|^4)$ where $\varepsilon(\boldsymbol{r})$, $E(\boldsymbol{r})$ are the local dielectric constant and electric field, E_0 is the average (macroscopic) electric field inside the film and ε_d the dielectric constant of the medium hosting the metal grains. E_0 can be different from the incident field amplitude and is dependent on polarization. For a Drude metal in the frequency range $1/\tau << \omega << \omega_p / \varepsilon_b^{1/2}$ the enhancement of the Raman scattering is independent of frequency and determined only by ξ^2 where ξ is the correlation length, i.e. the scale for the field inhomogeneity at resonance.

4.13 Phase Transitions

Optical measurements can be used to study also different types of transitions that can take place in solids, identifying both the critical parameter at which the transition occurs, as well as the changes in the electronic or phononic structure. This section outlines different types of transitions and the information about them that can be extracted by optical measurements.

The most common phase transitions are structural transitions, at which the structure of the material, and hence the set of IR- and Raman-active modes, change suddenly with temperature, pressure or other parameters. Such transitions occur in practically all materials and are a constant subject in optical investigations of solids. At phase transitions the volume of the primitive cell changes by a few times; an increase of eight times has been reported for $FeSi_2$, at the transition between the orthorhombic phase and the β-phase (Guizzetti et al., 1997). Such structural phase transformations have been observed also in polycrystalline materials. One example is $BaTiO_3$, which suffers a tetragonal to cubic transformation at an increase of the pressure up to 2 GPa and another phase transition at about 5 GPa. The presence of disorder due to grain boundaries and intergrain stresses, as well as the off-center positions of Ti atoms induce, however, a smooth transition between different phases, in the sense that the strong Raman lines present in the tetragonal phase survive well into the cubic phase (Venkateswaran et al., 1998). Phase transitions are well documented also with nonlinear optical methods. For example, the phase transition at 22 GPa in quartz has been inferred from the behavior of second harmonic generation (Pinnick et al., 1997): the second harmonic signal and its anisotropy were constant up to 22 GPa, where they suffered a sudden decrease, followed by a continuous reduction at higher pressures.

One type of transition is the ferroelastic phase transition. A recently studied material from this point of view is the rutile-type GeO_2 (Haines et al., 1998). Raman measurements of the dependence of the spectral position of the B_{1g} mode with pressure have revealed a continuous softening of this mode up to the ferroelastic transition pressure $P_c = 26.7$ GPa, followed by the transformation of this mode in the hard A_g mode (see Fig. 4.17).

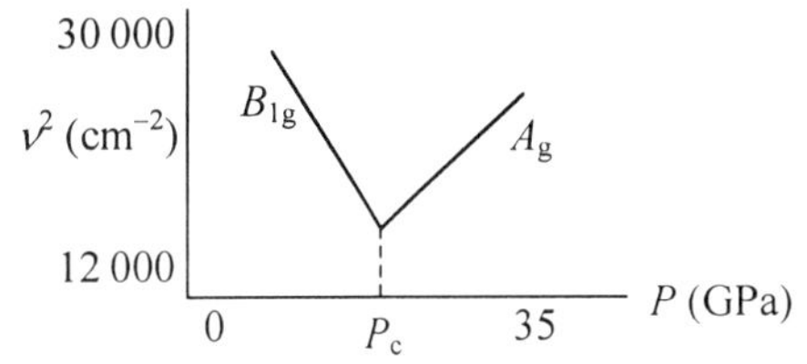

Fig. 4.17. Typical pressure dependence of the spectral position of the B_{1g} mode in GeO_2

The transition can be identified as a second-order phase transition since the squares of frequencies of the involved modes depend linearly on pressure. The change of the mode symmetry at transitions is an indication of the structural change in the material: the rutile structure – a paraelastic phase, stable at lower pressures, changes into the ferroelastic, orthorhombic $CaCl_2$-type structure at higher pressures. The unit cell constants of the orthorhombic phase, a and b, define the order parameter for the rutile ferroelastic transition, identical to the spontaneous strain $e_{ss} = (a-b)/(a+b)$. Another confirmation of the second-order-type transition comes from the calculation of the ratio κ_f / κ_p where κ_f, κ_p are the proportionality constants between ν^2 and P in the ferroelastic and paraelastic phases, respectively. The experimentally determined modulus of this ratio is 2.4 ($\kappa_f = 1503$ cm^{-2}/GPa, $\kappa_p = -631$ cm^{-2}/GPa), comparable with the value of 2 in Landau theory; also the calculated critical pressure is 27.8, close to the experimental value. Similar behaviors can be observed in SiO_2 and SnO_2. On the contrary, the critical pressure of the rutile-to-$CaCl_2$ ferroelastic phase transition in RuO_2 is determined by the observation of the splitting of the degenerate E_g component into two nondegenerate components in the Raman spectrum, with increasing pressure. This second-order transition at 11.8 GPa is accompanied by an abrupt change of the Grüneisen parameter $\gamma_i = (B_0 / \nu_i)(\mathrm{d}\nu_i / \mathrm{d}P)$, with B_0 the isothermal bulk modulus (= 270 GPa), for all phonons (Rosenblum et al., 1997).

Disorder-related effects, which appear at the transition from the ordered orthonormal to the disordered tetragonal phase of KSCN were observed in Raman scattering as the enhancement of the coupling between a sideband (a two-phonon process) and a one-phonon state belonging to a neighboring mode. This coupling is called Fermi resonance (Li and Hardy, 1997). The high sideband intensity suggests an anharmonic coupling between the CN internal stretching mode and the lattice vibrations, modeled by two coupled oscillations. The renormalized frequencies of the coupled sideband and CN modes are $\Omega_{s,CN} = (\omega_s + \omega_{CN})/2$

$\pm[(\omega_s+\omega_{CN})^2/4+V^2]^{1/2}$ where V is the coupling constant and ω_s, ω_{CN} the respective unperturbed frequencies. The coupling constant can be determined experimentally from the ratio of Raman intensities of the renormalized modes $r=[V^2+(\Omega_{CN}-\omega_s)^2]/[V^2+(\Omega_s-\omega_s)^2]$ and their difference $\Omega_s-\Omega_{CN}$. The coupling constant V increases almost linearly with temperature up to the transition temperature T_c, and decreases after it (see Fig. 4.18). The sideband shows a softening of about 12 cm^{-1} between 300 K and the phase transition temperature at 413 K; simultaneously, the CN mode shifts toward higher frequencies. Assuming that the sideband involves off-zone center lattice phonon and an internal mode with opposite wavevector, the softening suggests the possible off-zone center instability of lattice, which accompanies the phase transition.

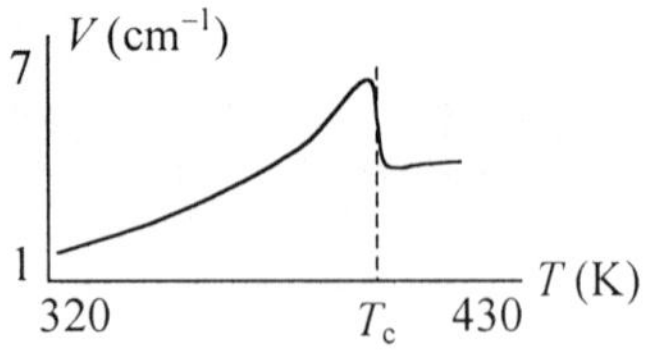

Fig. 4.18. Temperature dependence of coupling constant V in KSCN

Charge-ordering (CO) transitions can also be studied by optical means. In particular, in Fe_3O_4, the temperature dependence of the optical conductivity shows a notable spectral weight transfer in the 0–2 eV region (Fe 3d intersite transition region) above 121 K, the polaronic features being preserved. The highly diffuse charge dynamics produces at this transition temperature T_V (called Verwey transition) an ordering of Fe^{2+} and Fe^{3+} on the ferromagnetic octahedral B sites of the spinel structure, the t_{2g} conduction electrons on these sites being almost fully spin-polarized near T_V. The Verwey transition is associated with opening of an optical gap of about 0.14 eV (Park et al., 1998). CO transitions are also seen in many transition metal oxides with commensurate hole doping, the intersite Coulomb repulsion being one of the driving forces for these transitions. CO transitions are generally accompanied by lattice distortions and antiferromagnetic spin correlations. For example, in $La_{1/3}Sr_{2/3}FeO_3$, at the CO transition temperature T_{CO} = 198 K, an optical gap up to 2Δ = 0.13 eV is opened, the sequential 2:1 ordering of nominal Fe^{3+} and Fe^{5+} $(111)_c$ sheets inducing the appearance of additional optical (stretching and bending) phonon modes due to charge modulation (Ishikawa et al., 1998). This transition is of first-order, since the oscillator strength of the activated phonon modes S_A is discontinuous at T_{CO}, the optical gap increasing with decreasing temperature below T_{CO} (see Fig. 4.19). The temperature behavior of the optical gap implies a concomitant antiferromagnetic spin ordering.

The oscillator strength S_A of the activated Fe-O stretching phonon mode is obtained by fitting the optical conductivity spectrum between 0.06 and 0.1 eV, below T_{CO}, with two Lorentzians:

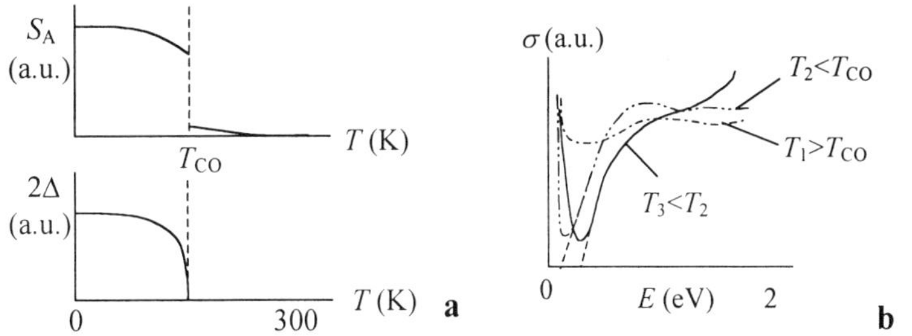

Fig. 4.19. (**a**) Typical temperature dependence of S_A and the optical gap, and (**b**) of the conductivity spectrum in $La_{1/3}Sr_{2/3}FeO_3$ below and above the CO transition

$$\sigma(\omega) = \sigma_0 /[1 + (\omega\tau_0)^2] + \sum_{j=E,A} S_j \omega_j^2 \gamma_j \omega^2 /[(\omega_j^2 - \omega^2)^2 + \gamma_j^2 \omega^2]. \tag{4.27}$$

These correspond to the Fe-O stretching phonon mode (at 0.07 eV) with oscillator strength S_E and to the activated Fe-O stretching phonon mode (at 0.08 eV) with oscillator strength S_A. The first mode exists also above T_{CO}, and its oscillator strength also has a jump at T_{CO}. The activated mode appears since the modulation charge density induces folding of the phonon branches along the corresponding direction. The optical gap is estimated by linearly extrapolating the onset part of σ to the line $\sigma \equiv 0$ (see Fig. 4.19). Above T_{CO} the reflectivity rises to approximately 1 and there is a clear Drude component below 0.1 eV, which indicates the presence of free carriers in this metallic phase. Concomitant metal-insulator (MI) and ferromagnetic-paramagnetic (FM-PM) transitions occur in $R_{1-x}A_xMnO_3$ ceramic samples such as $LaMnO_{3.1}$, and $La_{0.7}Sr_{0.3}MnO_3$.

CO transitions were also observed in 2D Mott insulators, such as $La_{1.67}Sr_{0.33}NiO_4$, where the doped holes tend to order in a stripe with the same role as a domain wall of antiferromagnetic domains. The electron-electron interaction then drives a charge and spin ordering, the CO ordering at 240 K being accompanied by the formation of a superlattice and lowering of the crystal symmetry. Due to the folding of phonon-dispersion branches, new modes appear in the reflectivity spectra and other modes split due to degeneracy lifting. The optical conductivity, calculated from the reflectivity applying the Kramers–Krönig relation, has a near parallel shift with temperatures for $T < T_{CO} = 240$ K (Katsufuji et al., 1996).

Incommensurate charge density wave (CDW) transitions have been observed in η-Mo_4O_{11} and γ-Mo_4O_{11} (McConnell et al., 1998). In the incommensurate phase there is no simple ratio between lattice parameters and the CDW wavelength, so that any phase shift of the CDW leads to an equivalent structure with the same energy, the vibrational spectrum of the incommensurate phase consisting of special excitations of vanishing frequencies which represent the uniform (long-wavelength) fluctuations of the modulation phase. The monoclinic η-Mo_4O_{11} shows two transitions: one at $T_{c1} = 105$ K with a CDW wavevector $\boldsymbol{Q} = 0.23\ b^*$, and another one at $T_{c2} = 35$ K with an in-plane nesting vector (0.42 b^*, 0.28 c^*), whereas the orthorhombic phase γ-Mo_4O_{11} has a single transition at $T_c = 100$ K with $\boldsymbol{Q} = 0.23\, b^*$. In both phases the CDW transition is a metal-metal transition, in which the electrons confined in the *b*-*c* plane have 2D properties, the optical conductivity being mostly suppressed along the *c* axis below the CDW transition. The optical gap is thus a partial optical gap, the conductivity along the *b* axis being significantly suppressed only at low frequencies. The change in the reflectance as the temperature decreases down to the CDW transition is at most 4%, the reflectance having a metallic character ($R > 0.9$) for all temperatures and wavelengths. On the contrary, in 1D systems as for example in $K_{0.3}MoO_3$, the CDW transition induces a change in the reflectivity of about 40%.

Normal (N)-incommensurate (IC)-commensurate (C) sequence of transitions at decreasing temperatures are reported in A_2BX_4 ferroic crystals, where A = Rb, K, NH_4, $N(CH_3)$, B = Zn, Mn, Cu, Co, X = Cl, Br (Kim et al., 1998). In the commensurate phase the phase of the ion-displacement pattern is locked to the electron gas, at the energetically most favorable value, possible only if the ratio between the lattice parameter and CDW modulation wavelength is a rational number. The vibrational spectrum of the C phase contains modes representing the fluctuations of the modulation phase. This sequence of transitions can be studied by measuring the linear birefringence, which is directly related to the order parameter of the N-IC phase transition, given by the amplitude Q of the incommensurately modulated polarization. The linear birefringence in this type of materials is proportional to $Q^2 \approx |T - T_i|^{2\beta}$ where T_i is the temperature of the N-IC phase transition. From the temperature dependence of the linear birefringence the coefficient β is found to be in the range 0.72–0.75, in agreement with theoretical predictions.

In the so-called Kondo insulators or Kondo semiconductors, the strong correlation between f-electrons leads to the occurrence of a metal-semiconductor transition at a critical temperature T^*. The materials in this class generally have a metallic character at high temperatures with local magnetic moments, and a semiconducting behavior at temperatures lower than T^*, when the magnetic susceptibility becomes much smaller and an energy gap develops simultaneously at the Fermi level. The appearance of this energy gap leads to a strong suppression of the optical conductivity in the far-IR for $T < T^*$. YbB_{12} is an example of a Kondo insulator; the measured optical conductivity in this material revealed a transition around 70 K from a metallic spectrum, with a reflectivity close to 1 for low energies, to a semiconducting spectrum in which $\sigma(\omega)$ shows a strong depletion below 40 meV (see Fig. 4.20). The energy gap inferred from optical measurements is about $2\Delta \cong 25$ meV (Okamura et al., 1998). This gap is due to many-body, as opposed to band-structure effects, a fact that can be demonstrated by comparing the optical bandgap to the gap measured by electrical means $2\Delta_{res} \cong 12$ meV $<< 2\Delta$. The same Kondo behavior was demonstrated in SmB_6, by Raman measurements. In this material, the energy scale below which the electronic states are renormalized by the gap development is 5 to 10 times larger than T^* (about 2 times larger in YbB_{12}), which indicates again the many-body origin of the energy gap in Kondo insulators.

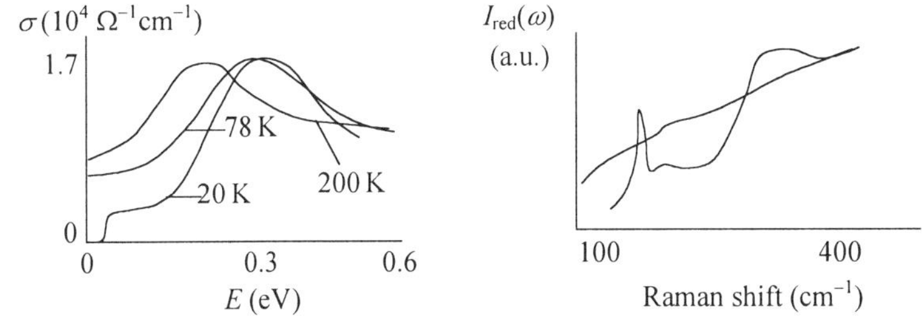

Fig. 4.20. Optical conductivity and reduced Raman spectrum of the Kondo insulator YbB_{12}

Raman studies in SmB_6 (Nyhus et al., 1997a) have revealed not only the energy gap development but also the energy and symmetry of low-frequency excitations inside the bandgap. Below the transition temperature of about 50 K, the electronic scattering intensity below 290 cm^{-1} is suppressed and the spectral weight is redistributed to higher energies; in particular, a higher-energy peak appears at 330 cm^{-1} and a sharp E_g symmetry peak appears inside the gap. The latter is about one order of magnitude larger than the transition temperature: $\Delta / k_B T^* \cong 8$. The in-gap resonances are found to decrease rapidly as the temperature increases towards the transition temperature, being strongly damped by the thermally excited conduction band d electrons. These in-gap resonances split in an external magnetic field, although the energy gap is independent of $\boldsymbol{B}$ for small fields ($|\boldsymbol{B}| = B \leq 8$ T). The dependence of the in-gap modes on B identifies them as non-phononic modes (phononic modes are insensible at B), and offers the possibility to calculate the effective g factor of these modes. For the E_g symmetry peak $g_{\mathrm{eff}} \cong 0.31$, the splitting levels being associated to crystal-field splitting.

The metal-semiconductor transitions in perovskite manganese oxides is seen in Raman scattering as a change from a diffusive electronic scattering in the semiconducting paramagnetic high-temperature phase to a flat continuum scattering in the metallic ferromagnetic low-temperature phase (Yoon et al., 1998). Mathematically, the transition is due to a crossover from a small-polaron regime at high temperatures to a large-polaron regime at low temperatures. In the small-polaron regime the optical conductivity can be written as

$$\sigma_{\mathrm{sp}}(\omega,T) = \sigma(0,T)\exp(-\hbar^2\omega^2/\Delta^2)\sinh(4E_b\hbar\omega/\Delta^2)/(4E_b\hbar\omega/\Delta^2), \qquad (4.28)$$

where E_b is the small-polaron binding energy, and $\Delta = 2\sqrt{2E_b E_{\mathrm{vib}}}$ with E_{vib} a characteristic vibration energy, equal to half the phonon energy at low temperatures and $k_B T$ at high temperatures. The effective number of carriers N_{eff} contributing to the conductivity below the frequency ω is given by $(m/m^*)N_{\mathrm{eff}} = (2mV_{\mathrm{cell}}/\pi e^2)\int_0^{\omega}\mathrm{Re}[\sigma(\omega')]\mathrm{d}\omega'$, where m, e are the bare mass and charge of the electron, respectively, and m^* is the renormalized mass. At high frequencies, the total Raman spectrum, which includes the phonon and electron contributions, is proportional to the optical conductivity: $I_R(\omega,T) = I_{R,ph}(\omega,T) + I_{R,el}(\omega,T) \approx \omega\sigma_{sp}(\omega,T)$. In EuB_6 the metal-semiconductor transition is favored by the formation of bound polarons, involving carriers bound to defects. In this material the Raman scattering in the semiconducting phase is typical for single-particle excitations in degenerate or doped semiconductors, i.e.

$$I_{R,ph}(\omega,T) \approx [1+\bar{n}_{ph}(\omega,T)]\Sigma_i A_i\omega\Gamma_i/[(\omega^2-\omega_i^2)^2+\omega^2\Gamma_i^2], \tag{4.29}$$

with $\bar{n}_{ph}(\omega,T)$ the Bose–Einstein occupation number for phonons. The electronic contribution to the Raman scattering is dominated by neutral charge density fluctuations at the anisotropic Fermi surface, which are not screened by Coulomb interaction:

$$I_{R,el} \approx \langle|\hat{\boldsymbol{e}}_f(1/\hat{m}_{eff} - \langle 1/\hat{m}_{eff}\rangle)\hat{\boldsymbol{e}}_i|^2\rangle, \tag{4.30}$$

where $\hat{\boldsymbol{e}}_f$, $\hat{\boldsymbol{e}}_i$ are the polarization vectors of the scattered and incident radiation, respectively, and $1/\hat{m}_{eff}$ is the inverse effective mass tensor. In the presence of an applied magnetic field, spin-flip Raman scattering is observed, energy shifted from the bound carriers due to Eu^{2+} magnetization (Nyhus et al., 1997b) with $\hbar\omega_0 \cong \bar{\chi}\alpha N_0\langle S_z\rangle$ where $\bar{\chi}$ is the concentration of defects contributing to the magnetization, αN_0 the exchange constant, and $\langle S_z\rangle$ the thermal average of Eu^{2+} spin, given by the Brillouin function for $J = 7/2$.

A metal-insulator Mott transition has been observed at a temperature $T_{MI} \cong 150$ K in V_2O_3, accompanied by a resistivity jump of over 7 orders of magnitude and a change from a triclinic structure above T_{MI} to a monoclinic structure below it (Misochko et al., 1998). The Raman scattering shows an abrupt change of both electron and phonon excitations at the transition temperature, the strong and broad electronic continuum in the metallic state disappearing at the transition. The coupling of the fully symmetric phonon modes to the electronic system is also changed, and found to be temperature dependent even above the transition. Above the Mott transition, the phonons superimposed on a strong continuum have an asymmetric shape with a Fano profile

$$I_R = (Q+\varepsilon)^2/(1+\varepsilon^2)+C, \tag{4.31}$$

where $\varepsilon = (\omega-\Omega)/\Gamma$ with Ω and Γ the renormalized frequency and bandwidth of the phonon line. The parameters Q, Ω and Γ, obtained from fitting the experimental data to this formula, have a jump at the metal-insulator transition, as seen in Fig. 4.21. Below T_{MI} the continuum disappears and the phonon structure is completely changed due to the structural change inside the material. A comparison between the results of Raman spectroscopy, which involves thermally excited, random-phase phonons, and femtosecond time-domain spectroscopy, involving coherently excited phonons with well-established phase relations, showed that the phonon frequencies are slightly displaced and have a different lineshape and linewidth in the two measurements, due to the different nature of the phonons involved. The linewidth of the phonons also have a different behavior with excitation: in Raman scattering the linewidth broadens with increasing excitation, due to an increase of the crystal lattice temperature, whereas in pump-probe spectroscopy the linewidths of the coherent phonons are almost independent of the degree of excitation.

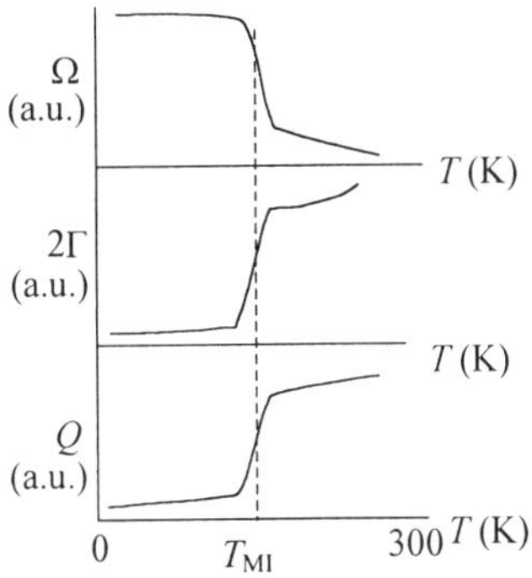

Fig. 4.21. Typical temperature dependence of Fano's parameters in V_2O_3 below and above the metal-insulator transition

A quite interesting type of phase transition is the spin-Peierls transition. Qualitatively, it appears due to the magnetoelastic coupling of the uniform, quasi-1D antiferromagnetic chains with the 3D phonon field. The magnetic atoms are displaced along the chains and form non-magnetic dimers, which involve pairs of opposite spins. The singlet ground state of the dimmers is separated from the excited triplet state by an energy Δ, called the spin-Peierls gap. At the spin-Peierls transition temperature T_{sp} the energy gained by splitting the degeneracy of the original 1D antiferromagnetic chains becomes higher than the lattice deformation energy. The second-order spin-Peierls transition has been mostly studied in $CuGeO_3$. Below the transition temperature the unit cell of this crystal is doubled along both a and c directions, the folding of the Brillouin zone producing the appearance of additional lines in the absorption spectrum (Popova et al., 1998). In $CuGeO_3$ the chains of spin $S=1/2$ Cu^{2+} ions couple by antiferromagnetic superexchange via oxygen orbitals. The exchange along the 1D Cu chains is expressed by

$$H = J\sum_i \{[1+\delta(-1)^i]\boldsymbol{S}_i \cdot \boldsymbol{S}_{i+1} + \alpha \boldsymbol{S}_i \cdot \boldsymbol{S}_{i+2}\}, \tag{4.32}$$

where J and α are, respectively the intrachain constant and frustration ratio, respectively, and the dimerization parameter δ vanishes at T_{sp}. In Raman scattering studies of $CuGeO_3$, the relevant operator in the A_{1g} symmetry is

$$H_R = \Sigma_i [1+\gamma(-1)^i] \boldsymbol{S}_i \cdot \boldsymbol{S}_{i+1}, \tag{4.33}$$

where γ is related to the exchange integral and is sensitive to the interionic distance. No Raman scattering occurs in the homogeneous state, for which $\delta = \gamma = 0$, unless $\alpha \neq 0$. In this case the spin-Peierls gap and the frequency ω_0 of the mode are found by fitting the calculated Raman intensity

$$I_R(\gamma = 0, \omega) = A\Theta(\omega - 2\Delta)\{1 - \tanh[2(\omega - \omega_0)]\} \tag{4.34}$$

with the experimental data. In the dimerized state, γ is obtained from fitting the Raman intensity to

$$I_R(\gamma, \omega) \cong \rho(\gamma, \omega) I_R(0, \omega), \tag{4.35}$$

where $\rho(\gamma, \omega_i)$ describes the relative weight of poles of frequency ω_i to the Raman spectrum (Muthukumar et al., 1996). The degree of interaction of magnetic excitations in a spin-Peierls compound can be estimated from the ratio between the gaps obtained from Raman measurements and from neutron scattering. A value close to 2 indicates a weak magnon interaction, whereas a value of 1.49–1.78 as in $CuGeO_3$ indicates the presence of an attractive magnon-magnon interaction (Gros et al., 1997). The value of this ratio is independent of the coupling constant *J*.

The elastic properties of $CuGeO_3$ above the spin-Peierls magnetic transition occurring at 14.1 K have been determined from the dependence of the Brillouin scattering (Jiménez Riobóo et al., 1998) on the incidence angles. The crossover between the longitudinal elastic constant c_{22} and the transverse elastic constant c_{44} observed at about 50 K was attributed to a translation-rotation coupling.

In the dimerized, non-magnetic phase, the magnetic excitation spectrum is formed from magnons with dispersion

$$\omega^2(k_b, k_c) = \Delta^2 + (\omega_c \sin k_c)^2 + [\omega_b \cos(k_b/2)]^2, \tag{4.36}$$

where $\omega_c = \pi J/2$, ω_b is a fitting parameter, and k_b, k_c are the wavevector components along the *b* and *c* directions. Applying a magnetic field *H*, higher than a critical value, the dimerized phase suffers a first-order transition to an incommensurate state in which the dimerized domains are separated by soliton-like walls of unpaired spins. Neglecting interchain interaction, the energy of magnetic excitations with wavevector $k = \pm(\pi - g\mu_B H/J)$ in the incommensurate phase is

$$E(H) = g\mu_B H + \sqrt{\Delta_{ic}^2 + (\pi J/2)^2 \sin^2(g\mu_B H/J)}, \tag{4.37}$$

or

$$E(H) = (1 + \pi / 2) g \mu_B H + (1/2)(\Delta_{ic}^2 / g \mu_B H), \quad (4.38)$$

for small fields (van Loosdrecht et al., 1996).

Another material in which the spin-Peierls transition was observed is α'- NaV_2O_5. In contrast to $CuGeO_3$, where $\alpha \cong 0.24$–0.36, this material is characterized by a negligible magnetic interaction between chains ($\alpha \cong 0$) due to their isolation by $V^{5+}O_5$ non-magnetic chains. The other parameters J, T_{sp}, Δ are however three-to-four times greater than in $CuGeO_3$, and the large value of $2\Delta / k_B T_{sp} = 4.8$–$6.6$ suggests the deviation from the weak coupling regime. The spin-Peierls transition temperature, inferred from the temperature dependence $(1 - T / T_{sp})^{2\beta}$ of the IR reflectivity is 34 K, with β = 0.25. The spin-Peierls Hamiltonian described above must be supplemented in this case with an interchain elastic coupling term that accounts for the 3D coherence of phonons and provides an additional elastic coupling $K_\perp$ between neighboring chains. If $K_\perp$ is not too small compared to $K_\parallel$, an effective elastic constant K can be defined such that the Hamiltonian is $H = J \sum_i [1 + \delta(-1)^i] \boldsymbol{S}_i \cdot \boldsymbol{S}_{i+1} + (1/2) N K \delta^2$, where N is the number of sites and the temperature-dependent δ parameter is obtained by minimizing the total free energy (Smirnov et al., 1998). The Raman spin excitation spectrum of dimerized spin ladders in α'-NaV_2O_5 show the existence of magnetic bound states at $T < T_{sp}$, due to lifting degeneracy of the singlet ground state (Lemmens et al., 1998). A similar bound state with zero total momentum has been observed also in the Raman spectra of $CuGeO_3$, as a δ-function (Gros et al., 1997).

Another magnetic transition, besides the spin-Peierls transition, is the phase transition that occurs in metals at quantizing magnetic fields (Gordon et al., 1997). This transition is due to the magnetic interaction between the conduction electrons; it is a cooperative orbital magnetic effect and not a spin magnetization. When a magnetic field is applied on the sample, the electrons feel the field $\boldsymbol{B}$, and not $\boldsymbol{H}$, and if the oscillation amplitudes in the de Haas–van Alphen experiment become comparable with the electron period, the magnetic interaction can lead to instabilities of the electron gas. The result is a diamagnetic phase transition, which can even lead to the stratification of the crystal into Condon domains. As shown by Gordon et al. (1997) the phase transition is, however, not necessarily accompanied by stratification; it is possible that the crystal forms a single-domain state. Since the transition occurs in each cycle of de Haas–van Alphen oscillations, a series of phase transitions are observed at discrete values of the magnetic field whenever the reduced amplitude of oscillations tends to 1.

The Brillouin spectra of several materials near the ferroelectric transition temperature T_c show a strong softening, due to the $\eta^2 \mu$-type coupling, where η is the order parameter and μ the strain. Examples of such materials are $CsTiOAsO_4$ (CTA) and $KTiOAsO_4$ (KTA), for which the LA [001] phonon peaks have a temperature dependence of the form $\omega = a\sqrt{T_c - T} + b$. The elastic constants of these materials have been measured by Tu et al. (1997).

Ferro- to paramagnetic phase transitions in $CoPt_3$ alloys can be induced in a sub-ps time scale, in pump-probe experiments (Beaurepaire et al., 1998). MOKE signals in polar geometry, with the magnetic field normal to the surface and p-polarized probe signal, show a non-hysteresis curve, typical for the paramagnetic phase for time delays of about 630 fs. When the probe precedes the pump, the Kerr signal has an s-like shape, with vanishing coercivity and a saturation value of about 500 Oe, whereas the signal in the absence of pump reveals a material with high perpendicular magnetic anisotropy and a coercivity of about 1 kOe. The behavior of the Kerr signal for different pump-probe excitations is depicted in Fig. 4.22.

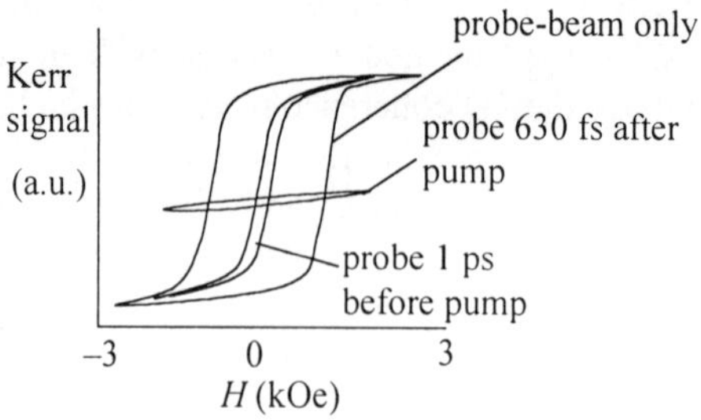

Fig. 4.22. Kerr signal at different pump-probe excitations in $CoPt_3$ alloys

These results suggest that during the spin dynamics a ferro- to paramagnetic phase transition takes place, the relaxation time of the polar Kerr signal being of about 500 fs. The polarization state of the pump beam does not affect the dynamics of the Kerr signal; only the excitation density influences it.

Magnetic transitions at the Néel temperature in antiferromagnetic materials such as $SmTiO_3$ have been observed in far-IR polarized reflectivity (Hildebrand et al., 1999). The onset of magnetic ordering in Ti and Sm sublattices at temperatures lower than T_N due to magnetic transitions is marked by the appearance of a peak near 120 cm^{-1} denoted by M when the electric field vector is polarized along the b axis. This relatively broad peak is due to a two-magnon process, the broad linewidth being caused by the $\boldsymbol{k}$-dependent magnon density of states. Not only the two-magnon peak indicates the magnetic transition but also the asymmetry of a nearby phonon mode with frequency ω_{ph} = 175 cm^{-1}, which has an asymmetric shape at temperatures lower than T_N due to interaction with the continuum of supposedly magnetic origin. σ_1 is then fitted with

$$\sigma_1 = C + A\omega + \frac{\omega}{4\pi}\mathrm{Im}\left(\frac{S_M^2}{\omega_M^2 - \omega^2 - i\omega\Gamma_M} + \frac{S_{ph}^2}{\omega_{ph}^2 - \omega^2 - i\omega\Gamma_{ph}}\exp(i\theta)\right), \tag{4.39}$$

where the linear term is the background conductivity, the oscillator strengths of the M and phonon modes are S_M and S_{ph} respectively, and the asymmetric line shape of the phonon mode is described by the phase factor exp(iθ). The phase of the phonon mode vanishes above T_N as can be seen from Fig. 4.23, which indicates a symmetric Lorentzian shape of the phonon mode.

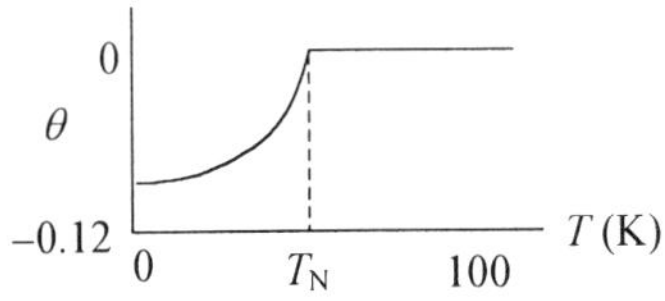

Fig. 4.23. The phase of the phonon mode in $SmTiO_3$ as a function of the temperature

References

Abelès, F. (1972) in *Optical Properties of Solids*, ed. F. Abelès, North-Holland, Amsterdam, p.95.

Abrashev, M.V, A.P. Litvinchuk, C. Thomsen, and V.N. Popov (1997) Phys. Rev. B **55,** 8638.

Arimoto, H., N. Miura, R.J. Nicholas, and N.J. Mason (1998) Phys. Rev. B **58**, 4560.

Basu, P.K. (1997) *Theory of Optical Processes in Semiconductors*, Clarendon Press, Oxford, U.K.

Beaurepaire, E., M. Maret, V. Halté, J.C. Merle, A. Daunois, and J.Y. Bigot (1998) Phys. Rev. B **58**, 12134.

Bellabarba, C., J. Gonzales, and C. Ricon (1996) Phys. Rev. B **53**, 7792.

Binet, F., J.Y. Duboz, E. Rosencher, F. Scholz, and V. Haerle (1996) Phys. Rev. B **54**, 8116.

Binet, F., J.Y. Duboz, J. Orff, and F. Scholz (1999) Phys. Rev. B **60**, 4715.

Blumberg, G., P. Abbamonte, M.V. Klein, W.C Lee, D.M Ginsberg, L.L. Miller, and A. Zibold (1996) Phys. Rev. B **53**, 11930.

Bonn, M., D. N. Dezler, S. Funk, M. Wolf, S. Svante-Wellerhoff, and J. Hohlfeld (2000) Phys. Rev. B **61,** 1101.

Boyen, H.G., R. Gampp, P. Oelhafen, C. Lauinger, and S. Herminghaus (1997) Phys. Rev. B **56**, 6502.

Brouers, F., S. Blancher, A. N. Lagarkov, A.K. Sarychev, P. Gadenne, and V. N. Shalaev (1997) Phys. Rev. B **55**, 13234.

Brouers, F., S. Blancher, and A. K. Sarychev, (1998) Phys. Rev. B **58**, 15897.

Campo, J., M. Julier, D. Coquillat, J.P. Lascaray, D. Scarllet, and O. Briot (1997) Phys. Rev. B **56**, 7108.

Cardona, M., ed. (1975) *Light Scattering in Solids*, Springer, Berlin.

Cardona, M. and G. Güntherodt, eds. (1999) *Light Scattering in Solids* VII, Springer, Berlin.

Cardona, M. and G. Güntherodt, eds. (2000) *Light Scattering in Solids* VIII, Springer, Berlin.

Castella, H. and R. Ziemmermann (1999) Phys. Rev. B **59**, 7801.

Cuscó, R., J. Ibáñez, and L. Artús (1998) Phys. Rev. B **57**, 12197.

Ebner, T., K. Thonke, R. Sauer, and F. Schaeffer, and H.J. Herzog (1998) Phys. Rev. B **57**, 15448.

Fleury, P.A. and R. London (1968) Phys. Rev. **166**, 514.

Fröhlich, D., W. Nieswand, U.W. Pohl, and J. Wrzesinski (1995) Phys. Rev. B **52**, 14652.
Fuchs, F., K. Kheng, P., Koidl, and K., Schwartz (1993) Phys. Rev. B **48**, 7884.
Fumagalli, P., A. Schirmeisen, and R.J. Gambino (1998) Phys. Rev. B **57**, 14294.
Ganikhanov, F., and F. Valeé (1997) Phys. Rev. B **55**, 15614.
Göbel, A., T. Ruf, M. Cardona, C.T. Lin, J. Wrzesinski, M. Steube, K. Reimann, J.C. Merle, and M. Joucla (1998) Phys. Rev. B **57**, 15183.
Göbel, A., T. Ruf, J.M. Zhang, R. Lauck and M. Cardona (1999) Phys. Rev. B **59,** 2749.
Gordon, A., M.A. Itskovsky, and P. Wyder (1997) Phys. Rev. B **55**, 812.
Goto, T., M.Y. Shen, S. Koyamo, and T. Yokouchi (1997) Phys. Rev. B **55**, 7609.
Grann, E.D., K.T. Tsen, D.K. Ferry, A. Salvador, A. Botcharev, and H. Morkoc (1996a) Phys. Rev. B **53**, 9838.
Grann, E.D., K.T. Tsen, and D.K. Ferry (1996b) Phys. Rev. B **53**, 9847.
Gros, C., W. Wenzel, A. Fledderjohann, P. Lemmens, M. Fischer, G. Güntherodt, M. Weiden, C. Geibel, and F. Steglich (1997) Phys. Rev. B **55**, 15048.
Guizzetti, G., F. Marabelli, M. Patrini, P. Pellegrino, B. Pivac, L. Miglio, V. Meregalli, H. Lange, W. Heurion, and V. Tomm (1997) Phys. Rev. B **55**, 14290.
Gurioli, M., F. Bogani, S. Ceccherini, M. Colocci, F. Beltram, and L. Sorba (1999) Phys. Rev. B **59**, 5316.
Hägele, D., R. Zimmermann, M. Oestreich, M.R. Hofmann, W.W. Rühle, B.K. Meyer, H. Amano, and I. Akasaki (1999) Phys. Rev. B **59**, 7797.
Haines, J., J.M. Léger, C. Chateau, R. Bini, and L. Ulivi (1998) Phys. Rev. B **58**, 2909.
Henn, R., A. Wittlin, M. Cardona, and S. Uchida (1997a) Phys. Rev. B **56**, 6295.
Henn R., T. Strach, E. Schönherr, and M. Cardona (1997b) Phys. Rev B **55**, 3285.
Herman, I.P. (1995) IEEE J. Selected Topics Quantum Electron. **1**, 1047.
Hess, S., F. Walraet, R.A. Taylor, J.F. Ryan, B. Beaumont, and P. Gibart (1998) Phys. Rev. B **58**, 15973.
Hildebrand, M.G., A. Slepkov, M. Reedyk, G. Amow, J.E. Greedan, and D.A. Crandles (1999) Phys. Rev B **59,** 6938.
Hirata, T., K. Ishioka, M. Kitajima, and H. Doi (1996) Phys. Rev. B **53,** 8442.
Hjelm, A. and J.-L. Calais (1993) Phys. Rev. B **48**, 8592.
Ho, C.H., Y.S. Huang, K.K. Tiong, and P.C. Liao (1998a) Phys. Rev. B **58**, 16130.
Ho, C.H., Y.S. Huang, P.C. Liao, and K.K. Tiong (1998b) Phys. Rev. B **58**, 12575.
Ishikawa, T., S.K. Park, T. Katsufuji, T. Arima, Y. Tokura (1998) Phys. Rev. B **58,** 13326.
Jiménez, J., E. Martin, A. Torres, and J.P. Landesman (1998) Phys. Rev. B **58** 10463.
Jiménez Riobóo, R.J., M. García-Hernández, C. Prieto, J.E. Lorenzo, and L.P. Regnault (1998) Phys. Rev B **58**, 8574.
Johannsen, P.G. (1997) Phys. Rev. B **55**, 6856.
Johannsen, P.G., G. Reiß, U. Bohle, J. Magiera, R. Müller, H. Spiekermann, and W.B. Holzapfel (1997) Phys. Rev. B **55**, 6865.
Joschko, M., M. Hassellbeck, M. Woerner, T. Elaesser, R. Hey, H. Kostial, and K. Ploog (1998) Phys. Rev. B **58**, 10470.
Katsufuji, T., T. Tanabe, T. Ishikawa, Y. Fukuda, T. Arima, and Y. Tokura (1996) Phys. Rev. B **54**, 14230.
Kim, D.Y., S.I. Kwun, and J.G. Yoon (1998) Phys. Rev. B **57**, 11173.
Klingshirn, C.F. (1997) *Semiconductors Optics*, Springer, Berlin.
Kono, J., A.H. Chin, A.P. Mitchell, T. Takahashi, and H. Akiyama (1999) Appl. Phys. Lett. **75**, 1119.
Konstantinović, M.J., Z. Konstantinović, and Z.V. Popović (1996) Phys. Rev. B **54**, 68.
Kovalev, D., B. Averboukh, D. Volm, B.K. Meyer, H. Amano, and I. Akasaki (1996) Phys. Rev. B **54**, 2518.
Kurita, S., M. Tanaka, and F. Lévy (1993) Phys. Rev. B **48**, 1356.
Kuroe, H., J.I. Sasaki, T. Sekine, N. Koide, Y. Sasago, K. Uchinokura, and M. Hase (1997) Phys. Rev. B **55**, 409.

Lami, J.-F. and C. Hirliman (1999) Phys. Rev. B **60**, 4763.

Lefebvre, P., T. Richard, J. Allégre, H. Mathieu, A. Combette-Ross, and W. Granier (1996) Phys. Rev. B **53**, 15440

Lemmens, P., M. Fischer, G. Els, G. Güntherodt, A.S. Mishchenko, M. Weiden, R. Hauptmann, C. Geibel, and F. Steglich (1998) Phys. Rev. B **58**, 14159.

Li, R. and J.R. Hardy (1997) Phys. Rev. B **56**, 13603.

Loulergue, J.C. and J. Etchepare (1995) Phys. Rev. B **52**, 15160.

Mahan, G.D. (1990) *Many-Particle Physics*, Plenum Press, New York.

Makino, T., M. Watanabe, T. Hayashi, and M. Ashida (1998) Phys. Rev. B **57**, 3714.

Manasreh, M.O. (1996) Phys. Rev. B **53**, 16425.

McConnell, A.W., B.P. Clayman, C.C. Homes, M. Inoue, and H. Negishi (1998) Phys. Rev. B **58,** 3565.

Michaelis, J.S., K. Unterrainer, E. Gornik, and U. Bauser (1996) Phys. Rev. B **54**, 7917.

Misochko, O.V., M. Tani, K. Sakai, K. Kisoda, S. Nakashima, V.N. Andreev, and F.A. Chudnovsky (1998) Phys. Rev. B **58**, 12789.

Moraldi, M., M. Santoro, L. Ulivi, and M. Zoppi (1998) Phys. Rev. B **58**, 234.

Morell, G., W. Pérez, E. Ching-Prado, and R.S. Katiyar (1996) Phys. Rev. B **53** 5388.

Morr, D.K., A.V. Chubukov, A.P. Kampf, and G. Blumberg (1996) Phys. Rev. B **54**, 3468.

Muthukumar, V.N., C. Gros, W. Wenzel, R. Valenti, P. Lemmens, B. Eisener, G. Güntherodt, M. Weiden, C. Geibel, and F. Steglich (1996) Phys. Rev. B **54,** 9635.

Nansei, H., S. Tomimoto, S. Saito, and T. Suemoto (1999) Phys. Rev. B **59**, 8015.

Nüsse, S., P.H. Bolivar, H. Kurtz, V. Klimov, and F. Levy (1997) Phys. Rev. B **56**, 4578.

Nyhus, P., S.L. Cooper, Z. Fisk, and J. Sarrao (1997a) Phys. Rev. B **55**, 12488.

Nyhus, P., S. Yoon, M. Kauffmann, and S.L. Cooper (1997b) Phys. Rev. B **56**, 2717.

Oestreich, M., S. Hallstein, and W. Rühle (1996) IEEE J. of Selected Topics in Quantum Electronics **2**, 747.

Okamura, H., S. Kimura, H. Shinozaki, T. Nanba, F. Iga, N. Shimizu, and T. Takabatake (1998) Phys. Rev. B **58**, 7496.

Oohashi, H., H. Ando, and H. Knabe (1996) Phys. Rev. B **54,** 4702.

Park, S.K., T. Ishikawa, and Y. Tokura (1998) Phys. Rev. B **58**, 3717.

Pinnick, D.A., S.A. Lee, X. Sun, and H.J. Simon (1997) Phys. Rev. B **55**, 8031.

Plekhanov, V.G. (1996) Phys. Rev. B **54**, 3869.

Popova, M.N., A.B. Sushkov, S.A. Golubchik, A.N. Vasil'ev, and L.I. Leonyuk (1998) Phys. Rev. B **57**, 5040.

Prabhu, S.S., A.S. Vengurlekar, and J. Shah (1996) Phys. Rev. B **53**, 10465.

Reedyk, M., D.A. Crandles, M. Cardona, J.D. Garrett, and J.E. Greedan (1997) Phys. Rev. B **55**, 1442.

Rönnow, D., P. Santos, M. Cardona, E. Anastassakis, and M. Kuball (1998) Phys. Rev. B **57**, 4432.

Rosenblum, S.S., W.H. Weber, and B.L. Chamberland (1997) Phys. Rev. B **56**, 529.

Samara, G.A., H.L. Tardy, E.L. Venturini, T.L. Aselage, and D. Emin (1993) Phys. Rev. B **48**, 1468.

Schwarz, U.T. and M. Maier (1997) Phys. Rev. B **55**, 11041.

Schwarz, U.T. and M. Maier (1998) Phys. Rev. B **58**, 766.

Scott, J.F. (1980) Rep. Progress in Physics **43**, 951.

Shah, J. (1999) *Ultrafast Spectroscopy of Semiconductors and Nanostructures*, Springer, Berlin.

Smirnov, D., P. Millet, J. Leotin, D. Poilblanc, J. Riera, D. Augier, and P. Hansen (1998) Phys. Rev. B **57**, 11035.

Stepniwski, R., K. Korona, A. Wysmolek, J.M. Baranowski, and K. Pakula (1997) Phys. Rev. B **56**, 15151.

Tang, P.J.P., M.J. Pullin, and C.C. Philips (1997) Phys. Rev. B **55**, 4376.

Tu, C.S., Y.L. Yeh, V.H. Smidt, R.M. Chien, R.S. Katier, R. Guo, R. Guo and A.S. Bhalla (1997) Phys. Rev. B **56**, 7988.

Ulivi, L., M. Zoppi, L. Gioé, and G. Pratesi (1998) Phys. Rev. B **58**, 2383.

Valeé, F., F. Ganikhanov, and F. Bogani (1997) Phys. Rev. B **56**, 13141.

Van Danzig, N.A. and P.C. M. Planken (1999) Phys. Rev. B **59**, 1586

Van Loosdrecht, P.H.M., S. Huant, G. Martine, G.Z. Dhalenne, and A. Revcolevschi (1996) Phys. Rev. B **54**, 3730.

Venkateswaran, U.D., V.M. Naik, and R. Naik (1998) Phys. Rev. B **58**, 14256.

Volm, D., B.K. Meyer, D.M. Hofmann, W.M. Chen, N.T. Son, C. Persson, U. Lindefelt, O. Kordina, E. Sörman, A.O. Konstantinov, B. Monemar, and E. Janzén (1996) Phys. Rev. B **53**, 15409.

Von Truchseß, M., A. Pfeuffer-Jeschke, C.R. Becker, G. Landwehr, and E. Batke (2000) Phys. Rev. B **61**, 1666.

Weber, D., W. Petri, U. Woggon, C.F. Klingshirn, S. Shevel, and E.O. Göbel (1997) Phys. Rev. B **55**, 12848.

Wegerer, R., C. Thomsen, M. Cardona, H.J. Bornemann, and D.E. Morris (1996) Phys. Rev. B **53**, 3561.

Wetzel, C., W. Walukiewicz, E.E. Haller, J. Ager III, I. Gazegory, S. Porowski and T. Suski (1996) Phys. Rev. B **53**, 987.

Yao, H., B. Johs, and R.B. James (1997) Phys. Rev. B **56**, 9414.

Yoon, S., H.L. Liu, G. Schollerer, S.L. Cooper, P.D. Han, D.A. Payne, S.W. Cheong, Z. Fisk (1998) Phys. Rev. B **58**, 2795.

Zhang, J.M., T. Ruf, M. Cardona, O. Ambacher, M. Stutzmann, J.M. Wagner, and F. Bechstedt (1997) Phys. Rev. B **56**, 14399.

Zhang, J.M., T. Ruf, R. Lanck, and M. Cardona (1998) Phys. Rev. B **57**, 9716.

Zimmermann, R., A. Euteneuer, J Möbius, D. Weber, M.R. Hofmann, W.W. Rühle, E.O. Göbel, B.K. Meyer, H. Amano, I. Akasaki (1997) Phys. Rev. B **56**, 12722.

5. Optical Properties of Interfaces and Thin Films

This chapter is devoted to the optical properties of interfaces and thin films. An interface is a region of separation between two materials with different properties, on which the otherwise bulk-like character of particle motion suffers a change. At the same time, due to fluctuations at the microscopic level, local potential minima, which can trap carriers or impurities, can form along the interface. Although we make references to such localized quasiparticles, we distinguish between surface, as mainly a restriction of movement, and intentionally spatial confined structures such as low-dimensional heterostructures. In this respect, thin films are considered here as regions where the motion of carriers is restricted in a certain plane, but with sufficiently large thicknesses such that the main characteristic of spatial confinement, i.e. quantization of energy levels, does not manifest itself.

At interfaces between two media Bloch's theorem does not hold for directions normal to the surface of separation, but only in directions parallel to the surface. In reality the optical properties do not have a sudden jump at the interface, but there is a surface region formed from the outer few atomic layers, in which the properties are different from the bulk material. Even the periodicity of the surface along the relevant directions may not be the same as in bulk. Since the symmetry properties at the surface differ from those of a bulk material, the wavevector can be decomposed as $\boldsymbol{k} = (\boldsymbol{k}_{\|}, k_z)$, the component normal to the surface, k_z, being able to take any value inside the bulk Brillouin zone. However, as k_z is varied, there might exist energy ranges for which there are no bulk states. In these energy ranges surface states can form, with the wavefunction decreasing exponentially from the surface. Also surface resonant states can develop, characterized by a larger amplitude in the surface region, which decays into a Bloch wave inside the solid. The crystal potential near the surface is no longer periodic along z, so that it can be expanded into

$$V(\boldsymbol{r}) = \sum_s V(\boldsymbol{K}_{s\|}, z)\exp(\mathrm{i}\boldsymbol{K}_{s\|} \cdot \boldsymbol{x}_{\|}), \tag{5.1}$$

where $\boldsymbol{K}_{s\|}$ is a two-dimensional (2D) reciprocal lattice vector, and $\boldsymbol{x}_{\|}$ a position vector in the surface plane. The wavefunction can also be expanded as

$$\psi(\boldsymbol{k}_{\|}, z) = \sum_s u(\boldsymbol{K}_{s\|}, z)\exp[\mathrm{i}(\boldsymbol{k}_{\|} + \boldsymbol{K}_{s\|}) \cdot \boldsymbol{x}_{\|}], \tag{5.2}$$

where u satisfies the Schrödinger equation

$$[-\mathrm{d}^2/\mathrm{d}z^2 + (\boldsymbol{k}_{\|} + \boldsymbol{K}_{s\|})^2 - E]u(\boldsymbol{K}_{s\|}, z) + \sum_t V(\boldsymbol{K}_{s\|} - \boldsymbol{K}_{t\|}, z)u(\boldsymbol{K}_{t\|}, z) = 0. \tag{5.3}$$

The electronic energy bands are calculated considering a film geometry instead of the semi-infinite geometry, such that some techniques can be borrowed from the bulk band structure calculation. Examples of techniques that can be adapted to surface problems are the tight-binding method and the linear augmented plane wave method (Callaway, 1991).

The modification of the electronic band structure at the interface has as a consequence a different sum rule. For example, for an interface between two different metals A and B, the new sum rule looks like

$$\frac{1}{4}(N_{\mathrm{A}} + N_{\mathrm{B}}) = N_{\mathrm{interf}} - \frac{\mathrm{i}N_{\|}\sigma}{(2\pi)^3}\int \mathrm{d}E\mathrm{d}k_{\|}\frac{\partial}{\partial E}\ln(\det S)\Big|_{E \le E_{\mathrm{F}}}, \tag{5.4}$$

where $N_{\mathrm{A,B}}$ is the number of electrons in metals A, B respectively, with normal velocity $v_z = 0$, N_{interf} is the number of localized interface electrons, $N_{\|}\sigma$ the area of the interface and S the scattering matrix. A more complicated sum rule holds for interfaces in magnetic heterostructures (Shaofeng et al., 1998).

5.1 Surface Effects

In some semiconductors, for example GaAs, an electric field appears near the surface due to the linear electro-optic effect. This electric field is induced by the strain/stress caused by changes in the crystalline structure at the surface. The effective electric field and its sign, as well as the depletion depth can be determined from the amplitude and lineshape resonances, respectively, in the reflection-difference spectroscopy (RDS) at the critical points E_1 and $E_1 + \Delta_1$. This method is also called reflection-anisotropy spectroscopy since it measures the polarization anisotropy of light linearly polarized along the two principal axes (two perpendicular directions in general) in the surface plane. The linear electro-optic part of the RDS, after eliminating the background, has the following shapes for n-GaAs and low-temperature grown GaAs (LT-GaAs) (see Fig. 5.1).

The RDS measures, in fact, the averaged difference of the dielectric function in the two directions over the light penetration depth inside the material $\langle\Delta\varepsilon\rangle = -2\mathrm{i}k\int_{-d}^{0}\exp(-2\mathrm{i}kz)\Delta\varepsilon(z)\mathrm{d}z$. For the n-GaAs sample where the depletion depth is quite wide (about 1500 Å), $\langle\Delta\varepsilon\rangle \cong \beta F_{\mathrm{s}}$, with β the linear electro-optic coefficient and F_{s} the value of the electric field at the surface. In the LT-GaAs sample for which $d \cong 10$ Å, $\langle\Delta\varepsilon\rangle \cong -2\mathrm{i}k\beta V_{\mathrm{s}}$, with V_{s} the surface potential. The presence of the factor i in the approximate formula of the dielectric function change in LT-GaAs explains the different shapes of the RDS curves for n-GaAs and LT-GaAs. The surface potentials in the two cases are 0.7 eV and 0.1 eV, respectively, the direction of the surface electric field being determined by the sign of the linear electro-optic component of the RDS (Chen et al., 1997).

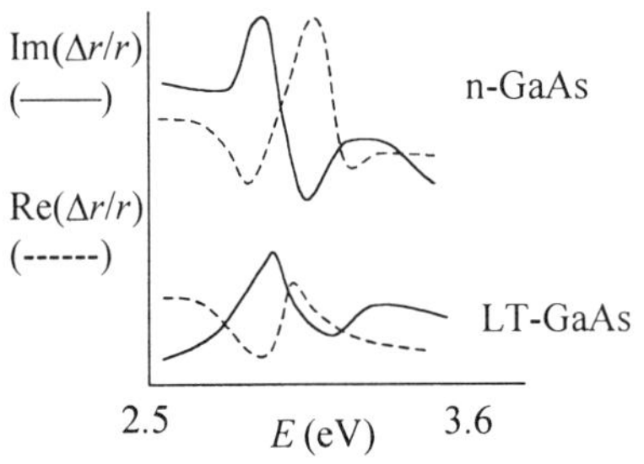

Fig. 5.1. The linear electro-optic part of the RDS signal in GaAs and LT-GaAs

In its turn, the surface electric field modifies the interband transition matrix elements through the piezoelectric effect, and induces a shift of E_1, $E_1+\Delta_1$ critical points in the strained crystal, with the same amount $\Delta E_0 = \gamma(1+2B^2 d'/d)$ but in opposite directions (the shift is towards larger energies for E_1). Here $\gamma = dd_{14}F/\sqrt{3}$ with d the deformation potential of the valence band, F the electric field, d_{14} the piezoelectric modulus, B a parameter and d' the deformation potential in the conduction band. The RDS can be related to this energy shift by

$$\frac{\Delta R}{R} = -\frac{1}{2R}\frac{\partial R}{\partial E}\Delta E_0 + \frac{4\gamma}{\Delta_1}(R'-R''), \tag{5.5}$$

where R', R'' are the Lorentzian lineshapes of R at the critical points E_1, $E_1+\Delta_1$ (Lastras-Martínez et al., 1999).

Charge transfer across interfaces can also take place. For example, at the Si/SiO_2 interface, the surface charge density σ varies as $\mathrm{d}\sigma/\mathrm{d}t = (\sigma_0-\sigma)/\tau_g - \sigma/\tau_d$ where τ_g, τ_d are accumulation and dissipation times, respectively, and σ_0 is the number of empty charge sites at $t=0$. The dissipation of surface charge density is due to the Coulomb repulsion among adsorbed O_2 molecules, due to the exchange between adsorbed and gas phases, and carrier recombination. The oxygen-assisted charging of the surface produces an electric field that enhances the second-harmonic generation (SHG). The intensity of the p-polarized SHG can be expressed as

$$I^{(2\omega)} \approx | a_0 + a_4\cos(4\varphi)|^2, \tag{5.6}$$

where a_0 is the isotropic part, due to interface-dipole and bulk-quadrupole electric-field-induced SHG contributions, and φ is the angle of sample rotation about the surface normal, which vanishes when the [011] axis is in the plane of incidence. Due to the electric-field-induced SHG a resonant enhancement of $|a_0|$ and a corresponding decrease of $|a_4 / a_0|$ is observed with the interface electric field, most pronounced near $2\hbar\omega$ (Mihaychuk et al., 1999).

The dependence of the SHG intensity on the applied or the internal electric field was used to estimate the internal electric field caused by the charge trapped at the interface between the oxide and the Si substrate in a metal-oxide-semiconductor field-effect transistor gate. The local electric field produced along the crystalline [100] direction by the electrons trapped in the oxide shift the SHG intensity curve, which is almost parabolic with respect to the applied field, with an amount equal to the local electric field. The latter parameter can thus be measured optically (Fang and Li, 1999).

The built-in field in a surface-intrinsic n^+-doped (s-i-n^+) GaAs can also be determined from the period of the Franz–Keldysh oscillations (FKO) in pump-and-probe reflectance measurements. FKO appear in the photoreflectance spectra for medium field strengths and for excitations above bandgap energies. Since the field F is uniform in the undoped layer, many FKO can be observed and beats in the FKO can be identified, especially at low pump powers, due to the HH and LH contributions with different reduced masses m_{ri} in the F direction. Performing a fast-Fourier transform analysis of the photoreflectance spectra, the frequencies of the HH and LH contributions are found, where $\omega_i = (2m_{ri})^{1/2}(2/3\pi e\hbar F)$, $i =$ HH, LH. The value of the electric field is determined by fitting (Wang et al., 1999).

Another effect of the surface is to favor transitions that are forbidden in bulk materials. For example, at the Si(100)-SiO_2 interface, interband transitions were observed, strongly resonant in the second-harmonic generation. These transitions are located between the E_1 and E_2 critical points of bulk Si, and are due to the Si atoms at the boundary between the Si and SiO_x transition region, which lack the T_d symmetry of bulk Si. Moreover, the E_2 transition is blueshifted at the interface. The second-harmonic intensity is fitted with a coherent superposition of critical-point-like resonances $I(2\omega) \approx |\sum_k A_k(\omega,\theta) f_k \exp(i\varphi_k)/(2\omega - \omega_k + i\gamma_k)|^2$ with θ the incidence angle (Erley and Daum, 1998).

Blueshifts of optical transitions due to the strain at interfaces, as large as 30 meV, have been observed also in GaAs deposited on top of GaP, the lattice mismatch in this case being about 3.6%. Surface-related optical transitions were measured in this case with chemical modulation spectroscopy, which monitors the change in the reflected light $\Delta R / R$ caused by the changing in the coverage of the last atomic layer. The results show that the blueshift of the transitions associated with the localized electronic states in the first surface layers (surface states) is smaller than that of the E_1 transition, for example, which shifts by 99 meV for an in-plane strain of 3.6% (Postigo et al. 1998).

Since Raman scattering is quite sensitive to stress, it can be used to obtain a 2D map of stresses that appear when a Si_3N_4 stripe, for example, is grown on a SiGe layer (Rho et al., 1999). This stress is caused by the difference in the thermal expansion coefficients of the two materials, the strain components being determined through the shift of the Raman peaks when the incident spot is moved across the investigated area. The spatial resolution attained with this method was 0.5 μm along the x direction in the plane of the SiGe layer and 0.25 μm along the z direction normal to this plane, *inside the sample*, for an incident laser wavelength of 514.5 nm. In a backscattering geometry, polarization selection rules can identify the corresponding LO or TO phonons along the x and z directions, whose frequency dependence on the applied strains are known. The values of the strain are then computed from the observed Raman peak shift. Experiments have shown that a compressive strain develops under the stripe, whereas tensile strain is observed outside the stripe edges. So, not only the value of the stress, but also its nature can be determined through nondestructive optical measurements. To increase the spatial resolution, and thus to probe shallower depths micro-Raman stress imaging can be performed with UV light (Holtz et al., 1999). Such a technique has been used to probe the stress in a patterned Si_3N_4/polycrystalline Si stack grown on a crystalline Si covered by oxide.

Self-induced optical anisotropy due to change in reflectivity from a surface, caused by different processes, has been measured by ellipsometric spectroscopy for Si(001)-2×1. Different mechanisms contributing to the reflectivity change, such as ion bombardment of the surface, etching with ion fluxes, annealing, and so on, are characterized by different temporal and thermal behaviors, so that these mechanisms can be identified from reflectivity changes as a function of temperature. The formulae that describe these mechanisms can be found, for example, in Wentink et al. (1997).

Adsorption kinetics of several types of atoms can be studied optically. Two different adsorption channels have been identified for H on Si(111)7×7 by measuring the temperature dependence of the surface differential reflectivity (Beitia et al., 1997). Supposing that H can bind on the adatom dangling bonds or can break the adatom back bond, and denoting by x and y the number of saturated bonds and broken bonds per cell which characterize the kinetics of these processes, their rate equations in terms of the H dose D are

$$x(D) = 12[1 - \exp(-\alpha D)], \tag{5.7}$$

$$y(D) = 36[1 - \beta \exp(-\alpha D)/(\beta - \alpha) + \alpha \exp(-\beta D)/(\beta - \alpha)]. \tag{5.8}$$

By measuring the surface differential reflectivity $\Delta R / R = (R^{\mathrm{Si}} - R^{\mathrm{H/Si}}) / R^{\mathrm{Si}}$ where R^{Si} is the reflectivity of the clean Si surface and $R^{\mathrm{H/Si}}$ of the hydrogenated Si, as a function of temperature, under an incidence angle of 60° and with p-polarized beams, the parameters α and β are fitted for every temperature value. The spectrum is approximated as $S = x(D)S_x / 12 + y(D)S_y / 36$, where S_x, S_y are the spectra corresponding to the two processes. The measured activation energies for the two processes, obtained from $\ln\alpha$, $\ln\beta$ as a function of $1/T$, are, respectively, a very small value, corresponding to practically no activation barrier, and 20 meV.

Adsorption and desorption kinetics on metallic surfaces have been studied by linear optical differential reflectivity, a method which measures with submonolayer sensitivity the changes for s- and p-polarized beams due to adsorbates, and can provide the amount of adsorbate, since for metallic substrates the change in reflectivity for p-polarized light is much larger than for s-polarized radiation. Another method of studying the surface kinetics employs an incident beam modulated with a frequency ω_m. The intensity of the reflected beam at second harmonic of the modulation frequency is

$$I(2\omega_m) = I_0(|r_p(\theta)|^2 - |r_p(0)|^2 |r_s(\theta)|^2 / |r_s(0)|^2), \tag{5.9}$$

with θ the coverage. $I(2\omega_m)$ increases almost linearly with θ, attaining a maximum for the saturation coverage θ_s. The kinetic of the isothermal adsorption is described by

$$\left(\frac{d\theta}{dt}\right)_{ads,T} = \frac{P}{N_s\sqrt{2\pi m k_B T_g}} S(\theta,T) - \left|\left(\frac{d\theta}{dt}\right)_{des,T}\right|, \tag{5.10}$$

where P is the pressure, N_s the surface density, T_g the temperature of the ambient gas, S the sticking probability and $(d\theta/dt)_{des,T} = -\nu\theta\exp(-E_d/k_B T)$ the desorbtion kinetics. The desorption energy E_d and the parameter ν are determined from fitting. A typical temperature variation of the coverage, measured for CO on Cu(110) in Jin et al. (1996), is shown in Fig. 5.2.

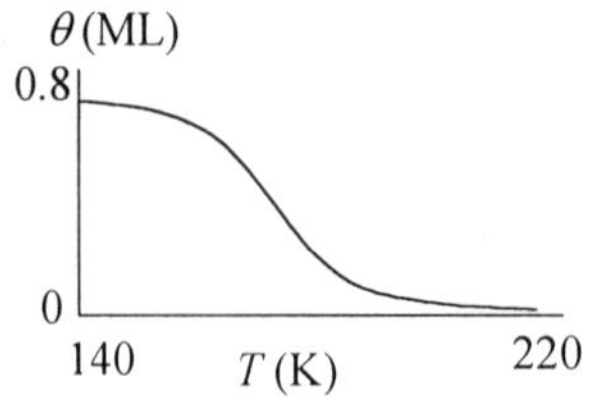

Fig. 5.2. The dependence on temperature of the coverage θ for CO on Cu(110)

Measuring the time evolution of the coverage at constant temperature through linear optical differential reflectivity, the adsorption and desorption contributions can be separated (the desorption being negligible for temperature less than 140 K), so that the sticking probability can be found. It is temperature independent for $\theta \le 0.4$ and can be modeled by $S = S_0(1 - \theta/\theta_c)$ (Langmuir kinetics) where the parameter $\theta_c = 0.6$ is determined from fitting. Taking into account the exponential temperature dependence of desorption, the equilibrium CO coverage on Cu is obtained in the isothermal Langmuir adsorption model for $\theta(T) = \theta_c/(1 + \theta_c A/P)$ where $A = N_S\nu(2\pi m k_B T_g)^{1/2}\exp(-E_d/k_B T)$. This relation is valid for $\theta \le 0.4$. For $\theta \ge 0.4$, $S(\theta)/S(\theta_x) = 1/[1 + k(\theta - \theta_x)/(\theta_s - \theta)]$, the fitting procedure giving

$\theta_S = 0.777$, $\theta_X = 0.4$, $k = 0.5$. k describes the surface mobility of adsorbed CO, its value indicating a precursor-mediated adsorption mechanism, the CO being first adsorbed to the surface, and then rearranging in more stable structures.

In situ monitoring of surfaces can be also made using nonlinear optical techniques such as sum-frequency generation or second-harmonic generation. The nonlinear optical methods have unique sensitivity to surfaces and interfaces, allowing growth control of Ge coverage on Si(001) surfaces (Parkinson et al., 1999). Recent experimental developments even allow a real-time growth control via second harmonic spectroscopy (Wilson et al., 1999).

Raman scattering is less sensitive to surface excitations. However, Raman scattering of terrace hydrogen vibrations on Si(111) has been measured; it has a ω^4 dependence in the 457.9–514.5 nm range, with a cross-section for the Si:H bond line at 448 nm of 8.37×10^{-28} nm^2/sr (Sano and Ushioda, 1996). A more detailed discussion on Raman spectroscopy on surfaces, interfaces and thin films can be found in Tsang (1989).

As for surfaces, stacking faults can also be viewed as planar discontinuities. They have, however, different roles, acting as accidental quantum wells, and so are able to confine carriers. 2D nonlocalized Frenkel excitons have been observed in stacking faults of BiI_3 (Mishina et al., 1993) their dynamics being studied with time-resolved PL. The confinement of the carriers in the fault has produced a multilevel exciton system as in Fig. 5.3, where the radiative transitions are indicated by solid lines and the nonradiative transitions by dotted lines.

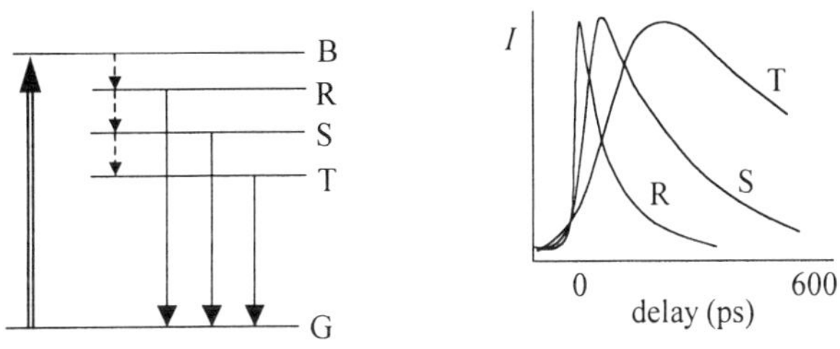

Fig. 5.3. Optical transitions and the time-resolved PL from stacking faults in BiI_3

The three-exciton transitions, R, S, T are observed in both absorption and PL spectra, with different intensities. In absorption the intensities increase slightly from T to R, whereas in PL the S line is an order of magnitude lower than the T line, and the R line is an order of magnitude smaller than S. This behavior, together with the relaxation curves in Fig. 5.3, suggest the existence of an internal, cascaded relaxation process of excitons. The decrease in the luminescence efficiency and the rapid decay of T excitons with increasing excitation density suggest a branching into a nonradiative recombination path of the excitons involving Auger exciton-exciton annihilation.

In the stacking-fault plane of BiI_3 the translational symmetry is maintained over about 100 μm, and the center-of-mass of the stacking-fault excitons has a quasi-2D motion. The propagation velocity of these excitons can be measured in space-resolved pump-probe experiments, where the pump and probe beams are spatially separated by a distance r (the spatial resolution is 30 μm). By exciting the sample with an energy equal to that of the S exciton, and measuring the PL at an energy equal to the T exciton, the radial PL intensity is approximated as $F(r) \approx \exp(-r / v\tau_{ex})$ where τ_{ex} is the decay time and v the exciton velocity. Knowing the value of the decay time as 0.8 ns, the exciton velocity can be obtained from experimental data. It is found that for large r values the exciton mean velocity is smaller due to scattering processes. Additionally, from pump-probe absorption measurements at the energy of the T exciton one can estimate the exciton concentration from the peak energy shift. The blueshift of the absorption peak is caused by mutual interaction between excited high-density excitons, and is proportional to the local exciton density (Kondo et al., 1998).

Optical phonons localized on buried interfaces can be observed with time-resolved second-harmonic generation (Chang et al., 1999). The access to the buried interface is facilitated if at least the medium on one side of the interface is transparent. From the peaks observed in the second-harmonic spectrum, that corresponding to the buried interface is identified on the basis of changes in phonon spectrum as a function of the pump laser intensity and during in situ oxidation. The mode is found to shift its frequency to lower values, from 8.48 to 8.29 THz for native oxide-covered GaAs(100), due to coupling with holes driven to the interface by the depletion field. The frequency of the phonons observed in time-resolved second-harmonic generation are obtained from the Fourier transform. The method used to identify the presence of the phonon associated to the buried interface, is the coherent time-domain analog of stimulated hyper-Raman spectroscopy.

Optical techniques, in particular the time-resolved PL imaging technique with high spectral, temporal and spatial resolution was successfully employed to measure the time- and carrier-density-dependent heterointerfacial band bending in $GaAs/Al_xGa_{1-x}As$ (Gilliland et al., 1998). This was possible by studying the evolution of the quasi-2D exciton system which self-localize at the heterointerfaces of wide, undoped structures, after laser excitation. The exciton density evolves due to radiative and nonradiative recombination and lateral transport along heterointerfaces, so that band bending at the heterointerface also changes in time. In fact, it was established that the spatially nonuniform band bending induced by laser excitation decays within 20 ns to the static, uniform bending. The exciton transport along the heterointerface is similar to exciton diffusion in a potential gradient induced by the photoexcited carriers, and applied externally. The transport is driven initially by the force induced by the spatially nonuniform band bending due to screening of the built-in heterointerface field by the photoexcited carriers. With increasing time exciton transport becomes asymptotically diffusive. The field screening caused by the photoexcited carriers is seen as a redshift with time of the broad, asymmetric PL peak due to recombination of the excitons localized at the heterointerface. The time dependence of the

PL was numerically reproduced by parametrical solutions of the Poisson and Schrödinger equations, which include the Coulomb interaction between electrons and holes. The actual transport of excitons could be monitored by spatially resolved PL techniques.

5.2 Thin Films

Thin films are also dominated by surface effects, their properties being influenced by the lack of translational invariance along the normal direction to the film surface. In this subsection we consider films sufficiently thick such that quantum confinement effects are not present. The opposite case is treated in Chap. 6.

For monolayer films grown on a material surface with a dielectric constant ε_m, the dielectric constant of the film can be calculated by taking into account the interaction between molecules and the material surface, as well as the local field, which acts on the polar molecule with a permanent dipole moment. The result is $\varepsilon = (S_0 / S)[1 + 11.0342 a^{-3} \alpha_p (2\varepsilon_m)/(\varepsilon_m + 1)]$ where α_p is the electronic polarizability, a the lattice constant, and $S_0 = (1 + \cos\theta_A)/2$ with $\theta_A =$ $\arcsin[(A/A_c)^{1/2}]$. $A = 3^{1/2} a^2 / 2$ is the molecular area, $A_c = 3^{1/2}(2l^2)$ the critical molecular area with l the length of the rod-like molecule, and $S =$ $\int_0^{\theta_A} \cos\theta f(\cos\theta) \sin\theta d\theta$ is an orientational order parameter, where $f(\theta) =$ $(1/Z)\exp[-W(\theta)/k_B T]$ with W the interaction energy among molecules and between the molecules and the surface of the material. $Z =$ $\int_0^{\theta_A} \exp[-W(\theta)/k_B T] \sin\theta d\theta$ is the single particle partition function (Iwamoto et al., 1996).

Thin films are used, for example, to study the different types of theoretically predicted surface waves. Quite recently the observation in the Brillouin spectra of peaks associated with guided longitudinal acoustic modes has been reported in elastically hard films on soft substrates (Chirita et al., 1999).

Optical constants of thin films are usually determined by measuring both amplitude and phase of the reflection coefficient. For example, for epitaxially grown semiconductor alloys with a thickness less than 20 Å, the complex dielectric function of the film is found from ellipsometric measurements of the reflection coefficient, if the dielectric function of the substrate and the film thickness is known. If the film is a semiconductor alloy, its composition can then be determined from the dielectric function of the film, in a nondestructive way. The method can even be used for a multilayer structure (Aspens, 1995). Inversely, real-time control of layer thickness in the epitaxial growth process can be achieved by monitoring the normal incidence reflectance. Since light interference occurs between the beams reflected at the two interfaces of the thin film, the reflectance has an oscillatory behavior as a function of the layer thickness. The eventual faults appearing during growth can also be detected by comparing reflectance signals from different oscillatory periods, and even the surface temperature during growth can be monitored by simultaneously recording the thermal emission (Breiland and Killeen, 1995).

In the case of thin metal films, the optical constants can be obtained by optical excitation of surface plasma waves in attenuated total reflection (Lopez-Rios and Vuye, 1979). Surface plasma waves are excited by illuminating the film through a prism, as in Fig. 5.4. The method to calculate the optical constants from the halfwidth and the reflection minimum in attenuated total reflection can also be extended to account for the influence of surface roughness. Since the algorithm is quite long, although straightforward, we invite the interested reader to consult the paper of Lee and Jen (1999).

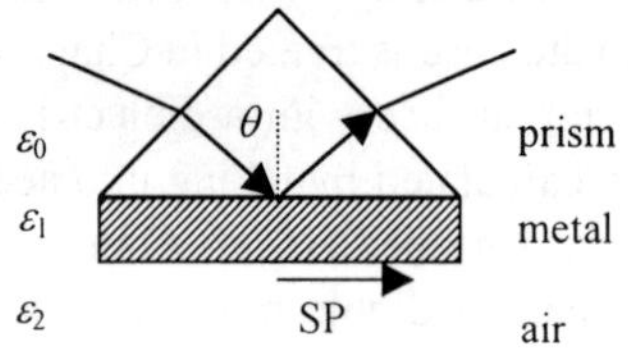

Fig. 5.4. Set-up for excitation of surface plasma waves in attenuated total reflection

Modes of the substrate can appear in the reflection spectra from thin films grown on it. For example, when a Fe_3O_4 film is deposited on a MgO substrate, a peak near the LO phonon frequency of MgO is observed in the reflectance spectra for oblique incidence with TM polarization. No peak is observed for the TE polarization (see Fig. 5.5).

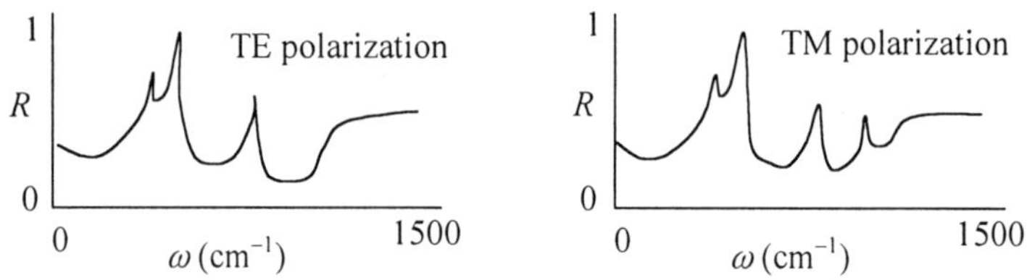

Fig. 5.5. Reflectance spectra of Fe_3O_4 on a MgO substrate for TE and TM polarizations

The appearance of this peak can be explained by the fact that the electric field component normal to the film induces coupling between charge carriers in the film and LO phonons in MgO, the peak appearing at the resonant frequency where the substrate dielectric constant equals $\sin^2\theta$, with θ the angle of incidence. This peak does not appear for thick films since the transverse electromagnetic wave cannot interact with LO phonons inside an infinite medium, but only with TO phonons. In a related, but different Berreman effect, an absorption associated with LO phonons of the film instead of those of the substrate is observed (Ahn et al., 1996).

The shift of Raman modes of epitaxial films from the value in bulk materials can be used to determine the value of the residual stress (Sun et al., 1997). For example, in ferroelectric $PbTiO_3$, the E(1TO) and A_1(1TO) modes are temperature and pressure sensitive, shifting in opposite directions in thin films and ultrafine powders due to the fact that the c/a ratio depends on the particle size. When the thickness of the film is much larger than the size of the powder particles and of the polycrystalline grains, the upward shift of the Raman modes is caused mainly by stress. Denoting by $\Delta p = P(2d_{31} + d_{33})$ the change of polarization due to the hydrostatic pressure P, the frequency in the stressed film is $\omega^2 = \omega_0^2[1 - \Delta p/(2d_{31} + d_{33})P_c]$ where ω_0 is the frequency in the unstressed film, d_{ij} are the piezoelectric coefficients, and P_c is the critical pressure above which the phonon frequency vanishes. Due to the geometry of thin films, the unit cell suffers only 2D in-plane stresses caused by film-substrate interaction. For in-plane stresses T, the change in polarization is $\Delta p' = 2d_{31}T$. For $PbTiO_3$ grown on (110)$NdGaO_3$ wafers the stress along (001) has an opposite effect on the spontaneous polarization compared to stresses along (100) and (010). The shift of the Raman mode is downward, respectively upward with increasing stress, the in-plane stress corresponding to a 7 cm^{-1} upward shift of the Raman line being 1.3%.

PL measurements can also be employed for the determination of the critical layer thickness (CLT) (Parker et al., 1999). For layers thinner than CLT the mismatch is accommodated by elastic strain, while for thicker layers strain, dislocations and three-dimensional growth appear as a result of the mismatch. To determine it optically, the PL is recorded for increasing thickness layers. The typical evolution of the PL peak is shown in Fig. 5.6; experiments have been performed on $In_xGa_{1-x}N$ films grown on GaN substrate for different x values.

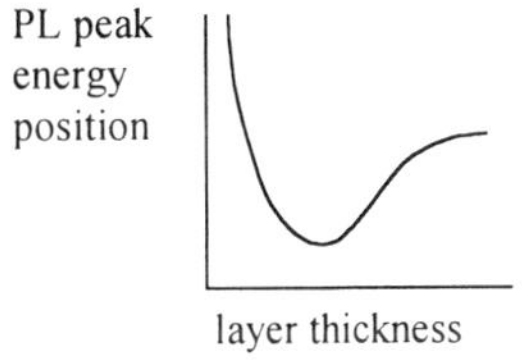

Fig. 5.6. Typical PL peak energy position as a function of the thickness of InGaN films

Three regions can be identified on this curve. For large layer thicknesses the PL emission occurs at about the same wavelength, and is due to band edge emission of relaxed films. For thin film regions the effective energy bandgap is larger than in relaxed thick films (its value corresponding to strained films under compressive stresses), whereas in the intermediate region the PL emission peak broadens, being dominated by deep-level defects. The CLT can be defined either as the thickness where the bandgap in the strained region equals that in the relaxed thick film, or as the thickness that corresponds to the onset of the transition region dominated by deep-level emission. For $In_xGa_{1-x}N$ both approaches led to similar values: 100 nm for $x = 0.08$, and 65 nm for $x = 0.15$.

5.2.1 Optical Properties of Magnetic Thin Films

Interesting phenomena take place when the film or the substrate is a magnetic ordered material, in which case the film thickness and temperature determines the direction of the easy axis of magnetization. A good review on Raman scattering in thin films and layered magnetic structures can be found in Grünberg (1989). In ultrathin films intrinsic anisotropies are always present, the surface anisotropy being able to induce an easy axis for magnetization perpendicular to the film. In thicker films, the in-plane magnetization is favored since magnetostatic anisotropy is dominant. For example, in the case of Tb/Co films, MOKE experiments in polar and longitudinal geometry show that the magnetization direction changes with temperature (Garreau et al., 1996) due to the competition between shape anisotropy described by $-2\pi M_s^2(T, d_{\mathrm{Tb}})\sin^2\theta$ and magnetocrystalline anisotropy $K_2(T, d_{\mathrm{Tb}})\sin^2\theta + K_4(T, d_{\mathrm{Tb}})\sin^4\theta$ where d_{Tb} is the thickness of the Tb film, θ the angle between the normal to the film and magnetization, and M_s the saturation magnetization. Magnetocrystalline anisotropy is caused by both long-range polar and short-range spin-orbit interactions. The equilibrium angle θ is that for which the two anisotropies cancel each other. Since M_s and the parameters K_2 and K_4 depend on temperature, three regions can be identified, as shown in Fig. 5.7.

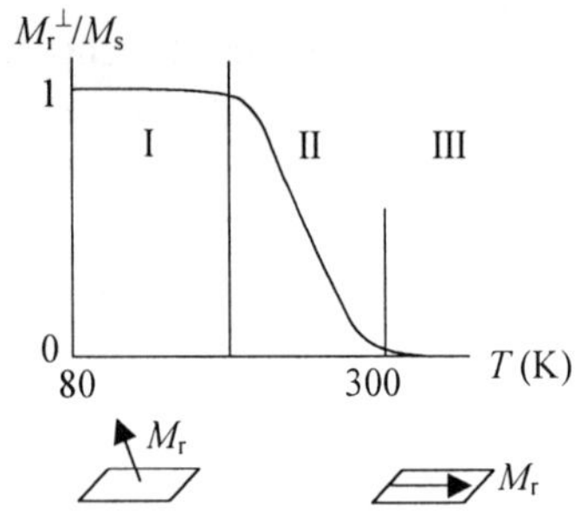

Fig. 5.7. Temperature dependence of the remanent magnetization in Tb/Co films

In region I, which extends between 80 and 210 K, the hysteresis loop has a square shape, with full remanence, the magnetocrystalline anisotropy being dominant. In region II, between 210 and 311 K, the component of the remanent magnetization $\boldsymbol{M}_r$ normal to the film decreases gradually to zero, while in region III, for temperatures higher than 311 K, the shape anisotropy dominates and the normal remanence vanishes, the easy axis being situated in the film plane. The normal component of the magnetization is seen in the polar geometry of MOKE, whereas the in-plane component is sensed in the longitudinal geometry of MOKE. When the normal component of the remanent magnetization vanishes, the hysteresis loop in the polar MOKE reduces to an S-like shape.

If the thin film is anisotropic, the magneto-optical response is asymmetric, apparently violating the invariance of the hysteresis loop under the transformation

$M \to -M$, $H \to -H$. The loops in MOKE for anisotropic thin films of Fe(110) and Co(110) look like those in Fig. 5.8.

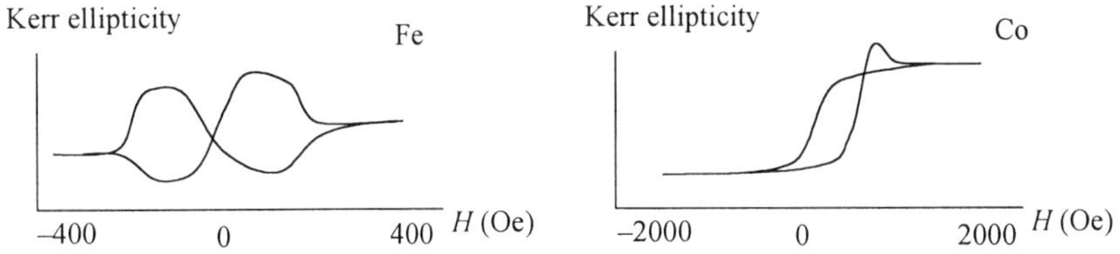

Fig. 5.8. MOKE signals for anisotropic thin films of Fe(110) and Co(110)

This strange behavior appears if there is a component of the magnetization normal to $\boldsymbol{H}$. In this case the magneto-optic response contains a term of second-order in the magnetization, proportional to the product of the longitudinal and transverse components of $\boldsymbol{M}$ (Osgood et al., 1997).

The exchange coupling across a ferromagnetic (FM)/antiferromagnetic (AF) interface can be optically modulated by illumination with ultrafast optical pulses. The exchange coupling across the interface produces an effective unidirectional anisotropy bias field H_{ex}, which shifts the hysteresis loop and increases the easy-axis coercivity of the FM layer. When illuminated, energetic electrons promoted via interband transitions tend to break the ground state electron-electron exchange, which in turn produces a nonequilibrium spin system, the relaxation of which can be studied with MOKE. The spin degrees of freedom in the thin FM film are optically manipulated before they transfer energy to the lattice. The transient Kerr rotation for the exchange biased bilayer NiFe/NiO is about five times larger than in the ferromagnetic NiFe material since the initial distribution of hot electrons has a net finite nonequilibrium spin polarization. Both the magnetization in the NiFe layer and the interfacial exchange energy $\delta S \cong \sum_{ij} J_{ij} \boldsymbol{S}_i \cdot \boldsymbol{S}_j$, where i, j are spins in monolayers across the interface on the FM and AF sites, are modulated by spin/electron heating (Ju et al., 1998).

Other studies on the FM/AF exchange coupling have indicated that in a wedged FM/AF system switching in the exchange coupling involves only two macroscopic domains, extending across the entire sample and separated by a 180° domain wall moving along the wedge direction. By applying an external field normal to the wedge direction, the form of the domains is changed, the hysteresis loop being shifted from the origin by an amount equal to the exchange field H_{ex}, with a simultaneous increase of the coercivity H_c. The exchange coupling is transmitted along the FM/AF interface (and also across FM/NM/AF interfaces, where NM is a non-magnetic material) due to the dependence of the exchange field on the thickness of the FM layer: $H_{ex} \approx 1/t_{FM}$.

Measurements on a system permalloy (FM)/FeMn (AF) have produced the results shown in Fig. 5.9; the evolution from b to d (for which MOKE is the same) involved the growth of one domain at the expense of the other. It was also found that at low temperatures $H_c \approx 1/t_{FM}^{3/2}$, this dependence being confirmed by measurements performed at different thicknesses of the FM material along the wedge (Zhou et al., 1998).

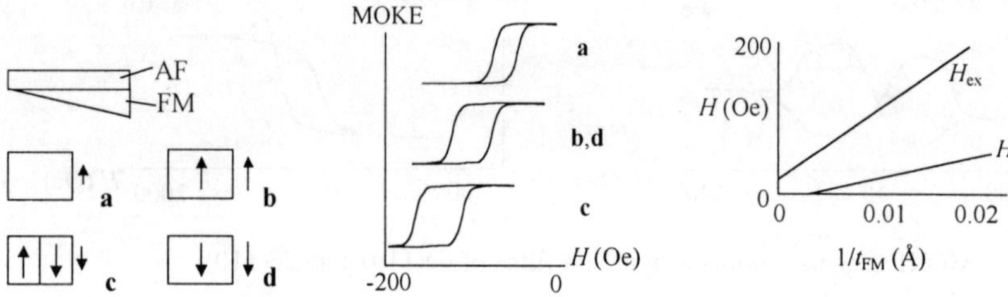

Fig. 5.9. MOKE behavior with H in a wedged permalloy (FM)/FeMn (AF) system

The magnitude of the bilinear and biquadratic interlayer coupling between Fe layers separated by Cr spacer layers can also be studied by longitudinal surface MOKE or Brillouin spectroscopy. It was found that the coupling between the Fe layers oscillates periodically with the spacer layer thickness between FM and AF, the period being inverse proportional to the distance between the extremal points of the Fermi surface normal to the layering direction. The energy of the bilayer system is $E = d_1 E_1 + d_2 E_2 - J_1 \boldsymbol{m}_1 \cdot \boldsymbol{m}_2 - J_2 (\boldsymbol{m}_1 \cdot \boldsymbol{m}_2)^2$ where d_i, E_i, $\boldsymbol{m}_i$ are the thicknesses, energy densities and magnetic moments of the two Fe layers, and J_1 and J_2 are the interlayer coupling constants. The energy densities can be written as $E_i = K_1[(1/3)\cos^4\theta_i + (1/4)\sin^4\theta_i] + K_u \cos^2\theta_i - HM_s\cos(\theta_i - \xi)$ where K_1, K_u are the cubic and uniaxial anisotropies. The equilibrium condition for the system is that E is minimum with respect to θ_1, θ_2, the condition from which the frequencies of the double-layer system, ω^+, ω^-, are also determined. These frequencies can be experimentally detected in Brillouin spectroscopy, the parameters K_1, K_u, J_1, J_2 and M_s being derived from fitting (Grimsditch et al., 1996).

The temperature dependence of the exchange coupling between two ferromagnetic films across a spacer layer can be studied with MOKE. For an amorphous ZnSe spacer layer between two Fe layers, the coupling was found to be antiferromagnetic with a positive temperature coefficient, if the spacer thickness was small 18–22 Å. For larger spacer thicknesses, 22–25 Å, a reversible transition has been observed from a ferromagnetic coupling at low temperatures to an antiferromagnetic coupling at higher temperatures, caused by localized defect states in the bandgap in the vicinity of Fermi energy (Walser et al., 1999).

Both second-order (fourth-power) and first-order (second-power) uniaxial anisotropy constants in E, with $E = -K_u^{(1)}\sin^2\theta - K_u^{(2)}\sin^4\theta - HM_s\cos\theta + (2\pi M_s^2)\sin^2\theta$, vary with the thickness of the spacer layer. For example, in ultrathin Co/Au/Cu(111) films the second-order perpendicular anisotropy increases monotonically with the Au thickness up to 5 monolayers (ML), the first-order anisotropy coefficient having a non-monotonic increase with a minimum at

1 ML. Both anisotropy constants saturate at 5 ML Au thickness, due to the saturation of the expansion of the in-plane Co lattice caused by the coherent growth of Co at the Co/Au interface. For larger Au thicknesses an increase of coercivity is observed in the polar-Kerr hysteresis curves, and a field-dependent broadening of the spin wave appears in the Brillouin spectrum, due to the atomic-scale transition of the underlayer material from Cu to Au. The dependence of the anisotropy constants on the thickness of the Au layer induces a thickness-dependent frequency of the spin waves ω_S, which can be observed in Brillouin spectroscopy (see Fig. 5.10) (Murayama et al., 1998).

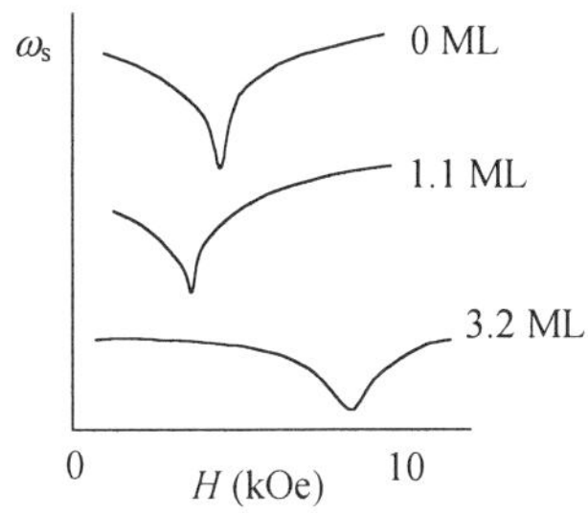

Fig. 5.10. Spin waves dependence on the applied magnetic field for different thickness of the ultrathin Au film in the Co/Au/Cu(111) system

When the layer that separates the two ferromagnetic films is antiferromagnetic, as for example in the CoFe/Mn/CoFe system, different antiferromagntic sublattices are exposed to interaction with CoFe spins, by varying the Mn thickness. A competition between parallel and antiparallel alignment of spins in the ferromagnetic films can occur if the exchange coupling at interfaces is strong and if the thickness variation of Mn layer occurs over a lateral length scale smaller than the exchange length in ferromagnetic films. This additional energy per interface area caused by the deformation of the antiferromagnetic order by coupling to the ferromagnetic films explained the upward frequency shift of the spin waves in the Brillouin spectra with increasing applied magnetic field. Actually, an in-phase and an out-of-phase ('acoustic' and 'optic') coupled surface spin-waves were observed, with large Stokes/anti-Stokes ratio, which crossed at a field between 1 and 2 kOe, the crossing point depending on the thicknesses of Mn and CoFe layers (Chirita et al., 1998).

The interlayer alignment in a trilayer system such as Ni/Cu/Ni(001) can change from ferromagnetic to antiferromagnetic by modifying the temperature, due to the variation of the magnetic anisotropic energy. The magnetic anisotropic energy dominates over the antiferromagnetic interlayer exchange at low temperatures, and fixes the sublayer magnetization at remanence in a ferromagnetic alignment. Measuring the Kerr ellipticity as a function of temperature at remanence and after applying a field of 100 Oe, a behavior such as that described in Fig. 5.11 has been observed (Poulopoulos et al., 1998).

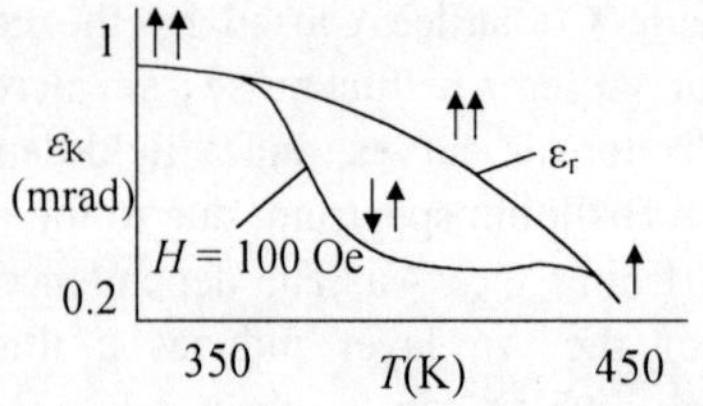

Fig. 5.11. Temperature dependence of the Kerr ellipticity in the trilayer system Ni/Cu/Ni(001), in the absence and presence of a magnetic field

Up to 360 K the Kerr ellipticities at remanence and at 100 Oe are almost equal, and the interlayer alignment is ferromagnetic. As the temperature increases, the interlayer alignment changes to antiferromagnetic when an applied field is present, and remains ferromagnetic at remanence. Above 450 K the magnetization of the bottom Ni layer vanishes after entering the paramagnetic phase, the signal originating entirely from the top Ni layer.

Exchange-coupled trilayers of amorphous rare-earth Co alloys, such as Y-Co, have a magnetization antiparralel to the applied field, for low values of H, which is reversed at a critical field $\mu_0 H \cong 3\,\text{mT}$. This is observed as a jump of the transverse MOKE signal, followed by a gradual increase and saturation of the signal for larger fields. The movement of Bloch walls is observed in longitudinal MOKE (Wüchner et al., 1997).

References

Ahn, J.S., H.S. Choi, and T.W. Noh (1996) Phys. Rev. B **53** 10310.
Aspens, D.E. (1995) IEEE J. Selected Topics Quantum Electron. **1**, 1054.
Beitia, C., W. Preyss, R. Del Sole, and Y. Borensztein (1997) Phys. Rev. B **56**, 4371.
Breiland, W.G. and K.P. Killeen (1995) J. Appl. Phys. **78**, 6726.
Callaway, J. (1991) *Quantum Theory of Solid State*, Academic Press, Boston, 2nd edition.
Chang, Y.M., L. Xu, and H.W.K. Tom (1999) Phys. Rev. B **59**, 12220.
Chen, Y.H., Z. Yang, R.G. Li, Y.Q. Wang, and Z.G Wang (1997) Phys. Rev. B **55**, 7379.
Chirita, M., G. Robins, R.L. Stamps, R. Sooryakumar, M.E. Filipkowski, G.J. Gutierrez, and G.A. Prinz (1998) Phys. Rev. B **58**, 869.
Chirita, M., R. Sooryakumar, H. Xia, O.R. Monteiro, and I.G. Brown (1999) Phys. Rev. B **60**, 5153.
Erley, G. and W. Daum (1998) Phys. Rev. B **58**, 1734.
Fang, J. and G.P. Li (1999) Appl. Phys. Lett. **75**, 3506.
Garreau, G., E. Beaurepaire, K. Ounadjela, and M. Farbe (1996) Phys. Rev. B **53**, 1083.
Gilliland, G.D., M.S. Petrovic, H.P. Hjalmarson, D.J. Wolford, G.A. Northrop, T.F. Kuech, L.M. Smith, and J.A. Bradley (1998) Phys. Rev. B **58**, 4728.
Grimsditch, M., S. Kumar, and E.E. Fullerton (1996) Phys. Rev. B **54**, 3385.
Grünberg, P. (1989) in *Light Scattering in Solids V*, eds. M. Cardona and G. Güntherodt, vol.66 in Topics in Applied Physics, Springer, Berlin, Heidelberg, p.303.
Holtz, M., J.C. Carty, and W.M. Duncan (1999) Appl. Phys. Lett. **74**, 2008.
Iwamoto, M., Y. Mizutani, and A. Sugimura (1996) Phys. Rev. B **54**, 8186.

Jin, X.F., M.Y. Mao, S. Ko, and Y.R. Shen (1996) Phys. Rev. B **54**, 7701.

Ju, G., A.V. Nurmikko, R.F.C. Farrow, R.F. Marks, M.J. Carey, and B.A. Gurney (1998) Phys. Rev. B **58,** 11857.

Kondo, H., H. Mino, I. Akai, and T. Karasawa (1998) Phys. Rev. B **58**, 13835.

Lastras-Martínez, A., R.E. Balderas-Navarro, L.F. Lastras-Martínez, and M.A. Vidal (1999) Phys. Rev. B **59**, 10234.

Lee, C.C. and Y.J. Jen (1999) Appl. Opt. **38**, 6029.

Lopez-Rios, T. and G. Vuye (1979) Surf. Sci. **81**, 529.

Mihaychuk, J.G., N. Shamir, and H.M. van Driel (1999) Phys. Rev. B **59**, 2164.

Mishina, T., H. Chida, and Y. Masumoto (1993) Phys. Rev. B **48**, 1460.

Murayama, A., K. Hyomi, J. Eickmann, and C.M. Falco (1998) Phys. Rev. B **58**, 8596.

Osgood III, R.M., B.M. Clemens, and R.L. White (1997) Phys. Rev. B **55**, 8990.

Parker, C.A., J.C. Roberts, S.M. Bedair, M.J. Reed, S.X. Liu, and N.A. El-Masry (1999) Appl. Phys. Lett. **75**, 2776.

Parkinson, P., D. Lim, J.G. Ekerdt, and M.C. Downer (1999) Appl. Phys. B **68**, 1.

Postigo, P.A., G. Armelles, and F. Briones (1998) Phys. Rev. B **58**, 9659.

Poulopoulos, P., U. Bovensiepen, M. Farle, and K. Baberschke (1998) Phys. Rev. B **57**, 14036.

Rho, H., H.E. Jackson, and B.L. Weiss (1999) Appl. Phys. Lett. **75**, 1287.

Sano, H. and S. Ushioda (1996) Phys. Rev. B **53**, 1958.

Shaofeng, W., Z. Wang, and S. Zou (1998) Phys. Rev. B **57**, 7305.

Sun, L., Y.F. Chen, L. He, C.Z. Ge, D.S. Ding, T. Yu, M.S. Zhang, N.B. Ming (1997) Phys. Rev. B **55**, 12218.

Tsang, J.C. (1989) in *Light Scattering in Solids V*, eds. M. Cardona and G. Güntherodt, vol.66 in Topics in Applied Physics, Springer, Berlin, Heidelberg, p.233.

Walser, P., M. Hunziker, T. Speck, and M. Landolt (1999) Phys. Rev. B **60**, 4082.

Wang, D.P., K.R. Wang, K.F. Huang, T.C. Huang, and A.K. Chu (1999) Appl. Phys. Lett. **74**, 475,

Wentink, D.J., M. Kuijper, H. Wormeester, and A. van Silfhout (1997) Phys. Rev. B **56**, 7679.

Wilson, P.T., Y. Jiang, O.A. Aktsipetrov, E.D. Mishina, and M.C. Downer (1999) Opt. Lett. **24**, 496.

Wüchner, S., J.C. Toussaint, and J. Voirou (1997) Phys. Rev. B **55**, 11576.

Zhou, S.M., K. Liu, and C.L. Chien (1998) Phys. Rev. B **58**, 14717.

6. Low-Dimensional Structures

In low-dimensional (low-D) structures the motion of carriers is spatially confined in at least one direction, the main consequence being the discretization of energy levels. Low-D structures are commonly (but not exclusively) made from alternating layers of different semiconductors, in which case they are called low-D semiconductor heterostructures. If the width of the layers of low-D semiconductor heterostructures is smaller than the mean free path of carriers it is possible to modify the carrier transport from incoherent, as in bulk materials, to coherent, ballistic-like, where no scattering effects are present. The modification of carrier transport is essential in advanced optoelectronic devices; for a recent review of the applications of low-D heterostructures in optoelectronics see Dragoman and Dragoman (1999). Depending on the number of degree of freedoms along which the motion of electrons and holes is restricted and the energy is quantized, low-D structures can be classified as quantum wells, quantum wires and quantum dots. A somewhat different category of low-D structures comprises the nanoparticles, which can have different shapes and are embedded in a surrounding medium with different properties. Although there is no formal border between nanoparticles and quantum wires or dots, we will refer in this chapter to nanoparticles as those low-D structures in which the motion of carriers is still confined but the quantization of energy levels is not evident, or not dominant in the optical response. Nanoparticles with sufficiently large dimensions have practically an identical behavior to bulk materials except for surface effects. Depending on material and size, the dimensionality of nanoparticles ranges from 0 (the dimensionality of a quantum dot) to 3 (the dimensionality of bulk materials). Nanoparticles can be produced from any materials, not necessarily semiconductors, and usually have a random distribution of positions and sometimes shapes and/or sizes, their optical spectra being generally isotropic, whereas the optical properties of low-D heterostructures are strongly anisotropic. Another difference between the optical properties of nanoparticles and low-D structures is that in the former surface effects have generally (with the exception of quantum dots) a larger influence on optical spectra, due to the larger surface-to-volume ratio. Since they have quite dissimilar optical responses, separate sections are devoted to low-D semiconductor heterostructures and nanoparticles.

6.1 Low-Dimensional Semiconductor Heterostructures

Depending on the number of dimensions along which the length scale of the structure is smaller than the mean free path of carriers L_{fp}, we are dealing with quantum wells, quantum wires or quantum dots.

The quantum well (QW) consists of a layer of width $L_z << L_{fp}$ placed between two, generally different materials, which act as barriers for charge carriers (see Fig. 6.1).

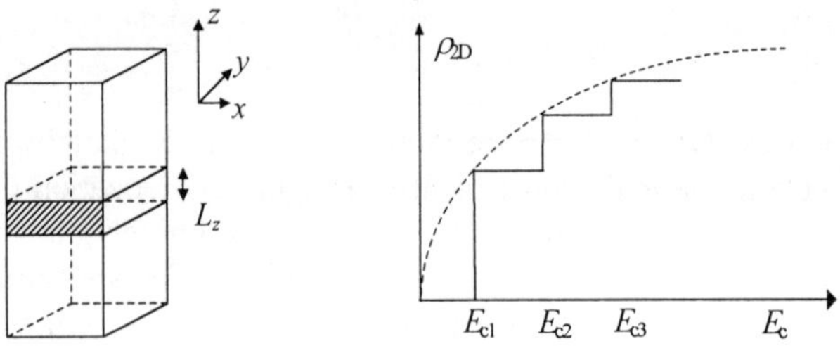

Fig. 6.1. The quantum well and its density of state

The motion of charge carriers is confined in the well region, i.e. in the *x-y* plane, the lack of translational invariance along the *z* direction implying that the energy levels of the carriers, as well as their momentum k_z along the *z* direction, are discrete and that the density of states is different compared to the bulk. For barriers of infinite height surrounding a well with a constant potential $V = 0$, the envelope of the electron wavefunction satisfies the time-independent Schrödinger equation in the conduction band, the discrete energy levels in the well measured from the bottom of the conduction band (bottom of the QW) being given by

$$E_{cj} = (\hbar^2 / 2m_e)(j\pi / L_z)^2, \tag{6.1}$$

with j an integer. Since $E_{cj} > 0$, the discrete energy levels are *inside* the conduction band. The corresponding discrete values of the momentum along the *z* direction are $k_{zj} = (2m_e E_{cj})^{1/2} / \hbar = j\pi / L_z$, and the total energy of the electron in the conduction band, obtained by adding to E_{cj} the kinetic energy of the electron motion in the confinement plane, is $E_c = (\hbar^2 / 2m_e)[k_x^2 + k_y^2 + (j\pi / L_z)^2]$.

The density of states for the electron, whose motion is restricted along the two-dimensional (2D) plane, is

$$\rho_{2D}(E_c) = (m_e / \pi\hbar^2 L_z)\sum_j \Theta(E_c - E_{cj}), \tag{6.2}$$

where Θ is the Heaviside step function. For $E_c = E_{cj}$, $\rho_{2D} = j(m_e / \pi\hbar^2 L_z)$ is identical to the electron density of states in a bulk semiconductor ρ_{3D}

$= (2m_e/\hbar^2)^{3/2}(E_c)^{1/2}/2\pi^2$, as can also be observed from Fig. 6.1 where ρ_{3D} is represented by a dashed line.

The quantum wire is a 2D confined QW (Fig. 6.2) for which L_z, $L_x << L_{fp}$, the one-dimensional (1D) motion of the carriers being allowed only along y.

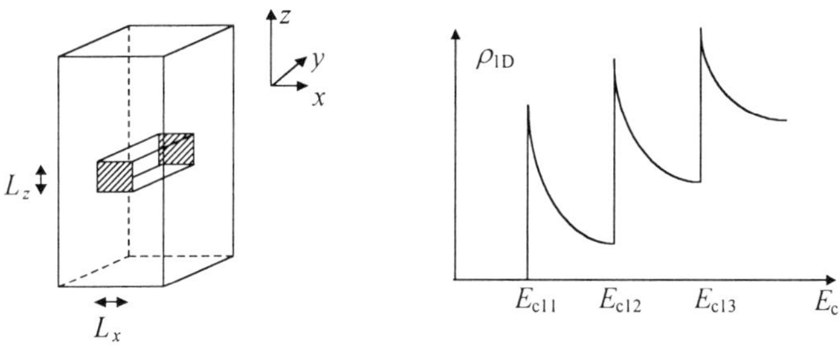

Fig. 6.2. The quantum wire and its density of states

The discrete energy levels for an electron in a quantum wire with infinite barriers can be written as

$$E_{cpq} = (\hbar^2/2m_e)[(p\pi/L_x)^2 + (q\pi/L_z)^2], \tag{6.3}$$

the quantized moments along the x and z directions being $k_x = k_{xp} = p\pi/L_x$, $k_z = k_{zq} = q\pi/L_z$. The total energy is now $E_c = (\hbar^2/2m_e)[k_y^2 + (p\pi/L_x)^2 + (q\pi/L_z)^2]$.

The density of states for the confined electron motion becomes

$$\rho_{1D}(E_c) = (1/\pi L_x L_z)(2m_e/\hbar^2)^{1/2}\sum_{p,q}(E_c - E_{cpq})^{-1/2}, \tag{6.4}$$

and is represented in Fig. 6.2. The exact form of $\rho_{1D}(E_c)$ depends on the shape of the quantum wire. For a square wire the levels (p,q) and (q,p) are degenerate for $p \neq q$, while for a rectangular wire with a cross-section $L_x \times L_z$ such accidental degeneracy is lifted unless L_x and L_z are commensurate.

The quantum dot is a three-dimensional (3D) confined QW (Fig. 6.3) for which L_x, L_y, $L_z << L_{fp}$; no degrees of freedom for electron motion exist. The moments are quantized along all directions $k_x = k_{xp} = p\pi/L_x$, $k_y = k_{yq} = q\pi/L_y$, $k_z = k_{zr} = r\pi/L_z$ and the discrete energy levels for infinite barriers are

$$E_{cpqr} = (\hbar^2/2m_e)[(p\pi/L_x)^2 + (q\pi/L_y)^2 + (r\pi/L_z)^2]. \tag{6.5}$$

The density of states is proportional to the Dirac function:

$$\rho_{0D}(E_c) = (2/L_x L_y L_z)\sum_{p,q,r}\delta(E_c - E_{cpqr}), \tag{6.6}$$

as displayed in Fig. 6.3.

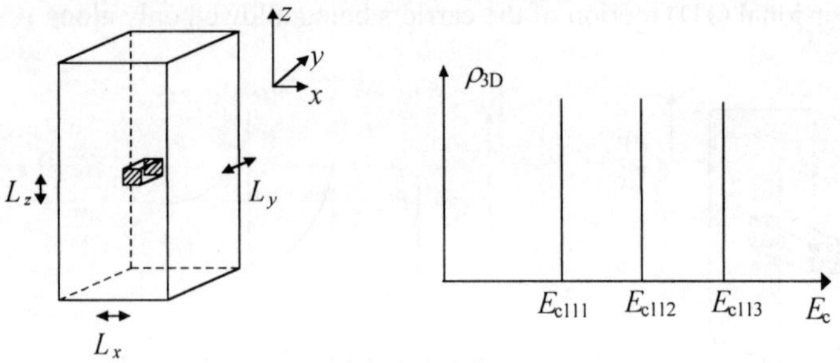

Fig. 6.3. The quantum dot and its density of states

The approximation of an infinite potential barrier height is not valid in real low-D structures, so that the energy values and momenta differ in reality from the expressions above. It is also possible to have wells with non-constant potentials $V(z)$, obtained by doping or by applying external fields (for example, the application of an electric field in the z direction tilts the energy bands).

Quantum confinement effects lead only to a redistribution of the existing density of states but do not create new energy states: for $E \to \infty$ the areas under all curves $\rho(E)$ become equal, even for the 0D systems. The sharp steps in the density of states of low-D structures are broadened by scattering processes, their discreteness being observable only if the energy spacing (the confinement energy) is larger than the collisional broadening of electron levels $\hbar/\bar{\tau}$, where $\bar{\tau}$ is a characteristic collision time. In particular, a QW with thickness L manifests as such only if $(\pi\hbar/L)^2/m_e >> \hbar/\bar{\tau}$.

The above oversimplified analysis of low-D structures is based on a single-particle picture and is valid only for structures with low density of electrons. For high electron densities the electric field produced by the inhomogeneous charge distribution across the structure must be included in the treatment, the energies E_n and wavefunctions being determined iteratively, in a self-consistent analysis.

The discrete energy levels in the valence band can be calculated in a similar manner. However, the computations are more difficult due to hole confinement which lifts the degeneracy between light hole (LH) and heavy hole (HH) bands, mixes the HH and LH bands, and induces mass anisotropy between the directions transverse to z (in-plane) and parallel to z (out-of-plane) for both HH and LH. Moreover, due to quantum confinement, the effective masses at small momenta become strongly nonparabolic, and anticrossing between different branches of the energy spectrum occurs at large momenta.

For narrow wells for which $L \leq \hbar/(2mE_g)^{1/2}$, the electron and hole states can no longer be treated separately, and a multiband $\boldsymbol{k} \cdot \boldsymbol{p}$ approach must be used. The interaction of electrons with photons and phonons, as well as the Coulomb interaction of carriers with each other and/or with charged impurities, can be described suitably with a generalization of the $\boldsymbol{k}\cdot\boldsymbol{p}$ band structure method. In the $\boldsymbol{k}\cdot\boldsymbol{p}$ formalism the wavefunction $\Psi(\boldsymbol{r})=\sum_l \psi_{l\boldsymbol{r}} u_l(\boldsymbol{r})$ is expanded in terms of $\boldsymbol{k}=0$ Bloch functions of the band extrema, $u_l(\boldsymbol{r})$, with l the band index, the expansion coefficients $\psi_{l\boldsymbol{r}}$ called *envelopes* combining at each point $\boldsymbol{r}$ in a vector $\psi_{\boldsymbol{r}}$ labeled by the band index. $u_l(\boldsymbol{r})$ are rapidly varying over the lattice period, whereas $\psi_{l\boldsymbol{r}}$ vary over distances much greater than the lattice period. Supposing an interface located at $z=0$, the envelope vectors away from the interface satisfy the equations (Vasko and Kuznetsov, 1999)

$$\begin{aligned} &[\hat{E}+\hat{\boldsymbol{v}}\cdot\hat{\boldsymbol{p}}+(1/2)\hat{\boldsymbol{p}}\hat{\boldsymbol{D}}\hat{\boldsymbol{p}}-E]\psi_{\boldsymbol{r}}=0, \quad z>0, \\ &[\hat{E}'+\hat{\boldsymbol{v}}'\cdot\hat{\boldsymbol{p}}+(1/2)\hat{\boldsymbol{p}}\hat{\boldsymbol{D}}'\hat{\boldsymbol{p}}-E]\psi_{\boldsymbol{r}}=0, \quad z<0, \end{aligned} \tag{6.7}$$

where $\hat{E}$ is the diagonal energy matrix whose elements E_l determine the positions of the band extrema, $\hat{\boldsymbol{v}}$ is the interband velocity matrix, and $\hat{\boldsymbol{p}}=-i\hbar\nabla$ is the momentum operator. The contributions of remote bands are accounted for by the introduction of the tensor of inverse effective masses $\hat{\boldsymbol{D}}$. These equations must be complemented by the conditions of continuity at the interface for the envelopes and the flux.

For A_3B_5 heterostructures the Kane model can be employed, the energy positions of band extrema being described by an 8×8 matrix. The Hamiltonian in the Kane model is $\hat{h}_K=\hat{E}+\hat{\boldsymbol{v}}\cdot\boldsymbol{p}+(1/2)\boldsymbol{p}\cdot m_h^{-1}\cdot\boldsymbol{p}$ with m_h the, generally position-dependent, HH effective mass. The Kane model can also accommodate states close to the conduction band extremum, for which the spinor (ψ^2,ψ^1) is determined from

$$\left[E_c(z)+\frac{1}{2}\boldsymbol{p}\cdot m_h^{-1}\cdot\boldsymbol{p}-\hbar\frac{\mathrm{d}M_s(z)^{-1}}{\mathrm{d}z}\sigma\cdot(\boldsymbol{e}_z\times\boldsymbol{p})\right]\begin{pmatrix}\psi_{lp}^2(z)\\ \psi_{lp}^1(z)\end{pmatrix}=E_{lp}\begin{pmatrix}\psi_{lp}^2(z)\\ \psi_{lp}^1(z)\end{pmatrix}, \tag{6.8}$$

with M_s the spin-orbit mass, or the presence of strain, when the electron energy levels are determined by solving the eigenvalue problem with a strain-induced contribution to the confining potential given by

$$V_c(z)=\begin{cases} C\sum_\alpha \varepsilon_{\alpha\alpha}=2C\varepsilon_\parallel(c_{11}-c_{12})/c_{11}, & |z|<L/2, \\ \Delta E_c, & |z|>L/2. \end{cases} \tag{6.9}$$

Here C is the deformation potential for this uniaxial stress, $\varepsilon_{xx}=\varepsilon_{yy}=\varepsilon_\parallel$ $=(a_b-a_w)/a_w$ is the in-plane stress caused by lattice constant matching requirements between well and barrier materials and $\varepsilon_{zz}=-2c_{12}\varepsilon_\parallel/c_{11}$ is the longitudinal strain with c_{ij} the elastic stiffness constants. A more complicated expression of the strain-induced potential energy must be considered for the valence band, which should include not only the rigid shift of the band but also the strain-induced LH-HH splitting.

The Kane or $\boldsymbol{k} \cdot \boldsymbol{p}$ theory can also describe the shallow electronic states that form at heterojunctions (with energies close to the conduction or valence band extrema), similar to the Tamm states localized at a perfect crystal surface. Apart from the shallow electronic states, an ideal abrupt heterojunction can also support deep localized interface states with binding energy comparable to the bandgap caused by the restructuring of chemical bonds at the interface due to lattice mismatch between the two materials. These deep states can only be analyzed numerically. Through filling of the localized states by carriers, internal electric fields can appear which modify the energy position of the levels and even the localization conditions. Confined electronic states at heterojunctions can also be created by external electric fields or by modulation doping.

Besides deep localized states, the $\boldsymbol{k} \cdot \boldsymbol{p}$ approach is not suitable for widegap materials, in which it can only treat states close to band extrema. To overcome the limitations of the $\boldsymbol{k} \cdot \boldsymbol{p}$ formalism, the tight-binding model, the empirical pseudo-potential model and the local density functional method are used.

The tight-binding model (see Sect. 1.3), also called linear combination of atomic orbitals LCAO, can also be applied to superlattices (SLs) if instead of the crystal unit cell the 'supercell' defined by the SL period is used. The tight-binding model works for any point in the Brillouin zone, which makes it particularly suitable for studying intervalley mixing, as well as the dependence of the energy spectrum in SLs on the crystallographic orientation of the growth axis and the semiconductor-semimetal transition in HgTe/CdTe SLs.

In the empirical pseudopotential method the Bloch amplitude of a state is expanded in a Fourier series in reciprocal lattice vectors, the spin-orbit interaction, deformation effects in strained SLs and the Coulomb interaction being also accounted for by including additional terms in the empirically determined free-ion formfactors, or by modifying the vector positions of atoms in the unit cell. The numerical calculations require usually 10^2–10^3 plane-wave terms in the expansion in order to describe the electronic states in widegap materials over the whole Brillouin zone, or the long-wavelength electronic states close to the gap in SLs based on narrowgap materials.

The local density functional (LDF) method does not use fitting parameters to select the orbitals or the pseudopotential, but is based on a self-consistent approach to calculate variationally the energy of the system expressed as a functional of the local electronic density. This approach can account for the many-body effects and is particularly suitable for the calculation of the structure of electronic bonds on the heterointerface.

A semiconductor heterostructure can be of type I or II, depending on the band structure of the constituent semiconductor layers. In *type I* heterostructures the semiconductor with the smaller E_g plays the role of well for both electrons and holes, as can be observed from Fig. 6.4a; the semiconductor with the higher E_g is then a barrier for both electron and hole motions. In this type of heterostructures the motions of electrons in the conduction band and holes in the valence band can be considered as independent. In *type II* heterostructures the motions of electrons and holes are coupled. In Figs. 6.4b and 6.4c the semiconductor with energy gap E_{g2} plays the role of well for electrons, while that with energy gap E_{g1} plays the role of well for holes. Examples of type I heterostructures are GaAs/AlAs, GaSb/AlSb, GaAs/GaP; type II heterostructures are InAs/GaSb, $InAs_{1-x}GaAs_x/GaSb_{1-y}GaAs_y$, etc. Sometimes the type II staggered and misaligned heterostructures are called type II and type III, respectively, and a new type of structure is introduced: type IV, which is a degenerate type I heterostructure in which the well is a gapless material (for example HgTe) (Vasko and Kuznetsov, 1999).

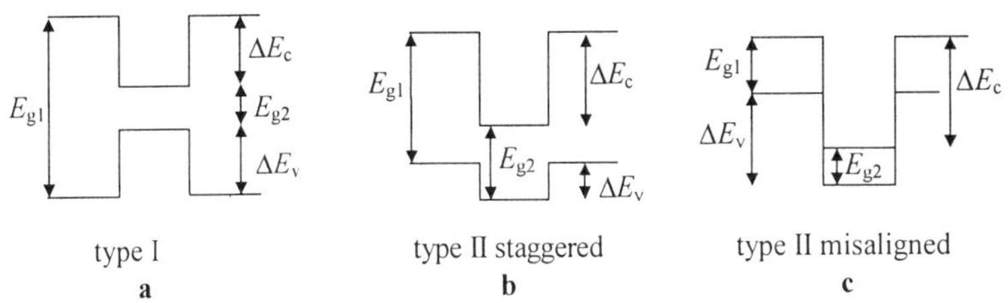

Fig. 6.4. Band energy diagrams for different types of heterostructures: (**a**) type I, (**b**) type II staggered, and (**c**) type II misaligned

Up to now we have considered only isolated QWs, and in general isolated low-D structures. If several QWs are placed in close proximity the electrons confined in one QW *may not or may interact* with the electrons confined in the neighboring wells. In the first case the electron wavefunction is *localized* in one well and the transport of electrons from one well to another in the presence of external fields, for example, is a sequential tunneling between adjacent wells. In the second case the QWs are *coupled*; if the structure is periodic, it is called a SL. The electron wavefunction extends now throughout the entire structure and the energy states are no longer discrete but form allowed and forbidden bands, just as discrete atomic levels transform into energy bands in crystals. For example, in a SL with a period Λ consisting of two non-identical QWs 1 and 2, with individual discrete electron energy levels E_{c1} and E_{c2} almost in resonance, i.e. $E_{c1} - E_{c2} = \Delta E_{12} << E_{c1},\ E_{c2}$, two minibands appear in the conduction band described by the dispersion relation $E_c(k) = E_{c1} - (\Delta E_{12}/2) \pm [(\Delta E_{12}/2)^2 + 4V^2 \cos^2(k\Lambda/2)]^{1/2}$ where V is the well-to-well interaction energy determined by the tunneling probability between the two wells. The same considerations apply also for the hole states in the valence band. However the SL differs from a crystal in that it is periodic in only one direction, the energy spectrum in the plane of the well remaining unchanged. Also, the energy bands in a SL are much narrower than in a 3D crystal (few meV versus few eV) since typical SL periods are 10–50 lattice constants.

Due to the periodicity, the electronic states in a SL are characterized by a new quantum number $p_\perp$ – the quasi-momentum in the z direction and by an effective mass defined by

$$m_{\mathrm{sl}} = \hbar^2 /(\partial^2 E_{\mathrm{c}} / \partial k^2)\Big|_{k=0} = (\hbar^2 / 2V\Lambda^2)[1 + (\Delta E_{12} / 4V)^2]^{1/2}, \tag{6.10}$$

which is a function of ΔE_{12}.

In the weak coupling case, i.e. when the tunneling through barriers is weak, the energy spectrum of a SL consists of a set of minibands, in which the total energy of electrons (adding the in-plane kinetic energy) is given by $E_{npp\perp} = E_n + p^2 / 2m + \Delta_n^{\mathrm{SL}} \cos(p_\perp \Lambda / \hbar)$. The center of the nth miniband, E_n, is close to the position of the corresponding level of an isolated well, the halfwidth being $\Delta_n^{\mathrm{SL}} = (8E_n / \kappa L_{\mathrm{w}}) \exp(-\kappa L_{\mathrm{b}})$ with $\kappa = [2m(U - E_n)]^{1/2} / \hbar$, U the height of the barrier, L_{w} the well width and $L_{\mathrm{b}} = \Lambda - L_{\mathrm{w}}$ the barrier width. For states above the barrier, for which $E > \Delta E_{\mathrm{c}}$, the periodic SL potential perturbs the continuum of states and leads to the formation of forbidden gaps for $p_\perp \cong N\pi\hbar / \Lambda$, $N = 1,2\ldots$ the width of the gaps decreasing with N. These narrow forbidden gaps appear also for strongly coupled SL with highly penetrable barriers.

The SL minibands significantly modify the density of states compared to an isolated QW. For a SL of total length $N\Lambda$, the density of states is

$$\rho(E) = (m_{\mathrm{e}} N\Lambda / \pi\hbar^2) \sum_n \int_{-\pi\hbar/\Lambda}^{+\pi\hbar/\Lambda} \frac{\mathrm{d}p_\perp}{2\pi\hbar} \theta[E - E_n - \Delta_n^{\mathrm{SL}} \cos(p_\perp \Lambda / \hbar)]. \tag{6.11}$$

This density of states gives van Hove singularities at the edges of each miniband where $\mathrm{d}\rho(E)/\mathrm{d}E$ diverges.

For weakly coupled QWs small *external fields can change the degree of localization* of the quantum states, from extended to localized. In particular, the position and separation of energy bands in a SL can be modified by illumination with photons with a suitable energy. This phenomenon is called the dynamic Wannier–Stark effect. For example, by illuminating the above-considered superlattice with an intense light beam with energy slightly lower (with δE) than the bandgap of the shallower well, the light will not be absorbed if δE is greater than the miniband widths in the non-illuminated superlattice, the net effect being the reduction of the miniband width. This is equivalent to a change of the electronic states from extended to more localized, accompanied by a change in the effective mass and the mobility of electrons.

When a dc electric field is applied on a SL, the continuum miniband spectrum transforms into a discrete set of equidistant levels (Wannier–Stark ladder) separated by the Bloch energy $E_{\mathrm{b}} = |e| F\Lambda$, which represents the potential drop between adjacent wells for an electric field F. The Wannier–Stark states are localized, their wavefunctions Ψ_z^ν being centered around the νth well and extending $2T / E_{\mathrm{b}}$ wells to the right and to the left, with T the tunneling matrix element. For weak electric fields $2T/E_b$ is large, each localized wavefunction extending over many wells, while for strong electric fields the levels become completely uncoupled. When a ladder of equidistant levels is formed, the electrons can oscillate in a dc field with the Bloch frequency $\omega_b = E_b/\hbar$. These Bloch oscillations are observed only when E_b is larger than the characteristic collisional broadening $\hbar/\bar{\tau}$. The formation of a discrete set of localized states in the presence of an electric field takes place also in crystalline solids, and in general in any periodic system, but are not observable since in a crystal the Bloch energy is only a fraction of 1 meV even for strong fields, whereas in SL it is 10–50 times larger.

The symmetry properties of the carrier wavefunctions in a SL depend on the type of the heterostructure (see Fig. 6.5). When the SL is formed in a type I heterostructure, both conduction and valence band wavefunctions are centered in the layer with a narrower gap, say A, while in type II structures the $p_\perp = \pi\hbar/\Lambda$ ($p_\perp = 0$) valence band wavefuntion is odd (even) with respect to the center of the A layer and the conduction band wavefunction is even. Therefore, the overlap integral between the conduction and valence band wavefunctions vanishes at the edge of a miniband in a type II SL, with implications in interband optical transitions.

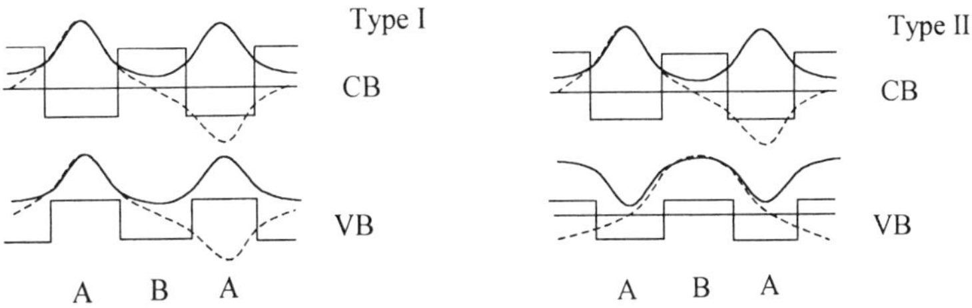

Fig. 6.5. Wavefunctions in type I and type II SL (*solid line* - wavefunctions at the bottom of a miniband $p_\perp = 0$, *dashed line* - wavefunctions at the top of the miniband $p_\perp = \pi\hbar/\Lambda$)

6.1.1 Quantum Well Excitons

An exciton in a QW satisfies the Schrödinger equation

$$[H_e(\boldsymbol{r}_e) + H_h(\boldsymbol{r}_h) + V(\boldsymbol{r}_e - \boldsymbol{r}_h)]\psi(\boldsymbol{r}_e, \boldsymbol{r}_h) = E\psi(\boldsymbol{r}_e, \boldsymbol{r}_h), \tag{6.12}$$

where H_e, H_h are the Hamiltonians of an electron and hole, respectively, when the interaction between them is neglected, $\boldsymbol{r}_e$, $\boldsymbol{r}_h$ their corresponding spatial coordinates and V the Coulomb interaction potential. Denoting by $\varphi_{cn}(z_e)$, $\varphi_{hm}(z_h)$ the eigenfunctions of H_e, H_h, characterized by a subband indices n, m respectively, the corresponding eigenvalues are $E_{cn}(\boldsymbol{k}) = E_{cn} + \hbar^2 k^2/2m_c$, $E_{hm}(\boldsymbol{k}) = E_{hm} + \hbar^2 k^2/2m_h$ where $\boldsymbol{k}$ is a 2D wavevector that describes the motion in the plane of the QW.

Since in optical transitions the total momentum must be conserved, we consider only the excitons for which $\boldsymbol{K} = \boldsymbol{k}_e + \boldsymbol{k}_h \cong 0$, or $\boldsymbol{k} = \boldsymbol{k}_e = -\boldsymbol{k}_h$. Then the excitonic wavefunction can be expanded as

$$\psi(\boldsymbol{r}_e, \boldsymbol{r}_h) = \sum_{nm} \sum_{k} \phi_{nm}(\boldsymbol{k}) | cn\boldsymbol{k}, hm\boldsymbol{k}\rangle, \tag{6.13}$$

where $| cn\boldsymbol{k}, hm\boldsymbol{k}\rangle = \varphi_{cn}(z_e)\varphi_{hm}(z_h)\exp[-i\boldsymbol{k}\cdot(\boldsymbol{x}_e - \boldsymbol{x}_h)]$ and the Wannier equation for excitons becomes

$$[E_{cn}(\boldsymbol{k}) + E_{hm}(\boldsymbol{k})]\phi_{nm}(\boldsymbol{k}) + \sum_q [V(\boldsymbol{q}) + \widetilde{V}(cn | hm)]\phi_{nm}(\boldsymbol{k}+\boldsymbol{q}) = E\phi_{nm}(\boldsymbol{k}). \tag{6.14}$$

Here $V(q) = -2\pi e^2 / \varepsilon_0 q$ (the 2D Fourier transform of the real-space Coulomb potential) is the long-range part of the Coulomb interaction and

$$\widetilde{V}(cn | hm) = -(2\pi e^2 / \varepsilon_0 q)\int dz \int dz' |\varphi_{cn}(z)|^2 |\varphi_{hm}(z')|^2 [\exp(-q|z-z'|) - 1] \tag{6.15}$$

is its short-range part that remains finite at $q \to 0$. For parabolic single-particle dispersion, and neglecting the short-range part of the potential, (6.15) becomes

$$(\hbar^2 \boldsymbol{k}^2 / 2m_r)\phi_{nm}(\boldsymbol{k}) - \sum_q (2\pi e^2 / \varepsilon_0 q)\phi_{nm}(\boldsymbol{k}+\boldsymbol{q}) = (E - E_{cn} - E_{hm})\phi_{nm}(\boldsymbol{k}), \tag{6.16}$$

with $m_r = m_c m_h / (m_c + m_h)$ the reduced mass. This equation is the Fourier transform of the 2D-hydrogenic Schrödinger equation that also describes impurity states. In particular the energy eigenvalues of this equation coincide with the values for the charged impurity with the difference that the effective Rydberg and the Bohr radius contain the reduced mass m_r instead of the conduction band mass m_c: $a_{ex} = \hbar^2 \varepsilon_0 / m_r e^2$, $R_{ex} = e^4 m_r / \hbar^4 \varepsilon_0^2$. For nonparabolic energy dispersion, the separation of the relative and the center-of-mass motion is not accurate.

The excitonic bound states for the angular momentum $l = 0$ (s-states) have discrete levels with energies $E_{nm}^j = E_{cn} + E_{hm} - R_{ex} / (j + 1/2)^2$ where j is the principal quantum number. The positive short-range potential lowers the binding energy. A small increase in the binding energy occurs due to the nonparabolicity of QW subbands caused mainly by mixing between the LH and HH subbands, whereas anisotropy lowers the binding energy of s-state excitons and increases the binding energy of excitons with $l > 0$. The intersubband coupling restores the 3D nature of the excitons in wells much wider than the Bohr radius, whereas in the opposite case, of the excitonic wavefunction in the z direction much less than the well width, the excitonic wavefunction is a coherent superposition of a large number of individual subband wavefunctions. In this case the relative motion of the electron and hole is bulk-like, but the center-of-mass motion in the z direction is quantized, the exciton behaving as a particle with mass $M = m_e + m_h$ and discrete energy levels $E_n = \hbar^2 \pi^2 n^2 / 2ML_w^2$.

The excitonic binding energy in a type I SL depends on the ratio between the well and barrier thicknesses, which controls the relative importance of quantum confinement and tunneling, whereas in type II SL, where the electrons and holes are confined in different materials, the binding energy of the indirect exciton increases with decreasing the barrier thickness, and depends also on m_c / m_h.

In quantum wires the Wannier equation is similar to the QW case but with the Coulomb matrix element replaced by $V(\boldsymbol{q}) = (e^2 / \varepsilon_0) \ln qd$. For this form of $V(\boldsymbol{q})$ the ground-state excitonic wavefunction should collapse into the origin, but this situation is avoided due to the contribution of the short-range potential. In 0D quantum dots with dimensions much less than the bulk Bohr radius the excitonic effects modify the already discrete two-particle energy eigenvalues by an amount that can be calculated perturbatively.

6.2 Optical Properties of Low-Dimensional Semiconductor Heterostructures

The confinement of carriers induces significant modifications in the optical properties of low-D heterostructures compared to the case of bulk semiconductors. An excellent recent review on the optical properties of low-D heterostructures is Vasko and Kuznetsov (1999). We sketch here briefly the characteristic features of optical spectra in low-D structures, leaving the interested reader to consult this review for more details.

6.2.1 Intraband Absorption

The absorption in 2D layers cannot be characterized by the usual absorption coefficient, defined per unit length, but by the absorption per layer α_{2D}. For radiation of frequency ω with a nonzero wavevector component in the growth direction, and which does not change appreciably across the 2D layer, the interaction with the electrons can be treated within the dipole approximation with an interaction Hamiltonian $\hat{W}_\omega = \mathrm{i}e\boldsymbol{E} \cdot \hat{\boldsymbol{v}} / \omega$. Then, the Fermi golden rule gives

$$\alpha_{\mathrm{2D}} = [(2\pi e)^2 / \omega c \varepsilon^{1/2} L^2] \Sigma_{\alpha\alpha'} |\langle \alpha | \hat{\boldsymbol{e}} \cdot \hat{\boldsymbol{v}} | \alpha' \rangle|^2 \, \delta(E_\alpha - E_{\alpha'} + \hbar\omega)(f_\alpha - f_{\alpha'}), \quad (6.17)$$

where $\hat{\boldsymbol{e}} = \boldsymbol{E} / E$ is the unit vector in the direction of polarization, $\hat{\boldsymbol{v}}$ the electron velocity operator and α a set of quantum numbers that label the eigenstates.

By denoting with $\Psi_{nn'}$, $\Phi_{nn'}$ the spin-independent overlap factors of electrons and holes in symmetric heterostructures, it is found that

$$\begin{aligned} |\hat{\boldsymbol{e}} \cdot \boldsymbol{v}_{n\sigma,n'\sigma'}(\boldsymbol{p})|^2 = s^2 \{ & (1 - \delta_{\sigma\sigma'}) [(\hat{\boldsymbol{e}} \cdot [\hat{\boldsymbol{n}} \times \boldsymbol{p}])^2 / p^2] \Psi_{nn'}^2 \\ & + \delta_{\sigma\sigma'} [(\hat{\boldsymbol{e}} \cdot \boldsymbol{p})^2 / p^2 \Psi_{nn'}^2 + (\hat{\boldsymbol{e}} \cdot \hat{\boldsymbol{n}})^2 \Phi_{nn'}^2] + \ldots \}, \end{aligned} \quad (6.18)$$

where the indices σ, σ' label the spin variables, s is the characteristic interband velocity and $\hat{\boldsymbol{n}}$ the normal to the 2D layer. The spin-flip processes, which vanish for $\hat{e} \parallel \hat{n}$, are described by the first term in (6.18), while the other two terms, present for any polarization, correspond to spin-conserving transitions. Since the QW has a preferential direction given by the normal $\hat{n}$ to the 2D plane, the optical properties can be anisotropic even for spin-independent and isotropic energy spectrum E_α and distribution function f_α.

The selection rules for transitions between quantum-confined subbands depend on the band offsets and on the relation between the well width L and the interband length $\hbar / ms$. When $L >> \hbar / ms$ the overlap factors reduce to $\delta_{nn'}$ since the wavefunctions of the confined states are almost identical, and the polarization dependence of absorption becomes the same as in a bulk crystal. For smaller wells, however, the carriers penetrate deeper into barriers, the transitions between states with different n and n' are no longer forbidden, and the matrix elements become dependent on the well width and the polarization direction. For even narrower wells only the parity selection rule, which forbids transitions between n and n' of different parity, remains.

In the Kane model the polarization dependence of interband transitions is determined by the different properties of HH and LH. In this model

$$w_{nn'}(\hat{e}) = \sum_{\sigma\sigma'} |\hat{e} \cdot \langle cn\sigma | \hat{v} | vn'\sigma' \rangle|^2 = P^2 {I_{nn'}^{cv}}^2 \begin{cases} 1-(\hat{e}\cdot\hat{n})^2, & v = h, \\ [1+3(\hat{e}\cdot\hat{n})^2]/3, & v = l, \end{cases} \tag{6.19}$$

with P the Kane matrix element and $I_{nn'}^{cv} = \int dz \varphi_n^c(z) \varphi_{n'}^v(z)$ the overlap integral between conduction and valence band states. According to (6.19), transitions from the HH band are excited only by the in-plane component of the electric field, while LH transitions are excited by both polarizations. However, HH and LH mixing violates the $\Delta n = 0$ selection rule and the energy spectrum becomes dependent on the orientation of the momentum in the 2D plane.

In particular, the in-plane anisotropy of the absorption coefficient, defined as $\rho(\omega) = \Delta\alpha_{2D}(\omega)/\langle\alpha_{2D}(\omega)\rangle = 2[\alpha_{2D}^x(\omega) - \alpha_{2D}^y(\omega)]/[\alpha_{2D}^x(\omega) + \alpha_{2D}^y(\omega)]$ for a growth direction z, was related to the Luttinger parameters, in the first order perturbation through $\rho_{HH} = [(\gamma_3 - \gamma_2)/2\gamma_2](1-\mu^2)(3\mu^2 - 1)[1 + 16\pi^2\gamma_2/(L^2\Delta)]$, $\rho_{LH} = -3[(\gamma_3 - \gamma_2)/2\gamma_2](1-\mu^2)(3\mu^2 - 1)$, where L is the well width, Δ the spin-orbit splitting, and $\mu = \cos\theta = n/(2m^2 + n^2)$ for a growth direction $[mmn]$ (Winkler and Nesvizhskii, 1996).

For an empty conduction band and a filled valence band, i.e. for $f_c = 0$, $f_v = 1$, and assuming a parabolic dispersion of these bands around the band edge, the relative absorption becomes

$$\alpha_{2D} = (e^2 / \omega c \varepsilon^{1/2}) \sum_{nn'} w_{nn'}(\hat{e}) \rho_{nn'}(\hbar\omega), \tag{6.20}$$

where the *joint density of states* is given by $\rho_{nn'}(E) = (m_{nn'} / \pi\hbar^2)\Theta(E - E_{nn'})$ with $m_{nn'}^{-1} = m_{cn}^{-1} + m_{vn'}^{-1}$ and $E_{nn'} = E_{cn} - E_{vn'}$ the transition energy between the bottom of n and n' subbands. Each pair of subbands contributes a step to the absorption, which is less than 1% per step per layer, if the transition between them is allowed. The transformation of a structureless spectrum in a wide QW into step-like spectra with decreasing well width is illustrated in Fig. 6.6.

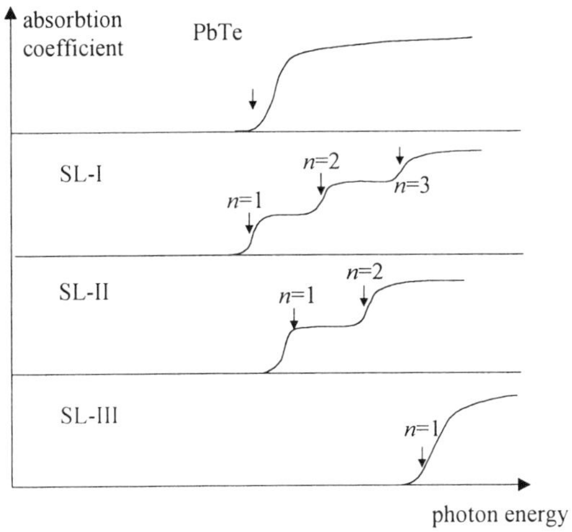

Fig. 6.6. Transformation of a structureless spectrum in a wide PbTe QW into step-like spectra with decreasing well width (the well width decreases towards the bottom)

Analogously, for an array of quantum wires formed in a 2D plane

$$\alpha_{1\mathrm{D}} = (e^2 n_\mathrm{w} / \omega c \varepsilon^{1/2}) \Sigma_{nn'} w_{nn'}(\hat{e}) \rho_{\mathrm{nn'}}^{1\mathrm{D}}(\hbar\omega), \qquad (6.21)$$

with $\rho_{nn'}^{1D}(E) = (1/\pi\hbar)\Theta(E - E_{nn'})[2m_{nn'}/(E - E_{nn'})]^{1/2}$ and n_w the number of wires per unit length, while for an array of quantum dots with n_d dots per unit area

$$\alpha_{0D} = (e^2 n_\mathrm{d} / \omega c \varepsilon^{1/2}) \Sigma_{nn'} w_{nn'}(\hat{e}) \delta(\hbar\omega - E_{nn'}) . \qquad (6.22)$$

The additional confinement in the quantum wire array splits each QW absorption step into a series of inverse-square-root singularities, while in the quantum dot each singularity is further split into a series of δ-function peaks.

The sharp absorption features in low-D systems are always broadened. The inhomogeneous broadening is caused by the variation from point to point in the 2D plane of the composition, well widths, etc., whereas the homogeneous, or lifetime broadening is due to the scattering of electrons on phonons, impurities and on each other. The inhomogeneous broadening is dominant in undoped samples at low temperatures, while homogeneous broadening becomes important with increasing temperature and impurity concentration.

6.2.2 Excitonic Absorption

In the Fermi golden rule expression for the absorption coefficient, it was assumed that the initial and final states of the optical transition are uncorrelated. This assumption is no longer valid for the wavefunction of an electron-hole pair in the presence of the Coulomb interaction; in this case the wavefunction is expressed not as a product but as a superposition of plane-wave electron and hole states, the relative excitonic absorption being

$$\alpha_X = [(2\pi e)^2 / \omega c \varepsilon^{1/2}] \Sigma_{nn'\nu} |\langle cn | \hat{e} \cdot \hat{v} | vn' \rangle|^2 C(\nu) \delta(\hbar\omega - E^{\nu}_{nn'}). \tag{6.23}$$

In (6.23) ν labels the excitonic eigenstates with eigenfunctions $\phi^{\nu}_{nn'}$ and eigenvalues

$$E^{\nu}_{nn'} \phi^{\nu}_{nn'}(\boldsymbol{k}) = E_{nn'}(\boldsymbol{k}) \phi^{\nu}_{nn'}(\boldsymbol{k}) - \Sigma_q [V(\boldsymbol{q}) + \widetilde{V}(cnn | vn'n')] \phi^{\nu}_{nn'}(\boldsymbol{k} + \boldsymbol{q}), \tag{6.24}$$

and $C(\nu) = |\Sigma_k \phi^{\nu}_{nn'}(\boldsymbol{k})|^2 = |\phi^{\nu}_{nn'}(\boldsymbol{x} = 0)|^2$ is the Sommerfeld factor, which gives the probability of finding the electron and hole in the νth excitonic state at zero distance from each other. The Sommerfeld factor renormalizes the transition matrix elements and is nonzero only for s-states. In the ideal 2D limit, for the jth bound state, where j is the principal quantum number, $C(j) = 2/(j + 1/2)^3$, while for s-states in the continuum

$$C(\Delta) = \frac{2\exp(\pi/\sqrt{2\Delta})}{\exp(\pi/\sqrt{2\Delta}) + \exp(-\pi/\sqrt{2\Delta})}, \tag{6.25}$$

with $\Delta = (E - E_{nn'})/R_{\text{ex}}$ the detuning from the subband edge in units of the exciton Rydberg. At the subband edge $E = E_{nn'}$ the Sommerfeld factor enhances the QW absorption by a factor of 2, the enhancement factor dropping to 1 for large detunings from the subband edge.

The optical absorption in QW can be modulated by the electric field to a much larger degree than in the bulk. When the electric field is applied in one of the quantum confined directions the transition energies and matrix elements are modified by the electric field, which enters through an additional term $-eFz$ in the Schrödinger equation. The net effect is that the electrons and the holes in the well are pushed in opposite directions, their energies and the overlap integral between them changing. In particular, the binding energies are slightly reduced and the excitonic peaks become more sensitive to the imperfections of the heterointerfaces. The oscillator strengths of the excitonic peaks allowed in the zero-field case change, and new excitonic peaks appear that were forbidden in the zero field. The reduction of exciton binding energies, called Stark shift, is a relatively minor effect in QWs with a smaller or comparable width with respect to the excitonic Bohr radius. Although this effect is present also in bulk, the excitonic resonances are destroyed in fields of the order 10^3 V/cm, too low to cause any appreciable Stark shift, due to large field-induced exciton line broadening caused by field ionization. On the contrary, in low-D systems the Stark shift is larger than the field-induced broadening and the fields required to ionize the exciton are orders of magnitude larger than in bulk.

6.2.3 Luminescence

For electrons and photons localized in the same plane, the problem of calculating the PL spectrum is a 2D counterpart of the bulk case, in which electrons and photons are characterized by a 2D wavevector. If 2D electrons emit 3D photons, the spontaneous emission becomes dependent on the angle between the photon wavevector $\boldsymbol{Q} = (\boldsymbol{q}, q_\perp)$ and the normal to the 2D layer, $\hat{\boldsymbol{n}}$. The rate of spontaneous emission associated with transitions between electron and hole states with 2D momentum $\boldsymbol{p}$ is then

$$I_{\mu\mu'}(\boldsymbol{Q}) = (e^2/\hbar^2\varepsilon\omega_Q)\Sigma_{nn'}\int \mathrm{d}\boldsymbol{p} f_{np}(1-f_{n'p})M_{\mu\mu'}(n\,|\,n')\delta(\hbar\omega_Q + E_{n'p} - E_{np}), \quad (6.26)$$

where $M_{\mu\mu'}(n\,|\,n') = (\hat{\boldsymbol{e}}_{Q\mu} \cdot \boldsymbol{v}_{nn'}(\boldsymbol{p}))(\hat{\boldsymbol{e}}_{Q\mu'} \cdot \boldsymbol{v}_{nn'}(\boldsymbol{p}))^*$. $M_{\mu\mu}$ is proportional to the squared modulus of the interband velocity matrix element, while $M_{\mu\mu'}$ with $\mu \neq \mu'$ depends on the phase correlation between the two-photon polarization states. The PL polarization is characterized by the Stokes parameters

$$\varsigma_1(\boldsymbol{Q}) = [I_{12}(\boldsymbol{Q}) + I_{12}(\boldsymbol{Q})^*]/I_0(\boldsymbol{Q}), \quad (6.27a)$$

$$\varsigma_2(\boldsymbol{Q}) = [I_{12}(\boldsymbol{Q}) - I_{12}(\boldsymbol{Q})^*]/I_0(\boldsymbol{Q}), \quad (6.27b)$$

$$\varsigma_3(\boldsymbol{Q}) = [I_{11}(\boldsymbol{Q}) - I_{22}(\boldsymbol{Q})]/I_0(\boldsymbol{Q}), \quad (6.27c)$$

where $I_0(\boldsymbol{Q}) = \Sigma_{\mu=1,2} I_{\mu\mu}(\boldsymbol{Q})$. Choosing $\hat{\boldsymbol{e}}_\mu$, which are perpendicular to $\boldsymbol{Q}$, at angles $\pm\pi/4$ to the plane formed by $\boldsymbol{Q}$ and $\hat{\boldsymbol{n}}$, linear polarization in the $\boldsymbol{Q}$-$\hat{\boldsymbol{n}}$ plane is described by $\varsigma_3 = 0,\ \varsigma_1 > 0$ and linear polarization in the plane parallel to the 2D layer corresponds to $\varsigma_3 = 0,\ \varsigma_1 < 0$.

The frequency dependence of the luminescence is given by

$$S_0(\omega) \approx \Sigma_{\mu nn'}\int \mathrm{d}\boldsymbol{p} f_{np}(1-f_{n'p})\,|\,\hat{\boldsymbol{e}}_\mu \cdot \boldsymbol{v}_{nn'}(\boldsymbol{p})|^2\delta(\hbar\omega + E_{n'p} - E_{np}), \quad (6.28)$$

which reduces in the case of weakly photoexcited undoped structures where the distribution functions of both electrons and holes are nondegenerate and of Maxwell–Boltzmann form with the same temperature T to

$$S_0(\omega) \approx \rho(\hbar\omega - E_g)\exp[-(\hbar\omega - E_g)/T]. \quad (6.29)$$

In this case the joint density of states can be extracted from the PL spectra; for an approximately constant density of states the carrier temperature can be directly determined from the slope of a log plot of PL intensity versus photon energy.

The luminescence and absorption spectra are related by the fluctuation-dissipation theorem as

$$S_0(\omega) \approx \alpha(\omega)/[\exp(\hbar\omega - E_g + \Delta\mu/T) - 1], \tag{6.30}$$

where $\Delta\mu$ is the sum of the chemical potentials of electrons and holes.

For doped structures, as well as under high photoexcitation, the carriers are degenerate and the spectral dependence of luminescence is given by

$$S_0(\omega) \approx \int \mathrm{d}E \rho_c(E)\rho_v(\hbar\omega - E) f_c(E)[1 - f_v(E)], \tag{6.31}$$

the PL peaks in this case being shifted compared to the undoped case.

The PL polarization is caused by the difference in the transition matrix elements for photons polarized in the plane or perpendicular to the plane of the heterostructure. For example, in the Kane model the transitions involving LH produce photons polarized in the $\boldsymbol{Q}$-$\hat{\boldsymbol{n}}$ plane while at transitions into HH states the photons are polarized in the 2D plane. On illumination with circularly polarized light the photoexcited carriers have a preferential spin orientation, subsequently destroyed by scattering processes. The PL has, in this case, a certain degree of circular polarization.

6.2.4 Scattering in Low-Dimensional Heterostructures

The additional quantum number in heterostructures, i.e. the subband index that characterizes the initial and final states i and f, induces additional scattering channels compared to the bulk case, corresponding to transitions that change the subband index. Since the wavevectors of the initial and final states coincide in these transitions, they can be described in the dipole approximation.

For scattering processes that conserve the subband index, a finite wavevector transfer $\boldsymbol{q} = \boldsymbol{k}_f - \boldsymbol{k}_i$ must be considered. An effective second-order electron-photon interaction is introduced to describe such processes, for which

$$(\mathrm{d}^2\sigma / \mathrm{d}\omega_f d\Omega)_\mu = r_e^2 (\omega_f / \omega_i) A(\omega_f) S(\boldsymbol{q}, \omega), \tag{6.32}$$

where $\omega = \omega_f - \omega_i$, $A(\omega_f)$ describes the resonant enhancement of scattering amplitudes due to virtual interband transitions, and

$$S(\boldsymbol{q}, \omega) = (1/2\pi)\int_{-\infty}^{+\infty} \mathrm{d}t \exp(\mathrm{i}\omega t)\langle \hat{n}_{\boldsymbol{q}}(t)\hat{n}_{\boldsymbol{q}}^+(0)\rangle, \tag{6.33}$$

is the structure factor, expressed in terms of Fourier components of the electron density operator $\hat{n}_{\boldsymbol{q}}$. When scattering occurs on fluctuations of spin density, for example, the spin density instead of charge density operators must be introduced. Scattering on spin fluctuations can be seen only for $\hat{\boldsymbol{e}}_i \perp \hat{\boldsymbol{e}}_f$, while scattering on density fluctuations occurs when $\hat{\boldsymbol{e}}_i \parallel \hat{\boldsymbol{e}}_f$. Raman scattering can be used for the determination of band offsets in strained low-D heterostructures (by fitting the positions of transition energies with calculated values), for studying the effects of confinement, or (when Raman scattering takes place on intersubband excitations in valence band) for evidencing the anisotropy of hole energy bands and for estimating the Luttinger parameters.

6.2.5 Intersubband Optical Transitions

At low frequencies ($\omega < 1/\bar{\tau}$) the optical response is dominated by free-carrier absorption. Besides this region, assuming the same effective masses for the well and barriers and neglecting the momentum dependence of matrix elements, the absorbance for transitions from the filled ground state $|1\rangle$ with electron density n_{2D} into an empty first excited state $|2\rangle$ is

$$\alpha_{2D} = [(2\pi c)^2 / \omega c \varepsilon^{1/2}] \cos^2 \theta \, | v(1,2) |^2 \, n_{2D} \delta(E_1 - E_2 + \hbar\omega), \tag{6.34}$$

where $v(n,n') = -i4nn'\hbar/[mL(n^2 - n'^2)]$, and θ is the angle between the optical field $\boldsymbol{E}$ and the unit vector $\hat{\boldsymbol{e}}_z$. The unavoidable collisional broadening of the intersubband absorption peak can be phenomenologically described by a Lorentzian $\delta_\Gamma(E) = \Gamma/[\pi(E^2 + \Gamma^2)]$.

Intersubband absorption can be used to investigate the Stark shift of subbands in an external dc field. Typical experimental curves as those in Fig. 6.7 show a quadratic dependence of the peak position on the applied field.

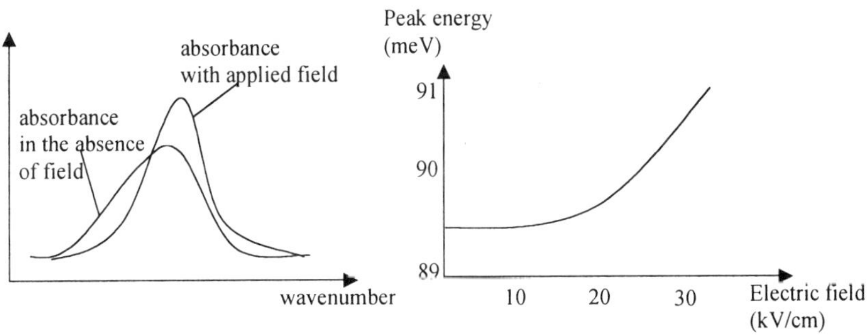

Fig. 6.7. Typical behavior of the intersubband absorption in the presence of the Stark shift of subbands in an external dc field

For high enough photon energies, such that transitions into the above-barrier delocalized states are possible, the absorption is dominated by resonant state effects at low transverse energy $E_\perp = E - p^2/2m$. For resonant well widths for which $L(2m\Delta E_c)^{1/2}/\pi\hbar$ is an integer, the absorption is proportional to $1/(\hbar\omega - E^*)^{1/2}$ where $E^* = \Delta E_c - E_1$ is the edge of the continuum absorption with E_1 the ground state energy. For other well widths the absorption varies as the square root of $\hbar\omega - E^*$, except for L half of the resonant values where the spectrum scales as $(\hbar\omega - E^*)^{-3/2}$.

6.2.6 Depolarization Shift and Coulomb Renormalization

Excitation of intersubband transitions by transverse electric fields creates an inhomogeneous charge density in the z direction that can change the resonant frequency of the transition. This effect is called depolarization shift and must be distinguished from the Coulomb renormalization, which is also a density-dependent energy shift. The depolarization shift of the transition frequency between two subbands 1 and 2, in the dipole approximation is

$$W_{\mathrm{P}}^2 = (8\pi e^2/\varepsilon_0)L_{21}\tilde{n}_{2\mathrm{D}}\tilde{E}_{21}, \tag{6.35}$$

where $\tilde{n}_{2\mathrm{D}} = N_{11} - N_{22}$ is the total population difference between the two subbands, $\tilde{E}_{21} = \tilde{E}_2 - \tilde{E}_1$ is the intersubband transition energy renormalized by Coulomb effects and

$$L_{21} = \int_{-\infty}^{\infty} \mathrm{d}z\,\varphi_2(z)\varphi_1(z)\int_{-\infty}^{z}\mathrm{d}z'\int_{-\infty}^{z'}\mathrm{d}z''\varphi_2(z'')\varphi_1(z''). \tag{6.36}$$

The depolarization effect can cancel the divergence of the absorption spectrum for continuum transitions in the 'resonant' well widths case, but does not significantly change the shape of the spectrum for nonresonant well widths except for very high densities, where the renormalized transition energy $\tilde{\omega}_{21} = [\tilde{E}_{21}^2 + W_P^2]^{1/2}/\hbar$ becomes comparable to the continuum edge frequency $E^*/\hbar$. In this case the intersubband absorption peaks merge with the continuum absorption.

6.2.7 Radiative Intraband Transitions in Heterostructures

Unlike the interband case, nonequilibrium population between subbands of the same band can be created by heating the electron distribution in a strong dc field in the plane of the layer, or by carrier redistribution in the tunneling transport. The nonequilibrium population can be more easily created between energy levels for which either $E_{21} < \hbar\omega_{\mathrm{LO}}$ or $>> \hbar\omega_{\mathrm{LO}}$. In the first case electrons from the excited state cannot relax back to the ground state through the emission of LO phonons, while in the second case phonon emission involves large momentum transfer for which electron-phonon coupling is inefficient. References to experimental studies that evidenced this possibility of creating nonequilibrium population can be found in Vasko and Kuznetsov (1999).

Another far-IR emission mechanism peculiar to 2D structures appears in a 2DEG (2D electron gas) for energies less than the lowest intersubband transition. If the properties of the 2D layer are periodically modulated in one direction, an electron moving with constant velocity will periodically oscillate and thus emit an electromagnetic radiation with frequency $\omega_q = \boldsymbol{q} \cdot \boldsymbol{p} / m$ where $\boldsymbol{q}$ is the wavevector of the periodic modulation. To observe this effect, the mean free path of electrons must be greater than the period of modulation. The effect cannot be observed when the contributions of individual electrons are averaged over a symmetric distribution function in the momentum space (see Vasko and Kuznetsov (1999) for references for experimental observation of this effect).

6.2.8 Collective Excitations

In low-D structures the free carriers behave as an ideal 2DEG. Inelastic light scattering on free carriers reveals both polarized spectra due to charge-density excitations, and depolarized spectra due to spin-density excitations. Charge-density excitations are collective modes with higher energies than single-particle transitions due to the effects of macroscopic electric fields (higher with the depolarization shift), while spin-density excitations have different frequencies than single-particle transitions due to many-body corrections (Pinczuk and Abstreiter, 1989).

Collective intrasubband excitations include plasmons. The plasma frequency of a single 2D layer with a density n embedded in a medium with a dielectric constant ε_0 is

$$\omega_{2\mathrm{D}}^2(q) = 2\pi n e^2 q / (\varepsilon_0 m_{\mathrm{eff}}), \tag{6.37}$$

while the plasma frequency of an infinite SL is $\omega_{\mathrm{p}}^2(q, q_z) = \omega_{2\mathrm{D}}^2(q) S(q, q_z)$ with

$$S(q, q_z) = \sinh(q\Lambda) / [\cosh(q\Lambda) - \cos(q_z\Lambda)], \tag{6.38}$$

the structure factor of the SL with period Λ. In a SL the modes are quasi-3D for $q\Lambda < 1$ and $q_z\Lambda = 0$, i.e. $\omega_{\mathrm{p}}^2 = 4\pi n e^2 / \varepsilon_0 m_{\mathrm{eff}} \Lambda$ does not depend on q, while for $q\Lambda < 1$ and $q_z\Lambda \neq 0$ they are acoustic-like, the frequency being linearly dependent on the wavevector.

In MQW (multiple QW) or SL with a finite number N of layers the plasmon modes are discrete; they are doublets in a finite SL (degenerate in an infinite SL) with amplitudes determined by admixtures of wavevector components q_β and $(2\pi / d) - q_\beta$, and frequencies $\omega_\pm^2(q, q_\beta) = \omega_{\mathrm{p}}^2(q, q_\beta) + \omega_{2\mathrm{D}}^2[f_1 \pm (f_2 f_3)^{1/2}]$, where (Pinczuk and Abstreiter, 1989)

$$f_1 = [1 - \exp(-Nq\Lambda)](1 - \cosh q\Lambda \cos q_\beta\Lambda) / [N(\cosh q\Lambda - \cos q_\beta\Lambda)^2], \tag{6.39}$$

$$f_{2,3} = [1 - \exp(-Nq\Lambda)][1 - 2\cosh q\Lambda \exp(\pm \mathrm{i} q_\beta\Lambda) + \exp(\pm \mathrm{i} z q_\beta\Lambda)] \times [2N(\cosh q\Lambda - \cos q_\beta\Lambda)^2]^{-1}. \tag{6.40}$$

In a finite SL the mode degeneracy is lifted due to the confinement of charge-density fluctuations within the finite width $(N - 1)\Lambda$.

When a high magnetic field is applied in a direction normal to the layer, the collective modes associated to Landau level transitions are called magnetoplasmons. Their frequencies are equal to the spacing between Landau levels $\hbar\omega_c = \hbar eB / m_{eff}$ at $q = 0$, while for $q << 1/l_m$ with $l_m = (\hbar / eB)^{1/2}$ the magnetic length the magnetoplasmon frequencies are $\omega_p(q,B) = [\omega_p^2(q) + \omega_c^2]^{1/2}$, with ω_p the corresponding plasma frequency for a single layer or a SL. For $q\Lambda > 1$ the coupling between layers is negligible and the SL magnetoplasmons are identical to those of a single layer. In high magnetic fields and polarized spectra the collective intersubband excitations and plasmon modes of MQW appear as additional peaks with respect to the intersubband transition present in depolarized spectra for $B = 0$. The collective excitations converge to the characteristic features of 3DEG with increasing carrier concentration.

6.2.9 Nonlinear Optics

In low-D semiconductors nonlinear effects appear at lower excitation intensities than in bulk (as low as 1 nW/cm^2), since the characteristic momentum is $\hbar / d$ instead of E_g / P. The magnitudes of nonlinear optical susceptibilities in low-D structures are, in general, of the same order as in the corresponding bulk materials, with the exception of susceptibilities that vanish in the bulk due to symmetry requirements. For example, in inversion-symmetric materials the bulk susceptibilities of even orders vanish, but the inversion symmetry can be broken in low-D structures by an asymmetric confinement potential. Thus, second-harmonic generation can be generated in asymmetric heterostructures even when in bulk $\hat{\chi}^{(2)}$ is zero. Also, new mechanisms of producing nonlinear optical responses can appear in low-D structures, due to the spatial nonuniformity of the structure. For example, in asymmetric double QW, the absorption spectrum is modified due to intensity-dependent redistribution of carriers between the wells and the related changes in the energies of the double QW levels. Also, low-D structures exhibit nonlinear properties in the frequency range corresponding to intersubband transitions that are absent in bulk materials. For example, in asymmetric double QW second harmonic generation is resonantly enhanced when the level spacing in the two QWs becomes equal to approximately twice the pump frequency.

6.3 Optical Characterization of Low-Dimensional Semiconductor Heterostructures

6.3.1 Confinement Effects

The absorption curve is a very strong indicator about *the confinement* of carriers. For example, the transition from 2D to 3D behavior can be seen in the electric-field-dependent absorption of ZnSe-based quantum wells where strong excitonic effects are present and the Coulomb interaction leads to intersubband coupling (Merbach et al., 1998).

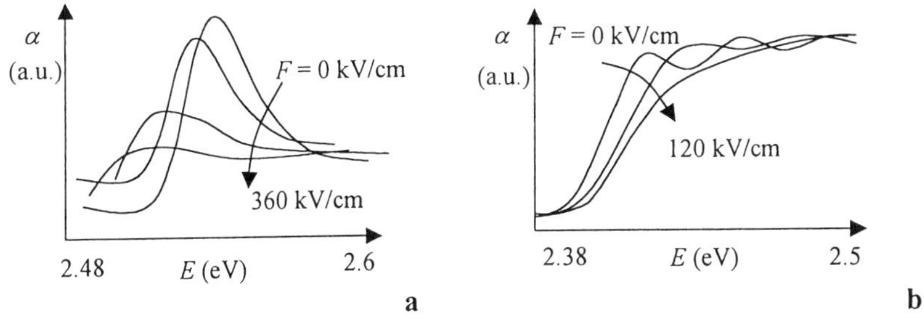

Fig. 6.8. Dependence of the absorption on electric field in (**a**) narrow and (**b**) wide ZnSe-based QW

The shape of the absorption spectra is dramatically influenced by the width of the QW. In Fig. 6.8a the well width, of 3 nm, is less than the exciton Bohr diameter and the absorption redshifts due to the quantum confined Stark effect. For well widths larger than the exciton Bohr diameter, as in the case of Fig. 6.8b where the well is 15 nm wide, the absorption shows a bulk-like behavior due to the quantum confined Franz–Keldysh effect. Namely, the exciton peak at $F = 0$ vanishes before any peak shift occurs, and the spectrum at 160 kV/cm is practically monotonic. The eventual transition from 2D to 3D center-of-mass motion in wider wells cannot be observed in absorption spectra since the momentum conservation law requires the center-of-mass motion to vanish.

The influence of interfaces on confinement effects can be studied in monolayer thick Ge quantum wells (Olajos et al., 1996). In Ge-Si heterostructures with QWs a few monolayers thick the transition energy is dictated by the competition between strain, which decreases the transition energy, and confinement effects that shift the transitions to higher energies. At the same time, the electronic structure is strongly influenced by interfaces, which enhance the probabilities of no-phonon transitions due to scattering at Si-Ge interfaces. Spectra of 2–4 monolayer thick Ge wells show that interface effects are quite strong, the observed PL originating from the no-phonon transitions and its phonon replica.

Not only electronic, but also photonic states can become confined. The confining structures are called photonic wells, wires and dots depending on the number of confined degrees of motion, analogously to quantum wells, wires and dots. The confinement of optical modes in photonic wires fabricated by lithography of planar microcavities can be investigated using angle-resolved PL (Kuther et al., 1998a). The electric fields in photonic wires are confined in the direction of the wire widths (y) due to the large difference in refractive indices at wire boundaries. The form of the electric field is then $E(x,y) \approx \exp(ik_x x)E(y)$ where k_x is the phonon wavenumber along the wire. The confined electric field is given by $E(y) \approx \cos k_y y$ or $E(y) \approx \sin k_y y$ where $k_y = (\pi / L_y)(n_y + 1)$, with $n_y = 0,1,2,...$. The energy of confined photon modes is then

$$E_{n_y}(k_k) = \sqrt{E_0^2 + (\hbar^2 c^2 / \varepsilon_r)[k_x^2 + \pi^2 (n_y + 1)^2 / L_y^2]}, \tag{6.41}$$

where E_0 is the energy of the fundamental cavity mode and ε_r the relative electric permittivity of the well material. The discrete character of the wavevector in the lateral direction can be evidenced in angle-resolved PL by changing the angles θ and φ (see Fig. 6.9). For the given quantum wire geometry $k_x = k \sin\theta \cos\varphi$ and $k_y = k \sin\theta \sin\varphi$. Experimental results showed that for $\varphi = 0$ only the $n_y = 0$ mode is observed in the PL, its spectral position shifting to higher energies as θ increases. On the other hand, for $\varphi = \pi / 2$ the energies of modes do not shift with θ, but the emission intensities depend on the detection angle. Higher confined modes appear at large angles, as can be seen in Fig. 6.9. The emission intensity of each mode becomes maximum when the detection direction is related to the confined wavenumber as $k \sin\theta = \pi / L_y(n_y + 1)$. The detected intensity can be written as $I(\theta, \varphi) = \delta(k_x - k_x')| \int E(y) \exp(ik_y y) dy |^2$.

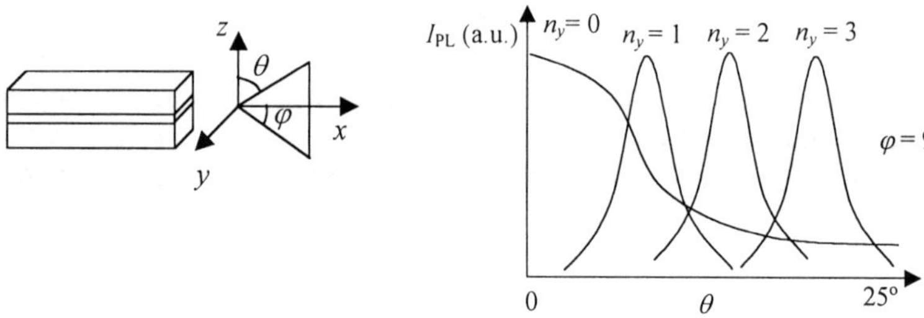

Fig. 6.9. The geometry of the quantum wire and the PL intensities of the confined modes

Similar angle-resolved PL results were obtained in photonic dots (Gutbrod et al., 1999). The energies of confined modes have in this case a form similar to (6.40), but with k_x replaced by $k_x = \pi(n_x + 1) / L_x$. For both quantum wire and quantum dots, the angular dependence of PL allows the determination of the electromagnetic field distribution in the dot plane.

The effect of QW confinement on the ultrafast carrier dynamics in an $In_{0.47}Ga_{0.53}As/In_{0.47}Al_{0.53}As$ heterostructure can be investigated with pump-probe experiments, when the excitation energy is (at 30 meV) above the band edge (Bolton et al., 1998). The results show that the rate of carrier thermalization is independent of the well width whereas the rate of carrier cooling to the band edge is strongly affected by confinement. The origin of this behavior resides in two phenomena (i) an increased number of thermalized electrons able to emit LO phonons in 3D rather than in 2D, due to the dimensionality dependence of the density of states, and (ii) a decrease of electron-LO phonon coupling due to the change in phonon density of states caused by ionic mass discontinuity at QW boundaries.

When exciting at energies above the bandgap, the time evolution of the nonthermal population follows the steps represented in Fig. 6.10. The initially sharply peaked nonthermal population (Fig. 6.10a) thermalizes in 0.3 ps to a hot Boltzman distribution through Coulomb-mediated scattering (Fig. 6.10b), and in about 20 ps reaches a Fermi distribution in equilibrium with the lattice through cooling accompanied by phonon emission (Fig. 6.10c). The band-edge absorption can be correspondingly attributed to nonthermal population, to screening by the hot thermal distribution and Pauli blocking, and to screening by cooled carriers and Pauli blocking. Time-resolved absorption experiments on samples with different well thicknesses, ranging from 100 to 6000 Å, but the same carrier density of 10^{16} cm^{-3} have shown that the thermalization in 2D and 3D have the same order of magnitude. These results contradict the common belief that 2D thermalization is faster than 3D.

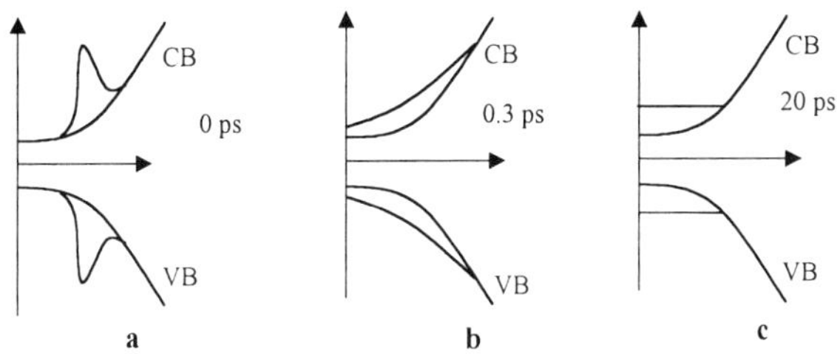

Fig. 6.10. Carrier dynamics for laser excitation above the energy gap

The differential transmission at band edge is strongly influenced by the screening of Coulomb-bound excitons and Pauli blocking, or phase-space filling, of exciton absorption. Pauli blocking is caused by the fermionic nature of carriers and is characterized by the difference in oscillator strengths in the presence and absence of carriers. In the presence of the carrier the absorption is $\alpha = \alpha_0(1 - f_\mathrm{e} - f_\mathrm{h})$, the change in the oscillator strength due to phase-space filling being

$$\Delta f / f = [1/\phi_n(r=0)]\Sigma_k (f_\mathrm{e} + f_\mathrm{h})\phi_n(k). \quad (6.42)$$

The electron contributions to phase-space filling in the 2D and 3D cases are, respectively, $S_\mathrm{e} = (\pi a_0^2 N_\mathrm{e}) 2\alpha_{\mathrm{e2D}} I_{\mathrm{2D}}$ and $S_\mathrm{e} = (8\pi^{1/2} a_0^2 N_\mathrm{e}) 2\alpha_{\mathrm{e3D}}^{3/2} I_{\mathrm{2D}}$, where I_{2D} and I_{3D} are integrals over the error function, and $\alpha_{\mathrm{e2D}} = (m/m_\mathrm{e}) 4E_0 / \mathrm{k_B}T$, $\alpha_{\mathrm{e3D}} = (m/m_\mathrm{e}) E_0 / \mathrm{k_B}T$ with N_e the carrier density and E_0 the excess energy. *An estimate of phase-space filling in 2D and 3D is obtained by comparing the differential absorption spectra at zero time delay with those at* 0.5 *ps*. The experiments show that phase-space filling is twice as effective in QW compared to bulk materials. However, the cooling process towards the Fermi–Dirac distribution was found to be much faster in bulk than in QW, the respective time constants being 7 ps and 21 ps. The longer cooling time in QW can be attributed to the change in electron-phonon scattering caused by confinement-induced changes in the phonon and electron density of states.

The *lateral confinement energy* in T-shaped $In_xGa_{1-x}As$ ($x = 0.17$) quantum wires (QWRs) can be obtained from PL spectra (Ayiyama et al., 1998). In T-shaped QWR a 1D electronic state forms at the intersection of the two parent QWs, i.e. QW_1 (with thickness a) for the 'stem' part of the letter 'T' and QW_2 (thickness b) for the arm part of 'T'. In GaAs T-QWRs with $Al_{0.7}Ga_{0.7}As$ barriers the lateral confinement energy of excitons E_{1D-2D} was found to be only 18 meV. This parameter was obtained as the difference between the QWR PL peaks and the lowest energy peak between QW_1 and QW_2. These three peaks appear simultaneously in the PL spectrum if the whole structure composed of the intersecting QW_1 and QW_2 wells is illuminated. E_{1D-2D} can be regarded as the stabilization energy of 1D excitons, and can be enhanced by changing the quantization energy of the parent QWs, i.e. by increasing the barrier energy or by decreasing the well energy. For example, E_{1D-2D} increases to 38 meV for AlAs barriers.

Quantum confinement effects appear irrespective of the crystalline or amorphous nature of the well material. For example, in amorphous Si/SiO_2 QWs quantum confinement effects on the electron wavefunction in the QW layer are evidenced through a strong blueshift of the PL spectrum with the decrease of the well width (Kanemitsu et al., 2000). However, the well-thickness dependence of the PL peak is different at room temperature and at low temperatures, of 10 K, due to the different mechanisms that predominate in the two temperature regions. At room temperature the PL is caused by radiative recombinations in deep states, and not in the near band edge states of amorphous Si. This is why the Stokes shift between absorption edge and the PL peak energy increases with increasing temperature, and is accompanied by a decrease in PL intensity. The PL spectrum from amorphous Si QWs is broad due to structural disorder in the well, fluctuations of the well thickness, and structural variations at interfaces. Not only does the PL intensity but also the PL lifetime, defined as $\tau_{PL} = (1/I_0)\int I(t)dt$, where I_0 is the intensity immediately after the laser pulse, depend on temperature. It shortens with increasing temperature, which suggests that PL decay is dominated by nonradiative recombination processes. Recombination processes at interfaces are also important, a fact demonstrated by the increases of τ_{PL} with the well thickness. This thickness dependence is not observed for PL energies higher than 2.3 eV, which means that the visible PL above 2.3 eV originates from defects in the SiO_2 layer or at Si/SiO_2 interface, as in Si nanocrystals in SiO_2 matrices. In the red spectral region τ_{PL} increases when the PL energy is decreased or the width of the well increases, which implies that the time-resolved PL decay reproduces the carrier diffusion and the relaxation of energy in the band-tail state. The photocarriers are rapidly thermalized in the extended states and recombine after the carriers are trapped in band-tail states.

Confinement effects contribute also to mixing of valence subbands states, leading to complicated far-IR (FIR) absorption spectra. In quantum dots with electrons, due to the parabolic form of the confining potential and of the momentum dependence of the electron kinetic energy, FIR absorption displays a single-particle-like spectrum, independent of the number of confined electrons (Darnhofer et al., 1995). In this case, the many-electron Hamiltonian separates the center-of-mass (COM) motion from the relative motion, and the FIR absorption couples only to COM quasiparticles. This property is a generalization of Kohn's theorem, which explains the analogous observation for cyclotron resonance (CR) in homogeneous electron systems. On the other hand, in quantum dots with holes the COM and the internal motion are coupled due to the degeneracy at the valence band edge where HH and LH bands coincide. The observed CR in the absorption spectrum deviates strongly in this case from the single peak structure and has a linear magnetic dependence that characterizes the CR in a 2DEG. The COM and internal motion are not coupled when there is no LH-HH mixing, in which case the two types of holes satisfy independently the generalized Kohn theorem.

The discrete electron spectra in quantum dots show very sharp features due to 3D confinement, the observed broadening being dominated by the size and shape inhomogeneities of QDs, which depend on the technological processes. The electron-phonon scattering process, which dominates the homogeneous lifetime broadening in QWs, and in which the finite lifetime electron state is scattered in other energy states by phonons, cannot occur in small QD because of the reduced scattering rate due to acoustical phonons. Instead, the homogeneous broadening in the single dot optical transitions can be explained by the lattice relaxation mechanism consisting of the coupling of an electron to acoustical phonons during optical transitions (Li and Arakawa, 1999). Lattice relaxation, treated in a weak-coupling approximation predicts that the homogeneous linewidth of an individual quantum dot with radius R has a linear temperature dependence $\Gamma_{\text{hom}} = \Gamma_0 + AT$, where $\Gamma_0 \approx 1/R$ and $A \approx 1/R^2$.

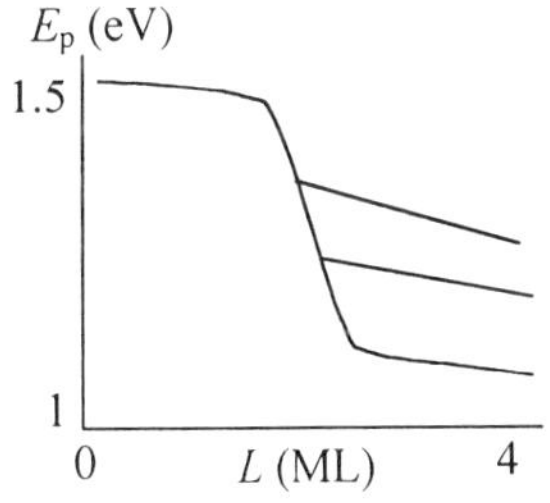

Fig. 6.11. The dependence of the quantum dot PL peak energy E_p on the thickness of deposited InAs

The blueshift of the PL peak in quantum dots with decreasing size is attributed to the lateral confinement and is characteristic of any mesosocopic structure, i.e. to quantum wells, wires and dots. Common trends in PL spectra of InAs quantum dots have been identified by comparing results obtained under different conditions (Grassi Alessi et al., 1999). It was found that the PL peak energy E_p depends on the thickness of deposited InAs as shown in Fig. 6.11. Around a critical thickness L_c a steep decrease of E_p with increasing L occurs independent of growth conditions, above this value the PL peak energy attaining a saturation value, which can assume only discrete values depending on growth condition, due to aggregation of quantum dots with different faceting. (In some quantum wire structures, for example $Cd_xZn_{1-x}Se/ZnSe$ the blueshift for small wire diameter is overcome in wires with larger diameter by the redshift due to strain relaxation).

In quantum dots not only can carrier wavefunctions become confined, but also optical or acoustic phonons. Moreover, due to the dramatic increase of the surface-to-volume ratio, surface phonons (SP) can appear in optical spectra. The SP can be detected in the frequency range between bulk LO and TO phonons, where the dielectric constant of the spherical nanoparticles $\varepsilon(\omega)$ is negative, and only SP can exist. The SP frequencies are obtained by solving the equation $\varepsilon(\omega) = -(l+1)\varepsilon_m / l$, with $l = 1,2\ldots$, where ε_m is the dielectric constant of the surrounding medium. The mode with $l = 1$, called the Fröhlich mode, is dipolar and corresponds to the uniform polarization of the sphere; the $l = 2$ mode is quadrupolar and, as the other higher-order modes, is more localized. The frequency of the SP modes, as well as the average particle size, can be determined from the first-order Raman spectrum $I(\omega) = I_c(\omega) + I_{SP}(\omega)$, the two terms corresponding to confined optical phonon modes and SP modes. From tight- binding calculations the Raman intensity of the confined optical mode j is $I_c^j(\omega) = A^j \int_0^{q_{max}} \mathrm{d}\boldsymbol{q}\,|C(0,\boldsymbol{q})|^2 / [(\omega - \omega^j(\boldsymbol{q}))^2 + \Gamma_c^2]$ where $\omega^j(\boldsymbol{q}) = \omega_0^j - \Delta\omega^j \sin^2(q/4)$ is the phonon-dispersion in bulk material, $|C(0,\boldsymbol{q})|^2 = \exp(-q^2 d_{av}^2 / 4a^2)$ is the Fourier coefficient of the phonon confinement function with d_{av} the average diameter and a the lattice constant. The contribution of SP is given by $I_{SP}^j = B^j \Gamma_{SP}^j / [(\omega - \omega_{SP}^j)^2 + (\Gamma_{SP}^j)^2]$. By fitting these theoretical formulae with experimental curves both d_{av} and ω_{SP}^j can be determined. Multiphonon Raman peaks can then be used to check the frequencies of SP modes. Such studies have been performed for example in CdS_xSe_{1-x} nanoparticles embedded in a glass matrix (Roy and Sood, 1996).

6.3.2 Optical Determination of Electronic Structure Parameters

In low-D structures, in particular in highly degenerate QWs the density of states can have singularities of a different kind from those encountered in bulk materials. For example, PLE experiments carried on highly p-doped $Al_xGa_{1-x}As/In_yGa_{1-y}As$ QWs revealed a second-order *van Hove singularity* in the joint density of states, which can appear when the effective mass of LH ground state is negative and equal to the electron effective mass near the Γ point. This singularity implies $\Delta_k E_c(\boldsymbol{k}) - \Delta_k E_v(\boldsymbol{k}) = 0$, a much restrictive condition than for the usual van Hove singularity, which demands $\nabla_k E_c(\boldsymbol{k}) - \nabla_k E_v(\boldsymbol{k}) = 0$. The joint density of states for the second-order van Hove singularity is given by $J(E) = A(E - E_0)^{-1/2} + B$

for $E > E_0$ and $J(E) = B$ for $E < E_0$, and shows $x^{-1/2}$ divergence at E_0. This singularity can be observed in PL and PLE experiments with circularly polarized light. Defining the polarization as σ^{ij}, where i, j refer to the emission and detection, respectively, and can take the values $-,+$ for left and right circular polarizations, respectively, it is known that PL and PLE for $\sigma^{-+}(=\sigma^{+-})$ are LH sensitive and are HH sensitive for $\sigma^{--}(=\sigma^{++})$. The divergence in PL corresponding to the second-order van Hove singularity was observed only in σ^{-+}; the PL peak could unambiguously be attributed to the singularity in the joint density of states, since high doping suppressed the alternative possibility of strong excitonic effects (Kemerink et al., 1996).

Optical experiments can also determine the main parameters of the electronic structure of QW, wires or dots. For example, the *valence band offset* defined as the ratio of the difference of absolute energy positions of the valence band maxima to the difference of the bandgaps of the isolated materials from which the heterostructure is made, can be determined in several ways: (i) based on an analysis of the splitting in energy between LH and HH excitons for different well widths, (ii) from optical transitions, which are highly sensitive to the valence band offset, in heterostructures designed to separately confine electrons and holes, or (iii) from the giant Zeeman splitting, i.e. from an analysis of the dependence of the energy position of QW excitons on the magnetic field or from the magnetic field splitting between the spin states of lowest electron levels in the conduction band of the well as a function of the well width, in heterostructures based on diluted magnetic semiconductors (DMS).

For example, in the type II $In_{0.52}Al_{0.48}As$/InP heterostructure, the observation of PL radiative recombination across the bandgap and between the InP conduction band and InAlAs valence band allows a direct measurement of the *change of valence band offset under pressure*. The bandgap hydrostatic deformation potential is usually determined from the hydrostatic pressure dependence of the bandgap measured by PL or optical absorption. However, individual band-edge potentials are more difficult to estimate since a pressure-insensitive reference level is very difficult to find.

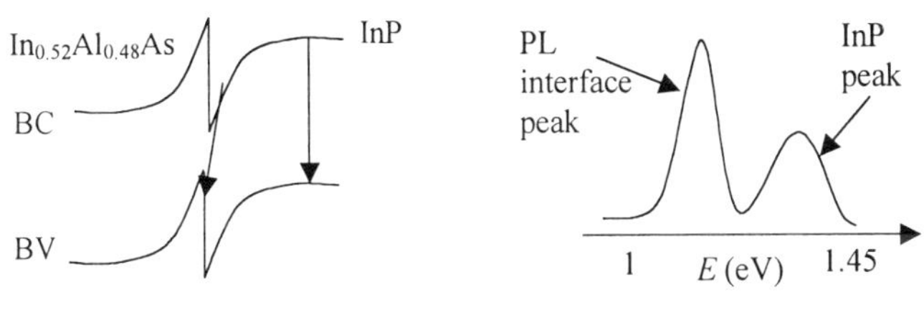

Fig. 6.12. Energy band diagram and the assessment of PL peaks in InAlAs/InP systems

The valence band offset is given by $\Delta E_{VBO} = E_g^{InP} - E^{inter} + E^{confin}$, where the first two terms are the energies of the transitions represented in Fig. 6.12, and the last term, usually negligible, represents the total confinement energies of electrons and holes. From PL measurements it was found that: $d\Delta E_{VBO}/dP = 0 \pm 0.4$ meV/kbar. The pressure dependence of the valence band offset can be expressed as $d\Delta E_{VBO}/dP = d/dP(\Delta E_v^{InAlAs} - \Delta E_v^{InP} + \Delta E_{is})$ where the first two terms are the valence band shifts under hydrostatic pressure and the last one is the interface strain; calculating the last term, and assuming that $\Delta E_{v,p} = 3a_v \Delta a / a_0$ with a_0 the unstrained lattice constant and Δa its change under pressure, the valence band hydrostatic deformation potential is found to be $a_v = -0.8$ eV (Yeh et al., 1995).

In self-assembled InAs/GaAs quantum dots the PL peak energy blueshift with increasing pressure up to 53 kbar, and then redshifts at higher pressure due to the $\Gamma - X$ crossover (Fig. 6.13a). At this critical pressure, a type I – type II band alignment crossover occurs, manifested in qualitative changes in the PL. The transition to the type II band alignment above the crossover pressure is accompanied by a decrease in the integrated line intensity and significant changes in the PL lineshape. Above the crossover pressure the PL energy is very sensitive to the optical intensity but no changes in the asymmetric lineshape are observed. The PL lineshape is very asymmetric, with a relatively sharp blue-edge (see Fig. 6.13b) above the crossover pressure, and becomes almost symmetric at low pressures. The $\Gamma - X$ crossover point can be taken as reference for electron and hole energy levels, which are found to be $E_h = 235$ meV and $E_e = 30$ meV after extrapolating to zero pressure the X and Γ energy levels (Itskevich et al., 1998).

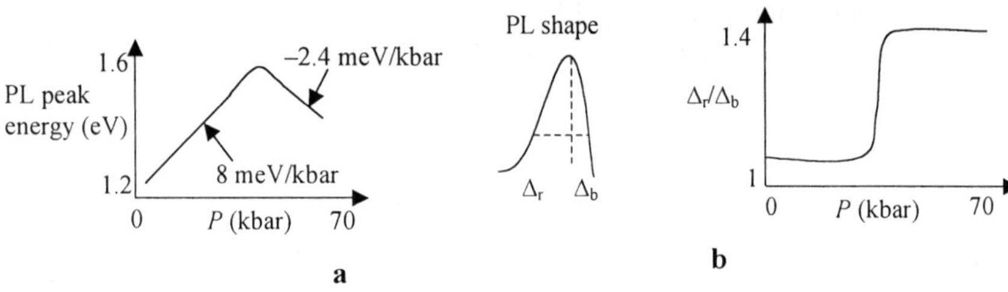

Fig. 6.13. (**a**) PL peak and (**b**) PL shape dependence on pressure in InAs/GaAs quantum dots

The situation is more complicated in CdTe/Cd$_x$Zn$_{1-x}$Te strained heterostructures, which show mixed type I/type II excitons depending on the type of holes. More precisely, the HH exciton is of type I, HHs being confined in the same layer as the electrons, and its energy is redshifted when an increasing electric field is applied. On the other hand LH excitons are of type II, the LH being confined in Cd$_x$Zn$_{1-x}$Te barriers whereas electrons are confined in the well; the type II exciton energy blueshifts with an increasing electric field. This type I/type II exciton mixing is only possible when the strain between wells and barriers is greater than the valence band offset. The valence band offset is determined from polarized PLE (PPLE), which measures the ratio of circular polarizations. A typical PPLE spectrum is shown in Fig. 6.14.

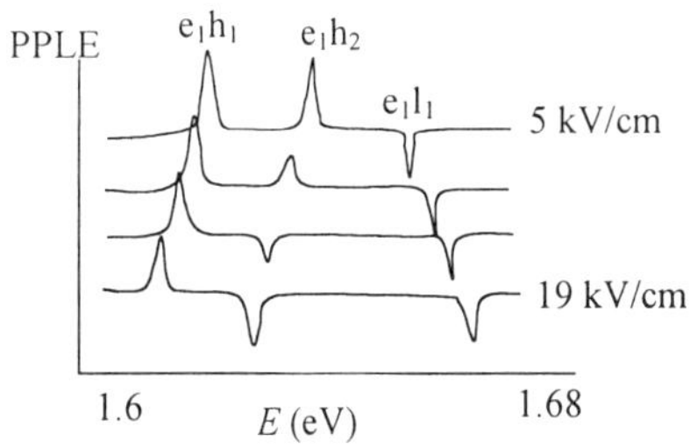

Fig. 6.14. PPLE spectrum in CdTe/Cd_xZn_{1-x}Te strained heterostructure

The experimental data show that at low electric fields the polarization ratio of e_1l_1 exciton peak is opposite to that of e_1h_1 and e_2h_2, reflecting the opposite behavior of LH and HH. At higher electric fields however e_1h_2 changes sign and becomes negative, as e_1l_1, slowing a mixing of HH and LH subbands for fields larger than a threshold of 12 kV/cm. The electric-field-induced mixing of LH and HH states is used to determine the valence band offset (Haas et al., 1997).

In ZnSe/$Zn_{1-x-y}Cd_xMn_y$Se QWs the excitons are of type I at zero magnetic field, but can become mixed exciton in the presence of a magnetic field normal to the layers. In this case the band edges split, such that the upper $\sigma^-(+1/2,+3/2)$ transition becomes of type II for high magnetic fields while the ground-state $\sigma^+(-1/2,-3/2)$ exciton is of type I at all field values. In general, in DMS-based heterostructures the large spin splittings in conduction and valence bands enhances the possibility of hole-spin population inversion via optical pumping and allows the modification of band offsets in magnetic fields. In particular, the lower transition becomes indirect (of type II) in the presence of the magnetic field because the $m_j = -3/2$ holes become confined in magnetic layers, whereas the upper $m_j = +3/2$ holes remain in the non-magnetic layers and the corresponding HH exciton is spatially direct (of type I). However, this is not a rule. In ZnSe/$Zn_{1-x-y}Cd_xMn_y$Se heterostructures the situation is reversed since for $B > 3$ T the $m_j = +3/2$ holes become confined in ZnSe barriers and the $m_j = 1/2$ electrons are localized in the same layer due to Coulomb attraction (Yu et al., 1997).

Magneto-optical methods can be used to determine the valence band offset in a CdTe/$Cd_{1-x}Mn_x$Te heterostructure (Siviniant et al., 1999). Measurements of the giant spin splittings of excitonic states for magnetic fields oriented either parallel or perpendicular to the growth axis showed that the spin splittings of the hole states confined in the QW are very sensitive to valence band offsets. The reason for this sensitivity is the possibility of controlling the potential profile in DMS through an external applied magnetic field. In particular, at zero magnetic field the electrons and holes in the $Cd_{1-x}Mn_x$Te/CdTe/$Cd_{1-x}Mn_x$Te structure are confined in the QW by the potentials $U_e(z)$ and $U_h(z)$, respectively. The barrier band energy can decrease or increase due to carrier-ion exchange interaction when a magnetic field is applied parallel to c axis, the corresponding energy shifts being comparable with the band offsets U_e and U_h. Knowing the value of the bandgap offset at zero magnetic field to be $\Delta E_g = 1574x$ meV, the experimental splittings observed in the Faraday configuration are compared with those obtained in the Voigt configuration and with the theoretical predictions. Calculations show that there is a unique set of parameters $Q_v = U_h/(U_e + U_h)$ and ΔL, the interface potential, which accommodate all data. These parameters have typical values 0.4 ± 0.05 and 1.6 Å, respectively.

The electron effective mass and the nonparabolicity also have different behaviors in low-D structures compared to bulk semiconductors. The effective mass is determined from the cyclotron mass measured in magnetoabsorption. In particular, the axial symmetry of QW induces an anisotropy of the effective mass, which has different values for the in-plane motion in the QW, $m_{\mathrm{eff}\parallel}$, and for the motion perpendicular to it, $m_{\mathrm{eff}\perp}$. The nonparabolicity, i.e. the dependence of the effective mass on energy, becomes important in narrowgap materials and narrow QW. In the last case the strong nonparabolicity of the conduction band originates in the penetration of subband electronic wavefunctions into the barriers caused by the fact that the quantization energies are comparable with the band offsets. In general, both in-plane and out-of-plane effective masses are energy dependent. In $Ga_{0.47}In_{0.53}As/InP$ QWs the cyclotron resonances observed in FIR magneto-absorption allow the determination of in-plane effective masses. In undoped samples the cyclotron mass $m_c = eB_{\mathrm{res}}/(\lambda/2\pi c)$ is determined at the bottom of the subbands, while for doped structures it is determined at the Fermi energy. To obtain the in-plane effective mass the cyclotron resonance peaks in the FIR absorption curve are extrapolated for $B = 0$. Experiments carried out for different well widths showed a 50% increase of the in-plane effective mass for the lowest well width of 15 Å compared to the bulk value of $0.044m_0$ (Wetzel et al., 1996). The increase of $m_{\mathrm{eff}\parallel}$ with the decrease of the well width is also predicted by the $\boldsymbol{k} \cdot \boldsymbol{p}$ model, which demonstrates that $m_0/m_{\mathrm{eff}\parallel} = [1 - 1.96E(\mathrm{eV})]/0.044$ where E is the energy of conduction band electrons.

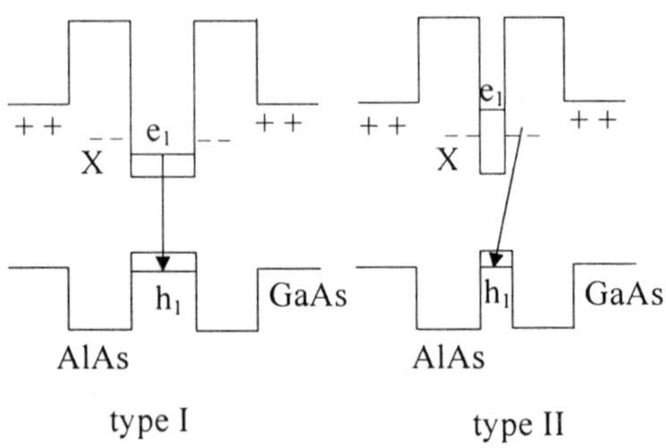

Fig. 6.15. Type I and II interband transitions in AlAs/GaAs heterostructures

The in-plane effective mass for the n-type modulation-doped $Al_{0.3}Ga_{0.7}As(Si)/AlAs/GaAs/AlAs/Al_{0.3}Ga_{0.7}As(Si)$ heterostructures was determined by magneto-PL. In this heterostructure the transitions can be of type I or II depending on the position of the e_1 level in the GaAs QW with respect to the X-valley minima in AlAs (see Fig. 6.15). In type I interband transitions the electrons are confined in the well, while in type II the electrons are located in the barriers and the transition is 'oblique'. The type of transition depends on the well thickness. Experiments in both cases have extracted the in-plane effective mass $m_{\text{eff}\|}$ from the slope $\mathrm{d}E/\mathrm{d}B$ of the PL peaks associated to the $l = 0$ and $l = 1$ Landau levels of the AlAs X valleys (Haetty et al., 1999).

Up-converted PL spectra of near IR-photons (near the E_g of GaAs) in visible photons (near the E_g of $GaInP_2$) in GaAs/(ordered)$GaInP_2$ type II heterostructures showed two peaks with energies linearly dependent on the applied magnetic field B. This linear dependence is only possible if the peaks correspond to free electrons-free holes recombination or for the recombination between a localized type of carrier and a delocalized, free carrier of the other type. The free electron-free hole recombination peak was excluded because the experimental value of the effective mass determined from the slope of the peak with respect to B did not agree with the theoretical value. The experimental values for the hole and electron masses, obtained from the peaks due to recombination of localized electrons and free holes and localized holes and free electrons, respectively, are $m_h = 0.24m_0$ and $m_e = 0.084m_0$ (Zeman et al., 1997). The PL up-conversion mechanism in this case is strongly correlated with the type II band alignment, and is a two-step, two-photon absorption process.

The cyclotronic resonance observed in transmission experiments through a 2DEG in Si metal-oxide semiconductor was investigated by Kaesen et al. (1996) in the extreme quantum limit, in high magnetic fields and for electron densities N_s satisfying $\nu = hN_s/eB << 1$, with ν the Landau filling factor. In this limit it was observed that the cyclotron mass has an oscillatory behavior with the filling factor for $\nu > 4$, with strong maxima at $\nu = 4n$ when the Landau levels are completely filled, and with weaker maxima at $\nu = 4n + 2$ when the spin-degeneracy is lifted (see Fig. 6.16). These data were obtained for a fixed magnetic field, the filling factor being modified by changing N_s through the gate voltage; the amplitude of the oscillation decreased with increasing B.

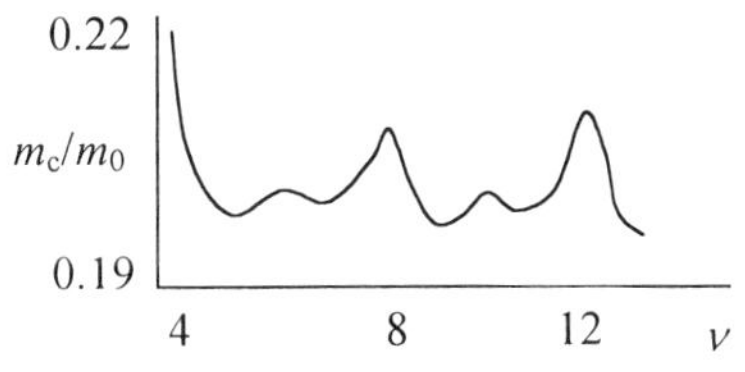

Fig. 6.16. The dependence of the cyclotron mass on the filling factor in a 2DEG Si MOS structure

These oscillations in the cyclotron mass were also observed in other semiconductor heterostructures at high filling factors and were attributed to the electron-hole interaction, the collective influence of the impurities or to nonparabolicity. For smaller filling factors, $1 < \nu < 4$, a splitting in the cyclotron resonance is observed, caused by the interplay of localization and electron-electron interaction. In all cases N_s and the filing factor were determined by measuring SdH oscillations. For $\nu > 4$ the lineshape of $1 - T(N_s)/T(0)$ was approximated with a Lorentzian profile and the cyclotron mass $m_c = eB/\omega_c$ was extracted from the resonance position that fitted the lineshape profile. The smallest width of this profile can also be used to determine the maximum scattering time, which gives the peak mobility μ_p. The peak mobility was found to range from 0.55 to 0.9 m^2/Vs in different samples.

The concentration of photogenerated electron-hole pairs can be determined from the excitation intensity dependence of PL. Such studies on narrow $\langle 001 \rangle$ and $\langle 111 \rangle$A grown $In_xGa_{1-x}As$/GaAs single QWs have revealed that the PL intensity has a linear dependence on the incident optical power density along these directions, while along $\langle 100 \rangle$ direction the blueshift of the PL peak and the band-filling effects, which increase with the excitation, suggest a strain-induced piezoelectric field of about 70 kV/cm. The PL linewidth for $\langle 100 \rangle$ depends on the excitation intensity as $\Gamma = \Gamma_0 + \beta I_{ex}^{2/3}$ due to the dependence of the quasi-Fermi level on the carrier density. From PL (along $\langle 100 \rangle$) and absorption measurements the density of photogenerated electron-hole pairs is given by $N \cong (\alpha\tau / E_{ex}) I_{ex}$ where $1/\tau$ is the carrier recombination rate (Sauncy et al., 1999).

When thermal quenching of the PL is due to nonradiative carrier loss from wells to barriers caused by thermal excitation, the change in PL intensity is modeled by an Arrhenius law in the 10–100 K temperature range followed by a strong quenching at 300 K (Vening et al., 1993). The PL intensity from each well i is then given by:

$$I_i = PR_i / \{(\beta_i + R_i / u_i)[R' + \textstyle\sum_j R_j / (\beta_j + R_j / u_j)]\}, \tag{6.43}$$

where P is the excitation rate in the barrier, $\beta_i u_i$ and u_i denote detrapping and trapping rate constants in the ith well, with $\beta_i = \exp(-E_i / k_B T)$ and E_i the depth of the confined state below the barrier band edge. R_i is the radiation rate constant from well i and R' the nonradiative recombination rate from the barrier. This expression for PL is obtained from the coupled-well rate equations. The *activation energies and the relative rates* R_i / u_i for each well are extracted from fitting the temperature dependence of the PL with the above analytical expression. Good fits are found in InGaAs/GaAs and GaAs/AlGaAs MQW where the thermal quenching of PL is due to emission of carriers to barriers, whereas in InGaAs/AlGaAs heterostructures the fitting procedure does not work well since the nonradiative loss mechanism is different (defect related).

To measure the carrier capture efficiency of electrons into QW, PL measurements have been performed on $GaAs/Al_xGa_{1-x}As$ QW doped with Be. The PL intensity ratio between the free-to-bound and excitonic transitions exhibited strong oscillations as a function of the QW width, which reflects the buildup of excess negative charge in the well due to different capture efficiencies of electrons and holes. The oscillations appear when virtual bound states are formed in the barrier continuum for well widths for which the highest QW level lines up with the band edge of the barrier, or when one of the QW levels is within one LO-phonon energy below the barrier band edge. In these conditions the matrix element of the electron-LO-phonon interaction, and thus the electron capture rate, is resonantly enhanced. The enhancement of the free-to-bound transition occurs only for excitations above the barrier bandgap, and the oscillation period becomes longer, the peaks shifting toward larger well widths, with decreasing x. The increase of the PL for free-to-bound transitions is accompanied by a decrease of the excitonic transition, the total PL intensity being almost constant. Simple calculations show that the PL intensity oscillations imply that the electron and hole concentrations oscillate with the well width, which can only occur when the charge neutrality is broken within the QW. This is what happens at resonant electron capture, i.e. charge neutrality breaks through preferential capture of electrons. The subsequent accumulation of negative charge in the QW leads inevitably to band bending and an increase of the hole capture rate until they become equal in the steady state (Muraki et al., 1996).

From the circular polarization of time-resolved PL one can determine the spin configuration of the ground state of a 2D-electron system, for different values of the filling factor ν. Measurements of the PL polarization degree $\gamma = (I_- - I_+)/(I_- + I_+)$ at long time delays such that the photoexcited holes reach equilibrium, show that the electron system becomes fully spin-polarized around $\nu = 1$. The PL intensities for σ^+ and σ^- polarizations, are given by the population of the corresponding levels: $I_+ \approx (n^{\mathrm{e}}_{+1/2} n^{\mathrm{h}}_{+1/2} + 3 n^{\mathrm{e}}_{-1/2} n^{\mathrm{h}}_{+3/2})$, $I_- \approx (n^{\mathrm{e}}_{-1/2} n^{\mathrm{h}}_{-1/2} + 3 n^{\mathrm{e}}_{+1/2} n^{\mathrm{h}}_{-3/2})$, the factor of 3 accounting for the difference in matrix elements for the optical transitions involving $3/2$ and $1/2$ hole states. The populations of the hole sublevels are in the ratio $n^{\mathrm{h}}_{-3/2} : n^{\mathrm{h}}_{-1/2} : n^{\mathrm{h}}_{+1/2} : n^{\mathrm{h}}_{+3/2}$ $= 1 : \exp(-\delta/T) : \exp[-(\delta+\Delta)/T] : \exp[-(2\delta+\Delta)/T]$, where δ, Δ are splitting energies (expressed in terms of k_{B}) proportional to B. At zero magnetic field, the Landau levels are fully occupied $n^{\mathrm{e}}_{+1/2} = n^{\mathrm{e}}_{-1/2}$ so that

$$\gamma_0 = \frac{3 + \exp(\delta/T) - \exp[-(\delta+\Delta)/T] - 3\exp[-(2\delta+\Delta)/T]}{3 + \exp(-\delta/T) + \exp[-(\delta+\Delta)/T] + 3\exp[-(2\delta+\Delta)/T]}, \tag{6.44}$$

whereas for $\nu = 1$, in the spin-polarized case $n^{\mathrm{e}}_{-1/2} = 0$ and γ_1 $= \{3 - \exp[-(\delta+\Delta)/T]\}/\{3 + \exp[-(\delta+\Delta)/T]\}$. These expressions can be further simplified for the limit $\delta/T \ll 1$, $\Delta/T \ll 1$. Measurements of the polarization degree of the PL in the two cases allow the determination of δ, Δ. The obtained values can be checked, by noting that the PL lines for the σ^+ and σ^- polarizations are shifted in energy with $\Delta E = \delta + \Delta$, which increases linearly with B (Kukushkin et al., 1997).

The radius of quantum dots can be estimated from light scattering measurements. Assuming that the optical vibrations are confined in spherical dots with radius R, the equivalent wavevectors are $q_n = \mu_n / R$, where μ_n is the nth node of the spherical Bessel function J_1. The negative bulk LO-phonon dispersion in the quadratic approximation is then $\omega_n^2 = \omega_L^2 - \beta_L^2 (\mu_n / R)^2$ with ω_L the bulk LO-phonon frequency at the Brillouin zone center. In resonant Raman spectra, a set of quantum dots with radii R_i are selected in either the incoming or outgoing resonances whenever $\hbar\omega_i = E_{\mu 1}(R)$ or $\hbar\omega_f - \hbar\omega_n(R) = E_{\mu 2}(R)$. The main contribution to the resonant Raman spectra for quantum dots with incoming or outgoing resonant radius R_e comes from dots with radii $|R_e - R| < \delta R$. Knowing the dot radii from absorption measurement of, for example, the $n = 1$ exciton state, δR is determined from the linewidth $\Gamma_\mu = (\partial E_\mu(R)/\partial R)\delta R$. For a Gaussian ensemble of quantum dots, an asymmetric broadening of the Raman line on the low-frequency side is expected. The dependence on the size distribution of the lineshape of Raman phonon modes in CdSe quantum dots was studied in Trallero-Giner et al. (1998). Alternatively, the phonon dispersion constant β_L can be determined from the shift of phonon frequency from the bulk value. Apart from the negative phonon dispersion discussed above, which causes an increased redshift with decreasing size, a blueshift of the Raman peak can occur due to lattice contraction. The compressive strain can even overcome the redshift due to phonon confinement. The peak shift due to lattice contraction is given by $\Delta\omega_c(R) = \omega_L[(1 + 3\Delta a(R)/a)^{-\gamma} - 1]$ with γ the Grüneisen parameter, and $\Delta a(R)/a$ the change in the lattice constant, determined from X-ray data. This contribution must be accounted for in CdSe dots embedded in glass matrices, for example, while in the GeO_2 matrix this contribution is negligible (Hwang et al., 1996).

6.3.3 Determination of Internal Electric Fields

The low-temperature PL dependence on optical excitation in single QWs allows the determination of internal electric fields. When an electric field is applied normal to the heterostructure, the absorption strongly shifts in energy and decreases in intensity due to the polarization of electrons and holes in QWs. This effect is called the quantum confined Stark effect (QCSE) and has important applications in the implementation of advanced optical modulators (Dragoman and Dragoman, 1999). Under intense optical excitation the photoinduced carriers screen the internal electric field, inducing an energy change of the QW states, seen in PL as a strong shift of the peak position with increasing optical intensity. In Fig. 6.17 this effect is shown schematically for the case of a single QW in a InGaAs/GaAs heterostructure.

This nonlinear behavior of the emission peaks of free and bound excitons is not observed in bulk GaAs or epilayers. The dependence of the internal electric field on the illumination P can be modeled as

$$F_s = \{F_{s0}^2 - (F_{s0}^2 / eV_{s0}) \ln[bPg(1-R)/\hbar\omega + 1)]\}^{1/2}, \tag{6.45}$$

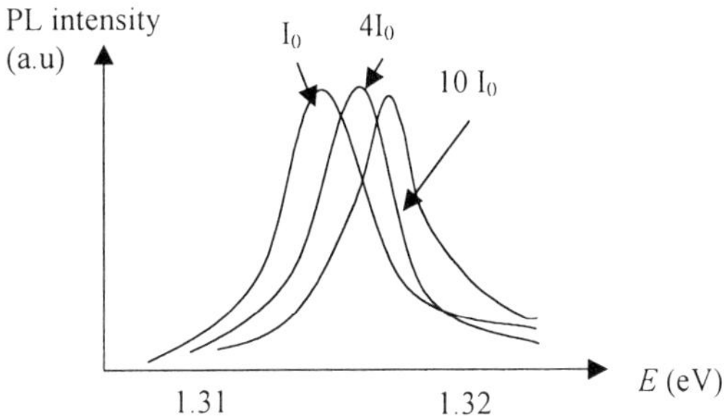

Fig. 6.17. QCSE effect on PL in InGaAs/GaAs heterostructures

where V_{s0} is the surface potential in the absence of photocarriers and F_{s0} the internal electric field in the same conditions, g is the quantum efficiency, R the reflectivity, $\hbar\omega$ the photon energy and $b = \exp(eV_{s0}/k_BT)(e/AT^2)$ with A the Richardson constant. The PL peak energy has a quadratic dependence on the internal electric field, its energy varying as $E_{PL} = E_0 - \kappa F_s^2$, where E_0 is the PL energy in the absence of an internal electric field, and κ is a constant. So, the internal electric field F_{s0} can be determined from the shift in the PL peak, experimental values for this parameter in $In_xGa_{1-x}As$/GaAs heterostructures at 12 K being around 44 KV/cm (Chatov et al., 1996).

Time-resolved PL experiments of excitons can equally well probe the internal electric fields. Such studies have been performed, for example, on GaN/(GaAl)N QWs (Lefebvre et al., 1999). The electric fields in QWs are determined by both the strain induced through the piezoelectric effect and by the competition between spontaneous and piezoelectric polarizations in the well and the barriers. The total electric fields in wells and barriers for an ideal heterostructure are given respectively by

$$E_w = [L_b/(\varepsilon_b L_w + \varepsilon_w L_b)][|P_w^{sp}| - |P_b^{sp}| - |P_w^{pz}| - |P_b^{pz}|], \tag{6.46a}$$

$$E_b = [L_w/(\varepsilon_b L_w + \varepsilon_w L_b)][|P_b^{sp}| - |P_w^{sp}| + |P_w^{pz}| + |P_b^{pz}|], \tag{6.46b}$$

where the superscripts sp and pz denote the spontaneous and piezoelectric polarizations, respectively. These fields favor the carrier transfer across thin GaAlN barriers, influencing the PL decay time.

In materials with large piezoelectric coefficients, such as for example in GaN quantum dots, the lateral confinement competes with the piezoelectric effect. In this case (see Fig. 6.18) for large quantum dots the PL peak can even be below the bulk GaN bandgap. The PL peak is due to intrinsic quantum dot emission, as evident from the constancy of its shape, linewidth and intensity over the 2–300 K temperature range. The emission energy depends on the piezoelectric field as $E = E_g + E_e + E_h - E_{pz}$, with $E_{e,h}$ confinement energies of electrons and holes. The fact that the PL peak energy can be even lower than in bulk is due to the giant

5.5 MV/cm piezoelectric field which appears in noncentrosymmetric materials with wurtzite structure. One method to confirm the piezoelectric origin of the redshifted PL peak compared to the bulk value is to perform studies for different excitation powers. When the piezoelectric field is present, the PL peak blueshifts with increasing excitation power due to its partial screening by the photogenerated electron-hole pairs. Such a blueshift of 70 meV between 60 and 450 W/cm^2 was also observed in GaN quantum dots (Widmann et al., 1998).

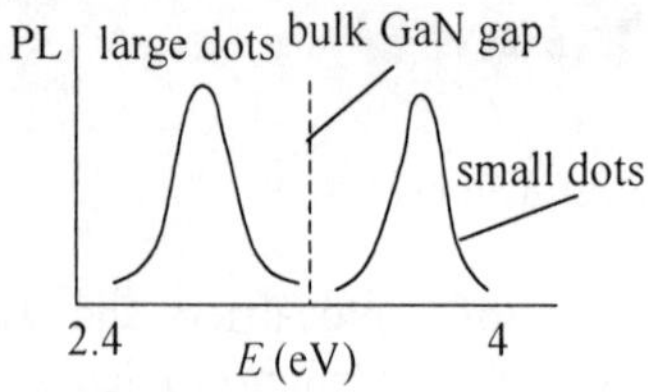

Fig. 6.18. PL spectra of GaN quantum dots with different size

6.3.4 Diffusion

PL can also be used to measure the vacancy diffusion coefficient in III-V materials repeatedly annealed at a given temperature. In In$_x$Ga$_{1-x}$As/GaAs MQWs the vacancy diffusion is evidenced by a shift in the PL peak to higher energies after each annealing, as the QW effectively narrows due to interdiffusion, followed by a reduction of the In concentration at the well center at later stages of diffusion (Khreis et al., 1997). The *diffusion length* after each annealing is determined from the shift in the PL peak position, assuming Fick's law. Then, plotting the square of the diffusion length versus the annealing time, the diffusion coefficient is deduced as the gradient of the graph. The diffusion length is given by $L_D^2 = 4D_v \int_0^t N_v dt$ where D_v is the diffusion coefficient for vacancies, and $N_v = N_{vS} + N_{vB}$ is the ratio of vacancies to the number of sites, with N_{vS} the vacancy concentration due to the diffusion source and N_{vB} the background thermally activated vacancy concentration. Writing

$$N_{vS}(x,t) = (N_0/2)\{\mathrm{erf}[(d/2 - x_0)/2(D_v t)^{1/2}] + \mathrm{erf}[(d/2 + x_0)/2(D_v t)^{1/2}]\}, \tag{6.47}$$

where N_0 is the initial concentration of vacancies in the layer with width x, and x_0 the depth of the layer from the surface, the parameters D_v, N_0 and N_{vB} are obtained by fitting the PL dependence on the annealing time. Repeating the measurements for different temperatures, the activation energy for vacancy diffusion can be obtained from the plot of the logarithm of the diffusion coefficient versus $1/T$ (the background vacancy concentration is temperature independent).

The effect of thermally induced compositional disordering on PL from 2D and 0D $In_{0.5}Ga_{0.5}As/GaAs$ structures was studied to determine the diffusive coefficient and the activation energy (Leon et al., 1998). The experimental results show that the PL peaks have much larger blueshifts in QDs compared to QWs under similar conditions. The diffusion coefficient and activation energies are extracted from PL shifts using a fitting procedure as above. Interdiffusion affects the lifetime in recombination processes through a decrease of the activation energy E_a for radiative recombinations. The latter parameter can be determined from the quenching of PL with temperature, which is described by $I / I_0 \approx 1/[1+\rho(-E_a / k_B T)]$, where ρ is related to the recombination lifetime.

6.3.5 Optical Properties of Excitons

The binding energy of excitons is approximately given by the Stokes shift between the absorption and PL line. In QW excitons can become localized due to potential fluctuations at the interfaces, on a length scale comparable to the exciton Bohr radius. As in 3D electronic states where there is a sharp boundary between localized and delocalized states, the so-called mobility edge E_m, excitonic states can be categorized in states with values of the center-of-mass localization length on the order of the diffusion length (delocalized) and those with much smaller values of this parameter (localized). The mobility edge in this case is situated in the low-energy part of the exciton absorption spectrum. It can be evidenced in time-integrated FWM by monitoring the homogeneous linewidth Γ of the exciton, when excited with ps pulses that shift across the absorption line: Γ is small below the absorption line center due to the localized character of excitons, and rises monotonically above the absorption line center. With increasing temperature the Stokes shift between the absorption and PL peaks decreases due to thermal activation of localized excitons to higher energy states (Braun et al., 1998a).

When the well width is smaller than twice the exciton Bohr radius, i.e. when $L_w \leq 2a_{ex}$, we are in the strong-confinement regime where the quantum confinement affects primarily the individual electron and hole wavefunctions. In the opposite case – the thin-film regime – the quantum confinement has negligible affect on individual carriers, but affects the exciton envelope function. In this case the center-of-mass motion of the exciton, in particular the exciton momentum, becomes quantized along the growth direction, and the exciton polariton becomes confined. The center-of-mass quantization is seen as multiple resonances in the optical spectra at energy levels equal to the quantized levels of a particle with mass equal to the effective exciton mass in a 1D QW: $E_n = E_0 + (\hbar^2 / 2M_{ex})(\pi / L_{eff})^2 n^2$ with L_{eff} the quantization length. This relation is strictly speaking valid for $L_w > 10a_{ex}$, when the relative electron-hole motion and the center-of-mass motions are decoupled, in the intermediate region $2.5a_{ex} < L_w < 5a_{ex}$ the QW exciton being treated with the effective mass approximation by variational envelope functions (Greco et al., 1996). The center-of-mass quantization has been observed for both HH and LH excitons in

ZnTe-(Zn,Mg)Te QW at $T = 2$ K, the two types of excitons being distinguished in piezomodulated reflectivity by their strain sensitivity (Lefebvre et al., 1997).

The binding energy and oscillation strength of quantum-confined excitons can be determined from the optical absorption. For example, in strained (Ga,In)Sb/GaSb quantum wells, distinct exciton resonances have been observed in absorption, although the electron-hole interaction is weak; since the absorption resonance was almost coincident with the PL peak, the excitons have been identified with free excitons. The value of the oscillator strength of the HH exciton was found to be $f = 13\times10^{-5}$ Å^{-2}, a value obtained from the area of absorption resonance, proportional to the oscillator strength. The exciton binding energy, given by $E_b = \pi\hbar^2 E_g f / 16 m_r E_P$, is then estimated as $E_b = 2.5$ meV, where m_r is the reduced exciton mass, and E_P is the Kane matrix element. With increasing excitation density, the PL peak energy shifts almost logarithmically, which suggests an initial saturation of localized states that dominate the emission at low excitation powers, followed by phase-space filling at higher excitation levels. Due to the thermionic emission of free excitons out of the quantum well, the PL intensity drops for temperatures higher than 50 K, the slope of the PL curve with temperature determining the activation energy of this process (Bertru et al. 1997).

For quantum dots occupied by single excitons, the PL linewidth decreases continuously with reducing diameter. This behavior, observed in $In_{0.14}Ga_{0.86}As$/GaAs quantum dots with sizes ranging from excitonic Bohr radius up to 2D structures, was related to compositional fluctuations. The PL exciton linewidth is narrow at low excitation powers, when the dot is occupied by a single photogenerated electron-hole pair, and broadens for higher powers due to simultaneous occupation of the dot by several electron-hole pairs. In this case biexcitons can form when the ground state is occupied by two excitons. At high excitation powers the dependence of PL linewidth on the dot diameter is opposite to the low excitation case: it increases continuously with reducing dot size (Steffen et al., 1996).

The shape of quantum dots can be determined by studying the quantized exciton levels. If site-selective spectral hole burning experiments are performed on quantum dots, the fundamental excitonic state, as well as satellite holes due to excited exciton states appear. These holes appear at excited exciton energies given by $E_{n_x,n_y,n_z} = E_{ex} + \hbar^2\pi^2(n_x^2 + n_y^2 + n_z^2)/[2M(L - a_{ex})^2]$ for cubic quantum dots with size L, Bohr radius a_{ex}, bulk exciton energy E_{ex} and translational mass M. (Here L was replaced with $L - a_{ex}$ to account for dead-layer correction.) On the contrary, if the quantum dots are spheres, the quantized exciton levels are at $E_{n,l} = E_{ex} + \hbar^2\pi^2\varsigma_{n,l}^2/[2M(R - a_{ex}/2)]$ with $\pi\varsigma_{n,l}$ the nth root of the spherical Bessel function of lth order. So, from the positions of the satellite holes it is possible to determine whether the quantum dots have cubic or spherical shape. In reality, the experimental data correspond neither to cubic or spherical shape, but to something in between (Sakakura and Masumoto, 1997).

Confined excitons are one of the most efficient probes of interfacial quality of MQW, due to their sensitivity to the structural aspects of interfaces on the atomic scale (de Oliveira et al., 1999). For example, in GaAs/$Ga_{0.7}Al_{0.3}$As both free and bound excitons peaks (FE and BE) were observed in PL spectra, only the FE peak showing a Gaussian inhomogeneous broadening due to structural patterns of the interfaces, whereas the BE peak associated to excitons bound to defects localized at QW interfaces showed a Lorentzian broadening. The total FWHM of the PL line is due to both alloy fluctuations and interface microroughness caused by defects with smaller extensions than that of the lateral dimension of the confined exciton: $\sigma_{tot} = [\sigma_{alloy}^2 + \sigma_{int}^2]^{1/2}$. The contribution of the interface microroughness is given by $\sigma_{int} = 0.59\delta_1\delta_2 / m_r L_z^3 R_{ex}^2$, where δ_1, δ_2 are, respectively, the height and lateral dimension of the interface islands, m_r and R_{ex} are the reduced mass and the lateral dimension of the exciton, respectively, and L_z is the quantum well width. The contribution due to alloy fluctuation is modeled by $\sigma_{alloy} = 2[1.4x(1-x)r_c^3 / R_{ex}^3 P_0^{ex}](\delta E_{ex} / \delta x)_{x=x_0}$ where x is the Al concentration and x_0 the nominal concentration, r_c is the radius associated with the volume per cation, $P_0^{ex} = 1 - L_z^3 / R_{ex}^3$ the fraction of exciton volume that penetrates into the barrier, and E_{ex} the excitonic transition energy. R_{ex} can be modified (decreased) by applying a magnetic field parallel to the growth direction, in which case it is obtained from the diamagnetic shift of the fundamental excitonic 1s state $\Delta E = (e^2 B^2 / 8m_{xy}) R_{ex}^2$ with m_{xy} the in-plane exciton mass. Fitting the experimental linewidth with the expressions given above, it was found that for the sample under study $\sigma_{tot} \cong \sigma_{int}$; the parameter δ_2, for example, can be obtained from fitting, if all other parameters are known (de Oliveira et al., 1999).

Interesting optical responses are observed in mixed type I-type II GaAs/AlAs MQWs. These structures consist of alternating GaAs narrow and wide wells separated by AlAs barriers and designed such that the 2DEG is confined in the wide well and the 2DHG (2D hole gas) is confined in the narrow well. The recombination time of electrons and holes is thus significantly longer than in usual MQWs, and the density n_e of the 2DEG can be varied without applying a bias, just by varying the photoexcitation intensity. Unlike the spectra of ordinary MQW, which are independent of the excitation intensity, the spectra of the mixed type QWs depend on the excitation intensity. In particular, the (e1:hh1) and (e1:lh1) excitonic transitions broaden and weaken with increasing n_e, but their energy is unchanged. Using the transfer matrix formalism to calculate the spectra of excitonic transitions and the free electron-hole interband transitions, and by comparing them with experimental data, the dependence of exciton energies on n_e, the onset energy of the free electron-hole interband transition, the exciton-photon interaction strength and the exciton damping can be determined (Harel et al., 1996).

The exciton formation time in Al_xGa_{1-x}As/GaAs quantum well structures has been studied by magneto-PL measurements at low temperatures (Zhu et al., 1995). Measurements for the applied magnetic field B parallel to the growth direction showed that the exciton formation time is dominated by the time during which electrons and holes move toward each other attracted by the Coulomb force. The excitons are formed after the carrier system is cooled down and their formation time is a function of the carrier density. Rate equations show that when $\tau_e < \tau(B)$, where τ_e and τ are the formation times of electrons and excitons, at small magnetic fields

$$I_{PL}^{-1} = A(2^{3/2}\pi\varepsilon/3e^2 n^{5/2})(m_e/\tau_{ef})[1+(\tau_e e/2\pi m_e)^2 B^2], \quad (6.48)$$

where A is a constant determined by the optical system, n is the electron density ($n = p$) and τ_{ef} is the mean free time of electrons. A similar relation exists for large B, with τ_{ef}/m_e replaced by τ_{hf}/m_h. The values of these ratios are obtained by fitting the PL curves at high and low magnetic field values, the formation times of excitons then being calculated from

$$r_0^3/3 = (e^2/8\pi\varepsilon)[(\tau_{ef}/m_e)/(1+\tau_{ef}^2/\tau_{ec}^2) + (\tau_{hf}/m_h)/(1+\tau_{hf}^2/\tau_{hc}^2)]\tau(B), \quad (6.49)$$

where τ_{ec}, τ_{hc} are the electron and hole cyclotron periods in the quantum well and $r_0^{-2} = 2n$. τ increases dramatically with increasing B. The relation between the formation time and the carrier density suggests that for small n the nonradiative recombination dominates over the carrier loss, changing the PL intensity with B due to the decrease of the carrier fraction that forms excitons (Zhu et al., 1995).

The dependence of the exciton decay time τ_{dec} on the kinetic energy of electrons in shallow $In_xGa_{1-x}As$/GaAs QW has been studied by PL. The decay time reflects the electron transfer process into the ground state due to different phonon mechanisms. The kinetic energy was varied by exciting the sample above the GaAs band edge, at different excitation energies. As expected from theoretical considerations, τ_{dec} had an oscillatory behavior as a function of the excitation energy, with maxima at $E_{max} = E_0 + (n+1/2)\hbar\omega_{LO}(1+m_e/m_{hh})$, n = 4,5,.. and minima at $E_{min} = E_0 + n\hbar\omega_{LO}(1+m_e/m_{hh})$, n = 5,6,... with $\hbar\omega_{LO}$ = 36.4 meV. This behavior is explained by the fact that the excess energy ΔE is shared by electrons and holes, under the requirement of energy and momentum conservation. The electron excess energy is lost rapidly in n steps due to LO-phonon scattering, such that finally $0 \le \Delta E_e \le \hbar\omega_{LO}$, the electron reaching the QW ground state after n LO-phonon emission and the radiative recombination of the subsequently formed exciton if $\Delta E_e = n\hbar\omega_{LO}$. For other excess energy values a further cooling is needed to reach the ground state, which manifests as a radiative delay of the excitons compared to the decay time when $\Delta E_e = n\hbar\omega_{LO}$ (Kovac et al., 1996).

The exciton binding energies increase for lower-dimension structures, such as quantum wires and dots. Measurements of PL peak shift with an applied magnetic field in the Faraday configuration showed that for $B \le 4$ T the peak has a diamagnetic shift to higher energies described by $\Delta E = \gamma_2 B^2$. Assuming that the exciton wavefunction is an ellipsoid of revolution characterized by three effective anisotropic Bohr radii a_{ex}^i and reduced exciton masses m_{ri} along the principal axes i = x, y, z, the perturbation theory gives $\gamma_2 = 4\pi\hbar^2\varepsilon^2/e^2 m_r m_{rx} m_{ry}$ with $1/m_r = (1/3)(1/m_{rx} + 1/m_{ry} + 1/m_{rz})$. By comparing the value of γ_2 in the quantum well with that in the quantum wire and dots (etched from the quantum well) the in-plane reduced masses m_{rx} and m_{ry} can be obtained. For quantum wires it was found that only the mass normal to the wire is modified due to confinement, whereas in quantum dots $m_{rx} \cong m_{ry}$. The confinement affects the excitonic properties for structure sizes up to about $10a_{ex}$. From these data two other quantities can be calculated: the expectation value of the exciton radius in the well plane, obtained from $\gamma_2 = (e^2/8m_r)\langle x^2 + y^2 \rangle$ with x, y the coordinates of the center-of-mass motion, and $E_b = e^4 m_r / 32\pi^2\hbar^2\varepsilon^2$ (Bayer et al., 1998).

Although the linear optical processes such as absorption and PL give us a hint about the changes in excitonic properties when the dimension changes between 1 and 2, the nonlinear, two-photon absorption (TPA) of Wannier excitons, is much more sensitive to the change in dimensionality. For example, the TPA spectrum of quantum wires has a strong anisotropy as regards the energy position of the peaks for directions parallel and normal to the wire direction, but is almost isotropic with regard to the strength of the peaks. As the cross-section of the wire increases (as the dimensionality increases) the TPA spectrum approaches an isotropic spectrum with respect to both peak position and strength. Similar behaviors are encountered in quantum wells and dots (Ogawa and Shimizu, 1993).

The behavior of exciton transitions in quantum dots in the presence of an electric field F was studied by Heller et al. (1998). They observed the well-known Stark effect, consisting in a shift to lower energies of the exciton peak proportional to F^2 and a drop in intensity due to electron and hole separation, determined by the strength of the lateral electric field. The study was performed on both the ground state of the localized exciton (that which gives the peak in PL) and excited state of the localized exciton (that which peaks in PLE). It was found that the Stark shift is stronger for the excited state, since higher states are more extended than ground states. The Stark shift is accompanied by an increase in the peak linewidth, corresponding to a reduced exciton lifetime in the applied electric field, due to tunneling of carriers out of the dot. For a particle in a rectangular potential with extent a and depth V_0 the linewidth increase due to tilting the potential in the electric field is modeled in the WKB approximation by $\Gamma \geq \Gamma_0 + (\hbar^2/8ma^2)\exp[-(4/3)(2m)^{1/2}V_0^{3/2}/eF\hbar]$ where Γ_0 is the linewidth when $F = 0$. By fitting with experimental data it was found that $V_0 = 15$ meV and $a = 90$ Å. When exciting the sample with high powers both the ground and excited exciton states shift to higher energies, due to field screening when the dot is occupied by several excitons. Since the blueshift of the excited state, which has a lower population, is smaller, it is inferred that the screening is due to carriers in the dot and not those in barriers, which would have produced the opposite effect.

The exciton dephasing can be studied with FWM (Wagner et al., 1997). Two components have been identified in the transient FWM: a prompt one caused by the response of spin-coupled exciton states, which dominates for cross-linear polarized fields, and a delayed photon-echo (PE) component due to the distribution of localized noninteracting excitons. The two components have different polarization selection rules: for $\uparrow\rightarrow$ polarization (where the first sign indicates the direction of linear incident polarization and the second the polarization of the scattered radiation) the PE component decreases drastically compared to the ↑↑ polarization, whereas the prompt component reduces only by a factor of 2. Both components are suppressed about ten times for the $(\sigma^+\sigma^-)$ polarization compared to $(\sigma^+\sigma^+)$, the signal in the latter case being almost the same as for the ↑↑ case. Studying the dependence of the homogeneous linewidth on the exciton density n_X, the exciton-exciton scattering parameter β_{XX} can be determined from $\gamma_{hom}(n_X) = \gamma_{hom}(0) + \beta_{XX}^{mD} a_{ex}^m E_b n_X$, where m is the dimensionality of the system and E_b the exciton binding energy. For QWs with different well widths, and with $\gamma_{hom} = h/(\pi T_2)$ determined from FWM, β_{XX}^{3D} and β_{XX}^{2D} have been determined in ZnS_xSe_{1-x}/ZnSe structures. The temperature dependence of the homogeneous linewidth for a given exciton density can be expressed by $\gamma_{hom}(T) = \gamma_{hom}(0) + \beta_{ac}T + \beta_{LO}/[\exp(E_{LO}/k_BT) - 1]$, the parameters β_{ac}, β_{LO} resulting from fitting. In $In_{0.135}Ga_{0.865}As$/GaAs quantum wires the exciton-exciton scattering rate was found to increase with decreasing wire width, due to the increase of the exchange part of the exciton-exciton interaction with increasing confinement (Braun et al., 1998b).

Charged states of excitons can also be observed in optical spectra. The negatively charged exciton for example, X^- is formed from two excitons bound to a hole, and appears in absorption or emission spectra as a line below the neutral exciton. In can be observed in those systems where an electron reservoir is present in the well (from doping, current injection or electron-hole spatial separation in mixed-type crystals), which can bind the photoexcited electron-hole pair to form the X^-. Analogously, X^+ can be created in p-type modulation-doped structures or by current injection of holes. In magnetic fields X^- has a diamagnetic shift similar to a neutral exciton, and a Zeeman splitting different from it, but similar to that of 2DEG. In strong magnetic fields the two electrons of the X^- can occupy a triplet state, a new line X_t^- appearing when the magnetic length is smaller than the diameter of X^-. The recombination of one electron and the hole in X^- can be accompanied by the transfer of the second electron to a higher energy state. This process is called shakeup and can be observed in optical spectra as a fan of discrete lines associated with different Landau levels when the higher energy state is a higher Landau level. The separation between these lines follows the dependence of $\hbar\omega_c$ on B (see Fig. 6.19).

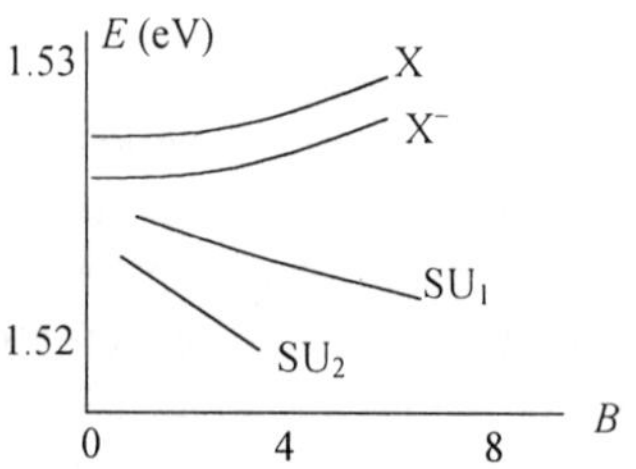

Fig. 6.19. Peak position dependence of magnetic fields in GaAs/AlGaAs systems

The two shakeup lines shown in Fig. 6.19 as SU_1 and SU_2 have different slopes; in fact SU_2 has twice the slope of SU_1. The energy difference between SU_1 and SU_2, and X^- and SU_1, of about 1.55 meV/T in GaAs/Al$_x$Ga$_{1-x}$As is similar to that expected for $\hbar\omega_c$ (Finkelstein et al., 1996).

Studies of X^- as a function of the electron density n_e, varied by applying a voltage on a CdTe/CdZnMgTe quantum well in fixed magnetic field have shown that the absorption intensity of X^- depends on the filling factor (Lovisa et al., 1997). More precisely, the triangular-shaped X^- absorption line, which is observed only in σ^+ polarization, increases almost linearly in intensity with n_e, reaches a maximum at Landau level filling factor $\nu = 1$, and then decreases linearly, disappearing at $\nu \cong 2$. This linear dependence is more accurate for high magnetic fields. The explanation of this behavior is that the X^- intensity is proportional to the number of electrons in the lowest Landau level (with degeneracy $2g$) when $0 < \nu < 1$, and proportional to the number of holes in this level, $2g - n_2$, when $1 < \nu < 2$. So, $I \approx n_e^+ n_h^-$ with n_e^+, n_h^- the number of electrons in the lowest and next lowest sublevels of the lowest Landau level, corresponding to $m = 1/2$ and $m = -1/2$.

Time-resolved PL studies of the decay time of charged excitons in GaAs QW showed that this parameter increases by an order of magnitude in high magnetic fields. This effect was explained by the fact that, unlike neutral excitons that move freely in magnetic fields, the center-of-mass of X^- is spatially confined to the cyclotron orbit, equivalent to an additional confinement in an effective quantum dot with a size determined by B. The inhibited recombination would then be a consequence of this additional confinement, a fact confirmed by the observation that in magnetic fields parallel to the QW (which do not confine X^-), the decay time of charged excitons is independent of B. For magnetic fields normal to the QW, the decay time was found to be proportional to B, changing from 100 ps at $B = 0$, to 1.2 ns at $B = 18$ T (Okamura et al., 1998).

The Overhauser effect in single GaAs/Al$_x$Ga$_{1-x}$As quantum dots was evidenced through the direct observation of spectrally resolved, polarization-dependent shifts in excitonic Zeeman splitting (Brown et al., 1996). The Overhauser effect originates in the dynamic polarization of lattice nuclei following optical pumping of circularly polarized light in a longitudinal external magnetic field. By absorption of photons the electron spins polarize and this polarization is then transferred to the nuclear system through the hyperfine interaction, preferentially orienting the nuclear magnetic moments. Finally, the static effective magnetic field proportional to the degree of nuclear orientation acts back on the electron system, shifting the energy levels; this is the Overhauser effect. In the PL of excitons localized by potential fluctuation in quantum dots, this effect was observed as a Zeeman splitting for σ^+ polarization at 0.5 T and for σ^- at 2 T, followed by a blueshift in energy of both levels with increasing magnetic field. Mathematically, the effect can be described by the Hamiltonian $H = g\mu_B BS_z + \alpha B^2 + \langle A \cdot I \rangle_z S_z$ where the first term describes the Zeeman splitting, the second the diamagnetic shift, with α a constant, and the third the hyperfine interaction, with S_z the exciton spin projection and I the nuclear spin. This Hamiltonian can be also written as $H = g\mu_B(B + B_N)S_z + \alpha B^2$, where B_N is the effective nuclear field. Its orientation is determined from the amount of exciton splitting for the two circular polarizations, for which this internal field either adds to or subtracts from the external magnetic field. Fitting the experimental data for a 4.2 nm wide well, it was found that $\alpha = 26\ \mu V/T^2$, $g = 1.1$, $B_N = 1.3$ T. To confirm the hyperfine origin of the effect, the strong dependence of the splitting on the excitation intensity was evidenced: for σ^+ the splitting varied from 143 to 192 μeV for excitation intensities between 1 and 20 mW, the splitting for σ^- varying between 87 and 12 μeV.

6.3.6 Biexcitons

Biexcitons are formed by pairing two excitons. The concentration of biexcitons is proportional to the square of the excitation power intensity, and thus proportional to the square of exciton concentration. The biexcitons are thus more important in the nonlinear regime of optical response, and can be directly excited by two-photon excitation. In II-VI quantum wells the exciton oscillator strength and binding energy are higher than in III-V quantum wells. Biexcitons can be of HH or LH type, denoted by XX_h and XX_l respectively, or of mixed type, formed by pairing a HH exciton with a LH exciton. These XX_m biexcitons are distinguished from XX_h biexcitons, for example, by the polarization selection rules. For example, in II-VI quantum wells, XX_h is equally strong in FWM experiments for $(\uparrow\uparrow)$ and $(\uparrow\rightarrow)$ polarizations and vanishes for (σ^+, σ^+) polarized light, whereas XX_m does not vanish for (σ^+, σ^+) polarization, being equally strong for $(\uparrow\uparrow)$ and $(\uparrow\rightarrow)$ polarizations. The intensities of these three types of biexcitons are in the relation $I_{XX_h} : I_{XX_m} : I_{XX_l} = 81:9:1$; XX_l biexcitons are usually not observed, being too weak (Wagner et al., 1999). The measured binding energies for XX_h and XX_m biexcitons were 4.8 meV and 2.8 meV, respectively.

The binding energy of biexcitons depends on the quantum confinement. It varies from 1.5 to 2.6 meV (as does the linewidth) when the well width of shallow $In_{0.18}Ga_{0.82}As/GaAs$ QWs increases from 1 to 4 nm, since at small quantum well widths the exciton wavefunction penetrates into barriers. Simultaneously, the ratio between exciton and biexciton binding energies increases from 0.23 to 0.3. From the temperature dependence of exciton and biexciton FWM signals it was inferred that exciton localization leads to an enhancement of the biexciton binding energy if the localization energy is higher than the binding energy of the delocalized biexcitons. Both exciton and biexciton binding energies are determined from the PL curves at low temperatures, and the dephasing times are determined from PL decay curves. The binding energies are related through $E_{XX} / E_X = \beta - (\eta^2 + \gamma^2)^{1/2} + \gamma$, where βE_X is the depth of the exciton-exciton interaction potential, ηE_X the harmonic oscillator energy $\hbar\omega_{XX}$ of exciton-exciton relative motion in the biexciton, and $\gamma = \hbar\omega_{loc} / E_X$ a localization parameter, with $\hbar\omega_{loc}$ the harmonic energy of the localization potential, determined from the inhomogeneous broadening. E_{XX} increases with increasing localization, a fact which can be seen from the quenching of the zero-point kinetic energy of exciton-exciton relative motion in the biexciton. The temperature dependence of the biexciton dephasing is deduced by comparing the FWM decay curves at a given temperature and at 5 K, to cancel the effect of the inhomogeneous broadening of E_{XX}, which is temperature independent and related to disorder. The intensity decay rate $\Delta\gamma$ due to phonon interaction is obtained from $\Delta\gamma_{XX} = \Delta\gamma - \Delta\gamma_{X}$, the temperature dependence of the exciton decay rate being $\gamma_X = \gamma_0 + aT + bN_{LO}$, where N_{LO} is the LO-phonon occupation number (Borri et al., 1999a).

Exciton pairs, not bound in a biexciton, but forming an excitonic molecule can have antibound states. Such states can be studied with spectrally resolved, subpicosecond transient FWM. In circularly polarized excitation, which suppresses the ground state of the exciton pair, a distinct peak was observed in ZnSe quantum wells at a 3.5 meV higher energy than the n = 1 HH exciton transition (see Fig. 6.20). This peak is distinct only for negative time delays, which identifies it as a manifestation of coherent four-particle correlated states within the biexciton continuum.

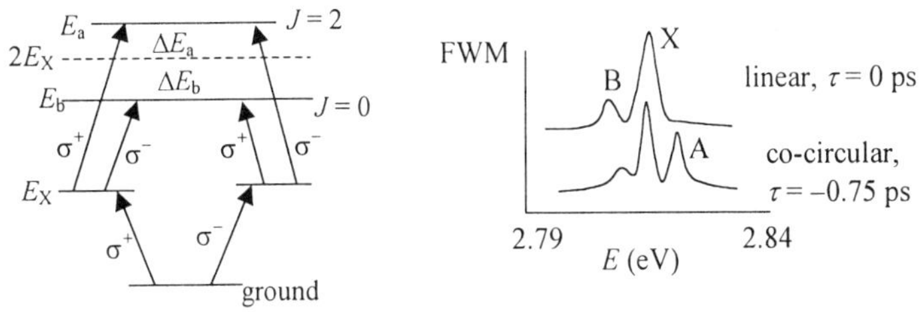

Fig. 6.20. Transitions involving two excitons in a ZnSe quantum well and the FWM spectrum showing the bound (B) and antibound (A) excitonic complexes

In contrast to the bound pair of excitons, which have a molecular binding energy ΔE_b = 5 meV in ZnSe and a total J = 0, being formed from HH excitons of opposite spins, the antibound state is characterized by a repulsive energy $\Delta E_a < 0$ and J = 2 (it is composed from two excitons with parallel spins). By illuminating with circularly polarized light, excitons are generated first, with J = 1 ($J_z = +1$ for σ^+ polarization and $J_z = -1$ for σ^-), the bound and antibound states of the exciton pair being reached then by the sequence of dipole-allowed transitions shown in Fig. 6.20; the antibound state is characterized by the peak A on the higher energy side of the HH n = 1 exciton (Zhou et al., 1998).

When the exciton density is sufficiently high, so that the interparticle spacing is comparable to the thermal de Broglie wavelength, the two components of the exciton-biexciton system confined in a QW can exhibit quantum statistical behavior. More precisely, Bose–Einstein condensation can occur for certain forms of the density of states (in particular when the density of states is proportional to $E^{1/2}$); this phenomenon consists in the fact that at a critical density, no more particles can be added to excited states and additional particles must go to the ground state of system. The signature of such a Bose–Einstein statistics would be the saturation of the exciton density and a continuing growth of biexciton density, as the photogenerated carrier pair density increases. Experimental results in GaAs QW showed that the onset for the Bose–Einstein condensation occurs for an exciton density of about 10^{11} cm^{-2} (Kim and Wolfe, 1998). Theoretical calculation demonstrated that the condensation is possible since both excitons and biexcitons generated from combining pairs of excitons, are composite bosons made from even numbers of electrons and holes. Although it was predicted that the quantum statistical regime is accessible before many-particle interactions alter the excitonic gas, a gradual broadening and blueshift of PL peaks was observed, indicating the onset of many-particle effects.

6.3.7 Magnetic Properties

The coupling between quantum well confinement and magnetic confinement is controlled by the angle θ between the magnetic field and the quantum well growth direction. The two confinement potentials are decoupled for $\theta = 0$ and $\theta = \pi$. In the first case the magnetic quantization enhances the geometric quantization, the length scale that determines the quantization changing from quantum well width to magnetic length with increasing B. For $\theta = \pi$ the magnetic, in-plane Landau quantization is independent of the geometric quantization, the spectral lines from intersubband transitions between Landau levels of different quantum well subbands crossing each other. In the intermediate case $0 < \theta < \pi$ the geometric and magnetic quantization mix and the electron levels become hybridized; the transitions between levels of different symmetry cross each other, whereas those involving levels of the same symmetry anticross. Additionally, the number of strongly allowed transitions increases near anticrossing due to the exchange character between the levels.

The rotational anisotropy of the QW energy level at the mutual alignment of the magnetic field and the momentum-dispersion of intersubband excitations, due to quasi-2D electron system asymmetry, is linear in both magnetic field and in-plane momentum, and is thus a tool for estimating the asymmetry of the 2D confining potential. The spectral positions of energy peaks associated with intersuband resonance in QW are not sufficient to accurately determine the shape of the confining potential. If the magnetic length is much smaller than the electric length determined by the gradient of the confining potential, an in-plane magnetic field B has a weak effect on the quasi-2D electron system, but modifies drastically the intersubband spectrum. In particular, the inelastic light scattering with $\boldsymbol{k}_{\parallel} \neq 0$ would be rotational anisotropic if the potential is asymmetric. For example, different Raman shifts for opposite directions of $\boldsymbol{B}$, for the intersubband collective mode due to spin-density excitation, and for the charge-density excitation mode in an $Al_{0.3}Ga_{0.7}As$/GaAs QW revealed the asymmetry of the confining potential in this structure (Kulik et al., 2000).

The g factors for both electrons and holes in QW can have different values as in bulk due to the change in band structure caused by confinement and strain effects caused by the lattice mismatch, due to carrier wavefunction penetration into barriers, and due to the reduced symmetry of the QW, which makes the g factors anisotropic. The g factors can be experimentally determined from either the shift between the excitonic PL peaks for σ^+ and σ^- polarizations, or from SFRS (spin-flip Raman scattering), the latter method having a higher accuracy. For CdTe/$Cd_{1-x}Mg_xTe$ single quantum wells the Stokes part of the spectrum looks like that in Fig. 6.21, where E and E_X are the electron and exciton spin-flip lines, respectively, and L denotes the laser line.

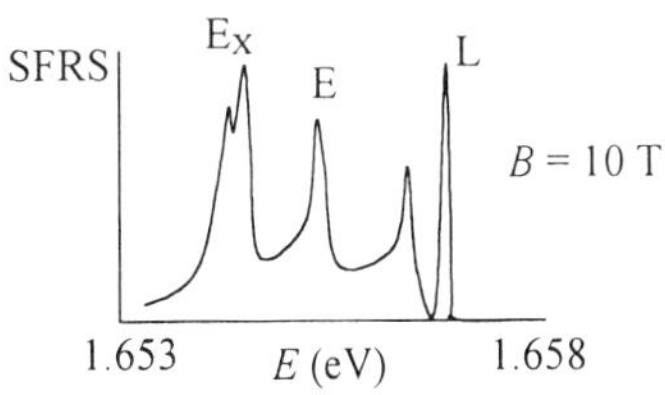

Fig. 6.21. Spin-flip Raman scattering spectrum of a CdTe/$Cd_{1-x}Mg_xTe$ single quantum well

In a magnetic field parallel to the growth axis, supposing that there is no mixing between LH and HH states, the Zeeman splitting of electron and HH exciton states are $\Delta_{e,ex} = |g_{\|e,ex}|\mu_B B$, where the g factors for the case when $\boldsymbol{B}$ is tilted with an angle φ with respect to the z axis are $g_e(\varphi) = -[(g_{\|e}\cos\varphi)^2 + (g_{\perp e}\sin\varphi)^2]^{1/2}$, and $g_{ex}(\varphi) = [(g_{\|e}\cos\varphi)^2 + (g_{\perp e}\sin\varphi)^2]^{1/2} + [(g_{\|hh}\cos\varphi)^2 + (g_{\perp hh}\sin\varphi)^2]^{1/2}$. Both parameters are strongly anisotropic, and, as in bulk, $|g_{\|hh}| >> |g_{\perp hh}| \cong 0$. From measurements one can determine both electron and hole g factors. It was found that for all single QW in which the ground state in the valence band is HH, $\Delta g = g_{\perp e} - g_{\|e} > 0$, and $\Delta g \cong E_P \Delta E_{lh-hh} / E_{ex}^2$, where E_P is the band parameter and ΔE_{lh-hh} is the splitting between LH and HH states (Sirenko et al., 1997).

It is interesting to note that the exciton PL peaks for σ^+ and σ^- polarizations can differ even in the absence of a magnetic field, in QW with unbalanced population of spin +1 and −1 HH excitons. Time-resolved PL experiments performed on undoped GaAs/AlAs QW with circularly polarized light showed that the splitting increases with the exciton density, and is connected with the polarization degree of PL. The splitting was assigned to exciton-exciton interaction, which can break spin degeneracy in 2D semiconductors (Viña et al., 1996). The frequency and linewidth of the QW exciton Zeeman components corresponding to σ^+ and σ^- polarizations in MQWs formed from DMS materials can also be determined from the MOKE rotation, which shows two peaks of opposite sign near the inflection points of the exciton magnetoreflectivity. The difference between the maximum and the minimum Kerr rotations is linearly increasing with the Zeeman splitting at low magnetic fields and saturates at higher values (Testelin et al., 1997).

Another way of determining both the magnitude and the sign of the conduction band electron g factors is by studying both the time-resolved Larmor beats in reflection, and the cw Hanle PL depolarization. This method has been applied for strained $In_xGa_{1-x}As/GaAs$ QW. $|g_e|$ is obtained from the Larmor beats in a magnetic field perpendicular to the incident laser beam, measured by a linearly polarized probe pulse which follows a circularly polarized pump beam. The pump generates excitons with spins polarized along the growth direction, their evolution being monitored via the rotation of the polarization plane of the probe pulse at reflection. The rotation is proportional to the relative spin population of excitons, and its decrease in time is due to spin relaxation and recombination. In an applied magnetic field the rotation of the probe polarization has an oscillatory behavior, due to spin precession about the field direction between the parallel and antiparallel orientations with respect to the pump beam. The frequency of Larmor precession for the geometry in this experiment is $\Omega_{ex} = (1/\hbar)(g_{ex}^2\beta^2B^2 + \Delta_{ex}^2)^{1/2}$, where $\Delta_{ex} \cong 10\ \mu eV$ is the electron-hole exchange energy obtained by fitting the signal with $A\exp(-t/\tau)\cos(\Omega_{ex}t + \phi)$, for different B values.

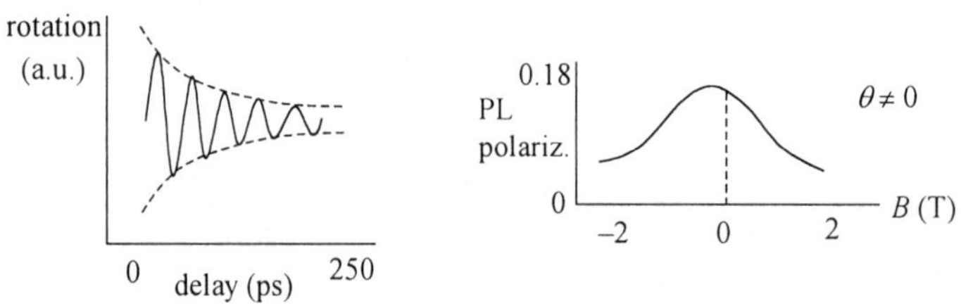

Fig. 6.22. Larmor beats (left) and Hanle effect in PL (right) in InGaAs/GaAs QW

In this configuration $g_{ex} = g_e$, the HH spin states having no contribution. The sign of the electron g factor is obtained from the nuclear Overhauser shift in the Hanle effect. The Hanle effect consists in measuring the circular polarization of the PL, as a function of B, under cw circularly polarized excitation along the growth axis, resonant with the HH exciton absorption peak. For the magnetic field normal to the growth axis the precession of electron spins produce a PL depolarization which follows a Lorentzian curve centered on $B = 0$. For $\theta \neq 0$ between the field and the normal to the growth axis, the electron spins precess on a cone centered on the applied field, inducing a steady resultant value of the electron spins antiparallel to B. These effective spins interact with the nuclear spins of the lattice via the hyperfine interaction, building a nuclear spin alignment $\langle \boldsymbol{I} \rangle$ which reacts back on the electron spins through an effective magnetic field $\boldsymbol{B}_{eff} = A\langle \boldsymbol{I} \rangle / g_e\beta$, where the hyperfine coupling constant A is positive for III-V materials. So, the sign of the effective field, which adds to the applied field and displaces the center of the Lorentzian with $\boldsymbol{B}_{eff}$ (see Fig. 6.22), depends on the sign of g_e (Malinowski et al., 1999).

The g factors for HH and LH have also been measured by magnetic circular dichroism (MCD), which measures the differential detection of left- and right-circularly polarized light. A typical MCD spectrum looks like that in Fig. 6.23.

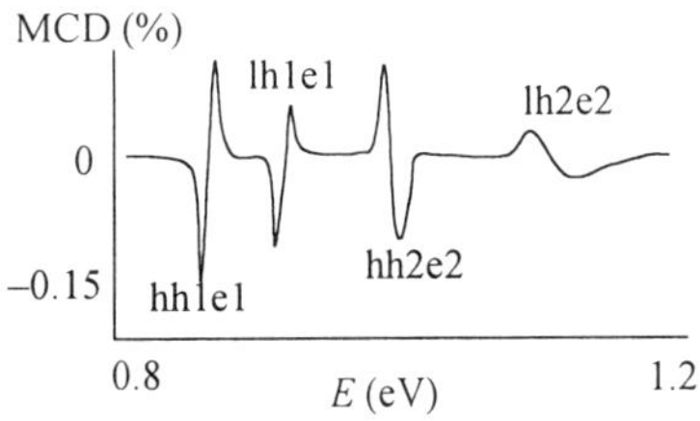

Fig. 6.23. Typical MCD spectrum in $In_xGa_{1-x}As$/InP QW

In $In_xGa_{1-x}As$/InP QW, where the exciton Zeeman splitting at low magnetic fields (< 1 T) is much smaller than the energy separation between maximum and minimum of the MCD, supposing that $I_\pm = I_0 \exp(-\alpha_\pm d)$ with $\alpha_\pm$ the Gaussian absorption lines for left and right circular polarizations, the MCD signal is $I_{\mathrm{MCD}} / \ln(I / I_0) = \Delta E / 2\sigma$. Here $I = I_+ + I_-$, $\Delta E = E_+ - E_-$ is the difference energy between the Zeeman exciton levels (between the maximum and minimum MCD) and σ is the width of the Gaussian absorption line. Since the I_+ light excites the $|+3/2,-1/2\rangle$ exciton state (first number is the hole spin, and the second the electron spin) in the HH subbands and the $|+1/2,+1/2\rangle$ excites the excitons in the LH subband, and I_- excites the $|-3/2,+1/2\rangle$ and $|-3/2,-1/2\rangle$ states, the left or right wing of the MCD is positive, depending on the relative order of these excited states. From MCD spectra it is not possible to determine the electron and hole g factors separately, but only combinations of them. For example, the first peak for the hh1e1 transition has $\Delta E = \mu_{\mathrm{B}} B(|g_{\mathrm{e}}| - 3|g_{\mathrm{hh1}}|)$, from which one obtains $g_{\mathrm{hh1}} = -0.68$ for a known $g_e = -3.3$. Similarly, g hole factors for lh1e1 and hh2e2 transitions can be obtained. These factors are closely related to the Luttinger parameters and the in-plane effective masses of the nth subbands: $g_{\mathrm{hh}n} = (2/3)(3\chi - \gamma_1 - \gamma_2 + 1/m_{\mathrm{hh}n})$, and $g_{\mathrm{lh}n} = 2(\chi + \gamma_1 - \gamma_2 - 1/m_{\mathrm{hh}n})$. Any one of these parameters can be determined if the others are known (Hofmann et al., 1997).

Electronic spin-flip Raman scattering (SFRS) in semiconductors of zinc-blende structures has been studied by Mal'shukov et al. (1997). Due to lack of inversion symmetry in these materials the spin degeneracy of the conduction band is lifted and the electron energy split $h(\boldsymbol{k})$ grows as k^3 with increasing electron energy. In narrow QW made from such semiconductors $h(\boldsymbol{k})$ can take large values. In the depolarized Raman geometry (with perpendicular incident and scattered light polarizations) the spin splitting of the conduction band is evidenced through low-energy peaks corresponding to transitions between pairs of spin-split electron states. The structure of the spin-splitting bands and the magnitude of spin splitting are determined from the peak positions at different sample orientations and different wavevector transfer. Additionally, when the incident and scattered photons are circularly polarized, interference of light inelastically scattered from different spatial components of spin-density fluctuations can offer information about the dynamics of electrons in spin-split bands due to phases of electron spin excitations. The interference term changes sign for non-gyrotropic materials in zero magnetic field when the direction of circular polarization is reversed. Detailed calculations show that if the growth direction z of MQW coincides with the spin quantization axis, the energies of the two spin-split subbands (SSSB) are $E_{\pm,k} = E_k \pm |\boldsymbol{h}(\boldsymbol{k})|/2$, and the intra-SSSB transitions produce a peak in the Raman spectrum at $\omega = v_F q$, with v_F the Fermi velocity. This result is similar to the single-particle intrasubband excitation in an electron gas, except that the peak intensity is now angular dependent due to spin splitting of electron energies. Inter-SSSB transitions give rise to two more Raman peaks at positions given (at $T = 0$ K) by the extremal points on the Fermi line where $\mathrm{d}/\mathrm{d}\phi(v_F \hat{\boldsymbol{e}}_i \pm |\boldsymbol{h}(\boldsymbol{k}_F)|) = 0$ with ϕ the angle between the 2D $\boldsymbol{k}_F$ at the Fermi energy and the x axis, and $\hat{\boldsymbol{e}}_i$ the polarization direction of the incident light. The polarization dependence of inter-SSSB peaks is given by $M_{\boldsymbol{k},\pm,\mp}(\omega,\boldsymbol{q}) = \gamma^2(|\boldsymbol{P}_\parallel \times \hat{\boldsymbol{n}}_k|^2 + |P_z|^2 \pm i\boldsymbol{P} \times \boldsymbol{P}^* \cdot \hat{\boldsymbol{n}}_k)$, where γ is a constant, $M_{\boldsymbol{k},i,j}(\omega,\boldsymbol{q})$ is the transition probability for single-particle excitations from state $\boldsymbol{k}$ in spin-subband j to state $\boldsymbol{k}+\boldsymbol{q}$ in spin-subband i, $\boldsymbol{P} = \hat{\boldsymbol{e}}_i \times (\hat{\boldsymbol{e}}_s)^*$ and $\hat{\boldsymbol{n}}_k = \boldsymbol{h}(\boldsymbol{k})/|\boldsymbol{h}(\boldsymbol{k})|$. The corresponding expression for intra-SSSB is $M_{\boldsymbol{k},\pm,\mp}(\omega,\boldsymbol{q}) = \gamma^2 |\boldsymbol{P}_\parallel \times \hat{\boldsymbol{n}}_k \pm iP_z|^2$. For linearly polarized incident and/or scattered light $\boldsymbol{P}$ is real, and the last, interference term in the inter-SSSB transition probability vanishes; for circularly polarized incident and/or scattered light, it vanishes only if the conduction band is not spin split, and it changes sign when both incident and scattered light are reversed. The two additional peaks in the Raman spectrum appear at $\boldsymbol{k} = \boldsymbol{k}_q = \boldsymbol{k}_F \cdot \boldsymbol{q}/q$ if $|\boldsymbol{h}(\boldsymbol{k})| < v_F q$, and at $\boldsymbol{k} = \pm\boldsymbol{k}_q$ in the opposite case. By using this interference effect in SFRS, it was possible to experimentally determine not only the magnitude, but also the sign of the bulk spin-splitting parameters a_{42} and a_{64}, which enter the expression of $\boldsymbol{h}(\boldsymbol{k})$ (Richards and Jusserand, 1999).

The magnetic-field dependence of exciton, hole and electron spin-relaxation rates in QW at low temperatures can be estimated from fitting the variation of the circular PL polarization with the magnetic field, with steady-state solutions of rate equations describing the population dynamics of exciton spin levels (Harley and Snelling, 1996). Unlike free electrons and holes, which have spin relaxation times of several hundred ps or even ns at low temperatures, excitons have much smaller spin-relaxation times, of only tens of ps. This is explained by an additional spin relaxation mechanism – the exciton spin relaxation – which involves simultaneous spin flipping of electron and hole, driven by the exchange coupling. The driving exchange interaction is enhanced by quantum confinement, and is a zero-phonon process at $B = 0$ which requires no energy interchange with the lattice, the exciton-spin relaxation being the dominant relaxation mechanism in QW in the absence of a magnetic field. The exciton-spin relaxation is strongly reduced when a magnetic field is applied due to the suppression of the zero-phonon relaxation by removing the degeneracy of optically allowed exciton states, and due to the enhancement of the hole-spin relaxation at level crossing fields, where the Zeeman energy cancels the exchange energy, such that hole-spin relaxation processes without energy interchange with the lattice are possible.

The electron density of states (DOS) as a function of B can be determined from PL measurements, which show an evolution of the broad zero-field feature into a series of peaks spaced by $\hbar\omega_c$ with increasing B. This behavior is caused by the transition from a constant 2D electronic DOS into a series of broadened Landau levels, the halfwidth of which has an oscillatory behavior as a function of the filling factor of Landau levels. Experiments on modulation-doped $In_xGa_{1-x}As$ quantum wells have confirmed these trends (see Fig. 6.24) (Harris et al., 1996).

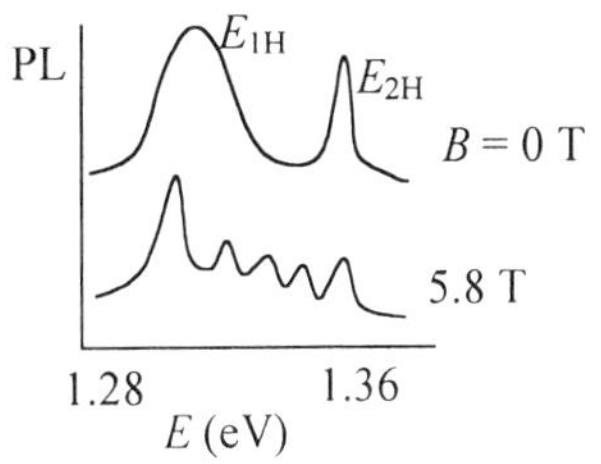

Fig. 6.24. PL dependence on the magnetic field in modulation-doped $In_xGa_{1-x}As$ QWs

For quantum dots placed in a magnetic field the excitonic PL shows a diamagnetic shift and a Zeeman spin splitting, depending on the size of the structure. The Hamiltonian in lateral confinement is $H = \boldsymbol{p}^2/2m_r + (1/2)m_r\omega^2 r^2 - e^2/\varepsilon r$ with m_r the reduced mass and $\omega = (\omega_0^2 + \omega_c^2)^{1/2}$, where ω_0 is the strength of the parabolic lateral confinement and $\omega_c = eB/2m_r$. The quadratic diamagnetic shift of the ground-state energy $\Delta = \langle 0|H_{\mathrm{rel}}|0\rangle_B - \langle 0|H_{\mathrm{rel}}|0\rangle_{B=0} = 3\varepsilon^2\hbar^4\omega_c^2/16m_r e^4$ is independent of ω_0, but dependent on the geometric dot size. In particular, experiments have shown that for geometric sizes smaller than 450 nm, the reduction in the lateral confinement is accompanied by an increasing blueshift of the whole spectrum with increasing B (Bockelmann et al., 1997).

In strong confined dots and parabolic quantum wells the excitonic effects play a minor role since the exciton wavefunction is a product of single-particle wavefunctions, the lack of degrees of freedom for the center-of-mass motion of excitons in quantum dots manifesting itself in the fact that the Zeeman and diamagnetic shift are expressed in terms of single-particle states (Rinaldi et al., 1998). More precisely, the Zeeman effect $\Delta E_Z = (e\hbar/m_0)(1/m_r^e + 1/m_r^h)Bm$ is observed only for states with quantum numbers $m > 0$, where m_0 is the electron mass and m_r^e, m_r^h are the in-plane mass of electrons and holes. The Zeeman splitting in strained quantum dots with different quantization energies shows a multifold splitting of π and Δ states, due to degeneracy lifting of $m > 0$ states. The diamagnetic shift is $\Delta E_d = (e^2/8m_0)(\langle r^2\rangle^e/m_r^e + \langle r^2\rangle^h/m_r^h)B^2$ where $\langle\rangle$ denotes expectation values calculated with wavefunctions corresponding to nonzero external fields. The diamagnetic shift is observed for all states. Experiments in

$In_{0.6}Ga_{0.4}As/GaAs$ self-assembled quantum dots have shown that the Zeeman splitting for biexcitons is the same as for excitons, whereas the diamagnetic shift is twice as large as for excitons. The identical Zeeman splitting for excitons and biexcitons, corresponding to a large g factor of about 3, suggests that the splitting originates from the final state of the exciton optical transition. From magneto-PL measurements, the binding energy of the biexcitons, as well as the lateral extensions of the electron and hole wavefunctions can also be determined. The first parameter, of about 3.1 meV, is much higher than the value in QW, of 0.13 meV, due to the strong confinement. The lateral extension of the wavefunctions was calculated assuming an equal extension for electron and hole wavefunctions, and was found to be $\langle r_{e,h}^2 \rangle^{1/2} \cong 12$ nm (Kuther et al., 1998b).

From measurements of spin-flip Raman scattering (SFRS) in quantum dots it is possible to obtain the electron g factor and the local sample temperature T^*. T^* is determined from fitting the ratio of Stokes and anti-Stokes low-frequency Raman lines, as a function of the Raman shift Δ with $I_S(\Delta)/I_{AS}(\Delta) = \exp(\Delta / k_B T^*)$. Experiments in CdS quantum dots show that the narrow Stokes and anti-Stokes SFRS lines due to acoustic-phonon-assisted electron spin-flip processes appear on top of the broad acoustic-phonon Raman scattering (APRS) line (see Fig. 6.25).

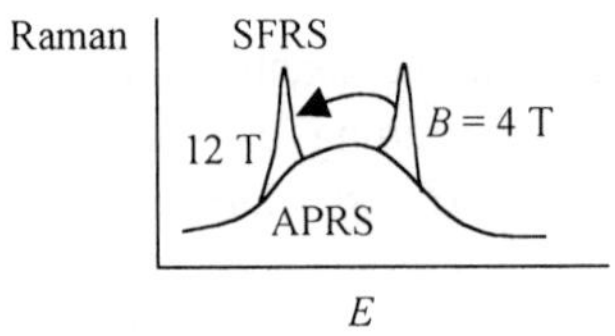

Fig. 6.25. Spin-flip Raman scattering in CdS quantum dots

The Raman shift of the electron spin-flip lines with magnetic field, due to Zeeman splitting can be fitted with $\Delta^e = | g^e | \mu_B B$, from which the electron g factor, g^e, and its dependence on the quantum dot size can be derived. If the CdS quantum dots are randomly oriented in the glass matrix, g^e is isotropic, in contrast to the results in bulk CdS or quantum wells. In contrast to APRS spectra which do not depend on B and have a circular and linear degree of polarization of $\rho_c = 0$ and $\rho_l = 0.3$, respectively, the intensity of the SFRS line depends on B and is strongly polarized, with $\rho_c = 0.75$ and $\rho_l = -0.5$. From the high-temperature dependence of the Stokes lines $\ln[I(T^*)/I(T^* = 12K)] = E_0 / k_B T^*$ the parameter E_0 can be also determined from fitting, and is found that it increases with decreasing the quantum dot radius (Sirenko et al. 1998).

When the low-D structures are fabricated from materials with magnetic properties Brillouin light scattering and MOKE can be used to characterize them. For example, the *magnetization and resonant frequency* of submicron Fe magnetic dot arrays were studied with both methods, a large in-plane anisotropy due to shape anisotropy of dots being detected with both investigation techniques

(Grimsditch et al., 1998). By measuring the magnon frequency as a function of the in-plane direction of the magnetic field with Brillouin light scattering, it was found that the maximum and minimum magnon frequency correspond to the magnetic field along the long and short axes of the elliptical dots, respectively. The predominant role of the shape anisotropy was also confirmed by MOKE measurements, which showed that the hard and easy magnetization axes did not coincide with the array axes but with the principal axes of the individual dots. For a magnetic field along one of the principal axes of the ellipsoid, chosen as the z direction, the magnon frequency, when the interaction between dots is neglected, is given by $\omega^2 = \gamma^2[H + 4\pi(N_y - N_z)M][H + 4\pi(N_x - N_z)M]$ where γ is the gyromagnetic ratio and N_i are the demagnetization factors. N_x, N_y and N_z ($N_x + N_y + N_z = 1$) can be determined by fitting the experimental dependence of the magnon frequency on H with this formula.

In QWs made from DMS, the exchange parameter between carrier spins and Mn is dependent on the width of the QW. In $Cd_{1-x}Mn_xTe/Cd_{1-x-y}Mn_xMg_yTe$ QWs with no discontinuity of Mn content at interfaces, a relative change of the exchange parameter of about 0.1 is determined from the Zeeman splitting of free-exciton states. The well width dependence of the Zeeman splitting is caused by the $\boldsymbol{k}$ dependence of the exchange parameter (Mackh et al., 1996), due to interface related effects between magnetic and non-magnetic materials.

The nearest-neighbor exchange constant J_{NN} in ZnSe/Zn(Cd,Mn)Se quantum wells can be determined from the magnetization steps in the Zeeman shift of the PL, which is a measure of the magnetization. The experiments were carried on $ZnSe/Zn_{0.8}Cd_{0.2}Se$ single quantum wells in which Mn^{2+} is 'digitally' incorporated as equally spaced fractional monolayer planes of MnSe. The Zeeman shift looks like the plot in Fig. 6.26.

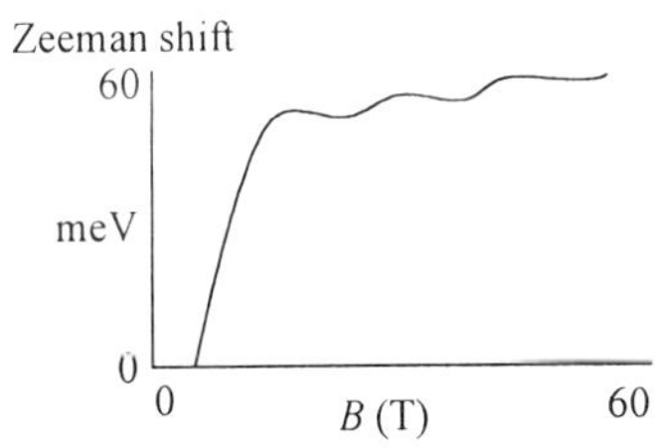

Fig. 6.26. Zeeman shift in $ZnSe/Zn_{0.8}Cd_{0.2}Se$ single QW

On increasing the magnetic field up to 60 T, a Brillouin-like saturation of the magnetization is observed at about 8 T, followed by three steps in the Zeeman shift at 19, 36 and 53 T, associated with a sharp drop of the PL linewidth. These magnetization steps at low temperature are due to the clustering of the Mn-Mn pair with a total spin S_{tot} =0, 1, 2.., whenever the applied field is commensurate with the exchange energy, i.e. when $H_n = 2nJ_{NN} / g_{Mn}\mu_B$, n = 1, 2,... In practice, however, the steps do not correspond exactly to the values of H_n due to the contribution of distant-neighbor exchange fields between Mn moments, so J_{NN} is obtained from the difference $H_{n+1} - H_n$. The value obtained experimentally is about −11.1 K (Crooker et al., 1999).

6.3.8 Collective Excitations

Collective excitations such as 2D intrasubband plasmons can be observed in Raman scattering (Vasko and Kuznetsov, 1999). 2D plasmons differ from their 3D counterparts in that the plasmon frequency ω_p is strongly dependent on q and vanishes for $q \to 0$, whereas for 3D plasmons ω_p is independent of $\boldsymbol{q}$. For 2D plasmons

$$\omega_p^{2D}(q) = [2\pi n_{2D} e^2 / \varepsilon m]^{1/2} q^{1/2}. \tag{6.50}$$

Similarly, for the 1D case the plasmon dispersion is

$$\omega_p^{1D}(q) = q[2n_{1D} e^2 \,|\ln(qd)|/ \varepsilon m]^{1/2}. \tag{6.51}$$

Plasmon scattering peaks are identified through their dependence on the angle of incidence in the backscattering geometry. A typical Raman spectrum for a low-frequency intrasubband 2D plasmon is shown in Fig. 6.27. The angular dependence of the Stokes shift is consistent with the 2D plasmon dispersion.

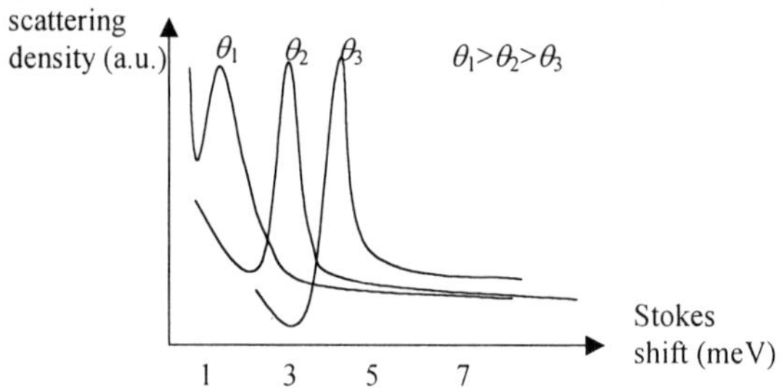

Fig. 6.27. Typical Raman spectrum for a low-frequency intrasubband 2D plasmon

Acoustical and optical plasmon modes can be observed in GaAs/$Al_xGa_{1-x}As$ double QWs by angle-resolved electronic Raman scattering. Acoustic plasmons (AP) occur in systems with two different types of free carriers: electrons and holes in semiconductors, electrons and ions in plasma, or carrier populations separated in real or momentum space. They can be observed, for example, in double QWs with multiple subband occupancy, or in single 2DEG drifting under the effect of in-plane electric fields. Another possibility of observing AP is in two parallel 2DEGs, sufficiently separated to avoid quantum mechanical interaction, but close enough to allow electromagnetic coupling between charge oscillations. AP are characterized by linear dispersion and by out-of-phase oscillations of similarly charged carriers and in-phase oscillations of oppositely charged particles. On the contrary, optical plasmons (OP) are in-phase oscillations of similarly charged particles and out-of-phase oscillations of oppositely charged carriers. In GaAs/Al_xGa_{1-x}As double QWs with no quantum mechanical interaction, the dispersion relations of AP and OP are

$$\omega_{\mathrm{AP,OP}}^2 = [e^2 q(N_1 + N_2)/4\varepsilon m]\{1 \mp [1 - 4N_1 N_2(1 - \exp(-2qd))/(N_1 + N_2)^2]^{1/2}\}, \tag{6.52}$$

where N_1, N_2 are the areal densities for the two parallel 2DEG. For $qd << 1$ with d the distance between the two 2DEG, the AP and OP dispersion relations scale as q and $q^{1/2}$, respectively. The OP or AP character of a plasmon mode can then be checked by angle-resolved Raman scattering with parallel incident and scattered polarizations, a configuration in which scattering by charge-density excitations (plasmons) is allowed. Experimental results showed that the acoustical and optical modes have different linewidths, due to the greater localization of electric fields of acoustic plasmons, which are less susceptible to impurity damping than the optical plasmons (Kainth et al., 1999).

Resonant Raman scattering has revealed that at low frequencies quasi-1D or quasi-0D laterally confined plasmons are excited in quantum wires or dots. However, at higher frequencies, additional modes appear at $\omega^2 \cong \omega_{\mathrm{2D}}^2 + \omega_{\mathrm{1D/0D}}^2$ where ω_{2D} is the frequency of vertical intersubband charge-density excitations. The polarization selection rules have identified these additional modes as collective charge-density excitations due to the coupling of lateral and vertical electron motion; they appear only in (s,s) or (p,p) polarizations (Biese et al., 1996). Coupling between motions in orthogonal directions can be achieved also by applying a magnetic field. If the magnetic field is tilted from the z direction perpendicular to a 2D electron system, the in-plane component of the magnetic field couples the electron motion in the perpendicular direction to the cyclotron motion in the confined 2D plane. As a result, an anticrossing of electron energy levels appears at $\hbar\omega_{\mathrm{c}} \cong E_z^{(2)} - E_z^{(1)}$, and the cyclotron-resonance absorption peak splits at this frequency. This phenomenon – called cyclotron-resonance-intersubband coupling – has been observed whenever the magnetic field and the lateral SL are tilted with respect to one another, but can also arise in untilted laterally corrugated SL in vertically applied magnetic fields since the corrugated interfaces are not perfectly vertical (see Sun et al. (2000) and references therein).

6.3.9 Coupled Excitations

Strong coupling between exciton and electromagnetic modes occurs in MQW whenever the MQW period is 1/2 or 1/4 of the light wavelength in the barriers. For example, in a double QW, in the absence of nonradiative damping, the symmetric and antisymmetric exciton modes are $E_{\pm} = E_{\mathrm{ex}}(k_{\|}) + \Gamma_0(k_z) \times[\pm\sin(k_z d) - \mathrm{i}(1 \pm \cos(k_z d))]$, where E_{ex} is the exciton energy for a single QW,

Γ_0 is the radiative coupling of exciton to radiation, and d the distance between the centers of the QWs. The splitting between symmetric and antisymmetric modes vanishes for $k_z d = \pi$, and is maximum for $k_z d = \pi / 2$. The oscillating behavior of splitting as a function of d can be observed in reflectivity: constructive interference of reflected light – Bragg condition – is observed for $d = \lambda / 2$, whereas destructive interference occurs for $d = \lambda / 4$. Similar results occur for more than two QWs (Merle d'Aubigné et al., 1996).

Resonant Fröhlich coupling between purely electronic states and mixed exciton-phonon states can form exciton-phonon complexes with a finite binding energy δ. These exciton-phonon quasibound states are intrinsically broadened due to the same overlap between exciton-phonon complexes and the electron-hole continuum which generates it. The quasibound states appear whenever the exciton binding energy $E_B \cong \hbar\omega_0$, and are identified through phonon sidebands, which occur below $E_0 + \hbar\omega_0$ with an amount $\delta \cong 0.1\hbar\omega_0$, where E_0 is the exciton energy at $\boldsymbol{q} = 0$ and $\hbar\omega_0$ the phonon energy. Such states have been evidenced in $CdTe/Cd_{1-x}Zn_xTe$ MQW as sharp peaks in absorption spectra; two such quasi-bound states associated to e1h1+LO and e1h2+LO sidebands were found to have binding energies of 1.1 meV and 0.3 meV, respectively (Pelekanos et al., 1997).

Coupled states of Wannier excitons and LO phonons have been observed in CuBr quantum dots by resonant hyper-Raman scattering. Resonances up to the Nth (N = 5) order have been observed, where $N \le (2\hbar\omega_i - E_{edge}) / \hbar\omega_{LO}$, with $\hbar\omega_i$, $\hbar\omega_{LO}$ the energies of the incident photon and LO phonon, and E_{edge} the one-photon absorption edge of the Z_{12}, s exciton. The relative intensities of the Nth LO phonons do not diminish appreciably when N increases, all signals from the N phonons vanishing when $2\hbar\omega_i > E_{edge}$ (Inoue et al., 1996). These exciton-phonon-coupled states are similar to molecules. Similar experiments in CuCl spherical quantum dots have shown that both LO and 2LO bands are resonant with the lowest energy 1s confined exciton state, the ratio of integral intensities of LO and 2LO bands increasing with increasing incident photon energy or with decreasing the radius of quantum dots (Baranov et al., 1997).

6.3.10 Quantum Microcavities

In semiconductor quantum microcavities (QMC) the exciton and photon modes are coupled into cavity polaritons. Polaritons can form in bulk materials between translational invariant excitonic bound states and electromagnetic modes. In 2D quantum wells the polariton concept disappears due to the lack of translational invariance, unless the electromagnetic mode is also confined. This happens in microcavities, where microcavity polaritons are formed. For a QMC with length L, the lowest photon mode has a quantized wavevector along the cavity axis $k_z = 2\pi / L$, and a total energy $E(k_\parallel) = (\hbar c / n)[(2\pi / L)^2 + k_\parallel^2]^{1/2}$ $= E_0 (1 + \hbar^2 c^2 k_\parallel^2 / E_0^2 n^2)^{1/2}$ where n is the effective refractive index of the QMC and E_0 the phonon energy for $k_\parallel = 0$. If the optical length of the QMC is an integer or half-integer multiple of the exciton wavelengh λ_{ex}, each in-plane photon mode couples with only one exciton with the same $k_{\|}$, the cavity polaritons having a strong in-plane dispersion. The polariton dispersion can be studied by monitoring the reflectivity spectra at 10 K as a function of the external angle of incidence θ, the photon in-plane wavevector being related to this angle through $k_{\|} = (E/\hbar c)\sin\theta$. In reflectivity spectra both cavity and exciton modes are observed, the cavity mode moving towards higher energies with increasing θ. For some range of θ values the cavity and exciton modes become coupled, at resonance their intensities being equal. Far from resonance the exciton intensity is weak. Experimental studies in GaAs QMC excited with either TE or TM waves have shown that exciton-polariton states in QMC have marked polarization dependences of energies, linewidths and intensities. At resonance, the spectra had an anomalous narrowing, and a broadening of the cavity mode due to interaction with exciton continuum states and QW excited states was observed (Baxter et al., 1997).

The coupling between excitons, with properties controlled by the QW embedded in the cavity, and photons controlled by the Fabry–Perot modes of the cavity, is called vacuum Rabi coupling. In magnetic fields the vacuum Rabi splitting increases since the shrinkage of the exciton wavefunction enhances the exciton oscillator strength. Reflectivity measurements in circularly polarized light reveal both cavity and exciton features which can be tuned by changing the temperature, and eventually made to anticross each other. The energy peaks are higher for σ^+ polarization than for σ^-, which indicates a positive g exciton factor.

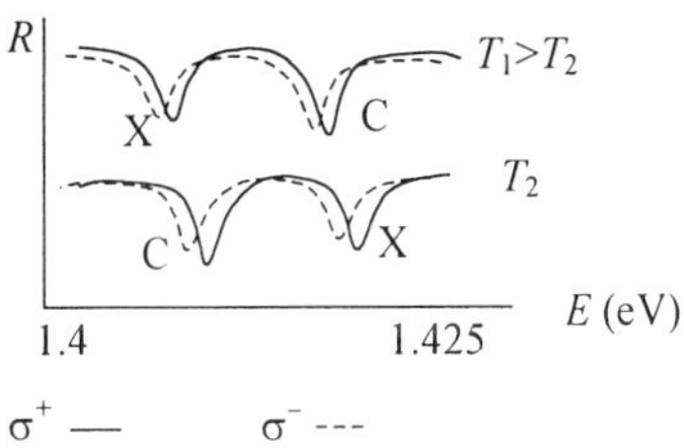

Fig. 6.28. Reflectivity spectra in quantum microcavities for two temperatures

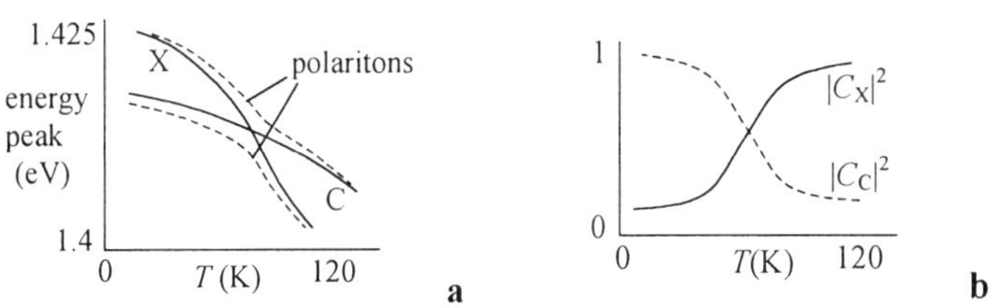

Fig. 6.29. (**a**) Dispersion curve and (**b**) exciton and cavity fraction of the lower polariton branch wavefunction

If the cavity is illuminated with circularly polarized light, the two fully decoupled exciton-spin components interact independently with the appropriately circularly polarized cavity mode. At resonance, the two coupled polariton branches have equal integrated intensities and are separated by the vacuum energy Rabi splitting Ω_v (see Fig. 6.28). The resonance is attained at 70 K for σ^- polarization and at 82 K for σ^+ in InGaAs/GaAs QMC inside $Al_{0.13}Ga_{0.87}As/AlAs$ DBRs (distributed Bragg reflectors) (Armitage et al., 1997). If Δ is the energy separation between the unperturbed components, the energy separation between the perturbed components (polariton branches) is $(\Delta^2 + \Omega_v^2)^{1/2}$, the state of the polariton being a combination of the exciton state and the cavity mode $|p\rangle = C_X|X\rangle + C_C|C\rangle$. The exciton and cavity fraction of the lower polariton branch wavefunction are represented in Fig. 6.29. For the lower polariton branch a motional narrowing of the linewidth occurs at resonance due to the strong difference of in-plane dispersion curves for exciton-cavity polariton compared to the uncoupled, bare excitons. Motional narrowing represents a decrease in the spectral line in disordered systems due to statistical averaging processes. In this case the motional narrowing is caused by the fact that the polariton wavefunction is much more extended than the exciton wavefunction since it has a much smaller effective mass, and thus an averaging over the disorder potential which localizes excitons occurs for polaritons. Also observed in experiments is the coupling between the first magnetoexciton ($N = 1$ Landau level) with the cavity mode, characterized by a smaller vacuum Rabi splitting than for the $N = 0$ ground state exciton, due to smaller oscillator strength.

Microcavity polaritons cannot form between confined electromagnetic modes and the continuum of unbound excitonic states at higher energies. However, in an applied magnetic field the continuum of unbound states is quantized in discrete Landau levels, and new Rabi splittings appear at sufficiently large magnetic fields applied normal to the quantum well. The magnetic field concentrates the oscillator strength in discrete magnetoexciton states, maintaining the in-plane translational invariance. The dip in the reflection spectrum corresponding to the resonance of the cavity mode with the continuum excitonic states split into two for higher magnetic fields. The critical field B_{cr} at which the new vacuum Rabi splitting appears differs for different excitonic modes labeled with n. This fact suggests that electron-hole correlation is not negligible. Experiments on InGaAs/GaAs QW in AlAs/GaAs DBRs show that B_{cr} increases with n, and that the vacuum Rabi splitting Ω_v, which increases with B, is proportional to the square root of the exciton oscillator strength $f^{1/2}$, for Rabi splittings larger than linewidths (Tignon et al., 1997). The exciton oscillator strength can be bleached at high excitation levels when the microcavity is injected with high electron-hole pair densities; this bleaching is much less effective in II-VI semiconductor microcavities than in GaAs microcavities, for example. When both HH and LH excitons can be excited, the cavity mode can be tuned towards one or the other by modifying the angle of incidence (Kelkar et al., 1997).

The cavity polariton splitting, irrespective of the homogeneous or inhomogeneous nature of the excitonic resonance, is given by $\Omega_V = [|V|^2 - (\gamma_C - \gamma_X)^2/4]^{1/2}$ where V is the exciton-photon coupling strength, and γ_C, γ_X are the half-linewidths of the empty cavity mode and excitonic resonance, respectively. In the strong coupling regime $2V > |\gamma_C - \gamma_X|$; the cavity polariton splitting is maximum when $\gamma_C = \gamma_X$ and the cavity and exciton modes overlap. When PL is used to measure the cavity polariton splitting, the obtained value is $\Omega_{PL} = \{2\Omega_V[\Omega_V^2 + (\gamma_C + \gamma_X)^2]^{1/2} - \Omega_V^2 - (\gamma_C + \gamma_X)^2\}^{1/2}$. From the temperature dependence of cavity polariton splitting in PL, the homogeneous exciton broadening can be determined from $\gamma_X(T) = \gamma_{inh} + \gamma_{hom} = \gamma_{inh} + \gamma_A T + \gamma_{LO}/[\exp(E_{LO}/k_B T) - 1]$ (Pratt et al., 1998).

In coupled QMC, the interaction between the symmetric and antisymmetric photon modes and the corresponding combinations of degenerate excitonic states, gives rise to bright excitons. If two identical coupled QWs are not placed in the QMC, the symmetric (S) and antisymmetric (AS) excitonic states have a small (about 1 meV) radiative splitting due to interwell dipolar coupling. This splitting is, however, washed out by disorder. When the two identical QWs placed each in a QMC, are coupled in the strong coupling regime, only one of the S and AS excitonic states is maximally coupled to light and observable. Analogously, if N identical excitons each inside a QMC are coupled, only one, 'bright' state is observed, the others being 'dark' states. In the reflectivity curve only one exciton peak with energy E_X is observed together with the S and AS photon modes. However, when the exciton is at resonance with the photon modes, the three dips become four since the degenerate exciton peak splits into two bright states. The optically induced lifting of the degenerate spatially separated QW excitons can be observed in reflectivity spectra by tuning the angle of incidence of the light. The four peaks of the coupled polariton states are solutions of the equations $(E_S - E)(E_X - E) = V_{XC}^2$, $(E_{AS} - E)(E_X - E) = V_{XC}^2$, where $E_{S,AS} = E_C \pm V_{opt}$ are the peak energies of the coupled cavity modes, with E_C the energy of the uncoupled mode and V_{opt} the optical coupling between cavities, and V_{XC} is the exciton-photon coupling (Armitage et al., 1998).

6.3.11 Relaxation Phenomena in Low-Dimensional Structures

The study of relaxation phenomena in low-D structures with time-resolved spectroscopy, in particular ultrafast spectroscopy, is a flourishing area of research. There are already many review papers and books (Shah, 1999; Vasko and Kuznetsov, 1999) focused on this subject; we present here mainly the recent developments in this area, not included in these reviews. There are no significant differences between carrier cooling in bulk and in low-D semiconductors. However, new phenomena can appear in the latter case, as for example the fact that in a coupled-quantum-well structure energy relaxation can be accompanied by real-space movement of carriers between the wells.

The tunneling dynamics of excitons and free carriers in semiconductor nanostructures such as asymmetric $Al_xGa_{1-x}As/GaAs$ double QWs can be studied using time-resolved PL (Emiliani et al., 1997). This method allows the determination of several parameters that reflect the competition between carrier tunneling and exciton formation, such as the exciton formation time and the tunneling time of photogenerated carriers before and after exciton formation. As an aside, we note that in time-resolved spectroscopy the tunneling time is defined as the reciprocal of the tunneling rate through a barrier (Shah, 1999). However, the tunneling time or traversal time is also defined as the time needed for a wavepacket to travel through a barrier (Dragoman and Dragoman, 1999). The two definitions refer to distinct physical times based on different physical processes, although they are encountered frequently in the literature under the same name. Returning to our asymmetric double QW, the exciton dynamics can be modeled with a system of coupled equations for the time evolution of the electron, hole and exciton densities, n, p and ν, respectively. The processes considered in the coupled system of equations are the tunneling processes of an electron or a hole from an excitonic state followed by the generation of a free carrier, described by the rates $\nu/\tau_{\mathrm{e,ex}}$ and $\nu/\tau_{\mathrm{h,ex}}$, the tunneling of electrons and holes in corresponding tunneling times τ_{e} and τ_{h}, and the exciton recombination characterized by the recombination time τ_{r}. The optical excitation is described by $G(t,\delta)$ where δ is the delay between two consecutive ultrashort pulses. For a fixed excitation density n_{ex}, the exciton density $\nu(t,\delta,n_{\mathrm{ex}})$ is proportional to the time-resolved PL, and the total PL intensity $I_{\mathrm{tot}}(\delta,n_{\mathrm{ex}})$ is obtained after integrating $\nu(t,\delta,n_{\mathrm{ex}})$ over time. The total PL intensity is found to be linearly dependent on n_{ex}, the dependence becoming quadratic when the tunneling of free carriers becomes the dominant process in the nanostructure. The time-resolved PL intensity is proportional to $I_{\mathrm{nl}} = I_{\mathrm{tot}}(\delta,n_{\mathrm{ex}}) - 2I(\delta=\infty,n_{\mathrm{ex}})$, where the last term represents the PL generated by a single pulse. The exciton recombination time is directly determined from the PL decay.

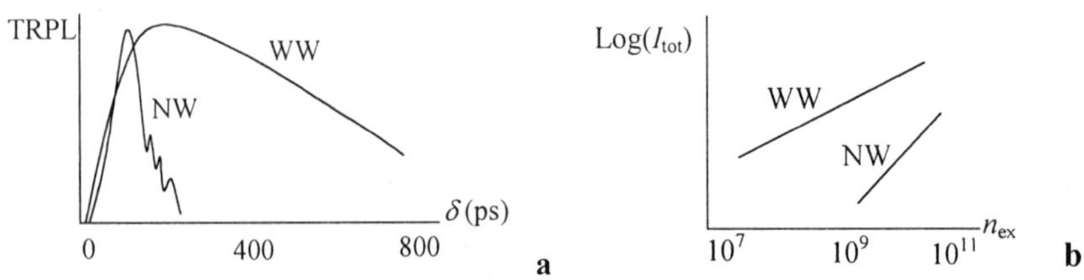

Fig. 6.30. (**a**) Time-resolved PL of the narrow (NW) and wide well (WW) of an AlGaAs/GaAs double quantum well and (**b**) the power dependence on excitation density

As shown in Fig. 6.30, the PL of the narrow well (NW) is very fast compared to the PL in the wide well (WW) for excitation at the fundamental exciton transition, suggesting that exciton tunneling as a neutral charge or particle is in strong competition with its radiative recombination. These considerations are confirmed by the power dependence of $I_{tot}(\delta, n_{ex})$ on n_{ex}, with a power coefficient 0.96 for the WW and 1.38 for the NW (see Fig. 6.30). The electron and hole tunneling times are obtained by fitting the time-resolved PL with the theoretical model; the hole tunneling time is found to be faster than the electron tunneling time. This last result contradicts the usual image of faster tunneling electrons with smaller effective masses, and is caused by valence band mixing.

The power-dependent temperature behavior of the time-resolved PL of GaAs QW can be explained considering that free carriers and excitons form an almost ideal gas in thermodynamic equilibrium. The time evolution of the temperature after exciting with a pump pulse can be measured from free-carrier recombination PL (Yoon et al., 1996). Considering that the band-to-band PL peak is due to recombination of free electrons and holes with the same temperature $T_e = T_h = T$, its intensity is $I(\hbar\omega) \cong \alpha(\hbar\omega - E_g)\exp[-(\hbar\omega - E_g)/k_B T]$, the carrier temperature being determined from the slope of the band-to-band PL on a semi-log plot. When illuminating with light pulses, for kinetic energy of carriers much larger than the homogeneous width of exciton recombination and the fundamental radiative recombination rate Γ_0, the time evolution of the PL is given by $I(T(t),t) = An_{ex}(T(t),t)2\Gamma_0 E_1 / 3k_B T(t)$ where $E_1 \cong 0.1$ meV and the exciton concentration n_{ex} depends on the incident photon density. Since the cooling of the gas of excitons and free carriers occurs through phonon emission, $T(t)$ is modeled as a sum of three exponentials corresponding to LO phonon cooling and slower acoustic phonon relaxations, the characteristic times for these processes being then determined from fitting.

Time-resolved PL was also employed to study the rapid carrier relaxation in self-assembled InGaAs/GaAs quantum dots (Ohnesorge et al., 1996). It allows the estimation of the rise time and barrier decay times as a function of temperature, excitation energy and excitation density, as well as the understanding of the accompanying physical mechanisms. The time-resolved PL of a dot is modeled as $I(t) \cong [\exp(-t/\tau) - \exp(-t/\tau_r)]/(\tau - \tau_r)$ where τ_r describes the carrier capture and relaxation from the excited to the ground state and τ is the excitation lifetime. Experimental data were then compared to expressions of these characteristic times corresponding to different mechanisms that dominate for different ranges of excitation intensity and energy. In this way the physical mechanisms that cause the PL decay are determined.

The influence of state-filling effects, Coulomb scattering and acoustic phonon scattering on the carrier relaxation between 0D states can also be studied using time-resolved PL (Grosse et al., 1997). Only interband transitions with $\Delta i = 0$ are allowed; the PL decays for higher transitions (higher i values) are faster than for the 1-1 transition due to carrier relaxation from higher-energy QD levels to lower-energy levels mediated by 0D-2D Coulomb scattering and acoustic phonon scattering. The inter-dot level relaxation rates, the 0D-2D scattering rates and the strength of the Coulomb interaction are determined by fitting the experimental data (time-resolved PL for 1-1, 2-2, 3-3 QD transitions) with a mathematical model of coupled rate equations for relaxation between levels i and $i-1$.

The relaxation phenomena in high-quality QWs showing homogeneous broadening and zero Stokes shifts are dominated by intrinsic excitons. In these QWs the photogenerated electron-hole pairs emit phonons with large in-plane wavevectors, the hot excitons relaxing to $\boldsymbol{k}_{\parallel} \cong 0$ through interactions among themselves and with phonons before recombining. In contrast, the QWs with interface steps are characterized by inhomogeneous broadening and Stokes shift in the optical spectra. The potential energy of excitons in these QWs follows the bandgap fluctuations generated by the thickness variations, so that the excitons can be localized in the minima of this potential energy and can move between these minima by emitting phonons. Time-domain PL can study the exciton dynamics in MQW that exhibit well-thickness fluctuations; such investigations were done on CdSe/ZnSe MQW by Yang et al. (1996). Experiments showed that the temporal behavior of PL is determined by three concurrent processes: (i) localization of excitons into local bandgap minima with a time constant of about 4 ps, (ii) tunneling of localized excitons through local minima accompanied by phonon emission, and (iii) recombination of localized excitons within a time constant of 470 ps. The redistribution of excitons in an inhomogeneous band of states is seen as a redshift in energy and a narrowing in the emission spectra. The exciton recombination lifetime, exciton rates for formation, cooling and localization, and the migration rate to lower energy states can be obtained from time-resolved PL experiments at different optical excitation energies, by fitting the data with a sum of exponentials describing these phenomena.

Time-resolved as well as nonlinear PL spectroscopy under various excitation conditions were used to investigate the recombination dynamics of localized excitons in one monolayer-thick highly strained CdSe QW in a CdSe/ZnSe/ZnS_xSe_{1-x} structure (Yamaguchi et al., 1996). Generally, in ternary and quaternary II-VI semiconductors the alloy composition has a statistical variation at a microscopic level, the alloy disorder inducing a tail in the density of states extending down exponentially into the forbidden bandgap. The excitons can easily localize on these states, the PL peak energy being situated at a lower energy than the absorption, the difference between these two peaks being known as the Stokes shift. In the CdSe/ZnSe QW the exciton localization is caused by interfacial disorder, the PL spectra being dominated by sharp lines corresponding to different localized excitons. Time-integrated PL experiments at 20 K and at various excitation powers revealed the presence of a localized exciton X_1 identified as a peak in the spectrum for low excitation density, another peak X_2 at 11 meV below it being observed at higher excitation densities. Varying the photon energy from the low-energy tail to the high-energy tail of the PL peak it was found that the lifetime of the localized exciton changes from 200 ps to 50 ps. The exciton lifetime can be expressed as $\tau(E)^{-1} = \tau_r^{-1}\{1+\exp[(E-E_{me})/E_0]\}$ where $E_0 \cong 8.5$ meV is the energy of the density of states, $\tau_r \cong 200$ ps is the radiative lifetime and $E_{me} \cong 2.72$ meV the energy for which the decay time is equal to the transfer time. For excitation energies near X_2 the time decay of the PL line has a slow and fast component and must be modeled by a double exponential time evolution. The dependence of the PL intensity on the excitation intensity has the form $I_{PL} = aI_{ex} + bI_{ex}^2$. Nonlinear PL experiments, made by focusing on the sample two excitation beams with different modulation frequencies Ω_1 and Ω_2, and detecting the $\Omega_1 + \Omega_2$ component, showed that at high excitation densities many-body effects of localized excitons are present. These cause the change of sign of the nonlinear signal corresponding to X_1, from positive to negative, due to band-filling effects of the localized exciton X_2.

The polarization decay of a homogeneous broadened single QW in transient FWM offers information about the dephasing in the quasi-2D exciton-biexciton system (Langbein and Hvam, 2000). From temperature and excitation density experiments it was inferred that in this quasi-2D system the main processes are radiative decay and phonon scattering. The radiative decay rate of biexcitons is comparable to that of excitons and the accompanying spontaneous photon emission from excitons and biexcitons are uncorrelated. On the other hand, the biexciton phonon scattering is twice as fast as the exciton-phonon scattering and is correlated to it, so that excitons and biexcitons interact with similar phonon modes. The FWM experiments performed between 5–80 K on a single, 25 nm wide $GaAs/Al_{0.3}Ga_{0.7}As$ QW, provide the possibility of discriminating between the transition from the ground state $|0\rangle$ to an active exciton state X, and that from X to a biexciton bound state XX or unbound state XX*, using the polarization selection rules. Theoretically, the selection rules are deduced from a five-level optical Bloch equation. For example, for the $\uparrow\uparrow$ polarization ($\boldsymbol{k}_1$, $\boldsymbol{k}_2$ directions) the FWM signal is resonant to the X-XX transition.

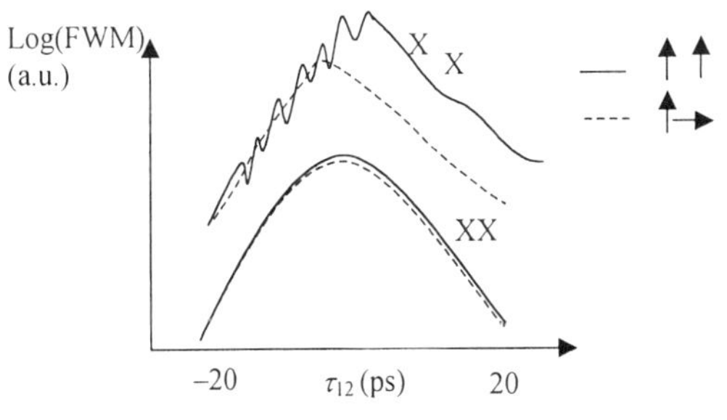

Fig. 6.31. FWM signal dependence on τ_{12} for different polarizations

The signal E_{XX} originates in phase-space filling for $\tau_{12} > 0$ and is due to two-photon coherence (TPC) when $\tau_{12} < 0$. Moreover, for this polarization the induced FWM signal resonant at E_X disturbs the signal resonant at E_{XX} due to the strong excitation induced dephasing. This disturbance can be avoided in the $\uparrow\rightarrow$ polarization. So, now the decay of E_{XX} for $\tau_{12} < 0$ is proportional to the linewidth γ_{XX_g} of the $|0\rangle$-XX transition due to TPC, while for $\tau_{12} > 0$ it is proportional to the linewidth γ_X of the transition $|0\rangle$-X. γ_{XX}, $\gamma_{XX}{}^*$, the linewidths of the X-XX and X-XX* transitions, respectively, are obtained from fitting the shift with temperature of the peak positions of E_{XX}, $E_{XX}*$ with a coherent superposition of two Lorentzians. The relation between γ_{XX}, γ_X, γ_{XXg} provides information about the scattering correlation R. If γ_X, γ_{XXg} are caused by a normal distribution of elastic Markovian scatterers, $\gamma_{XX} = \gamma_X + \gamma_{XXg} - 2R(\gamma_X\gamma_{XXg})^{1/2}$, where $|R| < 1$. Uncorrelated scattering corresponds to $R = 0$. Experiments have shown that the correlation increases with temperature. The linewidths are also temperature dependent; for example $\gamma_X = \gamma_0 + aT + b/[\exp(E_{LO}/k_BT) - 1]$, a similar relation holding also for γ_{XXg}. The FWM signal dependence on τ_{12} for both polarizations and for the excitation of either the X line or the bound biexciton XX line, is schematically represented in Fig. 6.31.

The dependence of exciton dephasing time on temperature and density for different QW thicknesses was studied in $In_xGa_{1-x}As/GaAs$ using degenerate FWM (Borri et al., 1999b). The exciton phonon-scattering contribution to the dephasing is obtained by extrapolating the dephasing rate to zero-exciton density. The temperature dependence of this rate gives the linewidth broadening coefficient for acoustical and optical phonons. These coefficients are defined through $\gamma_0 = \gamma_{00} + aT + b/\exp[(\hbar\omega_{LO}/k_BT) - 1]$. The coefficient a of acoustical phonons increases from 1.6 to 3 μeV/K when the well width varies from 1 to 4 nm because the exciton wavefunction penetrates into barriers in the case of thin quantum wells. No systematic dependence on well thickness is found for the optical phonon coefficient, which has almost the same value as in bulk GaAs. The acoustical phonon contribution to the zero-density dephasing rate is dominant below 70 K, the LO phonon scattering being predominant at higher temperatures.

The interaction and dephasing of center-of-mass quantized excitons in wide $ZnSe/Zn_{0.94}Mg_{0.06}Se$ QWs was also studied with the help of FWM (Wagner et al., 1998). For a homogeneous broadened exciton resonance T_2 is extracted directly from the spectral width of the FWM spectra as $\Gamma_{hom} = h/(\pi T_2)$. When there are several contributing mechanisms to the FWM, they can be discriminated through FWM polarization dependence. At negative delay times the FWM signal can originate from local field effects (LFE), exciton-induced dephasing (EID) and biexciton formation (BIF). The first two contributions decay twice as fast at negative delays than at positive time delays; the EID signal vanishes for the ↑→ polarization while LFE and BIF are not polarization dependent. So, from the FWM dependence on the angle θ_{12} between the linear polarization directions of the excitation pulses, $I(\tau_{12}) \approx \sin^2\theta_{12} + [1 + \sigma_{EID}/\gamma_2(n_x = 0)]^2\cos^2\theta_{12}$, the cross-section of the EID, σ_{EID}, and $\gamma_2(n_x = 0)$ can be determined. The BIF contribution can be observed for both negative and positive delay times and disappears in the $(\sigma^+\sigma^+)$ configuration. The signature of this contribution to the FWM is the spectral splitting of a part of the signal by the biexciton binding energy below the exciton energy; the biexciton formed by both simultaneously excited 11h and 11l excitons becomes visible in the ↑→ configuration and E_{bXX} can be determined. The BIF contribution to the FWM signal is greater than the LFE but smaller than the EID contribution. The spectral width of the FWM signal is different for coherent and incoherent exciton-exciton scattering, but in both cases its dependence on the exciton density n_X has the same form $\Gamma_{hom}(n_X) = \Gamma_{hom}(n_X = 0) + \gamma_X^{c,ic} n_X$. The corresponding interaction parameters for coherent and incoherent exciton-exciton scattering, γ_X^c and γ_X^{ic}, respectively, can be determined for different excitation conditions, with n_X measured from the absorption strength. The exciton-phonon scattering parameters are obtained by fitting the temperature dependence of the dephasing time at low n_X with $\Gamma_{hom}(T, n_x) = \Gamma_{hom}(T = 0, n_x) + \beta_{ac}T + \beta_{LO}[\exp(E_{LO}/k_B T) - 1]$ where β_{ac} and β_{LO} are the scattering parameters of acoustic and LO phonons.

In quantum wires the dephasing of excitons due to scattering by acoustical phonons is given by the temperature dependence of the time-integrated FWM. The phase relaxation time T_2 and the homogeneous linewidth of the FWM signal are related by $\Gamma_{hom} = 2\hbar / T_2$. Since the broadening depends on the temperature as $\Gamma = \Gamma_{hom}(0\,\mathrm{K}) + \gamma_{ac}T$, $\Gamma_{hom}(0\,\mathrm{K})$ and the acoustic-phonon scattering parameter γ_{ac} are determined from fitting. γ_{ac} scales with the wire width L as $1/L$ due to the lateral confinement of the exciton wavefunction.

The phase relaxation time T_2 at 2 K in CuCl spherical quantum dots embedded in glass, determined from transient FWM, is found to be very long, about 130 ps (Kuribayashi et al., 1998). This value is much longer than in the bulk since in QD all quasiparticles such as electron-hole pairs, excitons, acoustic phonons, are characterized by new quantum numbers and discrete energy levels, which change the selection rules and the interaction strengths between different quasiparticles. In spherical dots with radius $R_0 = 5.4$ nm in which the exciton Bohr radius is about 0.7 nm, the excitons are only weakly confined. PL experiments revealed that there is no Stokes shift for the Z_3S exciton in CuCl dots, which means that this is a free exciton. FWM experiments made in the Z_3S exciton transition region demonstrated that T_2 is governed by the energy relaxation time T_1 at low temperatures and becomes shorter with increasing temperature.

Both spectral hole burning and FWM have been employed to study the dynamics of excitons in CuBr nanocrystals (Valenta et al., 1998). Since the exciton-optical phonon interaction is strong in these nanocrystals, the holes consist of a zero-phonon line (at the excitation energy) and phonon sidebands. The Huang–Rhys factor S, which characterizes the strength of the electron-phonon interaction, is determined from the Stokes shift $\Delta_S = 2S\hbar\Omega$ between the maxima of excitonic absorption and emission, with Ω the energy of the coupled LO phonon. The lower limit of the exciton dephasing time T_2, found from the width of the zero-phonon holes, is about 2.3 ps. When the radius of the nanocrystals is decreased the spectral holes broaden and the persistent spectral hole burning becomes more pronounced. The temperature dependence of the width of the spectral holes can be modeled with $\gamma_{hom}(T) = \gamma_0 + AT + B\bar{n}(T) + C\bar{n}^2(T)$, where A, B and C describe the strength of acoustic-phonon scattering in the high-temperature approximation, and one-LO and two-LO phonon scattering, respectively, with $\bar{n}(T) = 1/[\exp(\hbar\omega_{LO}/k_B T) - 1]$. These parameters can be obtained from fitting.

6.3.12 Many-Body Effects

One of the many-body effects is FES (Fermi edge singularity), the coherent response of the Fermi sea to the creation or annihilation of the photocreated hole in the absorption or emission process. FES is observed in 2DEG, where it appears as a singularity at the Fermi level in the absorption or emission spectra. This singularity has a power-law dependence on energy $I_{\mathrm{FES}} \approx (E - E_0)^{-\alpha}$, with E_0 the HH or LH transition energy, which broadens and disappears as the temperature increases in a much sharper way than the excitonic transitions. The power-law exponent is related to the number of electrons that screen the photocreated hole, and depends on the electron density. Due to the attraction between the electron in the Fermi sea and the photogenerated hole, the multiple electron scattering in the collective charge motion involves only final states near the Fermi level, the FES being in fact a correlation enhancement. Theoretically, FES is observed in PL if the holes are localized and/or coupling of electrons at the Fermi level to empty conduction subbands takes place. Hole localization allows the indirect recombination of electrons at the Fermi level with $k = k_{\mathrm{F}}$ and the photogenerated hole with $k = 0$. FES disappears at high temperatures since the thermal activation of localized carriers induces a progressive recovery of momentum-conserving recombination. A typical FES curve in PL is shown in Fig. 6.32; the resonance-like structure at Fermi level is superimposed on an almost rectangular PL, which reproduces the step-like density of states in the 2DEG. The onset of the PL spectrum at the bandgap energy is due to band-to-band (wavevector-conserving) transitions involving recombination of conduction electrons with free holes. FES corresponds to band-to-acceptor (nonconserving of wavevector) transitions involving localized holes (on acceptors, for example) and electrons with wavevector between 0 and k_{F}. FES in PL and absorption is smeared out over the effective hole bandwidth $(m_{\mathrm{e}} / m_{\mathrm{h}})E_{\mathrm{F}}$.

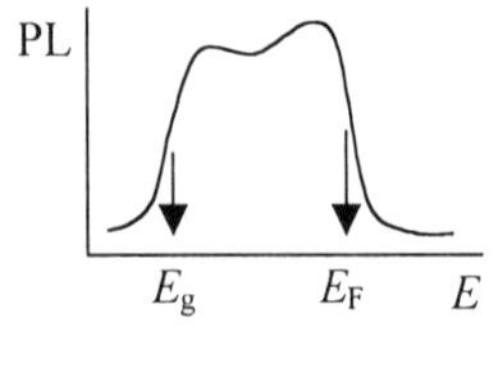

Fig. 6.32. Typical FES signature in PL

Experiments with II-VI modulation-doped QW, in which the holes are localized by compositional disorder, allowed the determination of the localization length of holes from the temperature dependence of FES (Coli et al., 1997). FES in PL can also appear for high-quality QW with delocalized, nonequilibrium population of holes at k_{F} (Brown et al., 1997). In p-type modulation-doped indirect-gap SiGe QWs FES has been also observed. In these structures the FES mechanism is different from that in direct-band QWs discussed above. Here, the radiative recombination itself has to be phonon-assisted to conserve the wavevector, the dominant mechanism being that of nearly resonant scattering between the electronic states at Fermi level and the next unoccupied subbands of the 2DHG (2D hole gas) (Buyanova et al., 1996).

FES can also appear in PLE, in high-density 2DEG with delocalized holes, if electron mobility is sufficiently high. PLE experiments performed on high-density, high-mobility 2DEG in GaAs QW have revealed FES behavior (sharp onset of PLE at the Fermi level) at both HH and LH transitions, with power exponents $\alpha = 0.3$ for the HH peak, and $\alpha = 0.38$ for the LH peak (van der Meulen et al., 1999).

In modulated-barrier quantum wires, if the excitons on one subband form an almost ideal gas in high magnetic fields, the renormalization of one subband is determined by the occupation of other subbands. The formation of an ideal exciton gas is possible due to the restoration of symmetry under continuous rotation in the electron and holes isospin space in high magnetic fields. This symmetry is lost if no magnetic field is present due to lateral confinement of carriers. The density and temperatures of the magnetoplasma can be calculated by fitting the PL spectra with theory. The PL peaks shift to lower energy values for increasing excitation, and their intensities increase; these changes in the fully quantized system, in high magnetic fields for which ω_c is larger than the lateral intersubband energy, occur discontinuously, when higher-lying states are significantly occupied with carriers (Bayer et al., 1997).

In high-density InAs/GaSb QW the 2D plasmon dispersion relation is $\omega_p^2 = \omega_{pl}^2 + (3/4)v_F^2 q^2$, where $\omega_{pl}^2 = N_s e^2 q/(2\varepsilon m)$ is the local plasmon frequency with N_s the 2D density of the 2DEG, ε the effective dielectric function of the medium surrounding the 2DEG and m the effective mass. The last term in ω_p^2 is the Fermi pressure in 2DEG, and has usually a very small contribution. The Fermi pressure is present also in 3DEG, the contribution being the same if the factor (3/4) is replaced by (3/5) and with a corresponding expression for the local plasmon frequency. The Fermi pressure describes the nonlocal dielectric response. Although the Fermi pressure has a very small value, measurements of the magnetoplasmon frequency $\omega_{mp}^2 = \omega_{pl}^2 + \omega_c^2$ in InAs/GaSb have succeeded to demonstrate it. At $B = 0$, a small difference between the calculated ω_{pl}^2 and the measured ω_p^2 has been observed, attributed to the Fermi pressure. In order to see this difference the plasmon wavevector q had to be artificially enhanced by applying periodic metal stripe grating couplers on top of the sample (Nehls et al., 1996).

In n-modulation doped QWs (n-MDQW) the carriers shift the absorption edge and renormalize (decrease) the E_g as a consequence of many-body exchange and correlation corrections due to the presence of free carriers. Bandgap renormalization (BGR) as a function of the 2DEG density can be easily studied in n-MDQW since the 2DEG density can be reduced by applying an in-plane electric field or by photon excitation with $\hbar\omega > E_g$. The density of 2DEG can, however, be controlled by illumination with photons with $\hbar\omega < E_g$; the blueshift of the magneto-PL spectra (MPL) with increasing optical power suggests a decrease of the 2DEG density caused by the drift of the holes created at barriers, to the GaAs QW by the built-in electric field, and their subsequent recombination with electrons in 2DEG (Plentz et al., 1993). The renormalized bandgap is obtained by extrapolating the MPL peak energy versus B to zero magnetic field. From the difference, ΔE, between the energies of the onset of fundamental absorption and PL peak at zero magnetic field, it is also possible to estimate the Fermi level from $E_F(1+m_e/m_{hh})=\Delta E$. The decrease of E_F with excitation intensity, from 28 meV at low excitations to 21 meV at high excitations, confirms the decrease of the 2DEG density. The decrease of E_g with excitation intensity is also observed in PLE spectra at zero field and different optical powers. Similar experiments can be made in p-MDQW, the BGR varying in this case with the hole concentration as $\Delta E_{BGR}(\text{meV}) = 2Cp^{\alpha}(\text{cm}^{-2})$, where $\alpha = 0.32$ and $C = 2.2\times 10^{-3}$ for GaAs wells and $Al_{0.32}Ga_{0.68}As$ barriers. Time-resolved PL spectra at various delays δ should therefore show a redshift of the PL peak since the 2D hole concentration varies in time as $\Delta p(\delta) = \Delta p_0 \exp[-\delta/\tau(\delta)]$, as is schematically represented in Fig. 6.33.

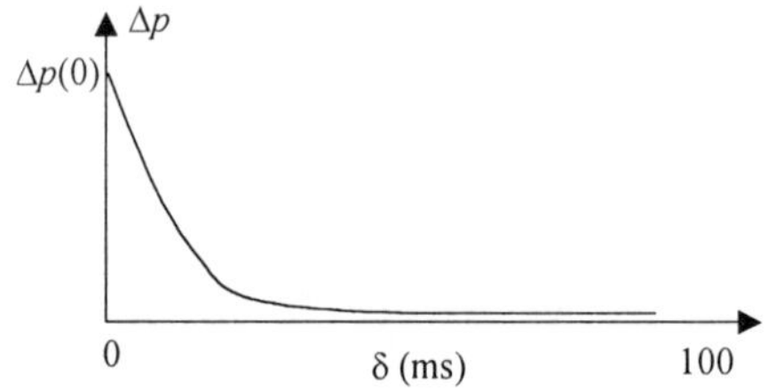

Fig. 6.33. Hole concentration decrease due to bandgap renormalization

6.3.13 Coupled Quantum Wells

Coupled quantum wells (CQW) are a succession of QW separated by thin barriers such that the wavefunction of the confined carriers in one QW can penetrate through the barrier into adjacent wells. Optical studied can reveal many properties of CQW.

The relative populations of the electron subbands E_1, E_2 in one QW can be determined from the relative PL intensity peaks due to electron recombination from these levels, when the lower level E_1 is in resonance with the electron level $E_1{}^*$ in the second QW (Li et al., 1998). The resonance between E_1 and $E_1{}^*$ can be achieved by applying a bias; at resonance the PL energy peak splits in two peaks, corresponding to the symmetric and antisymmetric states of the resonantly coupled QW. The electron wavefunction of the E_1 level becomes delocalized in the CQW structure, so that a population inversion between E_2 and E_1 levels occurs. Due to this, the two PL peaks observed at resonance involve both E_1 and E_2 levels, the ratio of the electron population being given by the ratio of the respective PL peak intensities (see Fig. 6.34). Experimental results showed that $n_2 / n_1 \cong 3$ at resonance. This value is smaller than the real population inversion since PL measures only a fraction of the E_1 level population, which is accessible to PL and relaxes to the subband minimum in N steps by sequential phonon emission $n_1^{PL} = n_1[\tau_1 /(\tau_1 + \tau_i)]^N$, with τ_1 the tunneling-out time from E_1 and τ_i the intrasubband LO-phonon emission time.

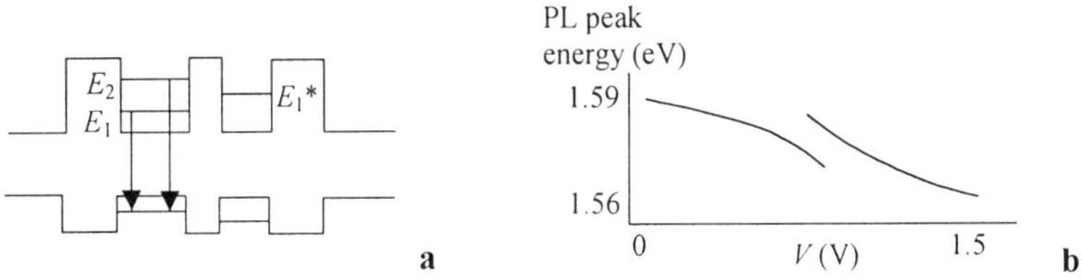

Fig. 6.34. (**a**) Energy levels in a coupled-QW system and (**b**) PL peak energy dependence on bias

Cyclotron resonance (CR) transitions in CQW have been studied by optical transmission in both quantum and semiclassical limits (Arnone et al., 1997). These two limits correspond to low and high filling factors $\nu = hN / eB$ with N the carrier density and B the magnetic field component normal to the 2DEG plane. At resonance, the density of carriers of strongly coupled QW separates into a symmetric and an antisymmetric subband, with carrier densities N_S, N_A, only the lower symmetric subband being occupied at B = 0. The transition from quantum to semiclassical limits can be made for example by tilting the applied magnetic field. The quantum limit corresponds to small parallel magnetic fields, when the electron motion along the confinement direction (z) is coupled to the cyclotron motion in the plane of the wells, and is characterized by absorption peaks corresponding to transitions between Landau levels in the same subband and anticrossing between Landau levels associated with different hybridized subbands. The semiclassical limit corresponds to large parallel fields when second CR absorption peaks appear in transmission curves, due to relaxation of the selection rules which allow transitions between different Landau levels associated with each of the electronic subbands. These transitions are enhanced when the subband separation in wells $\Delta = \hbar\omega_{ac}$ where $\hbar\omega_{ac}$ is the cyclotron energy at the anticrossing field $B_\perp = B_{ac}$. At this anticrossing field, the two CR peaks obtained by splitting the single CR peaks at B = 0, have minimum separation and equal peak intensities. Thus, by plotting the differential transmission $-\Delta T / T = (T_{ref} - T)/ T_{ref}$ as a function of the magnetic field, where T_{ref} is the reference transmission when both wells are totally depleted of electrons, one can identify the subband separation Δ in the wells.

The possibility of carrier tunneling between adjacent QW in CQW modifies the recombination kinetics and transport properties in these structures. The interaction between adjacent wells during the collection of carriers photogenerated in the barrier continuum can be studied by monitoring the PL and PLE spectra for excitation energy larger than the bandgap. A typical behavior for two CQW is shown in Fig. 6.35.

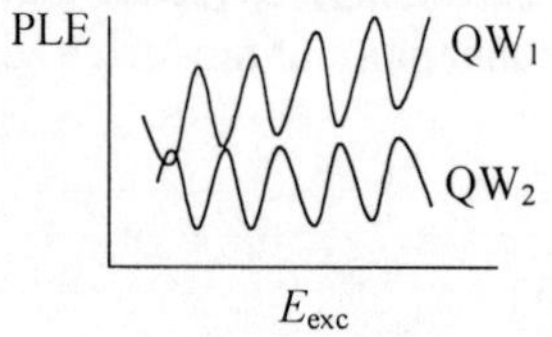

Fig. 6.35. Typical behavior of PLE of CQW for excitation energy larger than the bandgap

The oscillations in the PLE as a function of the excitation energy E_{exc}, at a detection energy equal to the HH exciton transition, are due to a modulation of the electron-collection efficiency caused by strong electron-LO phonon coupling. The average oscillation period for $In_xGa_{1-x}As/GaAs$ double QW is 0.42 meV, the oscillations corresponding to the two QW having opposite phases, which suggest a competition in the capture process of electrons in the continuum barrier states between the two QW. The carrier collection rates in the two CQWs become dependent on energy due to a phonon cascade relaxation process of the photoexcited electrons prior to their trapping in the QWs (Borri et al., 1997).

Optical studies can also reveal the exciton tunneling phenomenon in asymmetric double QW (ADQW). In III-V ADQWs the electrons and holes tunnel independently, not as an exciton, due to the small exciton binding energy. In widegap II-VI ADQWs, however, the exciton energy is an order of magnitude larger, and the Fröhlich electron-phonon interaction is enhanced due to the more ionic character of these compounds. In particular, in (Zn,Cd)Se/ZnSe ADQW the exciton binding energy is as large as 40 meV, comparable to electron and hole subband separation and the LO phonon energy ($\cong$ 31.7 meV). PL measurements of an ADQW formed from a narrow well (NW) and a wide well (WW), shows only the WW HH exciton recombination. No NW features appear since the tunneling-controlled lifetime of these excitons is very short. However, HH and LH NW excitons are present in PLE spectra, which demonstrates that excitons tunnel out of the NW into WW, where they recombine. The tunneling time has been estimated at about 1 ps from differential transmission measurements, which show a fast, complete recovery of the transmission change even for very high carrier density. This fast tunneling rate can be explained by the coupling of excitons and phonons through the Fröhlich interaction potential, the tunneling rate being

$$\tau^{-1} = (Me^2\omega_{\mathrm{LO}} / 2\pi\hbar^2)(1/\varepsilon_\infty - 1/\varepsilon_0)\int_0^{2\pi} \mathrm{d}\theta \int_{-\infty}^{\infty} \mathrm{d}q \,|\, V(Q,q)|^2 /(q^2 + Q^2), \qquad (6.53)$$

(Ten et al., 1996). The exciton does not tunnel as a whole by the emission of one LO phonon (which would correspond to a very slow process), but tunnels via an indirect state in a two-step process whose efficiency increases dramatically due to Coulomb effects.

The spatial character of the fundamental state exciton as well as its binding energy in coupled symmetric GaAs/$Al_{0.3}Ga_{0.7}As$ double QW can be modified by changing the AlAs barrier width. In wide barriers the symmetric and antisymmetric single-particle states are mixed by the Coulomb interaction, the exciton states having either a direct or an indirect character. In an applied magnetic field the direct exciton binding energy enhances more strongly than for indirect excitons and the oscillating strength is redistributed from indirect to direct excitons simultaneously with an increase of the splitting of indirect and direct excitons. The same behavior of the binding energy and oscillator strength with the magnetic field is observed also for narrow barrier structures, in which spatially direct and indirect character are strongly mixed in all exciton states. At resonance, the splitting between symmetric and antisymmetric single-particle states is $\Delta_{e/h} = E^{A}_{e/h} - E^{S}_{e/h} = (E_{e/h}/\pi)\exp[-(2m_{e/h}E_{e/h})^{1/2} L_b/\hbar]$, where $E_{e/h}$ is the energy of the electron/hole in the decoupled single quantum wells, and L_b the barrier width. In an applied magnetic field, the excitonic transitions transform into transitions between electron and hole Landau levels with increasing B, the transitions taking place between levels with the same Landau quantum number N. the transition energy is $E = E^{e}_{S/A} + E^{h}_{S/A} + \hbar(\omega^{e}_{c} + \omega^{h}_{c})(N+1/2) - E^{X}_{b}((N+1)_{shc})$ where $N = n-1$, with n the quantum number of the excitonic transitions, $E^{X}_{b}((N+1)_{shc})$ is the exciton binding energy of the $(N+1)_{shc}$ exciton, where e labels s for the s exciton, h for LH or HH, and c for L (lower energy) or H (higher-energy set of transitions). In high magnetic fields $\hbar\omega_c >> E^{X}_{b}$ the binding energies of pure direct (D) and indirect (I) excitons are given by $E^{X,D}_{b} \approx e^2/l_m$, $E^{X,I}_{b} \approx e^2/(l^2_m + d^2)^{1/2}$ with $d = L_b + L_w$ the effective spatial separation between quantum wells and $l_m = (\hbar/eB)^{1/2}$ the magnetic length (Bayer et al., 1996).

Exciton condensation, analogous to Bose–Einstein condensation of bosons, is expected to occur in CQW due to the long lifetime of interwell excitons. In zero magnetic field and dense electron-hole systems, the condensation temperature T_c is determined by the dissociation of condensed pairs. In a high magnetic field normal to the well plane, the center-of-mass motion and the internal structure of the excitons are coupled, the average in-plane spatial separation between electrons and holes being proportional to the exciton center-of-mass momentum. The critical temperature for condensation in a high magnetic field is $T_c = E_b(1-2\nu)/[2\ln(\nu^{-1}-1)]$ for $0 \le \nu \le 1$ with ν the Landau level filling factor, where the binding energy of the magnetoexciton is $E_b = (\pi/2)^{1/2} e^2/(\varepsilon l_m)$ with l_m the magnetic length (Butov and Filin, 1998). The exciton condensation, favored by the high magnetic field, is observed as an increase in the radiative decay rate and as a huge noise in the integrated exciton PL intensity.

6.3.14 Carbon Nanotubes

Carbon nanotubes are the latest members of the family of low-D structures. Unlike the structures we have referred to up to this point, they are not made from semiconductor materials, but from carbon. More precisely, carbon nanotubes are molecular wires consisting of giant fullerene molecules that can be modeled as a seamless cylinder having a diameter of a few nm and made by wrapping up a 2D graphite sheet with a hexagonal network. Due to the 'molecular' nature of carbon nanotubes, they have different properties compared to other semiconductor heterostructures, and therefore we have put them in a separate section. Carbon nanotubes can be single- or multi-walled, metallic or semiconductor in character. In a single-walled nanotube the edges of the cylinder are seamlessly jointed together and the ends of the cylinder are closed. Several concentric cylinders form multi-walled carbon nanotubes having a radius of 30–40 nm. Depending on the wrapping of graphene sheet (chiral angle) and radius, a single-walled carbon nanotube can be a 1D metal or a semiconductor with $E_g < 1$ eV. Electrons are confined in the radial and circumferential directions and are able to move only in the direction of the tube axis, so that the metallic carbon nanotubes behave like a quantum wire and therefore are sometimes called quantum molecular wires. The electronic movement in the tube axis direction is a ballistic propagation over a distance of a few microns. Closely packed carbon nanotubes (single-walled or multi-walled) form a 3D structure named a nanotube bundle. A recent review of carbon nanotubes can be found in Harris (1999).

The optical properties of carbon nanotubes are strongly dependent on the polarization of the incoming optical field due to the strong anisotropy of these nanostructures (Lin et al., 2000). The dielectric function of a multi-walled carbon nanotube bundle is featureless in the region $\omega < 2\gamma_0 = 6.06$ eV, with γ_0 the nearest-neighbor overlap integral, and decreases monotonically as a function of ω up to $\omega = 2\gamma_0$ where there is a peak followed by a plasmon-edge. Reflectance spectra show the same peak and display a large difference near the peak between the values corresponding to light parallel or perpendicular polarized (see Fig. 6.36). However, the shape of the reflection coefficient is almost the same for both light polarizations. The reflectance experiments made on single-walled and multi-walled nanotube bundles revealed important differences between them regarding their optical properties. In this respect, the reflectance of a single-walled nanotube bundle decreases much faster in the region $\omega < 2\gamma_0$, having in this region many peak structures and a weaker π-plasmon edge.

Strong visible light emission of multi-walled nanotubes at room temperature is observed in PL experiments in the band 2.05–2.3 eV, beyond E_g (Han et al., 2000). The PL has a Gaussian shape in this band with a FWHM of about 0.8 eV at 300 K, this shape being preserved over the whole temperature range 10–300 K. A strong nonlinear behaviour was also evidenced through the superlinear intensity dependence of the emission bands on excitation intensity. A two-step transition process ($\sigma - \pi$ electrons) is responsible for these PL features that demonstrate that carbon nanotubes could be a future candidate for optical displays, optical memories and other nano-optics devices.

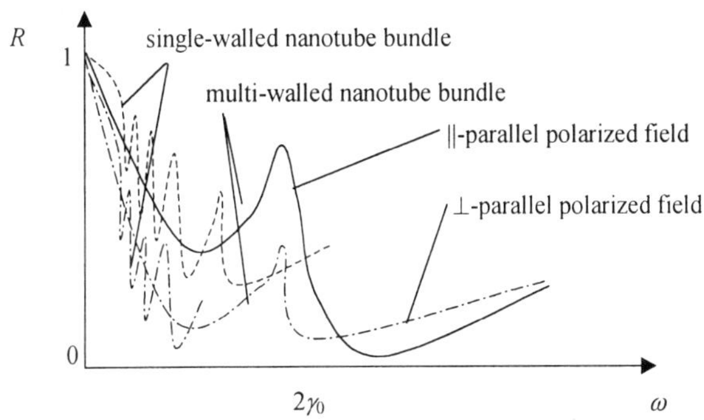

Fig. 6.36. Reflectivity of single- and multi-walled carbon nanotubes.

The properties of metallic carbon nanotubes can be evidenced using resonant Raman spectroscopy (Pinenta et al., 1998). In this respect, the Raman spectrum in a narrow region of laser energies (1.7–2.2 eV) shows special tangential phonon modes enhanced by electronic transitions between the first singularities in the 1D electronic density of states in the valence and conduction band ($v_1 \rightarrow c_1$). This enhancement (a broader band in the Raman intensity spectrum at the laser excitation energy of 1.9 eV accompanied by enhanced intensities of peaks) is a signature of metallic carbon nanotubes. If we have a bundle of metallic carbon nanotubes with a Gaussian distribution of diameters that varies between 1–1.6 nm, the overall enhancement of the intensity of the Raman modes is a sum of all contributions from each individual nanotube.

6.4 Optical Properties of Superlattices

One characteristic property of SLs, manifest in their optical spectra, is the folding of the Brillouin zone due to the increased periodicity along the SL growth axis. For example, if the primitive cell of the SL is equal to N_0 primitive cells of the original bulk crystals, each band of the original Brillouin zone gives rise to N_0 folded bands, the SL Brillouin zone becoming N_0 times smaller. Moreover, due to the induced superperiodicity in only one direction, the crystal symmetry is lowered. As a result, the selection rules are significantly altered and the appearance of the so-called folded modes, not present in the optical spectra of either bulk or QW components is favored.

When the SL is formed from materials with similar dispersion relations for acoustic phonons, the average dispersion relation of the constituents can be folded and the difference with the real SL can be treated by perturbation theory. The perturbation opens 'minigaps' in the folded dispersion relations at the center and edges of the SBZ. On the contrary, the dispersion relation of optical modes in the two constituents are in general well separated, so that the corresponding vibrations in the SL are confined to either one material or to the other, decaying rapidly and exponentially beyond the interface.

For small acoustic modulation the average frequency of the near-zone-center doublet in a SL with constituents A, B does not depend on the wavevector and is given by $\Omega_n = (n\pi/\Lambda)[\alpha/v_B + (1-\alpha)/v_A]^{-1}$ where $\alpha = L_B/\Lambda$, and $n = 0,1,2....$ This quantity is nearly independent of α, since the velocities v_A and v_B are nearly equal, its measurement enabling the determination of Λ from the distance between SL satellites that accompany the bulk reflexes (see references in Jusserand and Cardona (1989)). On the other hand, although the splitting $\Delta\Omega_n$ should depend strongly on the details of the structure, expressed through α, this dependence is not revealed in Raman experiments due to the finite wavevector of the phonons involved in scattering. The folded TA and LA modes are decoupled for vanishing in-plane wavevectors and their dispersion curves cross, since they have different symmetries. On the contrary, 'internal' acoustic gaps appear for finite in-plane wavevectors. Contrary to folded acoustic modes, which are mainly sensitive to the long range order, confined optical vibrations are sensitive to local properties, and therefore can provide information on the width and shape of the potential well in which they are confined, as well as on the strain between different layers in the SL. The observation of additional optic modes in SL is more difficult than for folded acoustic modes, because the optic phonon branches have lower dispersion.

The properties of Raman and Brillouin scattering in SLs on single or collective excitations were the object of volume V from the series Light Scattering in Solids, edited by M. Cardona and G. Güntherodt. We refer the interested reader to this excellent volume. One property of light scattering on free carriers that we mention here is the fact that in periodic multilayers, such as MQW and SL, the scattering amplitude from each layer has a phase factor $\exp(iq_\beta l\Lambda)$ where l is an integer that labels the layer, the total scattering intensity obtained by adding the contributions from all N layers being (Pinczuk and Abstreiter, 1989)

$$I(q,k_z) \approx |\textstyle\sum_{l=0}^{N-1}\exp[i(q-k_z)l\Lambda]|^2 = \{1-\cos[N(q-k_z)\Lambda]\}/\{1-\cos[(q-k_z)\Lambda]\}. \tag{6.54}$$

$I(q_\beta,k_z)$ has a peak at $q_\beta = k_z$ for large N values.

A consequence of folding in materials with indirect gaps is that the processes which start with absorption or emission of phonons, followed by the absorption of a photon (labeled by 1) can no longer be ignored. These processes were disregarded in bulk materials in which the processes involving the absorption of a photon followed by absorption or emission of phonons (denoted by a subscript 2 in Fig. 6.37) were dominant. The argument used to disregard them, namely that their probability is much smaller since E_{gk+q} at the X or L points at the boundaries of the Brillouin zone is larger than E_{gk} at the Γ point ($\boldsymbol{k} = 0$), is no longer valid in a SL since the extent of the Brillouin zone reduces from $\boldsymbol{q}$ to $\boldsymbol{q}'$ (see Fig. 6.37). This complicates the interpretation of the absorption spectrum.

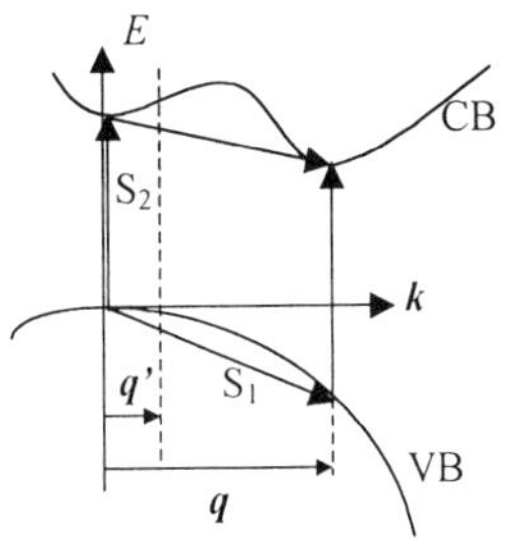

Fig. 6.37. Optical transitions in superlattice from indirect gap semiconductors

On the other hand, as we have already mentioned, the density of states in a SL is different from that in a QW, since the discrete states in the latter turn into minibands. As a result, the steps in the QW absorption broaden. More precisely, in type I SL where the conduction and valence band wavefunctions have the same symmetry, the absorption per well is similar to the QW absorption except that the Θ step functions in the joint density of states are replaced by (Vasko and Kuznetsov, 1999)

$$\Theta(\hbar\omega - E_{nn'}) \to \begin{cases} 0, & \hbar\omega < E_{nn'} - \Delta_{nn'} \\ \dfrac{1}{\pi}\arccos\left(\dfrac{E_{nn'} - \hbar\omega}{\Delta_{nn'}}\right), & -\Delta_{nn'} < \hbar\omega - E_{nn'} < \Delta_{nn'}, \\ 1, & \hbar\omega > E_{nn'} + \Delta_{nn'} \end{cases} \tag{6.55}$$

transitions between states of different parity being forbidden due to the parity selection rule. In type II SLs the conduction and valence band states have different transformation symmetry, transitions between the edges of the same parity minibands being forbidden while those at the zone center ($p_\perp = 0$) are allowed. The opposite rules hold for transitions between minibands of different parity. As a consequence, the absorption in type II SLs is different from the type I case, the arccosine in (6.55) being replaced with (Vasko and Kusnetsov, 1999)

$$\arccos\left(\frac{E_{nn'} - \hbar\omega}{\Delta_{nn'}}\right) + (-1)^{n+n'}\sqrt{1 - \left(\frac{E_{nn'} - \hbar\omega}{\Delta_{nn'}}\right)^2}, \tag{6.56}$$

Not only the expressions, but also the magnitudes of the absorption differ in type I and type II SLs; the absorption in type II SLs is much weaker than in type I since the overlap integral between the electron and hole wavefunctions is much reduced due to their localization in different materials.

The miniband absorption in SLs, which is observed in the IR spectral region, can be similarly treated; in this case the fact that the width of absorption peaks is determined also by the intrinsic width Δ^{SL} of the minibands involved in the transition must be explicitly accounted for. This can be achieved by performing an additional integration over $p_\perp$ in the general expression of ξ_ω using the appropriate expressions for the transition matrix elements between miniband states with a given $p_\perp$. The final result, after taking into account that the population difference factor is also dependent on $p_\perp$, is that the lineshape of interminiband absorption peaks is strongly temperature dependent. At temperatures lower than the miniband width the electrons buildup at the bottom of the SL miniband, the population difference factor favoring transitions from low $p_\perp$ and suppressing those from the top of the miniband.

The PL in SLs depends on the period; for example, in short-period $(GaAs)_n/(AlAs)_n$ SLs the PL is determined by the mixing of the low-lying Γ and X levels, whereas in long-period SLs it is dominated by direct transitions between Γ states in the conduction and valence bands. These transitions are also responsible for absorption processes, which cause the peaks in PL and PLE to almost overlap. On the contrary, in short-period SLs the absorption is controlled by transitions to higher-lying Γ states and luminescence occurs mainly from lower-lying hybridized Γ-X states, which creates a large shift between the absorption edge and the PL peak. The value of n that separates these two regimes, i.e. the type I SL from the type II with hybridized Γ-X states, is around 13–14.

6.4.1 Band Structure

From measurements of the optical absorption coefficient in Si_nGe_m ($n \cong m$) SLs, the bandgap and its dependence on the lattice period can be determined. The Si-Ge SLs have different band structures from those of the indirect band constituents; in particular the SLs have band-edge states at $\boldsymbol{k} = 0$, the optical matrix element between the top of the valence band and these states being several orders of magnitude larger than that of the lowest indirect transitions. The frequency dependence of the absorption coefficient $\alpha(\omega) \approx (\hbar\omega - E_g)^x$ is different for direct or indirect bandgap transitions, for which x = 1/2 and x = 2, respectively. For example, random Si-Ge alloys have indirect bandgap transitions, with x = 2, whereas for the pseudorandom alloy x = 3/2 and the energy gap is E_g = 0.76 eV. However, short-period Si-Ge SLs are characterized by none of these values, but rather by a linear dependence of the absorption coefficient on the photon energy, with x = 1, near the bandgap E_g = 0.71 eV, more precisely for $0.01\ \text{eV} < \hbar\omega - E_g < 0.1\ \text{eV}$. This linear dependence can be explained by the existence of a large number of conduction band states close to the bandgap energy throughout the Brillouin zone. The band-to-band absorption has thus both direct and indirect contributions (Pearsall et al., 1996). The exponent x in Si_nGe_m SL for energies close to the bandgap, depends also on the degree of interdiffusion across the Si-Ge interfaces: it has a value close to 0.5 for abrupt Si-Ge interfaces and close to 2 for highly diffused interfaces (Papadogonas et al., 1997).

Differences between random alloys and SLs optical properties have been also identified in laterally composition modulated GaP/InP short-period SL (SPS). A fundamental bandgap of 1.69 eV (for light polarized along the modulation direction y) has been obtained from electro-reflectance measurements of these SPS, with about 210 meV lower than the bandgap in GaInP random alloys with the same overall composition. In these SPS a spontaneous lateral composition modulation takes place. For off-resonance Raman scattering, the signal has contributions from both Ga-rich and In-rich regions, whereas for the incident photons in resonance with the fundamental bandgap of the lateral SL the Raman signal from In-rich regions is predominant, the wavefunctions of the ground-state electrons and holes being both localized in these regions. The phonon frequencies are therefore redshifted towards those for In-rich regions near resonance, the frequency shift determining the composition and strain in In-rich regions, which depend on both Ga and In compositions. In Raman spectra it was observed that the sharp GaP-like LO phonon, the InP-like LO phonon as well as the TO phonon move towards lower frequencies in the xx polarization as the excitation energy decreases, the intensity of the GaP-like LO phonon having a strong resonance near the fundamental bandgap. The dependence of the measured frequency ω_{meas} on the excitation energy E_{ex} can be explained by the introduction of a distribution $f(E)$ for the energy dependence of the local bandgap. Then, $\omega_{\mathrm{meas}}(E_{\mathrm{ex}}) = [(\omega_{\mathrm{nr}}^2 + |Cf(E_{\mathrm{ex}})\omega_{\mathrm{r}}(E_{\mathrm{ex}})|^2)/(1+|Cf(E_{\mathrm{x}})|^2)]^{1/2}$ with ω_{nr} the nonresonant phonon frequency, C the strength of the resonance, and $\omega_{\mathrm{r}}(E)$ the phonon frequency of the material with bandgap E; away from resonance $f(E_{\mathrm{ex}}) \cong 0$. From the measurement of the redshift of the GaP-like LO phonon in Raman spectra measured with incident and scattered photons polarized along the composition modulation direction, it was established that the average In composition in In-rich region is 0.7, while the average Ga composition in Ga-rich regions is 0.68 (Cheong et al., 1999).

Optical measurements can identify the conduction and valence band states involved in optical transitions. For example, piezomodulation spectroscopy of ultrashort-period GaAs/AlAs SL (2–8 monolayers) has shown that the SLs are indirect if grown on a GaAs substrate, and pseudodirect when grown on (In,Ga)As pseudosubstrates. GaAs/AlAs ultra-short period SLs are type II heterostructures, with the valence band maximum located at the Γ point in the GaAs layer, and the conduction band minimum located at the X valley states in the AlAs layer. The X valley states are split in X_z and X_{xy} subbands, the latter being higher in energy. When growing these SLs on an (In,Ga)As pseudosubstrate, the strain modifies the ordering of the X-like states, such that direct transitions between X_z electrons in AlAs and valence band in GaAs are favored (Rezki et al., 1998).

Fano interferences due to Fröhlich coupling between electronic and LO phonon states have been observed in inelastic light scattering in GaAs periodic δ-doped multilayers. This phenomenon appears when the modulation period is large enough such that the low-lying miniband states show a quasi-2D behavior, while higher-energy states display a nearly free-electron character. Also, charge-density oscillations in this intersubband continuum should coexist with Raman LO phonons polarized along the light scattering direction. The Fano interference with the profile $\sigma(\omega) = \sigma_0 + \sigma_1[(2q^2 - 1) + 2q\varepsilon]/(1+\varepsilon^2)$ can be observed in back-scattering Raman spectroscopy, where $\varepsilon = (\omega - \omega_{\mathrm{LO}} - \Delta\omega)/\gamma$ and σ_0 is a constant background superimposed on the Lorentzians $\sigma_i(\omega) = \sigma_i \sin^2(2\phi) \times[\gamma_{\mathrm{LO}}^2 + (\omega - \omega_{\mathrm{LO}})^2]^{-1}$ with ϕ the orientation of the [100] axis with respect to the linear polarization vector of the incident radiation. The two fitting parameters σ_1 and q are determined from experimental data obtained for different ϕ, the latter being related to the deformation potential a_{D} as $q = \mathrm{Re}(a_{\mathrm{D}})\sin(2\phi)/M_\gamma\gamma T$, with T the scattering rate amplitude associated with the intersubband continuum in the region of the LO-phonon, and M_γ the electron-phonon Fröhlich coupling (Bell et al., 1998). Different degrees of lineshape asymmetries are due to changes of the scattering amplitude of the deformation potential LO phonon relative to the scattering efficiency associated with the electronic channel.

The intraband optical absorption in SLs in the presence of an in-plane magnetic field was studied by de Dios Leyva and Galinda (1993). The theoretical model considers that the main contribution to the absorption comes from $0 \to \lambda$ vertical transitions ($\lambda \geq 1$) between parabolic bands, illuminated by radiation polarized parallel to the interfaces and perpendicular to the magnetic field, The features that appear in the absorption spectrum were discussed as a function of the ratio between the period of the SL, Λ, and the magnetic length $l_{\mathrm{m}} = (\hbar/eB)^{1/2}$.

The two lowest miniband widths and their dispersion in strongly coupled n-type GaAs/Al_xGa_{1-x}As SLs can be studied by monitoring the optical absorption (see Fig. 6.38).

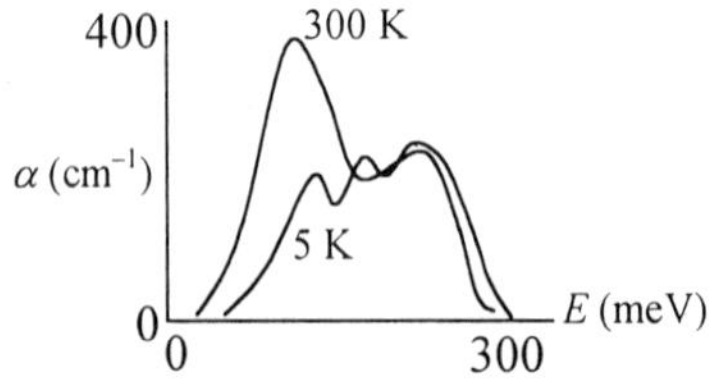

Fig. 6.38. Optical absorption at two temperatures in n-type GaAs/Al_xGa_{1-x}As SLs

The two peaks in the absorption spectrum at high temperatures are related to the critical points of the mini-Brillouin zone, and the asymmetric shape of the spectrum is caused by the variation of the oscillator strength over the mini-Brillouin zone. The absorption coefficient for transitions between the lowest minibands, i.e. between the first and second miniband, can be expressed as

$$\alpha = \frac{e^2 k_B T}{\varepsilon c \hbar^2 \pi m_{\mathrm{eff}} \omega} \int_0^{\pi/\Lambda} \mathrm{d}k_z \, |\langle 1 | p_z | 2 \rangle|^2 \ln\left(\frac{1+\exp[(E_F - E_1(k_z))/k_B T]}{1+\exp[(E_F - E_2(k_z))/k_B T]} \right) \times (\Gamma/\pi)/\{[E_2(k_z) - E_1(k_z) - \hbar\omega]^2 + \Gamma^2\}. \quad (6.57)$$

The two peaks in the absorption curve correspond to $k_z = 0$ and $k_z = \pi/\Lambda$ and are caused, respectively, by the higher occupation of states with finite in-plane wavevectors at $k_z = 0$, and from a stronger singularity in the joint density of states for $k_z = \pi/\Lambda$. At low temperatures, in heavily doped samples with metallic behavior but with the impurity band not yet merged with the conduction band, a third absorption peak appear due to 1s-2p$_z$ donor transitions, at an energy approximately equal to the separation between the lower edges of the first and second miniband. So, the width of the two minibands $\Delta_1 = 17$ meV and $\Delta_2 = 54$ meV are obtained from the energies of the three peaks, which correspond to $E_2(\mathrm{top}) - E_1(\mathrm{bottom})$, $E_2(\mathrm{bottom}) - E_1(\mathrm{top})$ and $E(2p_z) - E(1s) \cong E_2(\mathrm{bottom}) - E_1(\mathrm{bottom})$, (Helm et al., 1993).

The intervalley deformation potential for (GaAs)$_m$/(AlAs)$_m$ SLs can be determined from the temperature and pressure dependence of the type I and type II transitions in the PL. These types of transitions are shown in Fig. 6.39.

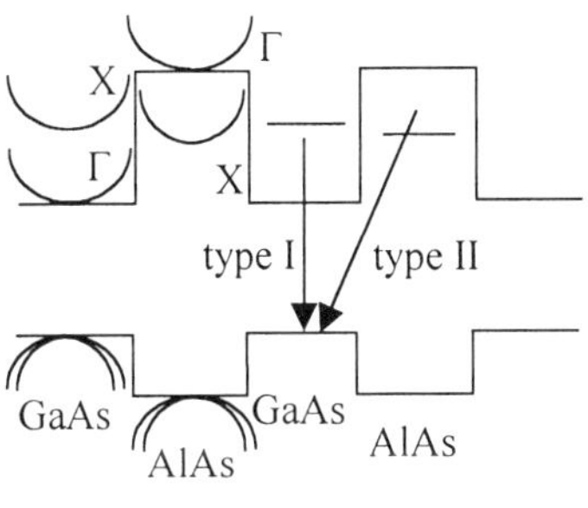

Fig. 6.39. The type I and type II transitions in a AlAs/GaAs SL

As a function of temperature and pressure, the dominant transitions can change from type I to type II. For example, for pressures higher than a critical value, the energy of electrons in the X well of AlAs become lower than that in the Γ well of GaAs and type II transitions are observed. Beyond type I-type II crossover, the linewidth of the PL increases as a function of temperature and pressure, the electron-phonon deformation potential for the Γ-X scattering being temperature dependent. The difference between the PL linewidth and the average linewidth below the crossover pressure can be defined as $\Delta = \pi |M_{\Gamma X}|^2 \rho_X(E) + \pi |M_{\Gamma L}|^2 \rho_L(E)$, where $|M_{\Gamma X,L}| \approx \Delta_{\Gamma X,L}$ is the temperature-dependent electron-phonon matrix for scattering between Γ and X, and between Γ and L, respectively, proportional to the deformation potential $\Delta_{\Gamma X}$, and respectively $\Delta_{\Gamma L}$. $\rho_{X,L}$ are the corresponding densities of states. From experimental data on the PL linewidth it is not possible to derive separately the deformation potentials $\Delta_{\Gamma X}$ and $\Delta_{\Gamma L}$, but assuming that $\Delta_{\Gamma L}$ is temperature-independent and equal to 6 eV/Å, both $\Delta_{\Gamma X}$ and its temperature dependency can be obtained (Guha et al., 1998). In the same type of GaAs/AlAs short-period SLs the transition from an indirect to a direct energy band can be induced by changing the distribution of the well and barrier thickness within the unit cell of the SL from a symmetric to an asymmetric one (Krylyuk et al., 1999). For a barrier thickness L_b as short as half the well thickness, L_w, the lowest state in the conduction band becomes the Γ point in the GaAs well, instead of the X point in the AlAs barrier, and the band structure becomes direct. The band structure is indirect for well thicknesses below 12 monolayers and equal to the barrier thickness, since the X minimum in the conduction band of AlAs is lower in energy in this case than the Γ state in GaAs. On the contrary in short-period SLs the X states of AlAs mix with the Γ states in GaAs, causing quasi-direct energy gaps in which the absorption or emission of electrons at the energy gap may occur without the participation of phonons. Moreover, if the ratio between the well and barrier thickness is larger than unity, direct energy-gap structure can be obtained, manifested in a significant increase of the PL. If this ratio is at least equal to 2, short-period SLs become direct for any well thickness. Due to the stronger coupling in the structures with $L_w = 2L_b$ (larger miniband width), the direct energy gap of these SLs lies below the direct gap of the structures with $L_w = L_b$.

In $(GaN)_m(AlN)_n$ SLs grown along the symmetry axis, as in any SL made of two binary materials with wurtzite structure and identical cation or anion, Raman measurements have confirmed the fact that the symmetry of the structure is different for odd and even $m+n$. The respective space groups are C_{6v}^4 and C_{3v}^1, the different point symmetries affecting non-monotonically the number of Raman-active modes as a function of $m+n$, whereas the number of IR-active modes depends monotonically on $m+n$. Therefore, Raman measurements can be used to distinguish the structures with even or odd $m+n$ (Kitaev et al., 1998).

SLs with very interesting properties are the finite-size mirror-plane SL (MPSL) formed from blocks of SL=$(GaAs)_{13}/(AlAs)_{18}$ and LS=$(AlAs)_{18}/(GaAs)_{13}$ arranged in a sequence $(SL)_{m/2}/(LS)_{m/2}$. Due to the interference of the scattering contributions from different quantum wells, the folded LA-phonon doublets in the MPSL suffer a pronounced splitting with satellites around the main lines, a phenomenon which cannot be observed in periodic $(SL)_m$ structures. This interference is due to the phase shift introduced by the mirror-plane symmetry, which can be modified by changing the width of GaAs buffers in $(SL)_{m/2}/(GaAs)_n/(LS)_{m/2}$ structures. The doublets observed in the Raman spectrum in the backscattering geometry along the growth direction z reflect the backfolded SL dispersion. In an infinite SL the sharp and pronounced doublets in the Raman spectrum are due to the crystal momentum conservation law, all the contributions of the scattering from GaAs layers adding constructively to the total intensity. In practice the doublet-lines intensity decreases due to interface roughness, which does not change the periodicity of the structure and therefore does not modify the width and position of the doublet lines.

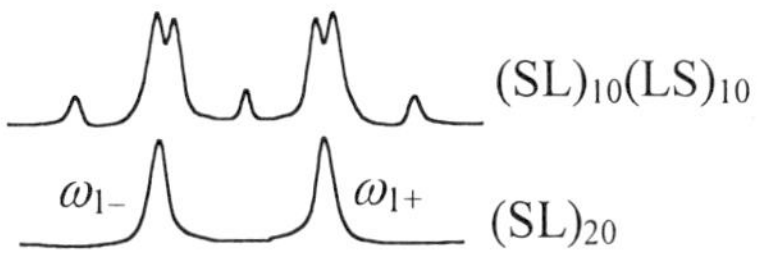

Fig. 6.40. Raman spectrum in a mirror-plane SL (top) and a SL with the same length (bottom)

Layer thickness fluctuations can introduce additional periodicity and thus induce a doublet fine structure in the Raman spectrum. In short-period SL, momentum conservation is relaxed due to the finite-size and weak satellites can appear around the folded-phonon peaks, caused by the standing waves of the whole SL. If L_{tot} is the total length of the SL, momentum conservation requires that the sum of the incident and scattered wavevectors, $k_i + k_f = q_z = \pi l / L_{\text{tot}}$, with $l = 0, \pm1, \pm2, \ldots$ the frequencies of the doublet peaks appearing at $\omega_{q_z} = |q_z| v_{\text{ac}}$ where the average sound velocity is $v_{\text{ac}} = (L_{\text{GaAs}} v_{\text{GaAs}} + L_{\text{AlAs}} v_{\text{AlAs}})/(L_{\text{GaAs}} + L_{\text{AlAs}})$. For SLs with a period $\Lambda = L_{\text{tot}} / m$ and $L_{\text{GaAs}} = L_{\text{AlAs}} = \Lambda / 2$, the average frequency of the folded-phonon doublets is $\omega_s = 2s\omega_{\text{Bragg}}$ with s the index of the zone-center doublet and $\omega_{\text{Bragg}} = v_{\text{ac}} \pi / \Lambda$, the frequency splitting of the doublet lines being $\Delta_{\pm} = \pm (q_z \Lambda / \pi) \omega_{\text{Bragg}}$. For MPSL, as can be seen in Fig. 6.40, the doublet frequencies are the same as in the SL with the same m, the interference effects inducing a frequency separation in the Raman spectrum between minor lines twice as large as in SLs with the same m (same thickness), the folded phonon doublets being split in two components. More complex interference patterns can be obtained by introduction of the buffer layers (Giehler et al., 1997).

SL can be made not only from layers of semiconductor materials, but also from arrays of metallic particles, for example. Silver nanoparticles can be arranged in well-organized 2D hexagonal array of isolated particles, which show in optical spectra, under p-polarized light, high-energy peaks corresponding to collective effects caused by the mutual interaction between particles. These high-energy peaks disappear for disordered and coalesced particles distributed randomly (Taleb et al., 1999).

Not only can metal particles spontaneously arrange in regular patterns, but also the semiconductor $GaInP_2$ disordered alloy. Spontaneous ordering lowers the symmetry of the point group from the zincblende cubic point group of the random alloy to rhombohedral point group with a threefold symmetry axis parallel to $\langle 111 \rangle$, doubling the real-space unit cell and producing folding states from L and Γ points. Alternating Ga-rich ($Ga_{1+\delta}In_{1-\delta}P_2$) and In-rich ($Ga_{1-\delta}In_{1+\delta}P_2$) monolayers appear normal to one of the four rotation axes [111], forming a SL. Brillouin-zone folding creates symmetry repulsion between the original and folded conduction band states at Γ, causing bandgap lowering with ΔE_g and valence band splitting with $\delta E_V \cong (1/5)\Delta E_g$. The latter parameter can be determined from the shift between the peaks of PL emissions polarized parallel and perpendicular to the electric field, which is applied in a direction normal to the ordering axis. The emerging of the ordering can be visualized in the Raman spectra through the appearance of the sharp folded transverse and longitudinal acoustic modes (FTA and FLA), superimposed on the broad disordered-induced longitudinal and transverse acoustic modes. FLA is observed only when the incident and scattered light have parallel polarizations, its intensity increasing with δE_V. FTA is observed in all geometries and its intensity has an opposite behavior with δE_V. Other modifications of the Raman spectrum due to SL formation include: the splitting of the GaP- and InP-like LO modes into doublets and the increase of the InP-like TO mode (Hassine et al., 1996).

The Raman selection rules of the spontaneously ordered phase in $GaInP_2$ are deduced from group theory. In $GaInP_2$ the bandgap of the ordered phase is by 100 meV less than that of the disordered phase. The effect of spontaneous $CuPt_B$ ordering on the lattice dynamics of $GaInP_2$ can be studied at room temperature using samples of various degrees of ordering. In the measurement on a (001) plane an extra peak at 354 cm^{-1} appears in the Raman spectrum corresponding to the LO mode due to the SL effect in the ordered alloy (Cheong et al., 1997). The origin of the extra peak is confirmed by selection rules for right-angle Raman scattering. The acoustic-phonon spectra range of (001) Raman spectra of the ordered samples show also extra peaks at 60 cm^{-1} and 205 cm^{-1} corresponding to folded transverse and longitudinal phonons, respectively. The Raman signal of $GaInP_2$ is presented in Fig. 6.41. In the $\bar{z}(y,z)x$ configuration only transverse modes are allowed, while both transverse and longitudinal modes are allowed in the $\bar{z}(x,z)x$ configuration.

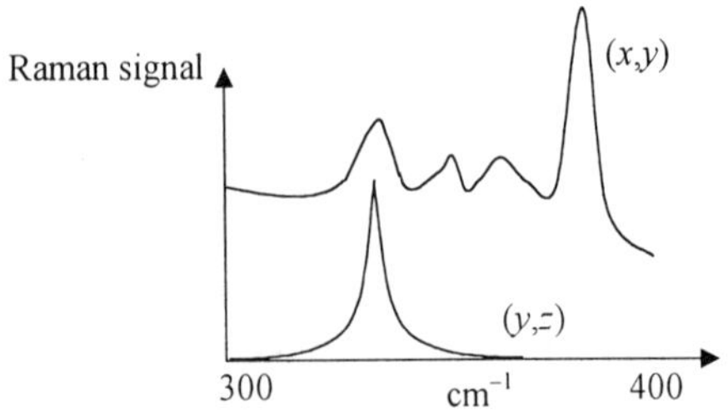

Fig. 6.41. Raman spectra for $GaInP_2$ at different scattering geometries

The selection rules and valence band crystal-field splitting of the ordered material are determined from edge-on polarized PL (Horner et al., 1993). In $GaInP_2$ the type A transition (between conduction band and A valence subband) exists only in the ordering-induced crystal field split-off band transition and is observed only for PL polarized along $x' = [1\bar{1}\bar{2}]$, $y' = [110]$ and $z' = [1\bar{1}1]$. Type B transitions (between conduction band and B valence subband) are observed for

x' and y' PL polarizations. The valence band crystal-field splitting is obtained as the difference between the two PL peaks for y' incident polarization, detected parallel to (solid line in Fig. 6.42), and normal to z', respectively (dotted line).

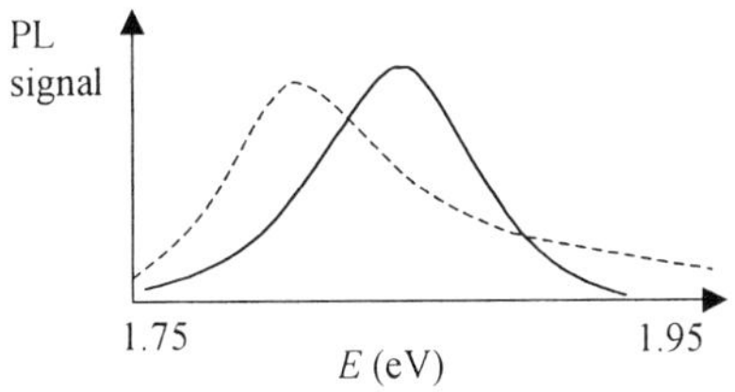

Fig. 6.42. PL of $GaInP_2$ for two incident polarizations: parallel (*solid line*) and normal to z' (*dashed line*)

Additional periodicity and thus additional folded phonons were observed in partially CuPt-ordered $Ga_{0.52}In_{0.48}P$ grown on GaAs (Kwok et al., 1998). The folded X-point zone-edge phonon mode was observed in addition to the folded L point mode in GaInP with two variants of CuPt ordering, due to antiphase boundaries between the $CuPt_A$ and $CuPt_B$ variants. This mode is resonantly enhanced for illumination with 2.25 eV.

Another example of an unusual folding pattern is the observation of folded phonons of all three acoustic branches for GaAs/AlAs corrugated SL grown along the [311] direction (Popović et al., 1993). In $z(x'y')\bar{z}$ Raman spectra the pure transverse acoustic-phonon doublet was observed shifted by 1.5 cm^{-1} from the quasitransverse doublet, accompanied by the weaker quasi-longitudinal doublet forbidden by the selection rules. In the $z(x'x')\bar{z}$ spectra the quasi-transverse and quasilongitudinal folded doublets showed almost the same strength, whereas in $z(y'y')\bar{z}$ the quasi-transverse doublet is about half of the quasi-longitudinal.

SLs from magnetic materials have interesting properties. For example, in a short-period biaxially strained $(CdTe)_n/(MnTe)_n$ SL, despite the antiferromagnetic ordering of the bulk Mn, the paramagnetic behavior dominates with increasing n due to the lack of spin ordering at CdTe/MnTe interfaces for wider CdTe wells. The paramagnetic contribution to the MOKE rotation is given by

$$\Delta\theta = m[B + (b/m)B_J(g_l J\mu_B B/k_B(T+T_0))], \tag{6.58}$$

with m, b fitting parameters, g_l (= 2) the g factor for localized Mn^{2+} electrons, and B_J the modified Brillouin function with J = 5/2. The small antiferromagnetic coupling of Mn ions responsible for the paramagnetic behavior characterized by T_0, increases slightly with decreasing n (of the layer thickness), whereas b/m increases with n (Pohlt et al., 1998).

The Wannier–Stark (WS) effect in GaAs/AlAs SL, i.e. the electric-field-induced localization of extended SL wavefunctions, was observed through acoustic-phonon Raman scattering (Sapega et al., 1997). Experimental results showed that whenever the incident energy was equal to the interband transitions between HH and electron Stark levels with $\Delta n = 0, \pm 1$, resonantly enhanced disorder-induced processes (continuous emission) take place. These processes were accompanied by oscillations of the scattering intensity caused by delocalization of localized excitons due to resonant interaction either with emerging Stark states of neighboring QWs in the same miniband or with Stark states from the next-higher electron miniband. The broad features in the Raman spectra are due to crystal-momentum nonconservation induced by the electric field. For low values of the electric field, the PL has only one peak due to quasi-2D excitons localized by interface roughness and layer thickness fluctuations. Measuring the scattered light intensity versus the electric field for a fixed Raman Stokes shift, it was observed that in high fields, for excitations around the $n = 0$ WS transition, a series of intensity dips are superimposed on the broad resonance, due to anticrossing with WS levels from the next-higher electron miniband of the $n = 0$ delocalized state. For even stronger electric fields, higher than 20 kV/cm, the energy separation between WS states is larger than the localization energy of electrons and the SL becomes equivalent to a set of single QW periodically distributed along the growth direction. The Raman spectra in this case for laser energy around the $n = 0$ WS level are similar to the resonance excitation of excitonic states in MQW. When quantum interference of WS states (Bloch oscillations) are induced in time-resolved transmission experiments, information on intraband dynamics of carriers is obtained (Cho et al., 1996). For example, in InGaAsP SLs beating was observed due to the coherent superposition of electronic WS(0) and WS(−1) states and a growth-induced localized electronic 2D state. The measured decay time of quantum beats gives the electronic intraband dephasing time constant, of about 0.5 ps, a parameter which cannot be determined in linear cw absorption since the linewidths of optical transitions are determined by excitonic interband dephasing time. The coupling of WS and 2D states, as well as the beat frequencies, were found to be dependent on the external bias.

6.4.2 Excitons in Superlattices

The excitonic response from a SL depends on the relation between the width of the minibands in SL, ΔE and the binding energy E_X^b of excitons in an isolated QW of the same thickness. If $\Delta E > E_X^b$ the interwell coupling is stronger than the Coulomb interaction, the electron and hole states are delocalized, and the exciton states have a 3D character. When an electric field is applied, the SL is no longer translational invariant, the minibands split in discrete localized levels, called a Stark (or Wannier–Stark) ladder, and excitonic peaks appear in the absorption spectrum. If $\Delta E < E_X^b$ the holes, particularly HH are confined in QWs due to their large mass, in contrast to electrons which are still delocalized. The Coulomb interaction can reduce the spread of the electron wavefunction, and so localize it in one well even in the absence of an applied field. The excitonic transition takes place at almost the same energy as in an isolated QW.

Both spatially direct and indirect excitonic transitions can take place in a SL. Spectrally resolved transient FWM in GaAs/$Al_xGa_{1-x}As$ SLs have shown that there are two HH excitonic states. One, denoted HHX, is identical to the HH exciton in isolated quantum wells, the other, IWX, corresponds to a new type of excitons – the interwell exciton, formed between electrons and holes in adjacent wells. The condition for IWX formation is $\Delta E < E_{\mathrm{IWX}} \cong E_{\mathrm{X}}^{\mathrm{b}}$, where $E_{\mathrm{IW}} = e^2/[\varepsilon(L_{\mathrm{w}} + L_{\mathrm{b}})]$ is the interwell Coulomb interaction energy. If this condition is satisfied for electrons and HH separated by more than one SL period, IWX excitons can still form, but have a smaller oscillator strength. The Coulomb interaction for IWX excitons is not so strong as for HHX since electrons and holes have a finite separation along the growth direction. Quantum beats were observed between HHX and IWX, the IWX excitons showing a dephasing time one order of magnitude faster than HHX, due to intensity-independent scattering mechanisms (Mizeikis et al., 1997a).

The excitons can be of either type I or type II, corresponding to the cases where the electrons and holes that form the exciton are localized in the same layer or in different layers, respectively. In SLs with layers made from DMS, a type I to type II exciton crossover can be observed in magneto-reflectivity in the Faraday configuration, by increasing the applied magnetic field. In CdTe/$Cd_{1-x}Mn_xTe$ with small x values, the σ^+ and σ^- polarized components show a strong asymmetric splitting due to the different binding energies corresponding to these polarizations (σ^+ light excites the transition between $m_{\mathrm{h}} = -3/2$ to $m_{\mathrm{e}} = -1/2$, while σ^- excites the $m_{\mathrm{h}} = 3/2$ to $m_{\mathrm{e}} = 1/2$ transition). For high x values and high magnetic fields, the shift of the σ^+ transition energy follows the valence band edge tuning of the barrier layer, a transition from type I to type II exciton transitions being observed at a critical field value B_{c} (see Fig. 6.43), which increases with increasing x. The first part of the curve in Fig. 6.43, up to B_{c} and common to both high and low x values, correspond to decrease in confinement. Type II excitons can only be observed when the type II band offset is larger than the binding energy of the exciton. In this material type II excitons are formed between electrons confined in non-magnetic CdTe layers and holes in magnetic $Cd_{1-x}Mn_xTe$ layers (Cheng et al., 1997).

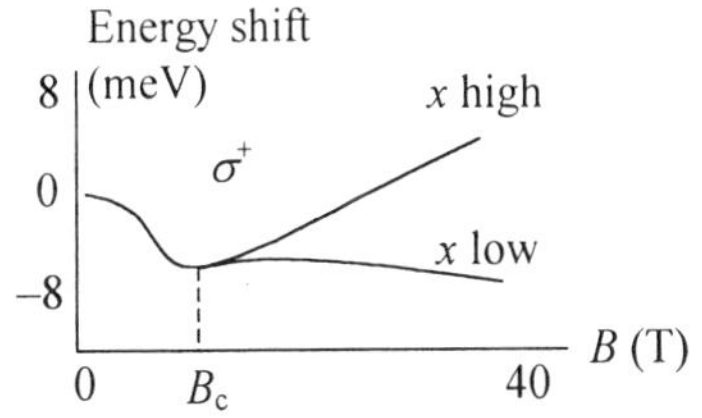

Fig. 6.43. Energy shift of the σ^+ transition in magnetoreflectivity of CdTe/$Cd_{1-x}Mn_xTe$ at different x values

A similar behavior is encountered in biexcitons, which appear in the nonlinear regime, when there are high concentrations of electron-hole pairs: in SL with miniband halfwidths smaller than the exciton binding energies, the biexciton binding energy $E^{\mathrm{b}}_{\mathrm{XX}}$ is almost the same as in isolated wells with the same width. In the opposite case $E^{\mathrm{b}}_{\mathrm{XX}}$ decreases sharply to the bulk value. Due to stronger confinement, the relative value of the biexciton binding energy with respect to the exciton binding energy is higher for the 2D case $E^{\mathrm{b}}_{\mathrm{XX}} / E^{\mathrm{b}}_{\mathrm{X}} = 0.228$ than for 3D where $E^{\mathrm{b}}_{\mathrm{XX}} / E^{\mathrm{b}}_{\mathrm{X}} = 0.1$. From optical measurements one can determine all the characteristics of exciton and biexciton states. For example, $E^{\mathrm{b}}_{\mathrm{X}}$ is given by the difference between the 1s peak and the band-to-band transitions in the linear absorption spectrum. The exciton binding energy depends on the dimension α: $E^{\mathrm{b}}_{\mathrm{X}} = 4E_{\mathrm{b}} /(\alpha - 1)^2$, where E_{b} is the bulk exciton binding energy. In quantum wells $\alpha = 3 - (1 - m_{\mathrm{r},0z} / m_{\mathrm{r},z}) \exp(-L_{\mathrm{w}} / 2a_{\mathrm{ex}})$, where $m_{\mathrm{r},0z}$, $m_{\mathrm{r},z}$, are the reduced mass along the z axis in bulk and SL, L_w is the effective width of the quantum well and a_{ex} the 3D excitonic Bohr radius. The exciton and biexciton energies in the ground state are, respectively, $E_{\mathrm{X}} = E_{\mathrm{g}} - E^{\mathrm{b}}_{\mathrm{X}}$ and $E_{\mathrm{XX}} = 2E_{\mathrm{X}} - E^{\mathrm{b}}_{\mathrm{XX}}$. $E^{\mathrm{b}}_{\mathrm{XX}}$ can be determined, for example, from transient FWM, when quantum beats are observed between exciton and biexciton resonances (Mizeikis et al., 1997b).

Exciton binding energy in quantum wire SL can be determined, for example, with magneto-optical methods. Such determinations have been performed in (Al,Ga)As serpentine SL quantum wire arrays by Weman et al. (1996). A signature of the 1D confinement of carriers in quantum wires is the linear polarization of the PL. Measurements of serpentine SL (SSL) have shown that at the PL peak energy the linear polarization is about 25%, the peak value of the polarization curve being redshifted with respect to the PL maximum, due to the distribution of quantum wires with different qualities in the peak. In magnetic fields the anisotropic diamagnetic shift of magnetoexcitons reveals the anisotropic nature of the electronic structure and its length scale. In the SSL the confinement is characterized by a/b (see Fig. 6.44).

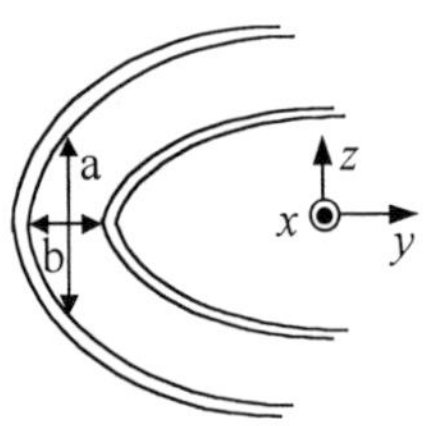

Fig. 6.44. The geometry of a serpentine SL

In a magnetic field the PL of quantum wire excitons has a diamagnetic energy shift depending on the direction of $\boldsymbol{B}$, which shows that the ground-state exciton has an anisotropic environment. In PLE at $B = 0$, four peaks are observed due to e_1hh_1, e_2hh_2, e_3hh_3 and e_4hh_4 transitions, which spin-split for $B \neq 0$ in Landau levels, the cyclotron orbit diameter being $d = 2(h/eB)^{1/2}(2n+1)^{1/2}$. In SSL the band edge is determined by the linear extrapolation of the higher (n = 2, 3, 4) Landau level transition to zero field. The binding energy of the delocalized ground-state exciton is obtained by subtracting the e_1hh_1 peak from the Landau level band edge. The exciton binding energy of around 11.5 or 13.5 meV (depending on the a/b value) is higher than the value in the 2D reference, of 8 meV. As a SL characteristic, the binding energy is found to be smaller in short-period SL, and higher in long-period quantum wire arrays. A maximum increase of 70% in the exciton binding energy with respect to the 2D reference, was found in a 160 Å-wide quantum wire array.

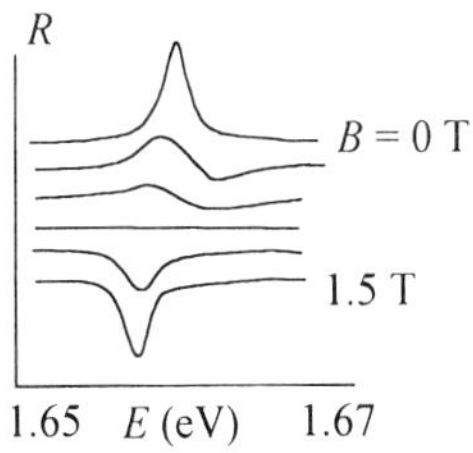

Fig. 6.45. The reflectivity dependence on magnetic field in SL made from DMS layers

Due to the giant spin splitting of electron bands, the magnetic field can drive a type I to type II transition in SL made from DMS layers. In a SL with thick CdMgZnTe barriers separated by thin equidistant quantum wells of magnetic CdMnTe and non-magnetic CdTe layers, placed alternatively, the magnetic field can tune the exciton energy across Bragg or anti-Bragg resonances (Sadowski et al., 1997). In Bragg structures with complex refractive index, the interference of light reflected by quantum wells drastically enhances the exciton reflectivity and its radiative recombination rate if the period Λ is equal to $\lambda/2$, with λ the light wavelength in the barriers corresponding to a photon with energy equal to the ground state exciton E_X in the quantum well. At the anti-Bragg condition $\Lambda = \lambda/4$ there is destructive interference. The Bragg and anti-Bragg resonant conditions can be obtained by simply changing the applied magnetic field. The observed reflectivity changes, respectively, from an emission-like to an absorption-like spectrum (see Fig. 6.45).

PL measurements can identify the type of interface on which the excitons are localized (Mashkov et al., 1997). For example, in type II pseudodirect GaAs/AlAs SLs the electrons and holes are confined in different layers, the electron-hole exchange interaction splitting the exciton state into a degenerative radiative doublet and two nonradiative states. The exchange Hamiltonian, which describes the electron-hole short-range exchange interaction and the additional splitting of radiative exciton states due to interface-induced coupling of HH and LH can be written as $H_{\mathrm{exch}} = -\sum_{i=x,y,z} 2c_{ii}S_{\mathrm{e},i}S_{\mathrm{h},i}$. Here $\boldsymbol{S}_{\mathrm{e}}$, $\boldsymbol{S}_{\mathrm{h}}$ are the electron and HH spins, c_{zz} is the electron-hole exchange energy, and the splitting energies Δ and δ between the radiative and nonradiative states are $\mathrm{i}(c_{xx}+c_{yy})$ and $c_{xx}-c_{yy}$, respectively. When an external magnetic field is applied the degeneracy of the radiative doublet is lifted, leading to two classes of excitons, with splittings of the same magnitude but opposite signs. These classes correspond to excitons localized either at the normal (AlAs-on-GaAs) or inverted (GaAs-on-AlAs) interface, the exciton population being unequally distributed between these classes. So, it is possible to identify the interface at which most of the excitons are localized by measuring the linear polarization of PL at exciton level anticrossings in a longitudinal magnetic field. The exciton splitting is due to either the random local deformation strain, which mixes HH and LH, caused by the presence of bonds of different nature (Al-As or Ga-As) on each side of the interface, or due to HH and LH mixing caused by the low C_{2v} symmetry of the interface. The splitting energy Δ between the radiative exciton states, of the order of a few μeV, can be experimentally obtained from the period of quantum beats (QB) in the PL decay time, when the two radiative sublevels are coherently excited by linearly or circularly polarized pulses. The QB in PL were observed during much longer times than the dephasing time T_2 of the excited state, the coherence loss between the exciton split states being caused by spin relaxation or by a spatially inhomogeneous distribution of splitting energies. The Zeeman contribution to the total Hamiltonian in the magnetic field is $H_{\mathrm{B}} = \mu_{\mathrm{B}} \sum_{i=x,y,z} (g_{\mathrm{e},i} S_{\mathrm{e},i} - g_{\mathrm{h},i} S_{\mathrm{h},i}) B_i$, the time-dependent degree of polarization of the PL being obtained by solving the equation of evolution of the density matrix $\rho(t)$: $\mathrm{d}\rho/\mathrm{d}t = \partial\rho/\partial t - (\mathrm{i}/\hbar) \times [H_{\mathrm{B}} + H_{\mathrm{exch}}, \rho]$ with the initial values $\rho(0)$ determined from the optical selection rules and the polarization and photon energy of the laser pulse (quasiresonant or band-to-band excitation). By varying the incident polarization and field value, one or several QB periods are observed. Apart from the exciton g factor for the radiative states, even the characteristic of nonradiative states, such as the splitting energy, can be deduced, since a magnetic field collinear with the [001] growth axis of SL couples the radiative states to the nonradiative ones.

6.4.3 Other Parameters

The thermal conductivity of the SL normal to its interfaces, $\kappa_{\perp}$, can also be optically measured using pump-and-probe techniques. For the $(\mathrm{GaAs})_n/(\mathrm{AlAs})_n$ SL this parameter was measured in the temperature range between 100 and 375 K, and was found to decrease with the reduction of the SL period. For example, at 300 K and for $n = 40$, $\kappa_{\perp}$ is three times less than the value in the bulk GaAs, whereas it is an order of magnitude less for $n = 1$. This behavior is to be expected since a SL modifies the phonon dispersion relation, and thus also the thermal transport properties. Due to its intrinsic anisotropy, the normal and parallel thermal conductivities with respect to the SL interfaces, $\kappa_{\perp}$ and $\kappa_{\parallel}$ have different values, but are both smaller than in bulk. In the pump-and-probe technique an outer Al film is deposited parallel to the SL interfaces and is illuminated with a strong pump pulse. The change in the film temperature as it cools by conduction into the SL determines a change in the reflectivity, measured by a time-delayed probe. The decrease in the reflectivity has a fast component, within the first 40 ps, caused by the redistribution of the heat within the Al film, followed by a slower component due to cooling in the SL. The cooling rate of the SL was found to increase with decreasing SL period. Assuming that the heat flow in the radial direction in the Al film and SL can be neglected on the basis that the thermal diffusivity of the SL is less than that of the Al film, and assuming a uniform illumination for the pump, the rate of change of the temperature in the film is $C_{\mathrm{Al}} d_{\mathrm{Al}} \partial T_{\mathrm{Al}}(t)/\partial t = -\sigma_{\mathrm{K}}[T_{\mathrm{Al}}(t) - T_{\mathrm{SL}}(z=0,t)]$ and the heat flow within SL is modeled by $C_{\mathrm{SL}} \partial T_{\mathrm{SL}}(z,t)/\partial t = \kappa_{\perp} \partial^2 T_{\mathrm{SL}}(z,t)/\partial z^2$. Here z is the direction normal to the film, $C_{\mathrm{Al,SL}}$ the heat capacities of the Al film and SL, $T_{\mathrm{Al,SL}}$ the corresponding temperatures, and σ_{K} the Kapitza conductance. These equations can be numerically solved to predict the change in temperature of the film due to a single pump pulse. For a series of pump pulses, with a large repetition rate such that the film cannot cool before the next pulse arrives, corrections must be made to account for the contributions of the previous pulses (Capinski et al., 1999).

Material coefficients for diffusion processes in different layers of SLs can be determined by optical means. For example, by Raman spectroscopy of the optical phonons in a $(^{70}\mathrm{Ge})_n(^{74}\mathrm{Ge})_m$ SL as a function of the annealing time, one can determine the parameters of the diffusion process in a nondestructive way. In this structure Ge atoms move through thermally activated jumps in neighboring vacancies, in a self-diffusion mechanism, the diffusion coefficient having the form $D = D_0 \exp(-H/k_{\mathrm{B}}T)$ where H is the self-diffusion enthalpy. The concentration profile of the material i near the jth interface in a SL is

$$\begin{aligned} C_i(z,t) &= (C_i^{0,1} + C_i^{0,\mathrm{sub}})/2 + (1/2)(C_i^{0,\mathrm{sub}} - C_i^{0,\mathrm{buf}})\mathrm{erf}[(z - d_{\mathrm{sub}})/2\sqrt{Dt}] \\ &\quad + (1/2)\Sigma_{n=1}^{2p}(C_i^{0,n+1} - C_i^{0,n})\mathrm{erf}[(z - d_n)/2\sqrt{Dt}], \end{aligned} \tag{6.59}$$

where $C_i^{0,j}$ is the initial concentration of material i in the jth layer, $C_i^{0,\mathrm{sub}}$, $C_i^{0,\mathrm{buf}}$ of the substrate and buffer layer, d_j is the distance between the surface and the jth interface, d_{sub} between surface and substrate, and p is the number of periods. Since the frequency of the optical phonons is an indication of the average mass, their evolution in time can be used for fitting the material parameters D, H (Silveira et al., 1997).

The temperature dependence of Auger recombination can be extracted from time-resolved PL up-conversion measurements in narrow-bandgap SLs (Jang et al., 1998). The PL decay rate increases with both lattice temperature and initial carrier density, and has two components: a fast one, occurring in a few hundred ps, due to Auger recombination, and a slower one due to Shockley–Read–Hall recombi-nation. From PL decay measurements as a function of time and carrier density N, the density dependent rates can be extracted as $R(N) = (\mathrm{d}N/\mathrm{d}t)/N = (1/N)(\mathrm{d}N/\mathrm{d}I_{\mathrm{PL}})(\mathrm{d}I_{\mathrm{PL}}/\mathrm{d}t)$. These rates comprise both radiative and nonradiative rates, the former being calculated from band structure and matrix elements. So, the nonradiative rates can be identified, and their dependence on the carrier density plotted. The Shockley–Read–Hall recombinations are density independent, their rate being obtained by extrapolating the curve of the nonradiative rates as a function of N for $N = 0$. The deviations from the zero slope line are associated with the Auger recombination, which increases with temperature, and has a subquadratic increase with the carrier density due to the degeneracy of conduction and valence bands.

Information about the carrier transport in SLs can be obtained from optical measurements. For example, the reverse transport of electrons due to electric-field screening by holes was evidenced by the delayed PL from the GaAs cladding layer in GaAs/AlAs short-period SL under high-intensity optical-pulse excitation (Hosoda et al., 1996). Due to the large mass difference, when an electric field is applied on a SL, the electrons tunnel away from the QW while the holes are left behind, creating a net opposite electric field when the hole density is sufficiently high. Under the influence of this field the electrons are pulled back and reinjected in the GaAs cladding layer, where they recombine with the remaining holes. This is the origin of the delayed PL, whose peak intensity has a monotonous variation with the applied voltage, with higher values for positive applied bias than for negative values. The delayed PL intensity depends strongly on the optical excitation intensity, being a consequence of the 'effective-mass filtering' effect, reflected electrically in nonresonant sequential tunneling times per barrier of 100 fs for electrons and 10 ps for HH.

6.5 Nanoparticles

As we have already specified, a nanoparticle is a spatially confined structure characterized by a dimension that ranges between 0 and 3. It is not a true low-D structure since quantum effects are not always present, or they are not predominant. At the same time, however, the optical properties of nanoparticles are not identical to those of bulk, and their geometry is certainly different from a thin film. Nanoparticles deserve therefore a special section; since, at least for some dimensions, quantum effects can be present, we choose to include nanoparticles in the chapter devoted to low-D structures.

Both the mean radius and the concentration of growing crystals can be determined by in situ absorption spectroscopy, by monitoring the exciton lines, which are a signature of crystallinity (Haselhoff and Weber, 1998). Such experiments have been performed for nanocrystals of CuCl growing in NaCl, the number of Cu^+ ions being estimated from the broad absorption peak. In the range of excitons the total absorption coefficient is described by $\alpha_{tot}(\omega) = \int_0^\infty (4\pi/3)R^3\alpha(\omega,R)f(R)\mathrm{d}R$, where $\alpha(\omega,R)$ is the contribution of nanocrystals with radius R and a distribution function over radii $f(R)$, the concentration of nanocrystallites being obtained by fitting this expression with the height of experimental exciton lines. Experiments at different temperatures have revealed two different growth mechanisms, which dominate in different temperature ranges. At a temperature of 300 K, the temporal evolution of the nanocrystal growth can be modeled by $x_{cr}/(x_i + x_{cr}) = p + q\ln(1/t_{ref})$, where x_i, x_{cr} are the crystal and ionic concentration, respectively, p, q are fitting parameters and t_{ref} is a reference time, which can be taken as 1 min. On the contrary, at 400 K the time evolution is described by $x_{cr}/(x_i + x_{cr}) = [(x_0 - x_s)/(x_i + x_{cr})][1 - \exp(-Dt/l^2)]$ with $x_0 = x_i + x_{cr}$ the total concentration, x_s the final Cu concentration in the equilibrium state of the sample, D the diffusion coefficient, and $2l$ the average distance between two particles.

Gold nanoparticles were studied with the help of the transient fs pump-probe absorption method (Inouye et al., 1998). The temporal modification of the electron temperature and effective damping constants are obtained by fitting the absorption spectra with the calculations based on Mie scattering theory. The gold nanoparticles show a nonlinear ultrafast response, the physical origin of which lies in the hot electron system heated by the incident optical pulse. The lattice temperature also affects the nonlinear response by the induced changes in the damping constant, such that on a long time scale of over 10 ps both electron and lattice temperatures have comparable contributions to the nonlinear response.

In noble metal nanoparticles the absorption spectrum has a peak located usually in the visible region and caused by surface plasmon resonance. For gold nanoparticles embedded in a SiO_2 matrix this absorption peak is situated near the edge of the band-to-band transition between the valence d band and the Fermi surface, a strong $\chi^{(3)}$ nonlinearity being identified around the surface plasmon resonance. The permitivity $\varepsilon(\omega) = \varepsilon_{intra}(\omega) + \varepsilon_{inter}(\omega)$ computed with the Mie scattering theory has both interband and intraband contributions, the former depending on the electron temperature and the damping constant in the band-to-band transitions. The intraband contributions dominate the low-energy spectrum, while the interband term is predominant in the high-energy region, both terms having comparable contributions at the surface plasmon band at about 2.3 eV. The surface plasmons are excited by the pump pulse, but due to electron-electron scattering lose their coherence within 100 fs and transform into a quasiequilibrium hot electron distribution. Within 10 ps this distribution turns into a quasi-equilibrium state between the electron and phonon systems in the nanoparticle due to electron-phonon interactions with the hot phonon system. For longer times, of over 100 ps, the heat transfer from the nanoparticle to the host matrix becomes predominant. By fitting the transient surface plasmon absorption peak with computations based on the Mie theory, it is found that the electron temperature increases first at 1700 K and then decreases at 400 K within 2.8 ps, cooling down in 120 ps. The electron phonon coupling, of 6×10^{16} $Wm^{-3}K^{-1}$, is more than twice that in gold films.

The behavior of different phonon modes when the size of nanoparticles decreases, was studied by Raman scattering in Bi nanocrystals embedded in amorphous Ge films. The two peaks observed in larger nanocrystals, which correspond to the optical A_{1g} and E_g modes, shift to lower and higher frequencies, respectively, when the size of the particles decreases. This behavior is attributed to the curvatures of the phonon-dispersion curves in Bi (Haro-Poniatowski et al., 1999).

Optical as well as acoustic phonons become confined in small-size nanocrystals. Due to the localization of acoustic vibrational motion, the coupling of excitons to LA phonons is strongly enhanced. The confined acoustic phonons can be coherently excited by the femtosecond pump-probe technique; the delay dependence of the differential transmission spectrum shows, in this case, a strong rapidly damped modulation, from which the frequency and the damping time can be obtained by fitting. The oscillatory behavior was assigned to coherent excitations of discrete confined acoustic phonons in nanocrystals, with frequencies varying with the nanocrystal radius R as $1/R$. This dependence is no longer accurate for small sizes due to an increasing influence from the surface of the nanocrystals and due to breaking of the assumption that the nanocrystal can be modeled as a continuum elastic body. For InAs nanocrystals, for example, with radii between 12 and 28 Å, the frequency of the discrete modes varies between 30 and 18 cm^{-1}. The damping coefficient of the modulation has a linear dependence with $1/R$, suggesting that the damping of coherent vibrations is due to coupling of the nanocrystal vibrations to the surrounding medium, mediated by the particle surface (Cerullo et al., 1999).

The low-frequency inelastic scattering from confined acoustic vibrations has different behaviors depending on the excitation wavelength. When the excitation domain is inside the absorption band the low-frequency inelastic scattering spectrum is unpolarized, and dependent on the excitation, while if the excitation domain is below E_g the spectrum is strongly polarized, and its position and shape are excitation independent, but shifts to lower energies as the mean particle size increases. This different behavior is explained by the different origin of the transitions. In the first case, the transitions are from carriers inside the nanocrystals, whereas in the second case it involves defects located at the surface of the nanoparticles. This behavior has been experimentally demonstrated in size-selective resonant Raman scattering in CdS nanocrystals embedded in a glass matrix (Saviot et al., 1998).

In CuCl nanocrystals embedded in a glass matrix, femtosecond degenerate FWM reveals that quantum beats appear between confined exciton sublevels for radii $R > 6.5$ nm, whereas for smaller radii quantum beats appear between confined exciton and LO-phonon-assisted transitions. The calculated LO-phonon energy (from the oscillation period) decreases with decreasing crystallite size, due to LO-phonon renormalization in the presence of excitons. The oscillation period is 830 fs for $R = 23$ nm, 220 fs for 8.6 nm and 165 fs for 5.3 nm, the simultaneous shift of the exciton peak to higher energies indicating quantum confinement effects (Ohmura and Nakamura, 1999).

The dynamics of excitons and biexcitons in CuCl nanoparticles embedded in NaCl were studied with time-resolved PL (Yano et al., 1997). Since electrons, holes and excitons are spatially confined, the optical nonlinearities in nanosize crystals are much greater compared to their bulk counterpart. When the ratio of the nanocrystal to exciton radii, β, is greater than one the exciton translational motion is confined. In the opposite case, when $\beta < 1$, the translational motion of electrons and holes are individually confined. When this last case is encountered, as in CuCl crystals, the exciton energy in the ground state is blueshifted with $\Delta E = \hbar^2\pi^2 / 2M(a^*)^2$, where the effective radius of the spherical nanocrystals is $a^* = a - 0.5a_{\mathrm{ex}}$, with a the real radius and a_{ex} the exciton Bohr radius. Upon irradiating the sample with high excitation powers, biexcitons are created, which are observed in PL as a band on the lower-energy side of the free exciton band that increases quadratically with the excitation intensity. In contrast, the PL peak of free excitons depends linearly on the light intensity. In the frequency-resolved PL three peaks are observed: one of them, BX, has a linear dependence on the excitation intensity, but cannot be associated to free excitons since its intensity decreases with increasing temperature and disappears beyond 20 K. Therefore, BX is identified as a bound exciton peak (this type of excitons is not encountered in bulk crystals). The other two peaks, M and another smaller peak BM, show a quadratic dependence on the excitation intensity and are attributed to the annihilation of the biexciton leaving a free exciton, and to the annihilation of a bound biexciton leaving a bound exciton which is the origin of BX, respectively. These assignments can be checked with time-resolved PL, or with the difference of optical density between spectra with and without pump light at different time delays. The results show that M has a zero rise time and a decay time constant of 125 ps, BM has a zero rise time and a two decay constants of 70 ps and 260 ps, while BX has a rise time of 70 ps an a decay time of 800 ps. Since the rise time of BX is equal to the decay time of BM, the bound exciton is created by the annihilation of the bound biexciton. Moreover, since the decay time of BX is larger than the decay time of free excitons in this material, which is about 400 ps, BX is indeed associated to bound excitons.

The dependence of the electron-LO-phonon coupling on the size of the semiconductor nanocrystals has been studied in CdS_xSe_{1-x} (Scamarcio et al., 1996). The strength of the electron-LO-phonon coupling is given by the ratio between two-phonon and one-phonon Raman scattering lines. In particular, from measurements of this ratio for the CdSe-like phonons, it was found that the strength of the electron-LO-phonon coupling increases with decreasing nanocrystallite size. The crystallite main radius R can be estimated from the blueshift of the first absorption band with respect to bulk material as $\Delta E(x) = \pi^2\hbar^2 / 2m_{\mathrm{r}}\overline{R}^2(x)$ with m_{r} the reduced mass $m_{\mathrm{r}}^{-1} = m_{\mathrm{e}}^{-1} + m_{\mathrm{h}}^{-1}$. Supposing a Gaussian distribution $f(R)$ for the radius in the sample, the total Raman cross-section for an n-phonon process is $\sigma_n = \int \sigma_n^{\mathrm{R}}(\omega) f(R)\mathrm{d}R$ with

$$\sigma_n^{\mathrm{R}}(\omega) = M^4 \left| \textstyle\sum_{m=0}^{\infty} \langle n | m \rangle \langle m | 0 \rangle / (E_0 + n\hbar\omega_{\mathrm{LO}} - \hbar\omega + \mathrm{i}\hbar\Gamma) \right|, \tag{6.60}$$

where M is the electric-dipole transition moment, m are intermediate vibrational levels in the excited state, E_0 the energy of the electronic transition, $\hbar\omega$ the incident photon energy and $\langle n | m \rangle = (n!/m!)^{1/2} \exp(-\Delta^2/2)\Delta^{n-m} L_m^{n-m}(\Delta^2)$, with

L_m^n Laguerre polynomials and Δ the dimensionless displacement of the harmonic oscillator in the excited state, related to the Huang–Rhys factor S as $S = \Delta^2$. Both Fröhlich and deformation potentials contribute to the electron-phonon interaction. Their relative contributions can be estimated in bulk samples by comparing the polarization properties of scattered light with the selection rules. When the nanocrystals are embedded in glass slabs, their position and orientation is random, so that statistical averages must be used to distinguish between Fröhlich and deformation potentials contributions. Calculations show that for CdS_xSe_{1-x} in cubic form, the ratio between the Fröhlich and deformation potentials Raman polarizabilities is $a_{\rm FP}^2 / a_{\rm DP}^2 = 3(3-4\rho)/5\rho$ where ρ is the depolarization ratio between scattered intensities polarized perpendicular and parallel to the incident radiation. From experimental measurements of ρ it follows that $a_{\rm FP}^2 / a_{\rm DP}^2$ increases by 1.7 times when the radius decreases from 28 to 20.5 Å, indicating that the main contribution at the electron-LO-phonon scattering in small-size nanocrystals is due to the Fröhlich potential.

Size effects can also be observed in femtosecond transient reflectivity measurements. For example, in metallic tin nanoparticles, embedded in a Al_2O_3 matrix, pump-and-probe cross-polarized reflectivity measurements with a pump wavelength of 390 nm and a probe wavelength 780 nm, have shown that the decay can be modeled with $1/\tau = 1/\tau_0 + v_{\rm F}/(\alpha R)$, where τ_0 is a size-independent time constant which includes all relaxation processes except the size-dependent electron surface scattering, v_F is the Fermi velocity and α the average number of collisions necessary to obtain electron-energy relaxation by surface interactions. Measurements with the same pump and probe wavelengths of 780 nm have evidenced a sharp risetime with a characteristic time constant of about 100 fs caused by the buildup of a hot Fermi distribution via electron-electron interactions (Stella et al., 1996).

Raman scattering in insulating Eu_2O_3 nanoparticles have revealed that the linewidth of the spectral hole burned in the $^7F_0 \rightarrow {}^5D_0$ transition in Eu^{3+} has a T^3 temperature dependence in nanoparticles, whereas in bulk it scales with $\approx T^7$. The linewidth depends also on the size of the nanoparticles as $\approx R^{-2.5}$. This behavior can be explained by a two-phonon Raman scattering process involving discrete phonon modes of homogeneous nanoparticles with stress-free boundary conditions (Meltzer and Hong, 2000).

The PL from nanocrystalline semiconductors depends also on the size of nanoparticles as $1/R^{\alpha}$ with an exponent determined by the disorder in the system. This variation is caused by two effects: i) a variation of the bandgap with size, which is described by an exponent $\alpha > 1$, and ii) the variation of the oscillator strength with size modeled by $5 \le \alpha \le 6$ (Ranjan et al., 1998). The PL peak energy in Ge nanocrystals, for example, was found to decrease with increasing size; the PL peak was at 0.88 eV for an average nanocrystal diameter of 5.3 nm, and at 1.54 eV for average diameter of 0.9 nm. Moreover, a drastic decrease of the PL intensity with decreasing size was observed, due to recombination of electron-hole pairs in nanocrystals (Takeoka et al., 1998).

References

Armitage, A., T.A. Fisher, M.S. Skolnick, D.M. Whittaker, P. Kinsler, and J.S. Roberts (1997) Phys. Rev. B **55**, 16395.

Armitage, A., M.S. Skolnik, V.N. Astrov, D.M. Whittaker, G. Panzarini, L.C. Andreani, T.A. Fisher, J.S. Roberts, A.V. Kavokin, M.A. Kaliteevski, and M.R. Vladimirova (1998) Phys. Rev. B **57**, 14877.

Arnone, D.D., T.P. Marlow, C.L. Foden, E.H. Linfield, D.A. Ritchie, and M. Pepper (1997) Phys. Rev. B **56**, 4340.

Ayiyama, H., T. Sameya, M. Yoshita, T. Sasaki, and H. Sasaki (1998) Phys. Rev. B **57**, 3765.

Baranov, A.V., S. Yamauchi, and Y. Masumoto (1997) Phys. Rev. B **56**, 10332.

Baxter, D., M.S. Skolnik, A. Armitage, V.N. Astrov, D.M. Whittaker, T.A. Fisher, J.S. Roberts, D.J. Mowbray, and M.A. Kaliteevski (1997) Phys. Rev. B **56**, 10032.

Bayer, M., V.B. Timofeev, F. Faller, T. Gutbrod, and A. Forchel (1996) Phys. Rev. B **54**, 8799.

Bayer, M., Ch. Schlier, Ch. Gréus, A. Forchel, S. Benner, and H. Haug (1997) Phys. Rev. B **55**, 13180.

Bayer, M., S.N. Walck, T.L. Reinecke, and A. Forchel (1998) Phys. Rev. B **57**, 6584.

Bell, M.J.V., L. Ioriatti, and L.A.O. Nunes (1998) Phys. Rev. B **57**, 15104.

Bertru, N., O. Brandt, R. Klann, A. Mazuelas, W. Ulrici, and K.H. Ploog (1997) Phys. Rev. B **55**, 4503.

Biese, G., C. Schüller, K. Keller, C. Steinebach, D. Heitmann, P. Grambow, and K. Eberl (1996) Phys. Rev. B **53**, 9565.

Bockelmann, U., W. Heller, and G. Abstreiter (1997) Phys. Rev. B **55**, 4469.

Bolton, S., G. Sucha, D. Chemla, D.L. Sivco, A.Y. Cho (1998) Phys. Rev. B **58**, 16326.

Borri, P., M. Gurioli, M .Colocci, F. Martelli, and M. Capizzi (1997) Phys. Rev. B **56**, 9228.

Borri, P., W. Langbein, J.M. Hvam, and F. Martelli (1999a) Phys. Rev. B **60**, 4505.

Borri, P., W. Langbein, J.M. Hvam, and F. Martelli (1999b) Phys. Rev. B **59**, 2215.

Braun, W., L.V. Kulik, T. Baars, M. Bayer, and A. Forchel (1998a) Phys. Rev. B **57**, 7196.

Braun, W., M. Bayer, A. Forchel, O.M. Schmitt, L. Banyai, H. Haug, and A.I. Filin, (1998b) Phys. Rev. B **57**, 12364.

Brown, S.W., T.A. Kennedy, D. Gammon, and E.S. Snow (1996) Phys. Rev. B **54**, 17339.

Brown, S.A., J.F. Young, Z. Wasilewski, and P.T. Coleridge (1997) Phys. Rev. B **56**, 3937.

Butov, L.V. and A.I. Filin (1998) Phys. Rev. B **58**, 1980.

Buyanova, I.A., W.H. Chen, A. Henry, W.X. Ni, G.V. Hansson, and B. Monemar (1996) Phys. Rev. B **53**, 1701.

Capinski, W.S., H.J. Maris, T. Ruf, M. Cardona, K. Ploog, and D.S. Katzer (1999) Phys. Rev. B **59**, 8105.

Cerullo, G., S. De Silvestri, and U. Banin (1999) Phys. Rev. B **60**, 1928.

Chatov, A., T. Baars, and M. Gal (1996) Phys. Rev. B **53**, 4704.

Cheng, H.H., R.J, Nicholas, D.E. Ashenford, and B. Lunn (1997) Phys. Rev. B **56**, 10453.

Cheong, H.M., A. Mascarenhas, P. Ernst, and C. Geng (1997) Phys. Rev. B **56**, 1882.

Cheong, H.M., Y. Zhang, A.G. Norman, J.D. Perkins, A. Mascarenhas, K.Y. Cheng, and K.C. Hsieh (1999) Phys. Rev. B **60**, 4883.

Cho, G.C., T. Dekorsy, H.J. Bakker, H. Kurz, A. Kohl, and B. Opitz (1996) Phys. Rev. B **54**, 4420.

Coli, G., L. Calcagnile, P.V. Giugno, R. Cingolani, R. Rinaldi, L. Vanzetti, L. Sorba, and A. Franciosi (1997) Phys. Rev. B **55**, 7391.

Crooker, S.A., D.G. Rickel, S.K. Lyo, N. Samarth, and D.D. Awschalom (1999) Phys. Rev. B **60**, 2173.

Darnhofer, T., U. Rössler and D.A. Broido (1995) Phys. Rev. B **55**, 14376.
de Dios Leyva, M. and V. Galinda (1993) Phys. Rev. B **48**, 4516.
de Oliveira, J.B.B., E.A. Meneses, and E.C.F. da Silva (1999) Phys. Rev. B **60**, 1519.
Dragoman, D. and M. Dragoman (1999) *Advanced Optoelectronic Devices*, Springer, Berlin.
Emiliani, V., S.Ceccherini, F. Bogani, M. Colocci, A. Frova, and S.S. Shi (1997) Phys. Rev. B **56**, 4807.
Finkelstein, G., H. Shtrikman, and I.B. Joseph (1996) Phys. Rev. B **53**, 12593.
Giehler, M., T. Ruf, M. Cardona, and K. Ploog (1997) Phys. Rev. B **55**, 7124.
Grassi Alessi, M., M. Capizzi, A.S. Bhatti, A. Frova, F. Martinelli, P. Frigeri, A. Bosacchi, and S. Franchi (1999) Phys. Rev. B **59**, 7620.
Greco, D., R. Cingolani, A. D'Andrea, N. Tommasini, L. Vanzetti, and A. Franciosi (1996) Phys. Rev. B **54**, 16998.
Grimsditch, M., Y. Jaccard, and I. K. Schuller (1998) Phys. Rev. B **58**, 11539.
Grosse, S., J. H. H. Sandmann, G. von Plessen, J. Feldmann, H. Lipsanen, M. Sopanen, J. Tulkki, and J. Ahopelto (1997) Phys. Rev. B **55**, 4473.
Guha, S., Q. Cai, M. Chandrasekhar, H.R. Chandrasekhar, H. Kim, A.D. Alvarenga, R. Vogelgesang, A.K. Ramdas, and M.R. Melloch (1998) Phys. Rev. B **58**, 7222.
Gutbrot, T., M. Beyer, A. Forchel, P.A. Knipp, T.L. Reinecke, A. Tartakovski, D. Kulakovski, N.A. Gippius, and S.G. Tikhodeev (1999) Phys. Rev. B **59**, 2223.
Haas, H., N. Magnea, and L. S. Dang (1997) Phys. Rev. B **55**, 1563.
Haetty, J., Kioseoglou, A. Petrou, M.Dutta, J. Pamulapati, and M.Taysing-Lara (1999) Phys. Rev. B **59**, 7546.
Han, H.X., G.H. Li, W.K. Ge,Z.P. Wang, Z.Y. Xu, S.S. Xie, B.H. Chang, L.F. Sun, B.S. Wang, G. Xi, and Z.B. Su (2000) eprint:cond-mat.000435.
Harel, R., E. Cohen, E .Linder, A. Ron, and L.N. Pfeiffer (1996) Phys. Rev. B **53**, 7868.
Harley, R.T. and M.J. Snelling (1996) Phys. Rev. B **53**, 9561.
Haro-Poniatowski, E., M. Jouanne, J.F. Morhange, and M. Kanehisa (1999) Phys. Rev. B **60**, 10080.
Harris, J.J., J.M. Roberts, S.N. Holmes, and K. Woodbridge (1996) Phys. Rev. B **53**, 4886.
Harris, P. (1999) *Carbon Nanotubes and Related Structures*, Cambridge University Press, Cambridge.
Haselhoff, M. and H.J. Weber (1998) Phys. Rev. B **58**, 5052.
Hassine, A., J. Sapriel, P. Le Berre, M.A. Di Forte-Poisson, F. Alexandre, and M. Quillec, (1996) Phys. Rev. B **54**, 2728.
Heller, W., U. Bockelmann, and G. Abstreiter (1998) Phys. Rev. B **57**, 6270.
Helm, M., W. Hilber, T. Fromherz, F.M. Peeters, K. Alavi, and R.N. Pathak (1993) Phys. Rev. B **48**, 1601.
Hofmann, D.M., K. Oettinger, Al.L. Efros, and B.K. Meyer (1997) Phys. Rev. B **55**, 9924.
Horner, G.S., A. Mascarenhas, R.G. Alonso, D.J. Friedman, K. Sinha, K.A. Bertness, J.G. Zhu, and J.M. Olson (1993) Phys. Rev. B **48**, 4944.
Hosoda, M., K. Tominaga, P.O. Vaccaro, and T. Watanabe (1996) Phys. Rev. B **53**, 13283.
Hwang, Y.N., S. Shin, H.L. Park, S.H. Park, U. Kim, H.S. Jeong, E.J. Shin, and D. Kim (1996) Phys. Rev. B **54**, 15120.
Inoue, K., A. Yamanaha, K. Toba, A.V. Baranov, A.A. Onushchenko, and A.V. Fedorov (1996) Phys. Rev. B **54**, 8321.
Inouye, H., K. Tanaka, I. Tanahahashi, and K. Hirato (1998) Phys. Rev. B **57**, 11334.
Itskevich, I.E., S.G. Lyapin, I.A. Troyan, P.C. Klipstein, L. Eaves, P.C. Main, and M. Henini (1998) Phys. Rev. B **58**, 4250.
Jang, D.J., M.E. Flatté, C.H. Grein, J.T. Olesberg, T.C. Hasenberg, and T.F. Boggess (1998) Phys. Rev. B **58**, 13047.
Jusserand, B. and M. Cardona (1989), in *Light Scattering in Solids V*, eds. M. Cardona and G. Güntherodt, vol.66 in Topics in Applied Physics, Springer, Berlin, Heidelberg, p.49.

Kaesen, K.F., A. Huber, H. Lorentz, J.P. Kotthaus, S. Bakker, and T.M. Klapnijk (1996) Phys. Rev. B **54**, 1514.

Kainth, D.S., D. Richards, A.S. Bhatti, H.P. Hughes, M.Y. Simmons, E.H. Linfield, D.A. Ritchie (1999) Phys. Rev. B **59**, 2095.

Kanemitsu, Y., M. Iiboshi, and T. Kushida (2000) Appl. Phys. Lett. **76**, 2200.

Kelkar, P.V., V.G. Kozlov, A.V. Nurmikko, C.C. Chu, J. Han, and R.L. Gunshor (1997) Phys. Rev. B **56**, 7564.

Kemerink, M., P.M. Koenraad, and J.H.Wolter (1996) Phys. Rev. B **54**, 10644.

Khreis, O.M., W.P. Gillin, and H.P. Homewood (1997) Phys. Rev. B **55**, 15813.

Kim, J.C. and J.P. Wolfe (1998) Phys. Rev. B **57**, 9861.

Kitaev, Yu.E., M.F. Limonov, P. Tronc, G.N. Yushin (1998) Phys. Rev. B **57**, 14209.

Krylyuk, S., D.V. Korbutyak, V.G. Litovchenko, R. Hey, H.T. Grahn, and K.H. Ploog (1999) Appl. Phys. Lett. **74**, 2596.

Kwok, S.H., P.Y. Yu, and K. Uchida (1998) Phys. Rev. B **58**, 13395.

Kovac, J., H. Schweizer, M.H. Pilkuhn, and H. Nickel (1996) Phys. Rev.B **54**, 13440.

Kukushkin, I.V., K. von Klitzing, and K. Eberl (1997) Phys. Rev. B **55**, 10607.

Kulik, L.V., I.V. Kukushkin, V.E. Kirpichev, K. von Klitzing, and K. Eberl (2000) Phys. Rev. B **61**, 1712.

Kuribayashi, B., K. Inoue, K. Sokoda, V.A. Tsekhomskii and A.V. Baranov (1998) Phys. Rev. B **57**, 15084.

Kuther, A., M. Bayer, T. Gutbroad, A. Forchel, P.A. Knipp, T.L. Reinecke and R. Werner (1998a) Phys. Rev. B **58**, 15744.

Kuther, A., M. Bayer, A. Forchel, A. Gorbunov, V.B. Timofeev, F. Schäfer, J.P. Reithmaier (1998b) Phys. Rev. B **58**, 7508.

Langbein, W. and J.M. Hvam (2000) Phys. Rev. B **61**, 1692.

Lefebvre, P., V. Calvo, N. Magnea, T. Taliercio, J. Allegre, and H. Mathieu (1997) Phys. Rev. B **56**, 10040.

Lefebvre, P., J. Allégre, B. Gil, H. Mathieu, N. Grandjean, M. Leroux, J. Massies, and P. Bigenwald (1999) Phys. Rev. B **59**, 15363.

Leon, R., D.R.M. Williams, J. Krueger, E.R. Weber, and M.R. Melloch (1998) Phys. Rev. B **56**, 4336.

Li, X.Q. and Y. Arakakawa (1999) Phys. Rev. B **60**, 1915.

Li, Y.B., J.W. Cockburn, J.P. Duck, M.J. Birkett, M.S. Skolnick, I.A. Larkin, M. Hopkinson, R. Grey, and G. Hill (1998) Phys. Rev. B **57**, 6290.

Lin, M.F., F.L. Shyu, and R.B. Chen (2000) Phys. Rev. B **61**, 14114.

Lovisa, S., R.T. Cox, N. Magnea, and K. Saminadayar (1997) Phys. Rev. B **56**, 12787.

Mackh, G., W. Ossau, A. Waang, and G. Landwehr (1996) Phys. Rev. B **54**, 5227.

Malinowski A., D.J. Guerrier, N.J. Traynor, and R.T. Harley (1999) Phys. Rev. B **60**, 7728.

Mal'shukov A.G., K.A. Chao, and M. Willauder (1997) Phys. Rev. B **55**, 1918.

Mashkov, I.V., C. Gourdon, P. Lavallard, and D.Yu. Roditchev (1997) Phys. Rev. B **55**, 13761.

Meltzer, R.S. and K.S. Hong (2000) Phys. Rev. B **61**, 3396.

Merle d'Aubigné, Y., A. Wasiela, H. Mariette, and T. Dietl (1996) Phys. Rev. B **54**, 14003.

Merbach, D., E. Schöll, W. Ebeling, P. Michler, and J. Gutowski (1998) Phys. Rev. B **58**, 9926.

Mizeikis, V., D. Birkedal, W. Langbein, and J.M. Hvam (1997a) Phys. Rev. B **55**, 7743.

Mizeikis, V., D. Birkedal, W. Langebein, V.G. Lyssenko, and J.M. Hvam (1997b) Phys. Rev. B **55**, 5284.

Muraki, K., A. Fujiwara, S. Fukatsu, Y. Shiraki, and Y. Takahashi (1996) Phys. Rev. B **53**, 15477.

Nehls, J., T. Schmidt, U. Merkt, D. Heitmann, A.G. Norman, and R.A. Stradling (1996) Phys. Rev. B **54**, 7651.

Ogawa, T. and A. Shimizu (1993) Phys. Rev. B **48**, 4910.

Ohmura, H. and A. Nakamura (1999) Phys. Rev. B **59**, 12216.

Ohnesorge, B., M. Albrecht, J. Oshinowo, A. Forchel, and Y. Arakawa (1996) Phys. Rev. B **54**, 11532.

Olajos, J., J. Engvall, H.G. Grimmeis, M.G. Gall, G. Abstreiter, H. Presting, and H. Kibbel (1996) Phys. Rev. B **54**, 1922.

Okamura, H., D. Heiman, M. Sundaram, and A.C. Gossard (1998) Phys. Rev. B **58**, 15985.

Papadogonas, I.A., A.N. Andriotis, and E.N. Economou (1997) Phys. Rev.B **55**, 4887.

Pearsall T., H. Polatoglou, H. Presting, and E. Kasper (1996) Phys. Rev. B **54**, 1545.

Pelekanos, N.T., A. Haag, N. Magnea, V.I. Belisky, and A. Cantarero (1997) Phys. Rev. B **56**, 10056.

Pinczuk, A. and G. Abstreiter (1989) in *Light Scattering in Solids V*, eds. M. Cardona and G. Güntherodt, vol.66 in Topics in Applied Physics, Springer, Berlin, Heidelberg, p.153.

Pinenta, M.A., A. Marucci, S.A. Empedocles, M.G. Bowendi, E.B. Hanlon, A.M. Rao, P.C. Eklund, R.E. Smalley, G. Desselhaus, and M.S. Dresselhaus (1998) Phys. Rev. B **58,** 16016.

Plentz, F., F. Meseguer, J. Sanchez-Dehesa, N. Mestra, and E.A. Meneses (1993) Phys. Rev. B **48**, 1967.

Pohlt, M., W. Herbst, H. Pascher, W. Faschinger, and G. Bauer (1998) Phys. Rev. B **57**, 9988.

Popović, Z.V., J. Spitzer, T. Ruf, M. Cardona, R. Nötzel, and K. Ploog (1993) Phys. Rev. B **48**, 1659.

Pratt, A.R., T. Takamori, and T. Kamijoh (1998) Phys. Rev. B **58**, 9656.

Ranjan, V., V.A. Singh, and G.C. John (1998) Phys. Rev. B **58**, 1158.

Rezki, M., A.M. Vasson, A. Vasson, P. Lefebre, V. Calvo, R. Planel, G. Patriarche (1998) Phys. Rev. B **58**, 7540.

Richards, D. and B. Jusserand (1999) Phys. Rev. B **59**, 2506.

Rinaldi, R., R. Mangino, R. Cingolani, H. Lipsanen, M. Sopanen, J. Tulkki, M. Braskén, and J. Ahopelto (1998) Phys. Rev. B **57**, 9763.

Roy, A. and A.K. Sood (1996) Phys. Rev. B **53**, 12127.

Sadowski, J., H. Mariette, A. Wasiela, R. André, Y. Merle d'Aubigné, and T. Dietl (1997) Phys. Rev. B **56**, 1664.

Sakakura, N. and Y. Masumoto (1997) Phys. Rev. B **56**, 4051.

Sapega, V.F., T. Ruf, M. Cardona, H.T. Grahn, and K. Ploog (1997) Phys. Rev. B **56**, 1041.

Sauncy, T., M. Holtz, O. Brafman, D. Fekete, and Y. Finkelstein (1999) Phys. Rev. B **59**, 5049.

Saviot, L., B. Champagnon, E. Duval, and A.I. Ekimov (1998) Phys. Rev. B **57**, 341.

Scamarcio, G., V. Spagnolo, G. Ventruti, M. Lugarà, and G.C. Righini (1996) Phys. Rev. B **53**, 10489.

Shah, J. (1999), *Ultrafast Spectroscopy of Semiconductors and Semiconductor Nanostructures*, Springer, Berlin, Heidelberg.

Silveira, E., W. Dondl, G. Allstreiter, and E.E. Haller (1997) Phys. Rev. B **56**, 2062.

Sirenko, A.A., T. Ruf, M. Cardona, D.R. Yakovlev, W. Ossau, A. Waag, and G. Landwehr (1997) Phys. Rev. B **56**, 2114.

Sirenko, A.A., V.I. Belitsky, T. Ruf, M. Cardona, A.I. Ekimov, and C. Trallero-Giner (1998) Phys. Rev. B **58**, 2077.

Siviniant, J., F.V. Kyrychenko, Y.G. Semenov, D. Coquillat, D. Scalbert and J.P. Lascaray (1999) Phys. Rev. B **59**, 10276.

Steffen, R., A. Forchel, T.L. Reinecke, T. Koch, M. Albrecht, J. Osinowo, and F. Faller (1996) Phys. Rev. B **54**, 1510.

Stella, A., M. Nisoli, S. De Silvestri, O. Svelto, G. Lanzani, P. Cheyssac, and R. Kofman (1996) Phys. Rev. B **53**, 15497.

Sun, H., J. Lei, and K.W. Yu (2000) Phys. Rev. B **61**, 2957.

Takeoka, S., M. Fujii, S. Hayashi, and K. Yamamoto (1998) Phys. Rev. B **58**, 7921.

Taleb, A., V. Russier, A. Courty, and M.P. Pileni (1999) Phys. Rev. B **59**, 13350.

Ten S., F. Hennberger, M. Rabe, and N. Peyghambarian (1996) Phys. Rev. B **53**, 12637.

Testelin, C., C. Rigaux, and J. Cibert, (1997) Phys. Rev. B **55**, 2360.

Tignon, J., R. Ferreira, J. Wainstain, C. Delalande, P. Voisin, M. Voos, R. Houdré, U. Oesterle, and R.P. Stanley (1997) Phys. Rev. B **56**, 4068.

Trallero-Giner, C., A. Debernardi, M. Cardona, F. Menéndez-Proupin, and A.I. Ekimov (1998) Phys. Rev. B **57**, 4664.

Valenta, J., J. Moniatte, P. Gilliot, B. Hönerlage, J.B. Grun, R. Levy, and A.I. Ekimov (1998) Phys. Rev. B **57**, 1774.

van der Meulen, H.P., I. Santa-Olalla, J. Rubio, J.M. Calleja, K.J. Friedland, R. Hey, and K. Ploog (1999) Phys. Rev. B **60**, 4897.

Vasko, F.T. and A.V. Kuznetsov (1999) *Electronic States and Optical Transitions in Semiconductor Heterostructures*, Springer, Berlin, Heidelberg.

Venig, M., D.J. Dunstan, and K.P. Homewood (1993) Phys. Rev. B **48**, 2412.

Viña, L., L. Muñoz, E. Pérez, J. Fernández-Rossier, C. Tejedor, and K. Ploog (1996) Phys. Rev. B **54**, 8317.

Wagner, H.P., A. Schätz, R. Maier, W. Langbein, and J.M. Hvam (1997) Phys. Rev. B **56**, 12581.

Wagner, H.P., H. Schätz, P. Maier, W. Longbein, and J.M. Hvam (1998) Phys. Rev. B **57**, 1791.

Wagner, H.P., W. Langbein, and J.M. Hvam (1999) Phys. Rev. B **59**, 4584.

Weman H., M. Potemski, M.E. Lazzouni, M.S. Miller, and J.L. Merz (1996) Phys. Rev. B **53**, 6959.

Wetzel, C., R. Winckler, M. Drechsler, B.K. Meyer, J. Scriba, J.P. Kotthaus, V. Härle, and F. Scholtz (1996) Phys. Rev. B **53**, 1038.

Widmann, F., J. Simon, B. Daudin, G. Feuillet, J.L Rouvière, N.T. Pelekanos, and G. Fishman (1998) Phys. Rev. B **58**, 15989.

Winkler, R. and A.I. Nesvizhskii (1996) Phys. Rev. B **53**, 9984.

Yamaguchi, S., Y. Kawakami, S. Fujita, Y. Yamada, T. Mishina, and Y. Masumoto (1996) Phys. Rev. B **54**, 2629.

Yang, F., G.R. Hayes, R.T. Phillips, and K. P. O'Donnel (1996) Phys. Rev. B **53,** 1697.

Yano, S., T. Goto, T. Itoh, and A. Kasuyo (1997) Phys. Rev. B **55**, 1667.

Yeh, C.N., L.E. McNeil, R.E. Nahory, and R. Bhat (1995) Phys. Rev. B **52**, 14682.

Yoon, H.W., D.R. Wake, and J.P. Wolfe (1996) Phys. Rev. B **54**, 2763.

Yu, W.Y., M.S. Salib, A. Petrov, B.T. Jouker, and J. Warnock (1997) Phys. Rev. B **55**, 1602.

Zeman, J., G. Martinez, P.Y. Yu, and K. Uchida (1997) Phys. Rev. B **55**, 13428.

Zhou, H., A.V. Nurmikko, C.C. Chu, J. Han, R.L. Gunshor, and T. Takagahara (1998) Phys. Rev. B **58**, 10131.

Zhu, J.B., H.I. Jeon, E.K. Suh, H.J. Lee, and Y.G. Hwang (1995) Phys. Rev. B **52**, 16353.

7. Optical Properties of Disordered Materials

In this chapter we present the optical properties of disordered materials, i.e. of materials that lack long-range order. These include glasses, amorphous and polycrystalline materials. All these different kinds of materials share the same characteristic: no sharp features in the optical properties, as encountered in their crystalline counterparts, due to disorder-induced violation of the momentum-conservation rule.

Depending on the growth method and conditions, a sample can be monocrystalline, polycrystalline or amorphous. The latter two phases have a high concentration of defects and imperfections, although these can also exist in the monocrystalline phase in much lower concentrations. The distinction between polycrystalline and amorphous materials is quite subtle. Polycrystalline materials are composed of grains surrounded by interconnective or boundary atoms, inside each grain the atoms being periodically arranged, as in the crystal. As the grains get smaller, the number of boundary atoms becomes larger, and at some point the distinction between grain surface and interior is lost. Even in this extreme case the material has not become amorphous. The disorder in the amorphous state is a topological disorder. A polycrystalline material does not have the amorphous state as a limiting case. Actually, there seems to exist a lower limit on crystalline size, and a smaller, upper limit of the correlation length in amorphous semiconductors. For example, in Si the upper limit on the correlation length in the amorphous phase is 12–15 Å, whereas the lower limit for crystallites is 30–40 Å (Brodsky, 1975). Between amorphous materials and glasses the distinction is not so clear; they differ mainly through their coordination. Therefore, these materials will be treated in the same section. Besides disordered materials, this chapter also deals briefly with disordered alloys, superlattices and surfaces, for a complete overview on the disorder-induced effects. We will start with the latter disordered structures since they are more clearly connected to the ordered structures encountered in the previous chapters.

Contrary to other chapters where we have shown how different parameters can be extracted from optical measurements, in disordered materials the situation is somewhat different. Optical measurements, among other types of measurements, help us understand the inner structure of disordered materials, and check the validity of different theories. In disordered solids, there is no single theory that can describe all materials, not even a category of disordered materials, because the optical properties are not only dependent on the material structure but also on the preparation conditions. The main concepts and models at the basis of the description of disordered materials were developed long ago; excellent reviews exist for either the theoretical work or its application to the analysis of optical data. In this chapter we will not insist on the theoretical formulation because there is no unitary treatment of the subject, but we will briefly describe the main concepts and models and refer to some recent work in the field.

7.1 Disordered Alloys

One type of disorder encountered in binary, as well as in ternary alloys, is compositional fluctuation, which violates the translational symmetry of the crystal, and thus modifies its vibrational spectrum. The crystal band theory is, however, still valid, since the deviation from perfect periodicity is small. However, local departures from periodicity cause transitions between Bloch states of different $\boldsymbol{k}$. In the weak disorder case, when the single-electron wavefunction is still extended over the whole crystal, the periodic description can be restored by introducing a coherent potential in the effective Hamiltonian such that there is no scattering on average and the electron wavefunction propagates coherently through the material. This method is suitable for describing metallic and many semiconductor alloys, substitutionally disordered systems in any concentrations and positionally disordered systems. The effective Hamiltonian is, in general, not Hermitian and energy dependent (Callaway, 1991). More precisely, a compositionally disordered semiconductor alloy can be modeled as an effective medium characterized by a self-energy $\tilde{\varepsilon}(\omega)$, with ω the phonon frequency with respect to the virtual crystal. In the coherent-potential approximation, the condition for zero average scattering at a single site in the medium is

$$\sum_{i=1}^{2}\frac{x_i[m_{\mathrm{VC}}\omega^2(1-\tilde{\varepsilon}(\omega))-m_{ri}\omega^2]}{1-[m_{\mathrm{VC}}\omega^2(1-\tilde{\varepsilon}(\omega))-m_{ri}\omega^2]G_M(0,0;\omega^2)}=0, \tag{7.1}$$

where x_i are the fractional concentrations of the constituents of the binary semiconductor and m_{ri} their reduced masses, $m_{\mathrm{VC}}=x_1m_{r1}+x_2m_{r2}$ and

$$G_M(0,0;\omega^2)=\frac{1}{Nm_{\mathrm{VC}}}\sum_{qj}\frac{1}{\omega^2(1-\tilde{\varepsilon}(\omega))-\omega_{qj}^2}, \tag{7.2}$$

with N the number of unit cells in the virtual crystal. The above equation is solved for $\tilde{\varepsilon}(\omega)$, the real part of which gives the shift in the phonon frequency, and the imaginary part – the linewidth broadening. The compositional disorder also changes the lineshape of vibrational modes.

At finite temperatures the compositional disorder is always accompanied by thermal anharmonicity. The temperature dependence of the line center position and linewidth of first-order optical modes at the Brillouin zone center, is modified due to cubic and quartic anharmonic terms. These changes can be described as $\omega(T)=\omega_0(T)+\Delta\omega(T)$ and $\Gamma(T)=\Gamma_0(T)+\Delta\Gamma(T)$, respectively, where the parameters labeled with 0 refer to the values in the harmonic approximation. The cubic and quartic contributions are

$$\Delta\omega(T)=C\left[1+\frac{2}{\exp(x)-1}\right]+D\left[1+\frac{3}{\exp(y)-1}+\frac{3}{[\exp(y)-1]^2}\right], \tag{7.3}$$

$$\Delta\Gamma(T)=A\left[1+\frac{2}{\exp(x)-1}\right]+B\left[1+\frac{3}{\exp(y)-1}+\frac{3}{[\exp(y)-1]^2}\right], \tag{7.4}$$

where $x=\hbar\omega_0/2k_BT$, $y=\hbar\omega_0/3k_BT$, A, C are related to three-phonon processes, and B, D to four-phonon process. At temperatures much lower than the Debye temperature T_D, three-phonon processes dominate.

The compositional disorder and thermal anharmonicity are nonadditive since they have a different character. At low temperatures the anharmonicity is mainly due to compositional disorder. For III-V ternary alloys $A_xB_{1-x}C$ with A, B anions or cations, the two contributions to anharmonicity in first-order Raman spectra can be separated by comparing it to Raman spectra of the binary systems. For low values of alloying, thermal anharmonicity is dominant. For example, in $GaAs_{1-x}P_x$ alloys (Ramkumar et al., 1996a) the degree of anharmonicity increases with disorder. More precisely, it increases for GaP-like LO phonons, and decreases for GaAs-like LO phonons with decreasing P concentration. For a given P concentration the degree of anharmonicity is higher for GaP-like LO phonons due to the mass difference between P and As. Since the density of phonon states for the 2TA(x) mode in GaP decreases with decreasing temperature, it follows that the composition-disorder-induced effect compensates temperature-induced anharmonicity in $GaAs_{0.1}P_{0.9}$. The lifetime of first-order modes decreases with increasing anharmonicity. Measuring the temperature dependence of the line center and linewidth of GaAs-like and GaP-like LO phonons, it is possible to obtain A, C, ω_0, and Γ_a/Γ_b, the ratio of halfwidths of the lower/higher energy sides of phonon peaks, which characterizes the lineshape asymmetry.

The resonant Raman spectra of ternary alloy semiconductors differ from those of binary semiconductors. For example, in GaP a resonant enhancement of all allowed phonons is observed, the spectrum maintaining its two-phonon structure. On the contrary, in the mixed crystal $GaAs_{1-x}P_x$ the two-phonon spectrum disappears, being replaced by an energy-shifted one-phonon spectrum, due to the enhancement of the $\boldsymbol{q}$-dependent Fröhlich interaction. The excitons, localized by the defect- and alloy-induced disorder, couple to the most energetic LO phonons, the exciton-LO-phonon coupled states becoming the intermediate states in Raman scattering. The coupling constant V determines the intensity of the 1LO, 2LO, 3LO modes and so on in resonant Raman scattering and can therefore be estimated from these intensities. Such a study has been reported for $GaAs_{1-x}P_x$ for different x values by Ramkumar et al. (1996b); the analytic dependence of the intensities of the 1LO, 2LO and 3LO modes on $V/\hbar\omega_{LO}$, where $\hbar\omega_{LO}$ is the energy of LO phonons, can also be found in this paper. Besides the determination of resonant Raman intensities, the disorder leads to the appearance of localized states in the bandgap, manifested through sharp resonances, and to an enhancement of the oscillator strength of the second-order scattering when the localized excitons interacted with LO phonons.

The effect of isotopic substitution can also be studied with the coherent-potential approximation. In an ordered medium, the frequencies and linewidths of phonons vary with the mass as $\omega \approx 1/\sqrt{M}$, $\Gamma \approx 1/M$. These relations are not valid in disordered materials. In diamond-like isotopically pure disordered samples, if phonon linewidth broadening is mainly due to anharmonic effects, the strength of the anharmonic three-phonon interaction is proportional to $M^{-3/2}$ and, since the two-phonon joint density of states is proportional to $\sqrt{M}$, $\Gamma \approx 1/M$ at low temperatures where all occupation numbers are 1/2. At higher temperatures $\Gamma \approx 1/\sqrt{M}$ due to the occupation number dependency on $k_B T/\hbar\omega \approx \sqrt{M}$. This relation is similar to that in a harmonic oscillator, so that isotopic mass dependency cannot distinguish the anharmonic from harmonic effects in disordered materials at high temperatures. Similar considerations apply to phonon frequencies, which depend on M as $\omega(M) \approx a/\sqrt{M} - c_1/M$ at low temperatures and as $\omega(M,T) \approx a/\sqrt{M} - c_2 T/\sqrt{M}$ at high temperatures. The first term in these expressions corresponds to the harmonic approximation (Wang et al., 1997)

The effective medium theory can be used to calculate the Faraday effect in binary composite materials consisting of a dielectric matrix with metallic inclusions. The calculations show that near the percolation threshold and in the dielectric region below it, the Faraday effect is greatly enhanced (Barthélémy and Bergman, 1998). The nonlinear optical properties of the same random metal-dielectric films received considerable attention by (Shalaev and Sarychev, 1998)

The effect on Raman spectra of crystal potential fluctuations, which destroy translational invariance, is in general a broadening and a change in asymmetry. Additionally, collective excitations (phonons or plasmons) can become localized if their coherency is destroyed. The study of phonons in disordered materials gives information about the structural quality; plasmons, in addition, provide information about the influence of crystal potential fluctuations on electron transport. In alloys, the fluctuations of crystal potential are due to random alloy potential or fluctuations of impurity potential. Structural disorder can also affect the coupled plasmon-LO phonon modes by changing the asymmetry of Raman lines. The asymmetry of these lines becomes opposite to that for optical phonons due to the sign difference in the dispersion curves of optical phonons and plasmons.

Free electrons in doped semiconductors couple electrically with LO phonons, forming coupled plasmon–LO-phonon modes. Since there are GaAs-like and AlAs-like LO phonons in both AlGaAs alloy and SLs, two types of coupled optical modes exist: ω_1^+ for GaAs-like LO phonons coupled with electrons, and ω_2^+ for AlAs-like LO phonons. When relaxation of the crystal momentum conservation due to crystal potential fluctuation is taken into account through the spatial correlation length ξ, the Raman intensity far from resonance is described phenomenologically as

$$I(\omega) \approx \int \mathrm{d}\boldsymbol{q} f_{\mathrm{sc}}(\boldsymbol{q}) \exp(-q^2\xi^2/4)/[(\omega-\omega(q))^2+(\Gamma/2)^2], \tag{7.5}$$

where $f_{\mathrm{sc}}(q)=[4\pi/(q^2+q_0^2)]^2$ is the 3D screening correlation function with q_0 the screening wavevector (Thomas–Fermi screening radius) and $\omega(q)$ is the dispersion of relevant collective excitations with damping constant Γ. ξ characterizes the volume where the collective excitations are localized, and thus the microscopic nature of crystal potential fluctuations. In bulk $Al_xGa_{1-x}As$ the dispersion of LO optical phonons is $\omega(q)=\omega_{\mathrm{LO}}(1-Aq^2)$; from experimental data it follows that for GaAs-like phonons $A(\mathrm{GaAs})=0.18(a/2\pi)^2$, whereas for AlAs-like phonons $A(\mathrm{AlAs})=0.05(a/2\pi)^2$ with a the lattice constant. On the other hand, the plasmon dispersion, in the random-phase approximation is $\omega(q)=\omega_{\mathrm{p}}[1+(3/10)(v_{\mathrm{F}}/\omega_{\mathrm{p}})^2q^2]$ where v_{F} is the Fermi velocity and ω_{p} the plasma frequency. This dispersion relation has opposite sign compared to that for phonons. For sufficiently high electron densities $\omega_{\mathrm{p}} \cong \omega_{\mathrm{LO}}$, the dispersion of coupled plasmon-LO-phonon modes being almost completely determined by plasmons, which have a much stronger dispersion than the LO phonons.

The different dispersions of phonons and plasmons lead to an opposite asymmetry of the plasmon-LO-phonon Raman line compared to that of LO phonons, observable in undoped samples. The asymmetry is influenced also by the localization length L. A typical behavior of the Raman line with increasing free-electron concentration N is shown in Fig. 7.1.

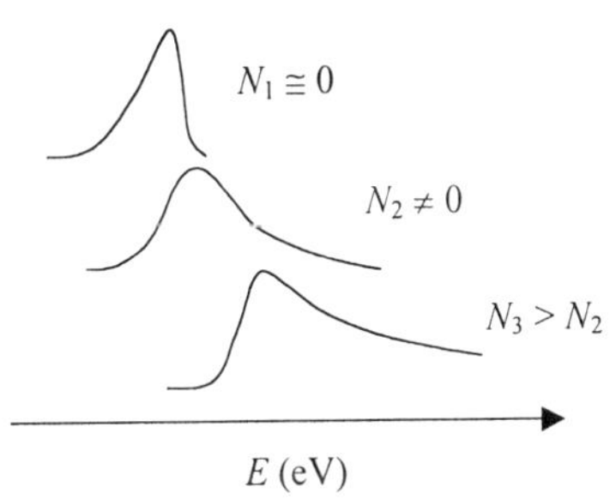

Fig. 7.1. Typical dependence of Raman spectrum on the free-electron concentration N

This effect has been evidenced experimentally in Si-doped $Al_xGa_{1-x}As$ alloys and doped GaAs/AlAs SLs (Pusep et al., 1998) in unpolarized backscattering Raman spectra at 8 K and 514 nm, for concentrations N of the order of 10^{18} cm^{-3}. When phonons couple to plasmons (in doped samples), a blueshift of the Raman line and a change in shape is observed, the line asymmetry becoming opposite to that of LO phonons. As N increases, L and Γ decrease (these are determined by fitting the experimental data with the expression above); typical values are L = 104 Å, $\Gamma = 20.4\,\text{cm}^{-1}$ for $N = 10^{18}\,\text{cm}^{-3}$, and L = 59 Å, $\Gamma = 10.4\,\text{cm}^{-1}$ for $N = 2\times 10^{18}\,\text{cm}^{-3}$.

The change in the width of the asymmetric Raman line is caused by the different range of contributions to scattering. More exactly, if L is small, collective excitations in a broad dispersion interval are involved in Raman scattering. For the undoped sample phonon states from almost the whole BZ contribute and the width of the line is determined by the phonon dispersion, whereas in doped alloys, the plasmon dispersion is well defined up to a critical value of the wavenumber q_c, plasmons with $q > q_c$ decaying into single-particle electron excitations. The halfwidths of plasmon-LO-phonon lines depend on free-electron concentration since q_c increases with N. The localization length is larger for AlAs-like than GaAs-like coupled modes due to the different atomic vibrations.

In some materials, instead of disorder, a spontaneous ordering takes place. One example is the $Ga_{0.51}In_{0.49}P$ alloy, which orders spontaneously in a CuPt-like structure when grown by organometallic vapor phase epitaxy on a GaAs (001) substrate. In the ordered alloy In-rich $Ga_{0.5(1-\delta)}In_{0.5(1+\delta)}P$ alternate with Ga-rich $Ga_{0.5(1+\delta)}In_{0.5(1-\delta)}P$ $\langle 111\rangle$ planes, where $0 < \delta < 1$ is the ordering parameter. In phonon Raman spectra folded acoustic-like phonons and LO-like phonons appear due to the changes in the Brillouin zone. Room-temperature polarized far-IR reflectance measurements, sensitive to the longitudinal A_1 and E vibrations, with the incident plane parallel to the growth direction in partially ordered GaInP show a polarization dependence (Alsina et al., 1997). This is not surprising since even in the disordered isotropic alloy, s and p-polarized spectra are different due to the angular dependence of the reflectivity. In partially ordered alloys, for s-polarization, only one feature related to the vibrations in the plane of ordering was found, whereas for p-polarization a whole structure in the range of InP-like vibrations was discovered. This structure is more evident as the order parameter increases, and is related to a resonance with longitudinal-optic character due to the anisotropy of longitudinal phonons. The appearance of these order-induced features can be explained using statistical arguments, which predict that any lattice property f of a partially ordered alloy is (Laks et al., 1992) $f(\delta) = \delta^2 f_{\text{ord}} + (1-\delta^2) f_{\text{dis}}$, where ord and dis refer to the values in the ordered and the completely disordered phase. In particular this relation can be applied to (Clausius–Mossotti relation) $f = (\varepsilon - 1)/(\varepsilon + 2)$.

Disorder can also change the nature of the bandgap from direct to indirect. For example, in long-range ordered $(Al_{0.5}Ga_{0.5})_{0.51}In_{0.49}P$, the fundamental edge PL shifts to lower energy by 120 meV at 11 K relative to the PL from Γ conduction band valley in the disordered alloy (Yamashita et al., 1997). In ordered alloys the bandgap is direct, the Γ PL appearing at a lower energy than the indirect PL from the X valley in disordered alloys.

7.2 Disordered Superlattices

As we have mentioned in the previous section, the structural disorder in SLs can change the asymmetry of the Raman lines of coupled plasmon-LO phonon modes. In SLs the fluctuations of the crystal potential are caused by the disordered character of the interface between GaAs and AlAs layers, and by the fluctuations of the impurity potential. As for the case of doped alloy, the asymmetry of Raman lines become opposite to that for optical phonons due to the sign difference in the dispersion of optical phonons and SL plasmons, i.e. plasmons polarized normal to the layers. Pusep et al. (1998) observed this effect in doped $(GaAs)_n(AlAs)_m$ SLs. Unlike alloys, discussed above, SLs are anisotropic systems so that $I(\omega)$ in (7.5) is to be understood as being averaged over all directions. Moreover, for thick enough barriers the quantum wells are isolated and the only contribution to Raman scattering is from plasmons polarized in the plane of layers; the 3D screening correlation function has thus to be replaced by the 2D function $f_{sc}(q) = 1/(q + q_0)$ where $q_0 = 2g_v / a_B$ with a_B the effective Bohr radius and g_v the valley degeneracy factor which gives the number of equivalent energy bands. The shape of the Raman line has the same behavior with increasing free-electron concentration as in Fig 7.1. When N reaches 3×10^{18} cm^{-3} the localization length L suffers a sharp decrease corresponding to the entrance of the Fermi level into a minigap. This localization of electrons serves as evidence of the metal-dielectric transition, which occurs with increasing electron density in SLs.

In a SL disorder can also be introduced by a random distribution of well widths. The electronic properties of disordered SL (d-SL) change compared to that of ordered SL (o-SL); at low energies the electronic states are localized, whereas they become extended at higher energies. In contrast to amorphous materials, the disorder induces a fine structure in the PL spectra and a redshift with temperature due to the reduction of the bandgap, described by $\Delta E_g(T) = -\alpha T^2 /(\beta + T)$. Moreover, the PL efficiency has a slower decrease with increasing temperature than in o-SL. Mathematically, the properties of d-SL can be calculated using the matrix-propagation technique with randomly chosen wells of thickness $L_{wi} = L_{w,min} + (i-1)\Delta L_w$ where $i = 1,\ldots, N_w$, N_w being the number of different widths that differ by ΔL_w. The degree of disorder can be modified by changing ΔL_w. In the presence of disorder, the matrix propagation technique shows that the miniband containing extended states in the o-SL (for $\Delta L_w = 0$), splits into N_w subminibands of localized states, the degree of localization being higher for lower-energy subminibands. Moreover, disorder can introduce localized states below the band of extended states, an effect more evident in strongly d-SL. The subminibands themselves exhibit a disorder-induced fine structure, which is more evident in weakly d-SL. A qualitative picture of the disorder-induced changes in the PL spectrum of an GaAs/(AlGa)As o-SL is shown in Fig. 7.2 (Capozzi et al., 1996). In this figure $X^{hh}(i)$ denotes the ground state HH excitonic recombination at well L_{wi}. The PL spectrum is shifted towards lower energies compared to the o-SL case and a fine structure can be seen for the HH excitonic recombination in the fourth well. Besides these changes in the lineshape and spectral position of the PL spectrum, its intensity is also modified due to disorder-induced localization. Since localized excitons have a smaller mobility, the nonradiative recombination probability decreases, at high temperatures (around 100 K) the d-SL PL spectrum becoming about three orders of magnitude more intense than the o-SL PL. At low temperatures, however, the PL intensity of the o-SL is more intense than for d-SL. Another quite unexpected behavior at low temperatures is that the PL at high energy grows as the temperature increases, due to the variable-range hopping between localized states of exciton propagation; the PL of localized states at higher energy increases with temperature. Moreover, in the high-temperature limit all PL lines show a steeper increase, with the Arrhenius dependence $\exp(-\Delta E / k_B T)$, suggesting the existence of a quasithermal equilibrium with activation energy ΔE. The dependence of the PL on the excitation intensity reveals no qualitative difference between o-SL and d-SL, with the exception of the disorder-induced fine structure, which shows a non-monotonic behavior caused by the different localization properties of the states that give these lines.

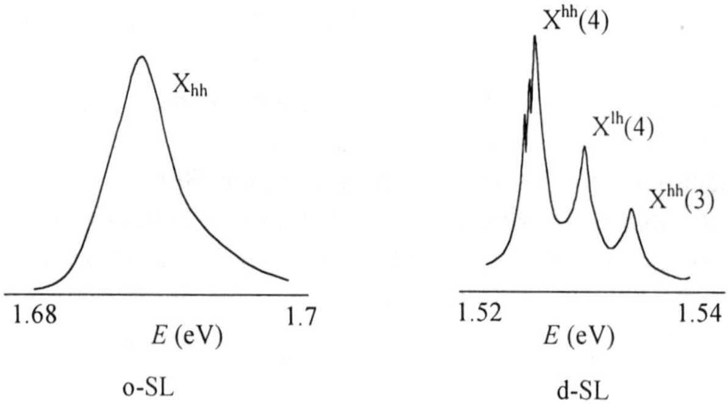

Fig. 7.2. Qualitative behavior of the disorder-induced changes in the PL spectrum of an GaAs/(AlGa)As o-SL

For disordered thin-layer GaAs/AlAs SLs expressions for the low-temperature PL has been obtained (Zhang et al., 1996) under the assumption that a disorder-induced set of localized energy levels in the bandgap exists, besides the excitonic levels. A good fit with experience is found if thermal relaxations from the conduction band to both exciton level and localized states are accounted for, as well as thermal excitation from the exciton level and localized states back into the conduction band, radiative recombinations of exciton states and localized states, and transitions from localized states to nonradiative centers.

7.3 Rough Surfaces

Not only SLs, but also surfaces can be randomized. The study of randomly rough metal surfaces has recently gained interest because it was observed that they favor second-harmonic generation (O'Donnell et al., 1997). For example, at a free air/silver interface, the roughness couples light incident at an angle θ_i to a counterpropagating surface plasmon polariton at the fundamental frequency ω. The angular distribution of the power spectrum of the second-harmonic light scattered from this metal surface with a weak random roughness has an almost rectangular form, centered on the surface plasmon polariton wavenumber at the fundamental frequency. The incident light produces a strong excitation of these surface waves if there are wavevectors k_r, k_r' in the roughness spectrum such that $k_{sp}(\omega) = k_i(\omega) + k_r$ or $-k_{sp}(\omega) = k_i(\omega) - k_r'$. These two cases correspond to excitation of counterpropagating surface plasmon polaritons (see Fig 7.3).

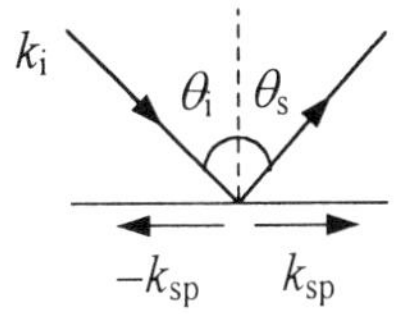

Fig. 7.3. Counterpropagating surface plasmon polaritons produced by the incident light in the roughness spectrum

The roughness spectrum for first-order scattering, $S(\omega)$, is centered on $k_{sp}(\omega)$ with a full width $\Delta k = (2\omega / c)\sin\theta_{max}$ where $|\theta_i| \leq \theta_{max}$. Analogously, the light scattered due to nonlinear interactions has a wavenumber that satisfies $k_s(2\omega) = k_{sp}(2\omega) - k_r$ or $k_s(2\omega) = 2k_i(\omega) \pm k_r$. So, the scattered distribution has two distinct peaks at angles consistent with the nonlinear interaction of the incident wave with the fundamental plasmon polariton. The mechanism of second-harmonic generation depends on light polarization: it is strongly related to the surface topography when the excitation light is s-polarized, whereas a local enhancement of the second harmonic, due to second-harmonic localization, takes place for p-polarized light (Smolyaninov et al. 1997).

The efficiency of second-harmonic generation at surface-plasmon polariton (SPP) excitation on corrugated metal surfaces is enhanced through the control of the surface-harmonic composition. If nanostructured, biperiodic gratings are fabricated, an enhancement of the reflectivity relative to a flat random silver surface has been observed, the enhancement taking place whenever counterpropagating SPP modes are exciting at the pump and second harmonic frequency. This double-resonance condition can be obtained by varying the grating wavevector (Pipino et al., 1996).

A backscattering enhancement of the diffusively scattered light from a weakly random silver surface due to excitation of surface plasmon polaritons was also observed (O'Donnell et al., 1998) (see the peak in the backscattered angular distribution of the diffuse scattered intensity in Fig. 7.4; the two curves correspond to different incident angles).

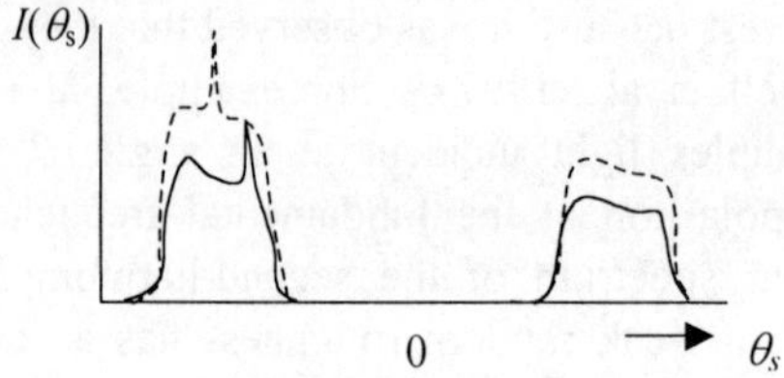

Fig. 7.4. Backscattered angular distribution of the diffuse scattered intensity from a weakly random silver surface

The backscattered peak which appears at $\theta_s = -\theta_i$ is caused by the constructive interference of multiple scattering processes within surface valleys for a surface roughness σ of the order of the wavelength of incident light λ, or by the excitation of surface-plasmon polaritons when $\sigma << \lambda$. In the latter case the peak appears when a double scattering occurs: the momentum conservation for both diffuse scattering $(\omega/c)\sin\theta_s = (\omega/c)\sin\theta_i + k_r$ and for excitation of plasmon polaritons $\pm k_{sp} = (\omega/c)\sin\theta_i + k_r$ must be simultaneously satisfied. This double-scattering process takes place when $(\omega/c)\sin\theta_s = -k_{sp} + k_r'$ and is described in the fourth order of the perturbation theory in (σ/λ). It can occur for $54° < \theta_s < 90°$, i.e. for large incident angles. A sixth-order perturbation theory describes the scattering process $(\omega/c)\sin\theta_s = k_{sp} + k_r''$, which can occur if $-90° < \theta_s < -54°$.

7.4 Polycrystalline and Nanocrystalline Materials

Polycrystalline materials are formed from grains of crystalline phases separated by boundaries. If the relative volume of the crystalline phase is high, these materials can be treated as a collection of separate crystalline regions, randomly oriented in space. The number of different possible crystalline arrangements depends on the material. It can reach quite large values, as for example in SiC, which has about 200 polytypes with different structures ranging from pure cubic to pure hexagonal stacking. Each of these polytypes differs with respect to the value of the energy gap and the location of the conduction band minima in the ***k***-space. In particular, for some SiC polytypes the optical and dielectric properties can be found in Adolph et al. (1997). The optical and dielectric properties depend not only on the polytype, but also on the size of the crystalline regions (clusters). Different models are used for different sizes. For example, in alkali-metal clusters the absorption profile can be calculated with a random-matrix model if the number N of atoms in the cluster is between 10 and 1000 (Akulin et al., 1997). The other two limiting models for small and large clusters are, respectively, non-adiabatic interaction of vibrational and electron motion, and weakly interacting quasiparticles, respectively. The first cannot be used for $N \geq 10$ because the adiabatic separation of electron and nuclear degrees of freedom cannot be used when the distance between neighboring electronic terms $\Delta E \approx 1/N$ becomes comparable to or smaller than the non-adiabatic interaction. Also, the weakly interacting quasiparticles model cannot be used for $N \leq 1000$ because the mean free path of electrons exceeds the typical cluster radius $R \approx N^{1/3}$.

Polycrystalline materials are macroscopically isotropic, so that the effective medium theory can be applied. If the material is a quasi-1D or a quasiplanar Drude metal, a part of its oscillator strength is pushed up in frequency to form a band of confined plasmon-like excitations. Assuming that the polycrystal is composed of spherical crystals with $\varepsilon_1 / \varepsilon_0 = 1 - \omega_p^2 / \omega^2$ along the x direction and $\varepsilon_2 = \varepsilon_0$ along the y and z directions, the optical sum rule can be written as

$$\int_0^\infty \mathrm{Im}\, \varepsilon_e(\omega)\omega \mathrm{d}\omega = (\pi/2) p \omega_p^2, \quad \int_0^\infty \mathrm{Im}[\varepsilon_e(\omega) - \varepsilon_{av}(\omega)]\omega^3 \mathrm{d}\omega = (\pi/6) p(1-p)\omega_p^4, \tag{7.6}$$

with $p = 1/3$. The effective dielectric constant is defined as $\varepsilon_e = \varepsilon_{av} - (1/9\varepsilon_{av})\sum_{i=1}^3 (\varepsilon_i - \varepsilon_{av})^2$ when the strong isotropy condition is satisfied, where in the same approximation the average dielectric constant is $\varepsilon_{av} = \langle \varepsilon(\boldsymbol{r}) \rangle = (1/3)\sum_i \varepsilon_i$, with ε_i the principal dielectric constants and $\langle\,\rangle$ the ensemble average. The strong isotropy condition is defined as $g_{lk}(\boldsymbol{r}, \boldsymbol{r}') = g(\boldsymbol{r} - \boldsymbol{r}')\delta_{lk}$, where $g_{lk}(\boldsymbol{r}', \boldsymbol{r}) = \sum_{i=1}^3 \langle [\varepsilon(\boldsymbol{r}') - \varepsilon_{av}]_{li} [\varepsilon(\boldsymbol{r}) - \varepsilon_{av}]_{ik} \rangle$. For ellipsoidal crystallites, which do not satisfy the strong isotropy condition, only the first rule is valid. These optical sum rules are valid also for 2D metals described by the Drude dielectric function along y, z, and a frequency-independent dielectric function along x, if p takes the value $2/3$ (Stroud and Kazaryan, 1996).

A particular situation arises when the crystalline regions are nanometer-sized. Since the atoms at the boundaries are displaced from lattice sites of adjacent crystals, the average over many boundaries shows very little short- or long-range order. The atomic structure is therefore different from either the crystalline or the amorphous state. It differs also from low-D structures in that the confining potential is not due to other, surrounding material, but to the disorder in the same material. If various sized nanoclusters are simultaneously present in the material, one can identify the peaks of each of them from the Fourier transform of the IR optical spectra and from the comparison with the ground nanostructured compacts. An interesting observation is that, besides the crystalline components, surface components and disordered interfacial components appear in the IR spectrum with decreasing grain sizes. Engineered materials, with a different structure on the molecular level compared to the bulk can be fabricated, using for example nanostructure oxides (Ying et al., 1993).

The character of localized and nonlocalized modes loses its meaning if the phonon correlation length is larger than the size of the nanoparticle. This affirmation can be exemplified with borosilicate glasses doped less than 1% by CdS_xSe_{1-x} nanocrystals with diameters smaller than 100 Å. In S-rich samples it is expected that CdSe-like modes are localized and CdS-like modes propagate. However, both types of LO modes are observed, with frequencies given by the solution of the equation $x\varepsilon_1(\omega)+(1-x)\varepsilon_2(\omega)=0$ where ε_1, ε_2 are the dielectric constants of CdS and CdSe, respectively (CdS and CdSe have non-overlapping restrahlen bands). The strength of these modes does not depend on x, and for isotropic cosine dispersions in both materials they scale as $\omega_0(\mathrm{CdSe})/\omega_0(\mathrm{CdS}) = [2\omega_{\mathrm{TO}}^2(\mathrm{CdSe})+\omega_{\mathrm{LO}}^2(\mathrm{CdSe})]^{1/2}/[2\omega_{\mathrm{TO}}^2(\mathrm{CdS})+\omega_{\mathrm{LO}}^2(\mathrm{CdS})]^{1/2}$. In $CdS_{0.55}Se_{0.45}$ crystals, disorder activates the zone-edge LO phonons, not present otherwise in first-order Raman spectra (Ingale and Rustagi, 1998).

Polycrystalline semiconductors and semiconductor compounds with nanometer-sized crystals are commonly called porous. The most studied material from this category is porous Si. As in many other nanocrystal assemblies, porous Si has a strong nonlinear optical response. Since porous Si has an indirect gap, the radiative decay time near the bandgap is long (ms–μs) and easily observable in PL spectra. These long-lived electron-hole pairs cause the photodarkening effect, i.e. PL quenching in nanocrystals that contain more than one electron-hole pair or an unpaired charge, due to the highly efficient Auger effect. Another effect in porous Si, as well as in other disordered materials such as C, is linear polarization memory, caused by the nonspherical shape of the nanocrystal. Due to this form anisotropy, the incident light selectively excites those crystals with the largest dimension parallel with its polarization direction; the emitted light is therefore polarized along the same direction. For an ellipsoidal nanocrystal the dependence of the optical excitation probability on the angle between the polarization vector of the exciting light $\boldsymbol{e}_{\mathrm{ex}}$ and a unit vector $\hat{c}$ along the major ellipsoidal axis is (Efros et al., 1997)

$$P(\omega_{\mathrm{ex}})=1+\kappa(\omega_{\mathrm{ex}})(\hat{\boldsymbol{c}}\cdot\boldsymbol{e}_{\mathrm{ex}})^2, \tag{7.7}$$

where

$$\kappa(\omega)=\delta(\omega)(1-3\eta^{(c)})[4+\delta(\omega)+\eta^{(c)}]/[4(1+\delta(\omega)\eta^{(c)})^2], \tag{7.8}$$

with $\delta(\omega)=[\varepsilon_{\mathrm{Si}}(\omega)/\varepsilon_{\mathrm{m}}(\omega)]-1$, $\varepsilon_{\mathrm{Si}}$ and ε_{m} being the dielectric constants of Si and the effective medium, respectively, and $\eta^{(c)}$ the depolarization factor along the $\hat{c}$ axis. For an ellipsoid with axes of length a, b, c, $\eta^{(c)}<1/3$ if $a=b<c$ and $\eta^{(c)}>1/3$ if $c<a=b$. The depolarization factors along the other two directions are $\eta^{(a)}=\eta^{(b)}=(1-\eta^{(c)})/2$. Since the probability of photon emission is proportional to $P(\omega_{\mathrm{ex}})$ and the photon detection probability is $P(\omega_{\mathrm{de}})=1+\kappa(\omega_{\mathrm{de}})(\hat{\boldsymbol{c}}\cdot\boldsymbol{e}_{\mathrm{de}})^2$, where $\boldsymbol{e}_{\mathrm{de}}$ is the polarization vector of the detected light , the intensity of the PL spectrum is

$$I_{\mathrm{PL}}\approx\langle N_0(\hat{c})P(\omega_{\mathrm{ex}})P(\omega_{\mathrm{de}})\rangle, \tag{7.9}$$

where $N_0(\hat{c})$ is the probability that the crystal contains no free electrons and/or holes. The PL, as well as the polarization spectra, shows very strong nonlinear effects even at low excitation intensities, being very sensitive to the detection energies, as can be seen from Fig. 7.5.

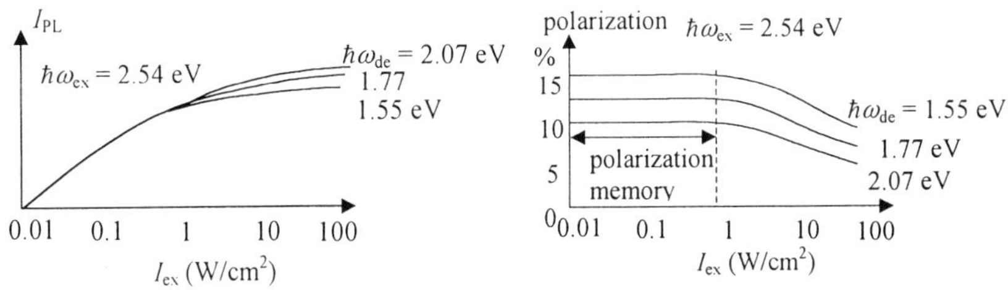

Fig. 7.5. PL and polarization spectra of porous Si

If $\hat{c}$ is aligned at an angle γ relative to the [100] growth direction of the porous Si and if it makes with the emission and detection directions the azimuthal angles φ_e and φ_d, respectively, then

$$I_{PL} \approx (2/\tau_r W_e \kappa_e)[(2+\kappa_e)(2+\kappa_d)I_0(b_e) + (2+\kappa_d)I_1(b_e)] \\ + \cos[2(\varphi_e - \varphi_d)][(2+\kappa_e)\kappa_d I_1(b_e) + \kappa_e\kappa_d I_2(b_e)], \tag{7.10}$$

where $\kappa_{e,d} = \kappa(\omega_{ex,de})\sin^2\gamma$, $I_0(x) = 1/(x^2-1)^{1/2}$, $I_1(x) = 1 - xI_0(x)$, $I_2(x) = -xI_1(x)$, and the parameter of nonlinearity is $b_e = [2 + \tau_r W_e(2+\kappa_e)]/(\tau_r W_e \kappa_e)$ with the average probability for exciting a nanocrystal $W_e = I_{ex}\alpha(\omega_{ex}, a)$. Here $\alpha(\omega_{ex}, a)\hbar\omega_{ex}$ is the effective absorption cross-section of a crystal with average size a and τ_r is the radiative lifetime of a single electron-hole pair. For an effective absorption cross-section 3.9×10^{-17} at an excitation energy 2.54 eV, $\eta^{(c)} = 0.08$, $\tau_r = 82$ μs, $\sin\gamma = 0.57$, the experimental curves are well fitted by the formula above.

The depolarization factor can be calculated from experimental Raman data, as for example in GaP (Kuriyama et al., 1998). In the Raman spectrum of porous GaP the formation of microcrystallites is evidenced by the appearance of a shoulder at the surface phonon frequency ω_s (see Fig. 7.6).

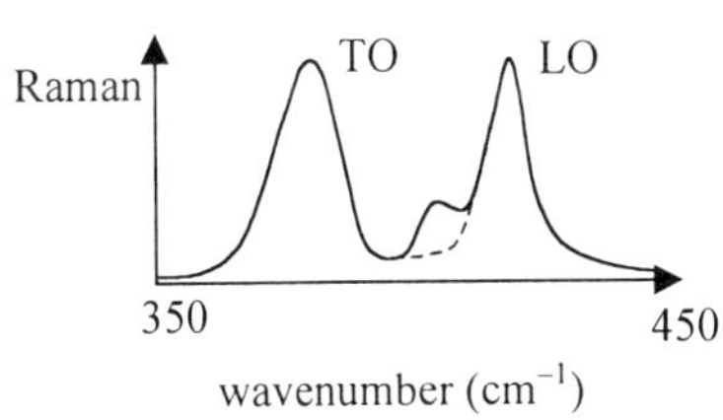

Fig. 7.6. Raman signal for porous (*solid line*) and crystalline (*dashed line*) GaP

The depolarization factor η is calculated from $(\omega_s^2/\omega_{TO}^2) = [\varepsilon_0 + \varepsilon_m(1/\eta - 1)] \times [\varepsilon_\infty + \varepsilon_m(1/\eta - 1)]^{-1}$ where ε_m is the dielectric constant of the surrounding medium and ε_0, ε_∞ are the static and high-frequency dielectric constants; the value of the depolarization factor gives us a hint about the shape of microcrystals. For example, if the experiment gives $\eta = 0.57$, we can infer that the microcrystals have a cylindrical shape (for which, theoretically, $\eta = 0.5$) with the polarization perpendicular to the rotational axis of the cylinder.

Another observed nonlinear optical effect in porous Si is the two-photon-excited PL with pulsed lasers. The spectrum and lifetime of this PL is almost identical to the one-photon-excited PL, but the degree of polarization ρ is significantly higher and depends on the orientation of the input polarization with respect to the crystalline axes of the sample, due to the selective excitation of ellipsoidal nanoparticles by the linearly polarized light and due to the intrinsic anisotropy in two-photon excitation of crystalline Si (Diener et al., 1998). The degree of polarization is maximum when the input polarization is along the [110] direction in the surface plane of porous Si from the Si(100) wafer, and minimum when it is along [010]. For two-photon excitation the PL intensity can be written similarly as above: $I_{PL}^{2\hbar\omega} \approx \langle P_e^{2\hbar\omega}(\omega_e) P_d(\omega_d)\rangle$ (for one-photon excitation $I_{PL} \approx \langle P_e(\omega_e) P_d(\omega_d)\rangle$) where the emission probability $P_d(\omega_d)$ is the same, but the excitation probability $P_e(\omega_e)$ is more complex than for one-photon excitation. It depends not only on the orientation of the ellipsoid, but also on the orientation of the crystalline structure inside the ellipsoid. For spherical particles photoexcited far above E_g, most of the emission occurs near E_g after successive nonradiative decays and tends to be unpolarized. On the contrary, photoemission at frequencies close to that of the pump is preferentially polarized along the pump polarization since nonradiative decays that scramble the polarization memory have only a small effect. When the particles are ellipsoidal, the local field favors emission at the bandgap with a polarization parallel to the long axis of the ellipsoid. However, the ellipsoidal particles are randomly oriented, so that the degree of polarization is dependent on the photoexcitation properties. If the photoexcitation is azimuthally isotropic with respect to the sample rotation about the surface normal, due to the random distribution of crystalline orientations in the crystals, ρ is finite but azimuthally isotropic. Otherwise, when the photoexcitation is azimuthally anisotropic about the surface normal (this can happen when the crystal axes are along fixed directions although the particles are randomly oriented), ρ is finite and azimuthally anisotropic.

The broad luminescence lineshape in porous Si can be described statistically. If we assume that porous Si is formed from spherical crystals with a Gaussian distribution of diameters L around a mean value L_m the PL intensity can be written as $I_{PL}(\lambda) \approx \rho(\lambda) W_r/(W_r + W_{nr})$ where $\rho(\lambda)d\lambda = \rho(L)dL$ is the size distribution $\rho(L) = (1/\sigma\sqrt{2\pi})\exp[-(L-L_m)^2/2\sigma^2]$ with root mean square σ. The wavelength λ emitted by the sphere of diameter L embedded in an infinite potential decreases with decreasing L. In some simulations it is assumed that the emitted wavelength is related to L as $\lambda \approx L^{1.4}$ (see Pellegrini et al. (1995) and the references therein). The radiative recombination rate W_r is considered independent of L, whereas the nonradiative recombination rate $W_{nr} >> W_r$ is proportional to $\exp(-C/\lambda)$ with C constant. This model of spheres embedded in an infinite potential also allows the introduction of effective masses inside and outside the sphere, with values determined from the boundary conditions at the sphere/surrounding medium interface (Fishman et al., 1993). This model explains the blueshift and the intensity decrease of the PL after optical excitation assuming that L decreases after excitation due to an electrochemical process. PL intensities as well as PLE spectra in porous Si are extremely dependent on the fabrication procedure. For example, in photochemically etched porous Si the PLE peak can shift by up to 1 eV with increasing etching time, first towards lower energies, then towards higher energies due to two independent and competing excitation mechanisms: direct-gap absorption and direct excitation of localized states (Koyama et al., 1996).

The PL intensity of porous Si can be enhanced more than one order of magnitude, and its linewidth can be narrowed by up to 18–25 eV when the emitting porous Si is placed in a microcavity consisting of two Bragg reflectors, fabricated from alternating layers of porous Si with different refractive indices (different porosity) (Pellegrini et al., 1995). The enhancement of spontaneous emission occurs when the emitting states are in resonance with the optical mode of the Fabry–Perot resonator, the radiative decay from all other states being inhibited. This behavior is modeled with the transfer matrix theory, supposing again a Gaussian size distribution of Si nanocrystals.

PL degradation under visible light illumination can be empirically described using a stretched exponential function, with temperature-independent stretching parameter and relaxation-time constant. In order to explain this behavior, it was assumed that there are two metastable defect states, A and B, the carriers having to overcome the barrier potentials E_{AB}, E_{BA} in order to move from A to B and from B to A, respectively (see Fig. 7.7) (Chang et al., 1993).

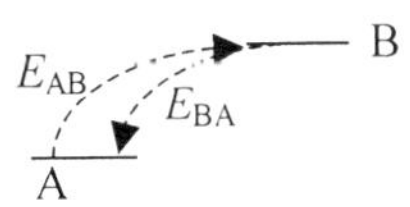

Fig. 7.7. Mechanism of PL degradation: carrier displacement between the two metastable defect states, A and B, where E_{AB}, E_{BA} are the barrier potentials

On illumination, state A traps the photoexcited carriers, which are then transferred to state B. The latter are nonradiative recombination centers, so that the PL intensity is proportional to the inverse of the number of defects in state B, $N_B(t)$: $I_{PL}(t) \approx I_{ex}/[N_B(\infty) - (N_B(\infty) - N_B(0))\exp(-(t/\tau)^{\beta})]$, where τ is the degradation time constant, β the stretching parameter (< 1) and $N_B(\infty)$, $N_B(0)$ are the number of nonradiative recombinations in the initial and saturated states, respectively (the PL saturates at a value proportional to the excitation intensity, the saturated nonradiative recombination centers being independent of the excitation intensity). On illumination, there is a broad distribution of activation energies between states A and B, but states A transfer first to states B since they have the smallest activation energy. The degradation profile follows the filling of states B. Although this mechanism is more proper for amorphous Si, it can nevertheless explain the PL decay in porous Si. The parameters β and τ are independent of temperature, but decrease as the excitation intensity increases. A lower β means faster degradation. To check this hypothetical mechanism, porous Si was illuminated by either an Ar laser at 488 nm, or by an Argon laser and a He-Ne laser. It was found that by illuminating with both lasers the PL intensity increases and the degradation is slower since both He-Ne and Ar-lasers pump carriers to excited states, which are then trapped by A, the energy needed to transfer carriers from A to B coming from the capture of photoexcited carriers. The PL intensity increases and the degradation slows down when the sample is illuminated with both lasers since in this case the liberated energy from carriers photoexcited by the He-Ne laser is smaller than E_{AB}, and hence the nonradiative recombinations associated to states B are less numerous. At room temperature the PL decay in porous Si has, however, a fast and a slow component. The first ($\cong 10^{-10}$ s) is due to bimolecular recombination of free carriers in the core of nanometer-sized crystals, whereas the slow component ($\cong 10^{-4}$ s) is caused by recombination of carriers rapidly trapped in surface localized states (Malý et al., 1996).

A stretched exponential describes well the PL decay also in oxidized Si nanocrystals at low temperatures. The following figures show the absorption curves for oxidized Si crystallites with average crystalline (c-Si) core diameters of 3.7 nm (a), porous Si (b), indirect-gap crystal bulk Si (c) and amorphous Si:H (d) (Kanemitsu, 1996).

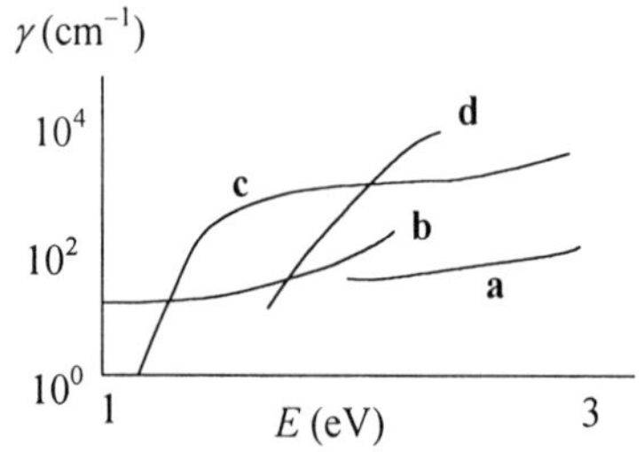

Fig. 7.8. Absorption curves for (**a**) oxidized Si with average crystalline (c-Si) core diameters of 3.7 nm, (**b**) porous Si, (**c**) indirect-gap crystal bulk Si and (**d**) amorphous Si:H

From Fig. 7.8 it follows that for curves a and b the absorption spectra are exponential in the near-IR and visible, described by $\gamma \approx \exp(\hbar\omega / E_0)$ with E_0 the Urbach energy. The Urbach tail reflects the nanoscopic disordered nature of the sample, and is about 250 meV in Si nanocrystals, and 40–50 meV in a-Si:H. The

PL peak energy is practically insensitive to the crystallite size, showing only a very small blueshift with decreasing size, whereas the PL intensity is sensitive to both crystallite size and surface oxidation. The mechanisms which determine the PL are different at high and low temperatures, as follows from the different behavior of PL intensity with temperature: it increases with decreasing temperature up to 100 K, then it starts decreasing for a further decrease of the temperature. The temporal evolution of the PL in the temperature range 10–300 K can be described by $I(t) = I_0(\tau/t)^{1-\beta}\exp[-(t/\tau)^{\beta}]$, where at high temperatures $\tau^{-1} \approx \exp(\hbar\omega_{PL}/E_1)$ with $\hbar\omega_{PL}$ the PL photon energy and E_1 a temperature-dependent constant. The radiative decay rate depends also on temperature as $\exp(-E_r/k_BT)$ where the thermal activation rate for oxidized Si is $E_r \cong 71$ meV. In porous Si the same constant is about 3–17 meV while in bulk c-Si it is 61 meV. The temperature dependence of the radiative decay rate shows that PL does not come from one state in a single crystallite but from multilevel light-emitting states in single crystallites, with a Boltzmann distribution. At low temperatures the PL lifetime is very long, of the order of μs–ms, due to tunneling and thermally activated transport of carriers to nonradiative recombination centers through barrier heights of the order 0.2–0.3 eV.

As in highly doped semiconductors and amorphous materials, the Urbach band tail in absorption is caused by the exponential extension of both conduction and valence band states into the bandgap. One explanation of the Urbach tail, valid for both monocrystalline and polycrystalline materials, assumes that it is caused by strain and bulk and grain boundary carrier trapping. E_0 is the sum of interactive and disorder contributions, depending on carrier concentrations, temperature and structural disorder. In polycrystalline materials it is supposed that both the grain boundary, which is bidimensional and amorphous, and the relatively organized crystalline region contribute to the potential fluctuations, their contributions being recently separated (Iribarren et al., 1999). Considering the grain boundary characterized by structural disorder caused by ionized traps, its contribution to the band-tail parameter is $E_{0,gb} = 8\pi^2 e^3 Z^2 m_{eff} L_D^2 \langle F_{gb}\rangle /(9\sqrt{3}\varepsilon_s \hbar^2)$ with Z the impurity charge, N the free-carrier concentration, m_{eff} the carrier effective mass, ε_s the static dielectric constant, $L_D = [\varepsilon k_B T/(e^2 N)]^{1/2}$ the Debye length, and $\langle F_{gb}\rangle = (F_s L_D/R)[1-\exp(-R/L_D)]$ where F_s is the field at the boundaries of the spherical grains with radius R. The band-tail parameter is thus proportional to the averaged electric field. In the crystalline region the band tail is caused by defects and imperfections such as wrong bonds and coordination, chemical and stoichiometric defects, dislocations, voids, etc. These defects create localized energy levels which trap carriers and can ionize them, causing the appearance of an average electric field $\langle F_d\rangle = eN_{def}L_D/\varepsilon_s$ where N_{def} is the trap-effective charge distribution, which includes all types of bulk defects. The contribution to the band-tail parameter of bulk defects is $E_{0,def} = 8\pi^2 e^3 Z_t^2 m^* L_D^2 \langle F_d\rangle /(9\sqrt{3}\varepsilon_s \hbar^2)$ with Z_t the trap ionization. This depends on the density of defects and the temperature. Besides these contributions, the band tail is also dependent on the strain through a term $E_{0,s} = (3/8)YkT_D$ where T_D is the Debye temperature, and $Y \approx 1/\sigma_0$, with σ_0 the Urbach-edge parameter. Y is related to the proper strain and elastic properties, to the structural lattice distortions in monocrystalline and amorphous materials. In polycrystalline materials, it is determined by the lattice distortions in grains, and by the influence of the surroundings. In both monocrystalline and amorphous materials $E_{0,gb} \to 0$, and $E_{0,\text{def}} \neq 0$, but $E_{0,\text{def}}$ has larger values in amorphous materials. In polycrystalline samples the disorder contribution to the band tail is typically about 4 times larger than the interactive one, whereas in monocrystals it is only about 1.5 times higher for the same carrier concentration.

For excitation energies above the bandgap, such as for amorphous materials, $\sqrt{\alpha\hbar\omega}$ depends almost linearly on $\hbar\omega$. From this Tauc plot, however, it is not possible to extract the bandgap of low-D nanostructures like porous Si (Datta and Narasimhan, 1999). Although no single model explains all optical data of porous Si, this particular energy dependence of the absorption coefficient can be explained by modeling porous Si as a pseudo-1D material system with a distribution of bandgaps. The assumption that the absorption coefficient of porous Si has the same form as in 3D bulk c-Si: $\sqrt{\alpha\hbar\omega} = D(\hbar\omega - E_g)$, holds only if the density of states $g(E) \approx \sqrt{E}$ with E measured from the band edge. Porous Si is, however, a nanostructured material with an enhanced bandgap due to quantum size effects; if we assume that porous Si is an assembly of pseudo-1D quantum wires with a parabolic band structure, $g(E) \approx 1/\sqrt{E}$. Moreover, porous Si has a distribution of quantum wire sizes d, so the energy upshift is different for each nanostructure. This quantum-size effect explain why the PL peak in porous Si has a greater energy than the bandgap of c-Si, and why this peak can be tuned through the visible spectrum by changing the preparation conditions (Canham, 1990). The absorption of porous Si can be calculated if it is assumed that it consists of 1D parallelepipeds of square cross-section of side d and constant length L. The absorption takes place in bulk under the hypothesis that $\boldsymbol{k}$ is not conserved in optical transitions, that the oscillator strength depends on the size of nanostructure in which absorption occurs, and that the distribution of bandgaps influences the optical absorption. The porosity P can be determined from transmission measurements by calculating first the refractive index under the assumption that porous Si is formed from air and c-Si. In the effective-medium approximation $P = 1 - [(2\varepsilon_{\text{PS}} + \varepsilon_{\text{Si}})(\varepsilon_{\text{PS}} - 1)/3\varepsilon_{\text{PS}}(\varepsilon_{\text{Si}} - 1)]$ where ε_{PS}, ε_{Si} are the dielectric constants of porous Si and bulk c-Si, respectively. The observed reduction of absorption at low temperature with respect to room-temperature absorption, and the shift of these two curves can be explained by accounting for the indirect bandgap of porous Si. The distribution of sizes that fit best the experimental data is the lognormal distribution $P(d) \approx (1/d)\exp[-(\ln d - \ln d_0)^2/2\sigma^2]$, where d_0 is the mean size and σ the standard deviation.

Micro-Raman scattering in porous Si layers can also be used for optical measurements of the thermal conductivity. The method is based on the fact that the surface layer is first heated by the focused laser beam, the subsequent rise in temperature producing a shift of the Raman peak, from which the local temperature rise is found. If P is the heating power, a the heating source diameter, and $\Delta T = T_l - T_b$ the difference between the local and the bulk temperature, the thermal conductivity is defined as $\kappa_s = 2P/(\pi a \Delta T)$. Measurements have shown that the Raman peak position is linearly decreasing with temperature, the thermal conductivity in 50 μm thick porous Si layers of 1 W/K m being much lower than the value of 63 W/K m in crystalline bulk Si (Périchon et al., 1999).

7.5 Continuous-Network Structures

Thin metal films grown by vacuum evaporation occur not only in amorphous or polycrystalline structures, but also as a continuous-network-structure (CNS) in which the deposited metal is separated by long, irregular and narrow channels. The thin CNS metal films have a very irregular shape and are intermediate between continuous and metal island films. The optical absorption characteristics of such CNS in Ag and Ir films can be found in Anno and Tanimoto (1999). In continuous metal films the Drude-type absorption due to conduction electrons appears at low photon energies and is in close relation with the structure of grains and boundaries of the film. In metal island film, the conduction electrons contribute to optical absorption as plasma-resonance absorption caused by plasma oscillations of conduction electrons in metal particles. For CNSs both optical absorptions can be found, in different proportions.

7.6 Optical Properties of Amorphous Materials

Amorphous materials lack long-range periodic order and exhibit continuous polymorphism, as opposed to the discrete polymorphism of the crystalline state. Amorphous materials can exist metastably within a range of densities, the densification process involving not only the elimination of free volume but also structural changes. Amorphous solids can consist of close-packed atoms as in amorphous metals, or of covalently bonded atoms arranged in an open network as in amorphous semiconductors. The latter are the most studied amorphous materials from the point of view of optical properties, hence we focus upon them.

There are two main categories of amorphous semiconductors: tetrahedrally coordinated amorphous semiconductors (TCAS) and chalcogenide glasses, distinguished by a particular short-range ordering, which differ mainly in coordination. In TCAS the short-range order in the amorphous phase is similar to that in the crystal up to small bond-angle distortions, but quite different in the long range. In glasses, the disorder is due to the coordination variation from site to site, although the average coordination is fixed in the amorphous material. A quite different type of disorder can appear in compounds – the chemical disorder. Several models of short-range order can be found in (Lucovsky and Hayes, 1979). Glasses, in general, differ from other solids in that they can be obtained directly by rapid cooling of the melted material below the glass transition temperature. The rapid cooling of a non-glass material results generally in a polycrystalline structure, the transition from the liquid to the solid phase occurring with latent heat. One can view the amorphous semiconductor as a metastable state of a solid, in contrast to a glass, which can be imagined as a liquid with high viscosity.

7.6.1 Theory of Electronic States in Amorphous Semiconductors

A theory that describes the electronic states/band structure of a general, or even a particular, amorphous material does not exist. There are, however, plenty of theories that help us understand different aspects of the behavior of real amorphous semiconductors. These theories are well reviewed in (Kramer and Weaire, 1979). We will only sketch here the basic concepts.

All theories of amorphous semiconductors are based on the notion of localization. In the absence of diffusion from a given site, the Anderson criterion for the existence of localized states in a system can be expressed in terms of the retarded Green's function, which gives the probability amplitude for an electron to move from the site j to the site j' during the time interval t:

$$G_{jj'}(t) = -i\langle 0 | c_j(t) c_{j'}^{+}(0) | 0\rangle. \tag{7.11}$$

$|0\rangle$ is the electronic vacuum state and $c^{+}(0)$, $c(t)$ are electronic creation and annihilation operators at times 0 and t, respectively, the subscripts in (7.11) indicating the electronic sites at which they refer. The evolution of the probability amplitude is determined by

$$-i\frac{\partial G_{jj'}(t)}{\partial t} = \Sigma_k H_{jk} G_{kj'}(t) + \delta_{jj'}\delta(t), \tag{7.12}$$

together with the initial conditions. The electron is localized if the probability of finding it on the initial site after time t satisfies the relation

$$\lim_{t\to\infty} | G_{00}(t) |^2 \neq 0. \tag{7.13}$$

For a disordered system for which the Hamiltonian $H = H_0 + V$ can be separated in a disorder-independent part H_0 and a disorder-dependent part V, $G(t)$ is identical to the Fourier transform of the operator

$$G(z) = G_0(z)\Sigma_{n=0}^{\infty}[VG_0(z)]^n, \tag{7.14}$$

with $G_0(z) = (z - H_0)^{-1}$. In the thermodynamic limit localized states are associated to dense distribution of poles of $G_{00}(z)$, and extended states correspond to branch cuts of the same function along the real energy axis (Economou and Cohen, 1972). The decomposition of the Hamiltonian in disorder-dependent and disorder-independent parts depends on the system: for structurally disordered systems H_0 is the electron kinetic energy and V the superposition of atomic potentials, for many-impurity problems H_0 describes the unperturbed crystal and V the superposition of impurity potentials, and so on. The simplest Hamiltonian that exemplifies the Anderson localization is of the type

$$H = \sum_j \varepsilon_j \mid j\rangle\langle j \mid + \sum_{j,j'} V_{jj'} \mid j\rangle\langle j' |, \tag{7.15}$$

where the sum in the second term is performed over sites j, j' corresponding to nearest neighbors, and the disorder is contained only in the first sum, ε_j being randomly distributed with a probability $P(\varepsilon)$ of width W. In two dimensions or higher the Hamiltonian in (7.15) can have extended or localized eigenstates depending on W (strength of disorder) (Anderson, 1958). Both theory and experiments suggest that there is an 'Anderson transition' between localized and extended states. Since the electron movement is qualitatively different for localized and extended states, corresponding to hopping and free-electron conduction, respectively, the energy of the transition from localized to extended states is called the mobility edge. The effect of disorder depends also on the symmetry of the state: for example, the energy of the s states is less sensitive to fluctuations of the local potential than the energies of the p or d states.

Since the optical properties are dependent on the density of states, much effort has been devoted to the calculation of this parameter. Both theories and experiments show a remarkable degree of similarity of the overall distribution of the density of states in crystalline and amorphous phases. This is due to the fact that the energy spectrum in an amorphous semiconductor consists of valence and conduction energy bands as in a crystalline semiconductor, with top and bottom edges E_v and E_c, respectively. These, however, do not terminate abruptly as in the crystalline case, but have tails into the energy gap (Fig. 7.9).

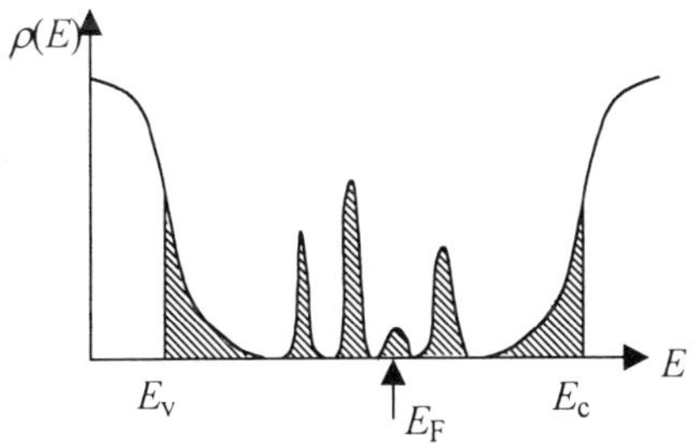

Fig. 7.9. Density of states in an amorphous semiconductor

The distribution of the energy levels in the energy gap can be described with several models: Cohen–Fritzsche–Ovshinsky model, Davis–Mott model, small polaron model, etc. (a review of the models for states in the gap of amorphous semiconductors can be found in (Davis, 1979; Nagels, 1979)). The existence even in amorphous semiconductors of energy bands is quite amazing, since they are a consequence of the translational symmetry in crystalline materials. Their survival in the amorphous state is due to the predominant role of short-range interactions; the interatomic forces in amorphous materials are commonly considered to be the same as in their crystalline counterparts.

7.6.2 General Optical Properties of Amorphous Semiconductors

The optical properties can be deduced from the general expression of the imaginary part of the dielectric constant (Connell, 1979)

$$\varepsilon_{\mathrm{im}}(\omega) = \frac{2}{V}\left(\frac{2\pi e}{m\omega}\right)^2 \sum_{i,f} |\langle f\,|\,P\,|\,i\rangle|^2\, \delta(E_{\mathrm{f}} - E_{\mathrm{i}} - \hbar\omega), \tag{7.16}$$

where V is the volume of the sample, and P is the momentum operator. In crystalline materials the sum is further evaluated by taking into account the translational symmetry, which leads to the rule of momentum conservation during optical transitions. The main difference in the amorphous phase is the breakdown of this rule. The short-range order, which is preserved in the fully coordinated system, independent of the topological disorder, is responsible for those features in the optical response of amorphous materials that are similar to the corresponding features in crystals, whereas the differences are determined by the topology of the amorphous material.

In fully coordinated materials the wavefunctions of the valence and conduction bands, $|\,i\rangle$ and $|\,f\rangle$, respectively, are expanded in terms of orthonormal, localized functions, centered on different atoms n:

$$|\,i\rangle = \textstyle\sum_n a_{inv}\,|\,nv\rangle, \qquad |\,f\rangle = \textstyle\sum_n a_{fnc}\,|\,nc\rangle. \tag{7.17}$$

When the initial and final states of the optical transition are localized, one of the coefficients a dominates, otherwise the amplitudes and phases of these coefficients vary randomly. Supposing that the ensemble average of the square modulus of the matrix element is $\langle|\,\langle f\,|\,P\,|\,i\rangle\,|^2\rangle_{\mathrm{ensemble}} = |\,P_{\mathrm{am}}(\omega)\,|^2$,

$$\varepsilon_{\mathrm{im}}(\omega) = \frac{2}{V}\left(\frac{2\pi e}{m\omega}\right)^2 |\,P_{\mathrm{am}}(\omega)\,|^2 \int_0^{\hbar\omega} \mathrm{d}E\, g_i(-E)\, g_f(\hbar\omega - E), \tag{7.18}$$

where $g(E) = (1/V)\sum_n \delta(E - E_n)$ is the one-electron density of states. The difference between the amorphous and crystalline solids is that, although the energy is conserved in both cases, the critical points associated with the momentum conservation do not appear in the amorphous case. Another important difference is that the optical response depends on the direction of the electric field in the crystalline material, but is independent of it in amorphous solids. $P_{am}(\omega)$ depends generally on the type of transitions we are referring to: delocalized to delocalized or delocalized to localized.

In general, the optical properties of amorphous materials depend strongly on the preparation procedure, mainly because the optical properties are very much influenced by defects, the concentration and types of which are determined by the fabrication procedure and the thermal history of the probe. Optical absorption is an example of a process that has the same form irrespective of the type of the amorphous material; a typical absorption curve is shown in Fig. 7.10.

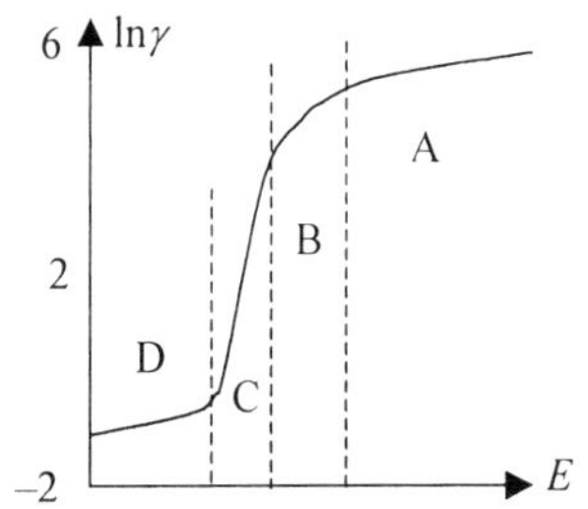

Fig. 7.10. Typical absorption curve in an amorphous material

It can be generally separated into four regions (three regions can be found in other references (Taylor, 1989)). Region A corresponds to interband transitions between delocalized valence and conduction states, separated by energies higher than the mobility bandgap. The absorption in region B is due to transitions between valence and conduction band states situated close to the respective mobility edges. This region can be used to estimate the optical bandgap. In region C, called the Urbach edge region, the absorption depends exponentially on the photon energy, the transitions involving defect states in the tails of the density of states. Finally, in region D the absorption depends less dramatically on energy than in the Urbach region, being caused by different processes. In chalcogenide glasses the absorption in this region is probably due to impurities, whereas in TCAS it is caused by the presence of unsatisfied bonds. Another difference between the two types of amorphous materials is that in chalcogenide glasses the optical absorption edges are highly reproducible and insensitive to preparation conditions, whereas in TCSC (most of which are fabricated only in thin film form), optical absorption edges depend dramatically on preparation conditions.

The gross features of the optical response can be obtained if, for excitations above the absorption edge, the conduction band density of states is taken as a step function at energy E_1 above the valence band mobility edge. The valence band density of states is then $g_i(E_1 - \hbar\omega) \approx \mathrm{d}[\omega^2 \varepsilon_{im}(\omega)/|P_{am}(\omega)|^2]/\mathrm{d}\omega$ where $|P_{am}(\omega)|^2$ can be obtained by averaging the corresponding term in the crystalline phase, or can be approximated for energies above the Penn gap $\hbar\omega_g$ by $|P_{\text{Penn}}(\omega)|^2 = \hbar k_F(\omega_g/\omega)^2$ (Penn, 1962) where k_F is the wavenumber at the Fermi level.

The densities of states just beyond the mobility edges, if the energies are measured from the valence band mobility edge, are usually approximated by power laws: $g_i(-E) \approx E^p$, $g_f(E) \approx (E - E_g)^q$ so that in this region (region B) $\omega^2 \varepsilon_{\text{im}}(\omega) \approx (\hbar\omega - E_g)^{p+q+1}$. This relation reduces to $\omega^2 \varepsilon_{\text{im}}(\omega) \approx (\hbar\omega - E_g)^2$ for parabolic valence and conduction bands. The optical absorption depends then on energy as $\alpha\omega \approx (\hbar\omega - E_g)^2$, a relation from which the Tauc optical bandgap $E_g = E_c - E_v$, equal to the difference between the conduction and valence band localization edges, can be determined (Tauc, 1974). The proportionality constant in the Tauc absorption law contains information on the convolution of valence and conduction band states and on matrix elements of optical transitions. An interesting observation is that the same quadratic absorption dependence on energy occurs for transitions between the mobility edge in the valence bands and localized states in the conduction band tail if $g_f(E) \approx (E - E_g)$ with E_g the energy between the mobility edge in the valence band and the tail of the conduction band. The Tauc definition of the optical bandgap is not unique, although widely used; it was not linked with any topological disorder or band-edge modification. In other definitions the bandgap is the energy corresponding to an absorption coefficient of $10^3\,\text{cm}^{-1}$ or $10^4\,\text{cm}^{-1}$. Optical absorption measurements show that the bandgap decreases with increasing temperature.

At energies below E_g (in region C) the absorption coefficient in almost all materials has the form $\alpha(\omega) = (\omega/nc)\varepsilon_{\text{im}}(\omega) \approx \exp[-\beta(E_0 - \hbar\omega)]$ with n the refractive index and β a constant parameter in the range 10–25 eV^{-1}. E_0 and $\sigma = \beta k_B T$ are also constant at high temperatures. This region is called Urbach since the behavior is similar to that first observed by Urbach (1953) in alkali halides, where it was found that $\beta \cong 0.8/k_B T$ near 300 K. The origin of this exponential law (Urbach absorption tail) has not been clarified up to now. Several mechanisms have been proposed as an explanation, including an exponential density of states in band tails, disorder-induced potential fluctuations, strained bonds, strong electron-phonon interactions, a Franz–Keldysh effect induced by the microfield caused by lattice vibrations, disorder-induced local electric field effects on the center-of-mass and relative motion of excitons, and so on. These early theories are briefly reviewed in (Connell, 1979). Which of them occurs depends mainly on the particular type of material.

Recent explanations of the Urbach tail include the model for the parameter σ valid in ferroelectric materials with order-disorder-type phase transition (Noba and Kayanuma, 1999). This model identifies the exciton absorption as the cause for the Urbach tail, and takes into account the thermal effect of exciton-lattice interaction and the internal Stark effect by local electric field due to displacement of protons. The random distribution of the local electric field induces a random polarization of the lowest exciton state that in turn induces an additional randomness of the exciton band in the disordered phase. At a temperature T the amplitude of the lattice distortion at the *j*th unit cell, q_j, has a thermal distribution $P(q_j) \approx \exp(-q_j^2 / 2k_B T_{eff})$ where T_{eff} is an effective temperature that incorporates the zero-point vibrations of phonons. Since this temperature is given by $k_B T_{eff} = (1/2)\hbar\omega \coth(\hbar\omega / 2k_B T)$, replacing for T_{eff} in the expression for Urbach tail gives $\sigma = \sigma_0 (2k_B T / \hbar\omega) \tanh(\hbar\omega / 2k_B T)$ with σ_0 a constant. At high temperatures σ becomes almost identical to σ_0.

Another recent model for the Urbach tail has been developed by John et al. (1986) starting from the calculation of the density of states in the case when the de Broglie wavelength is comparable to the correlation length ξ. The correlation length ξ describes the spatial extent of short-range order, which is typically of the order of the interatomic spacing. When the de Broglie wavelength $\lambda = \hbar / (2m|E|)^{1/2}$ is long compared to any correlation length ξ of the disorder (in the Halperin–Lax (H–L) limit), the kinetic energy of localization determines the scale of the most probable potential fluctuations for tail states near the band edge (Halperin and Lax, 1966). The 3D density of states in this limit scales exponentially with the square root of energy or, generally, in d dimensions, (Cardy, 1978), for a Gaussian white-noise potential, $\rho(E) \approx |E|^{d(5-d)/4} \times \exp(-\text{const.} \times |E|^{2-d/2})$. In the other case $\rho(E) \approx |E|^d \exp(-\text{const.} \times |E|^2)$. Neither of these expressions accounts for the universally observed Urbach tail. To model it, it is assumed that the electron involved in the optical transition is in a static random potential caused by the disorder due to lattice vibrations, impurities, etc. The Fourier components of the potential have several essentially Gaussian probability distribution contributions from various forms of disorder. The contribution to the electronic density of states at a certain energy $-|E|$ of such potential fluctuations is determined by the solution of a variational problem for the class of Gaussian potentials parameterized by a certain depth and a range a. The Gaussian potential is asymptotically exact for the deep tail, whereas there are small deviations from a Gaussian shape of the most probable potential fluctuations in the shallow tail, which gives an approximate linear exponential density of states for a range $0.1 < |E| / \varepsilon_L < 2$ (the regime accessible experimentally) with $\varepsilon_L = \hbar^2 / 2m\xi^2$. For $|E| / \varepsilon_L >> 2$ this distribution crosses over to the Gaussian density of states, whereas for $|E| / \varepsilon_L << 0.1$ the H–L density of states is recovered. The Gaussian-white-noise approximation is based on the constraint that the scale of potential fluctuation is large compared to the spatial correlation length ξ. In 3D (calculations have been performed also for lower dimensions) the Urbach edge is a consequence of the broad crossover between H–L and Gaussian tail regimes. Whenever the crossover region is narrower, as for example in the screened Coulomb-impurity model of H–L, which has longer-range correlations, the linear approximation is not so precise. This derivation of the Urbach tail from a quantitative estimation of the electronic density of states in correlated Gaussian random potentials gives an insight into the universality of this behavior.

It is interesting to note that in 1D amplifying random systems, the amplification suppresses the transmittance just as the absorption, so that we cannot distinguish from the transmitted intensity between transmittance and absorption.

To distinguish between absorption and transmission all moments of the transmitted intensity must be analyzed. Within the random-phase approximation all moments $\langle T^m \rangle$ with $m \geq 1$ are infinite in amplifying systems, whereas for absorption $\langle T \rangle \approx \exp(-\text{const.} \times L)$ decreases to zero when the length L of the medium increases to infinity (Freilikher et al., 1997). The localization is enhanced due to amplification.

TCAS and glasses have also similar properties in the far-IR region, i.e. in the region 1–100 cm^{-1} (Taylor, 1989). At lower energies in this interval both absorption and reduced Raman spectra are temperature dependent and scale with frequency as ω^2, whereas for higher energies these spectra are temperature independent and vary more rapidly with the frequency than ω^2. These energies regions are separated by 10 cm^{-1} in the glassy As_2S_3, for example. The frequency dependence of far-IR absorption in the temperature-independent region is the same as that of the density of states, which implies that the matrix elements that describe the photon-phonon coupling vary slowly with frequency in this range (Strom et al., 1974). They are almost constant for glasses, but vary with preparation conditions in TCAS.

Due to the disorder-induced breakdown of the selection rules, any IR and Raman modes of an amorphous solid become optically active, not only those allowed by group-theoretical selection rules. Both IR and Raman spectra exhibit the continuous, general features of the one-phonon density of vibrational states, modulated by a term determined by the symmetry of the local atomic environment (see, for example, Solin (1976)). More precisely, the IR absorption coefficient is $\alpha(\omega) \approx f(\omega)g(\omega)$, where $f(\omega)$ includes the frequency dependence of the matrix elements and $g(\omega)$ is the one-phonon density of states. By comparing the IR absorption and reduced Raman scattering spectra, we can infer whether $f(\omega)$ is strongly frequency dependent or not (it is if the spectra are very different). From the three types of modes: low-frequency acoustic modes, intermediate-frequency bond-bending modes, and high-frequency bond-stretching modes, the latter offer the most direct structural information, since their frequencies are determined mainly by nearest-neighbor interactions, and thus reflect the local molecular symmetry. Especially in binary alloys and compound systems vibrational spectra can help to discriminate between different bonding models.

In amorphous materials the nonresonant Stokes component of the Raman intensity at a frequency shift ω is given by the Shuker–Gamon approximation

$$I_{\alpha\beta\gamma\lambda}(\omega) = \sum_b C_b^{\alpha\beta\gamma\lambda}(1/\omega)[1 + \bar{n}(\omega, T)]g_b(\omega), \tag{7.19}$$

where $\bar{n}(\omega, T)$ is the Bose–Einstein distribution function, and $C_b^{\alpha\beta\gamma\lambda}$ is the coupling constant of the b band of vibrational states, the superscripts $\alpha\beta$ and $\gamma\lambda$ describing the polarizations of incident and scattered photons (Shuker and Gamon, 1971). The density of states $g_b(\omega)$ can be determined for each band b from the reduced Raman spectrum $I_{\text{red}}(\omega) = \omega(\omega_{\text{L}} - \omega)^{-4}[\bar{n}(\omega) + 1]^{-1} I(\omega)$ (for Stokes lines), where ω_{L} is the laser frequency. The approximation of a constant coupling is questionable when applied to all vibrational bands together, but proves useful for narrow bands. For anti-Stokes lines the reduced Raman spectrum is $I_{\text{red}}(\omega) = \omega(\omega_{\text{L}} + \omega)^{-4}\bar{n}(\omega)^{-1}I(\omega)$.

Also important in the characterization of amorphous materials is the depolarization ratio of Raman scattering, i.e. the ratio between the intensity of the scattered light polarized in the scattering plane to that polarized perpendicular to it. This is the only parameter that characterizes the symmetry properties of vibrational modes in media with a random orientation of individual scatterers. Like the Raman spectrum, the depolarization ratio is also a continuous function of ω (Taylor, 1989).

The vibrational Raman spectra in chalcogenide glasses and TCAS are not even qualitatively similar. In glasses, the spectra are to first order an approximation to the phonon densities of states, Raman spectra in glasses exhibiting in general sharp peaks (not as sharp as in crystalline materials). The reason is that glasses are dominated by a local 'molecular' structure. On the other hand, TCAS have generally very broad Raman features.

The PLE spectrum often tracks the absorption coefficient dependence on energy for energies below the bandgap. The PL spectra in amorphous materials is generally composed of a single broad peak, centered at least a few tenths of eV below the band edge, and a few tenths of eV wide. However, several broad PL peaks can appear when there are many intrinsic defects. The time-resolved PL has a broad distribution of decay rates with power-law decay with time. The PL (and absorption) intensities and lineshapes change under laser irradiation with bandgap light, due to rearrangements of electrons and atoms. As a function of temperature and applied pressure the PL modifies as: i) the PL peak position has an anomalous behavior (both blueshift and redshift) with temperature, ii) the PL intensity has a maximum as a function of temperature, iii) the PL decay time falls by more than an order of magnitude as T increases, iv) the PL intensity falls sharply when an external pressure P is applied, v) the PL peak position redshifts with P, an initial blueshift being sometimes discernible, and vi) the PL decay time depends systematically on the emission energy. A model for this behavior has been proposed (John and Singh, 1996) for porous Si, which, however, explains the temperature, pressure and emission energy dependence of the PL in all amorphous materials, chalcogenide glasses or TCAS. This is possible because no microscopic mechanism is assumed in the model, which is based solely on the hypothesis that the PL is the result of a competition between activated radiative processes and Berthelot-type nonradiative process. The PL decay time is then given by $1/\tau = R_{\text{r}} + R_{\text{n}} = \nu_{\text{r}}\exp(-T_{\text{r}}/T) + \nu_{\text{n}}\exp(T/T_{\text{n}})$ where R_{r}, R_{n} are the radiative and nonradiative rates, respectively. Supposing that the nonradiative processes are due to carrier tunneling across a barrier of width S vibrating with a frequency Ω, the parameters of the nonradiative rate are $\nu_{\text{n}} = \Gamma_0\exp(-2\alpha S)$, $T_{\text{n}} = M\Omega^2/2\alpha^2 k_{\text{B}}$, where Γ_0 is the jump frequency, $1/\alpha$ describes the extent of the carrier wavefunction, and M is the inertia of the vibrating system. In porous Si the vibrating barrier can be a defect such as a Si-O complex, a dangling bond, etc.;

in glasses, other mechanisms can be imagined as the source for a vibrating barrier. The values of the parameters introduced above assume quite a broad range in disordered systems, and can also be determined from non-optical measurements. Typical values for porous Si, obtained from non-optical experiments, are $1/\alpha = 2-7$ Å, $S = 8-20$ Å, $T_n = 50-150$ K, $\nu_r = 10^4\ \mathrm{s}^{-1}$, $T_r = 25-125$ K. The time-integrated temperature-dependent PL is

$$I(T) = I_0 R_r/(R_n + R_r) = I_0/\{1+\nu_0 \exp[(T/T_n)+(T_r/T)]\}, \qquad (7.20)$$

where I_0 is the initial intensity and $\nu_0 = \nu_n/\nu_r$. It has a maximum as a function of temperature at $T_m = (T_r T_n)^{1/2}$. If T_r is low, the PL intensity is approximately constant at temperatures lower than T_m (this behavior being attributed to Auger recombination). At high temperatures, $T > T_n > T_r$ a Berthelot behavior of the PL intensity is observed: $I(T) \cong I_0/[1+\nu_0 \exp(T/T_n)]$. A more complete expression of the temperature dependence of the PL,

$$I(T) = I_0/[1+(\Gamma_0/\nu_r)\exp(T_r/T)\exp(-2\alpha S)\exp(2\alpha^2 k_B T/M\Omega^2)], \qquad (7.21)$$

also explains the shift of the PL peak position with temperature. The peak position as a function of α has a maximum at $\alpha_p = (SM\Omega^2)/(2k_B T) \approx 1/T$, which implies a temperature dependence of the PL peak of the form $\hbar\omega_p = V + E_g - c_1 T - c_2/T^2$. Typical values for porous Si are $c_1 = 2.3\times10^{-4}$ eV/K, $c_2 = \hbar^2/2m(SM\Omega^2/2k_B)^2 = 3\times10^3$–$3\times10^5$ eV K^2. An initial blueshift followed by a redshift can occur, the crossover taking place at $T_c = (2c_2/c_1)^{1/3}$.

If the barrier width S has a linear dependence with pressure, $S(1-KP)$, with P the isothermal compressibility, the experimentally observed pressure dependence of the PL is explained by the relation

$$I(P) = I_0/[1+(\Gamma_0/\nu_r)\exp[(T_r/T)+(T/T_n)]\exp[-2\alpha S(1-KP)]], \qquad (7.22)$$

which implies also a quadratic pressure dependence of the PL peak. If the bandgap depends linearly on P as $E_g(P) = E_g - \eta P$, then $\hbar\omega_p = (V + E_g - c_2/T^2) + P(2Kc_2/T^2 - \eta) - P^2K^2c_2/T^2$.

7.6.3 Tetrahedrally Coordinated Amorphous Semiconductors

This class of materials includes amorphous Si (a-Si) and Ge (a-Ge), as well as III-V compounds, a-Si being usually considered the representative material of this class. Since the interatomic forces are the same as in crystals, in a-Si each Si atom is surrounded by four others, situated at the same distance as in the crystal and forming a regular tetrahedron. The first coordination sphere is thus almost identical to that of c-Si, the amorphous behavior being determined by the properties of the second coordination sphere; for example, the second-order neighbors have a deviation of the angles between bonds of about ±5° from the crystalline structure. The progressive increase of the disorder at large distances is confirmed by electron diffraction experiments that show that the diffraction maxima corresponding to neighbors increasingly far away is less and less discernible in the diffraction figure.

An ideal TCAS can be described by the Weaire–Thorpe Hamiltonian (Weaire and Thorpe, 1971)

$$H = \sum_{j\nu\nu'} V_1 \mid j\nu\rangle\langle j\nu' \mid + \sum_{jj'\nu} V_2 \mid j\nu\rangle\langle j'\nu \mid, \tag{7.23}$$

where V_1 describes the interaction between different orbitals at the same site, $\mid j\nu\rangle$ being the sp^3 orbital associated with site j, and V_2 describes the intersite interaction (see Fig. 7.11).

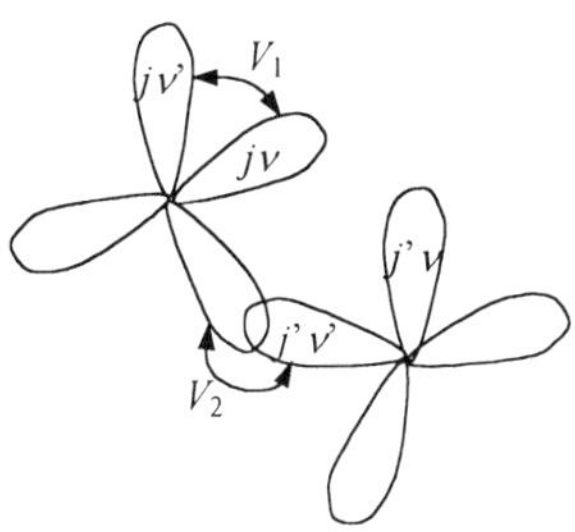

Fig. 7.11. Schematical representation of the interaction between two orbitals

The energy spectrum of this Hamiltonian consists of two bands (valence and conduction bands) separated by a bandgap of minimum width $2|V_2|-4|V_1|$. For $|V_1/V_2|=1/2$ the energy gap vanish; for $|V_1/V_2|<1/2$ the four states of each atom are arranged two in each band, the upper and lower energy bands being called antibonding and bonding bands, respectively. V_2 determines the separation of bonding and antibonding bands and V_1 characterizes their width. For $|V_1/V_2|>1/2$ three states in each atom are grouped in the upper energy, p-like band, and the remaining one in the lower energy, s-like band. This model is valid only when the interactions are confined to very short ranges, and does not account for small departures from perfect tetrahedral coordination. More refined models have to be used to describe the properties of real TCAS. These include tight-binding models, which involve atom clusters of a suitable structure, pseudopotentials which describe the relaxation of the long-range order, and muffin-tin models suitable for the description of liquid metals and clustered amorphous solids with short-range order (reviewed in Kramer and Weaire (1979)).

In TCAS the valence and conduction bands are formed from bonding and antibonding states, respectively. In compound semiconductors a new type of bond appears: like bonds, which in amorphous tetrahedral III-V compounds can be seen as antisite defects. For example, in GaAs, As-like bonds act as deep donors and Ga-like bonds act as deep acceptors. These like-bonds produce potential fluctuations and absorption tails well below the power-law edge E_g. The change in the potential due to like-bonds originates in the Coulomb interaction caused by the large ionic contributions to bonding. On the contrary, in chalcogenide glasses the local potential changes are small, since the bonding is mainly covalent.

Irrespective of the degree of sophistication of the Hamiltonians that describe ideal amorphous solids, their optical properties cannot be understood without taking into account the defects. In particular, the pinning of the energy level near the bandgap, in both types of amorphous materials (TCAS and glasses), which leads to a general insensitivity of properties with respect to impurification, is believed to occur due to the presence of discrete energy levels in the gap associated with specific defects (see the discussion in Davis, 1979).

For TCAS two kinds of defects must be considered: 'dangling bonds' which appear when an electron cannot be covalently paired with another, and 'lone-pairs' when an additional covalent bond forms between a normally coordinated atom and another through electrons which do not normally bond (as for example in the case of interstitial H doping). Dangling bonds are characterized through energy levels in the energy band.

From the point of view of the optical response, the most important defects in TCAS are the voids. These defects make the material inhomogeneous and the light propagates through multiple scattering. Whenever the complex dielectric constants of the medium and the voids, ε_m, ε_v are such that $0.05 < |\varepsilon_m(\omega)/\varepsilon_v(\omega)| < 20$ and the characteristic dimensions of the voids are smaller than the light wavelength, an effective medium theory can be used to describe the effects of voids on the optical absorption; other theories assume that small voids are accessible parts of an approximately homogeneous network (Phillips, 1971). A quantitative description of void effects on the optical response is in all cases difficult to predict. The spins can also affect the optical properties. For example, defect states with spin act both as deep donors and deep acceptors, the latter lying higher in energy by about 0.1–1 eV. Bipolaron formation can reverse the order of these levels, the doubly occupied state having a lower energy.

A comparison of the imaginary part of the dielectric constant in amorphous and crystalline tetrahedral materials is shown in Fig. 7.12, the dotted line representing a typical curve for the amorphous state and the solid line a typical curve in the corresponding crystalline phase (see Connell (1979) and the references therein). The two peaks in the crystal have different origins: the lower-lying one corresponds to transitions along directions coincident with the directions of bonds, and is less sensitive to amorphization. This is consistent with the fact that the peak of the absorption in the amorphous phase is close to E_1. The E_2 peak is due to transitions in directions that are not parallel to the bonds and is more sensitive to amorphization, being not present in the amorphous state. Estimating the Penn gap from the experimentally determined dielectric constant for a-Ge, one obtains a reduction of only about 5% from the crystalline to amorphous phases, which indicates again the preservation of short-range order. A calculation of the effective number of electrons per molecule contributing to absorption up to an energy E, $n_{\mathrm{eff}}(E) = (4m/h^2e^2N)\int_0^E E' n_{\mathrm{re}}(E') n_{\mathrm{im}}(E')\mathrm{d}E'$, confirms this conclusion. Here N is the number of molecules per unit volume, and n_{re}, n_{im} are the real and imaginary parts of the refractive index.

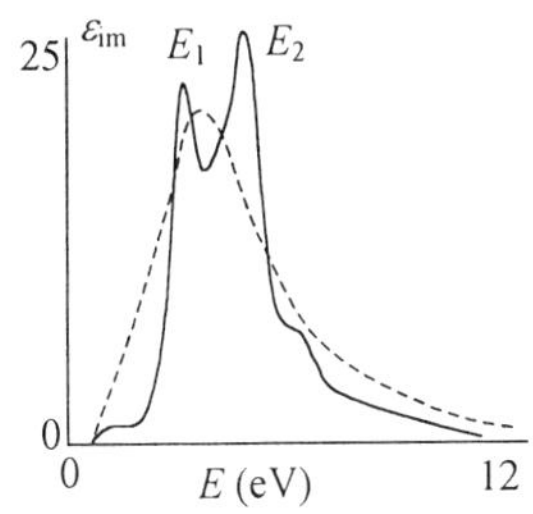

Fig. 7.12. Typical spectrum of the imaginary part of the dielectric constant in amorphous (*dashed line*) and crystalline (*solid line*) tetrahedral materials

The effective number of electrons in amorphous and crystalline phases contributing to a certain absorption band is an indication of the degree of similarity between the two phases; for example, for a-Ge this number is approximately 1.85 electrons per atom at 4.5 eV in both crystalline and amorphous phases (Connell et al., 1973). The dependence in a-Ge of the height and width of the peak of $\varepsilon_{\mathrm{im}}(\omega)$ on the preparation conditions is entirely explained by the volume and distribution of voids. In particular, the absorption edge position and shape are determined by voids. The voids can be almost completely eliminated by H incorporation: H atoms preferentially bond with dangling bonds on the void surface, transforming the material into a hydrogenated, fully bonded structure. The influence of H incorporation on the absorption spectra of Ge is reviewed in Connell (1979).

Analytical expressions for the optical constants in TCAS can be obtained from the Tauc joint density of states under the assumption of a collection of noninteracting atoms (Jellison Jr. and Modine, 1996):

$$\varepsilon_{\mathrm{im}}(E) = \begin{cases} 0, & E \le E_{\mathrm{g}}, \\ AE_0C(E-E_{\mathrm{g}})^2/\{E[(E^2-E_0^2)^2+C^2E^2]\}, & E > E_{\mathrm{g}}. \end{cases} \tag{7.24}$$

The above formula correctly describes the fundamental band-to-band transitions, which are characterized by the power-law $(E-E_{\mathrm{g}})^2/E^2$. When there are several valence bands connected to several conduction bands through optical transitions, each band contributes a term similar to (7.24) to the total dielectric function, which is obtained by summing all relevant terms. This summation procedure has proven very accurate in describing the experimental results for a-Se (Innami et al., 1999), a material in which the short-range order produces five local bands: an upper conduction band involving s atomic states, a lower conduction band of antibonding p states, an upper valence band associated with lone-pair (nonbonding) p states, a mid-valence band of bonding p states and finally, a lower valence band involving s states.

Although $n_{eff}(\omega)$ is approximately the same for the crystalline and amorphous phases, the density of states in the amorphous phase is quite a broadened version of that in crystals, and the optical matrix elements have a non-negligible dependence on frequency, as can be seen by comparing the IR and Raman curves. Figure 7.13 refers to a-Si at room temperature, but the dependencies are typical for other TCAS (Brodsky, 1975).

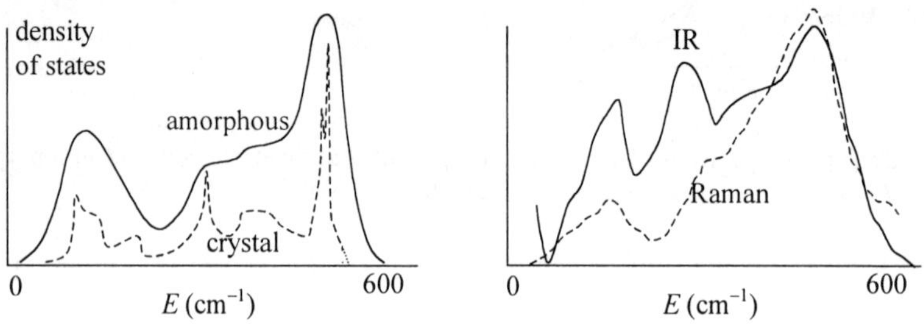

Fig. 7.13. Density of states in amorphous and crystalline TCAS, Raman and IR signals

Since TCAS include elements or compounds with indirect bandgaps (such as Si) the optical absorption in the amorphous state can be larger near the indirect bandgap than in the crystalline phase. Optically induced changes in absorption are of lower importance in TCAS compared to chalcogenide glasses, since the electron-lattice interaction plays a much smaller role. However, films of a-Si:H, when excited with light with an energy greater than the bandgap, show an increase of the below-gap absorption, which is metastable at 300 K (Stabler and Wronski, 1977). Most probably the optically induced absorption is caused by the dangling bonds (Amer et al., 1983). Transient-induced absorption can also be observed in TCAS, with a slope and time delay smaller than in chalcogenide glasses, and dependent on the film preparation method. Although it also has the same power law decay and the same temperature dependence of the exponent in the power law as in chalcogenide glasses, the mechanism in the two types of amorphous materials is clearly different, since transient-induced absorption can be excited with light above bandgap in TCAS, but with light below bandgap energy in glasses. The decay of the optically induced absorption is faster at 300 K than at 80 K due to the temperature dependence of hopping or multiple trapping of carriers; the decay is in general rapid, about a few ps, in the thermalization stage of hot photoexcited carriers, followed by a slower decay supposedly caused by the geminate recombination of trapped carriers (Taylor, 1989).

The amorphous state of a semiconductor, especially when it is porous, is modeled as a mixture of a crystalline phase and a pure amorphous phase. The ratio of the amorphous to the crystalline phases in thin Si films can be estimated as the ratio between the integral areas of a broad amorphous-like Raman peak to a narrow crystalline peak of TO phonon (Bustarret et al., 1988). Although the procedure is not very accurate, scattering data are used for the estimation, Raman data being sensitive to the degree of structural disorder. Raman spectra in the amorphous phase consist from an amorphous-like TO peak (at 480 cm^{-1} in porous a-Si) and a broad boson peak (at 150 cm^{-1}), which is absent in a crystal (Jäckle, 1981). The boson peak frequency ω_b is related to the correlation length ξ through $\omega_b \cong v/\xi$ where v is the sound velocity and expresses the vibration density of states in excess of the sound waves in the low-frequency region 20–100 cm^{-1}. In glasses, the structural correlation length is larger, so that the boson peak frequency is smaller than the frequencies of TA and LA modes, and is typically 1/5–1/7 of the Debye frequency. Experimentally it was observed that when crystallization begins, the amplitude of the boson peak decreases abruptly by a factor of 2, no change in amplitude and width of the amorphous-like TO peak being observed. Simultaneously, a small peak corresponding to the TO phonon in crystal appears in Raman spectra at 520 cm^{-1}. Qualitatively, this can be explained by a stronger tendency of localization for the optical compared to acoustic vibrations, the degree of violation of the wavevector selection rule being different for optical and acoustic modes. Mathematically, the coupling coefficient $C(\omega)$, which enters into $I(\omega) = C(\omega)g(\omega)[\bar{n}(\omega)+1]/\omega$ is proportional to ω, i.e. to $1/\xi$ for acoustic modes, no such localization dependence being found for optical modes. The different dependence of the coupling coefficient in these two cases is due to the fact that the coupling mechanism with light is via the deformation tensor in the acoustic case, whereas the center of weight of the elementary cell does not change on light excitations for optical vibrations (Ovsyuk and Novikov, 1998). So, the fraction of the amorphous component in solids that contain mixed phases should be determined optically using the boson peak, which is more sensitive to the structural disorder, instead of the amorphous-like optical mode.

For pure amorphous materials, such as a-Ge and a-Si, both theory and experiment show that the coupling coefficient $C(\omega)$, as well as the density of states, is proportional to ω^2 for small frequencies (less than 65 cm^{-1} for Si and less than 35 cm^{-1} for Ge) (see references in Brodsky (1975)). Raman spectra of compound TCAS show like-bonding modes and can detect small traces of crystallinity in elementary amorphous materials.

Luminescence is not commonly exhibited by TCAS due to the many alternative nonradiative recombination channels; the most notable exceptions are Si and Si_xC_{1-x} alloys (Fischer, 1979). A typical PL spectrum is shown in Fig. 7.14 for Si. The luminescence band is quite broad and without sharp features. Its shape and the spectral position of the peak depend generally on the fabrication procedure. For example, in glow-discharge Si films the most important parameter which determines the PL spectrum is the substrate temperature (Fischer, 1979): the spectrum moves to higher energies for increasing substrate temperature up to 250°C, and after that remains unchanged (as intensity, spectral position and shape). Three broad bands can be identified in PL, all three decreasing with temperature, suggesting the existence of a competing nonradiative temperature-activated recombination process. The three bands are due to network defects, dominant at high temperatures (Engemann and Fischer, 1977).

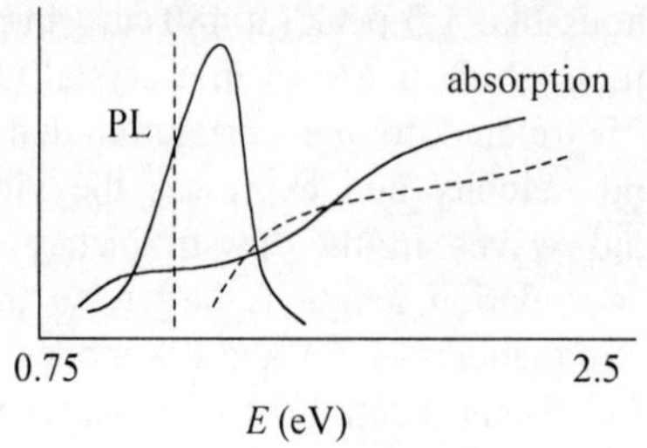

Fig. 7.14. PL and absorption of amorphous (*solid line*) and crystalline (*dashed line*) Si

The influence of preparation conditions on the properties of TCAS also includes the dependence of these properties on the film thickness. An example is the disorder-induced vibrational frequency shift dependence on the film thickness in hydrogenated a-Si films (Ossikovski and Drévillón, 1996). Due to the increased structural disorder at the film-substrate interface, the effective dynamical charges of Si-H bond vibrations in different configurations differ from those in a perfect crystal (see Fig. 7.15). They are also different from the static charge and are caused by the changes in charge distributions as the atoms vibrate.

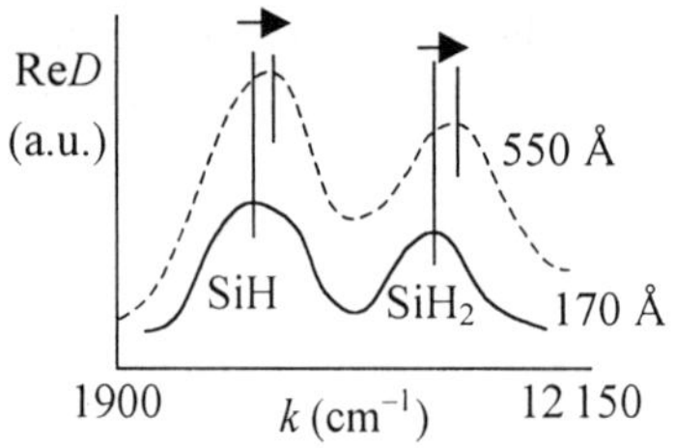

Fig. 7.15. The disorder-induced vibrational frequency shift dependence on the film thickness in hydrogenated a-Si films

This effective dynamical charge e^* can be determined from measurements of the real part of the optical density $D = \ln(\bar{\rho}/\rho)$ where $\rho = \tan\psi\exp(i\Delta)$ is the change of polarization state upon reflection, $\bar{\rho}$ being the same quantity referring to the substrate before deposition. $\mathrm{Re}\,D$, $\mathrm{Im}\,D$ are proportional respectively to $\mathrm{Im}\,\varepsilon$, $\mathrm{Re}\,\varepsilon$. In the polarized medium model, the material is assumed to consist of Si-H bonds uniformly embedded in a-Si homogeneous matrix, the frequency shift due to these bonds being equal to $\Delta\omega = -(\varepsilon-1)e^{*2}/[m_r\omega_0 R^3(2\varepsilon+1)]$ where ω_0 is the unperturbed frequency, determined by fitting the dielectric constant with $\varepsilon = \varepsilon_\infty + F/(\omega_0^2 - \omega^2 - i\Gamma\omega)$. R is the radius of the cavity (considered spherical) containing the dipole whose reduced mass is m_r, and ε is the IR dielectric constant of the nonabsorbing host matrix. If the cavity is not spherical, the frequency shift also depends on a geometric factor $A \in (0,1)$, which is equal to 1 for a slab and to 0 for a rod. For a general ellipsoidal cavity with semiaxes *a*, *b*, *c*, $\Delta\omega = -3e^{*2} A(1-A)(\varepsilon-1)/[2m_r\omega_0 abc(\varepsilon + A(1-\varepsilon))]$; the case of a spherical cavity is obtained for $A = 1/3$. The consideration of ellipsoidal, instead of spherical geometry corresponds to taking into account the disorder at the interface for Si:H isotropically distributed in the Si matrix. The thickness dependence of the disorder is modeled with $A = \Delta A \exp(-z/d_0) + (1/3)$ where d_0 is the characteristic disorder length and ΔA describes the disorder at interfaces with respect to 1/3. From experimental data the effective dynamical charges of Si-H bonds are $0.41e$ for the SiH and $0.24e$ for SiH_2.

Since PL is quenched by an externally applied electric field, it was inferred that in a-Si the photoexcited electrons and holes are correlated, with the possibility of forming free or bound excitons when trapped on the same center, which can be a Si-H bond (the electric field destroys the correlation). If the electron and hole are bound by Coulomb interaction, the binding energy $E_b = e^2/(4\pi\varepsilon r) \cong 0.13$ eV corresponds to an electron-hole distance r of about 10 Å. Since the excited state has such a small spatial extension, the surrounding network is deformed via electron-phonon coupling, and small polarons are formed. In optical spectra this can be seen as the Stokes shift of the luminescence band from the absorption edge. This shift is difficult to determine quantitatively due to the lack of sharp structures in both absorption and PL spectra. Another confirmation of the electron-hole pairing came from time-decay measurements of the luminescence intensity at low temperatures, which showed that the radiative recombination rate $1/\tau_r$ is proportional to the density of pairs and that the decay time of radiative recombination τ_r (of about 28 μs) is independent of the excitation intensity (Tsang and Street, 1978). At higher temperatures a temperature-activated radiationless process dissociates the photogenerated pairs into uncorrelated photocarriers. The dissociation rate has an exponential dependence on temperature: $1/\tau_d = (1/\tau_{d0})\exp(-\Delta E/k_B T)$ with ΔE the pair binding energy (Engemann and Fisher, 1976). The PL temperature dependence has the form $I(T) \approx 1/[1 + (\tau_r/\tau_{d,0})\exp(-\Delta E/k_B T)]$.

Doping influences the PL spectrum through several mechanisms: it can cause the filling of localized states in the bandgap with electrons or holes, creating internal electric fields, it can create donor and acceptor levels which generate additional luminescence bands, or the impurities which are not in the substitutional or interstitial sites as donors and acceptors can form network defects. For example, B, P and Li create additional bands in PL, which are the only ones to survive at high temperatures. However, for high doping concentrations the luminescence decreases due to either adding radiationless recombination channels or due to the internal electric field of ionized impurities. Oxygen contributes to the compensation of centers for radiationless recombination by saturating the dangling bonds. Hydrogen can bond covalently with Si, influencing the PL spectra through the electronic level (which probably merges with the valence and conduction bands) and localized vibrational modes (which *do not* provide channels for radiationless recombination), as well as by providing trapping sites for electron-hole pairs. H generally increases the PL intensity.

The PL spectra of a-Si:H has generally two bands, the higher-energy peak, situated between 1.2–1.4 eV caused by radiative transitions between conduction band tail states and valence band tail states, and the lower-energy peak, situated between 0.7–0.9 eV, which involves radiative transitions between electrons in the conduction band and Si dangling bonds near the middle of the gap. The higher-energy PL peak usually dominates at low temperatures, whereas the other is predominant at higher temperatures. A decrease of a few per cent in the amplitude of the higher energy peak and a concomitant increase in the amplitude of the lower- energy peak is observed after prolonged irradiation with light whose energy is greater than the optical bandgap. As in glasses, the peaks of PL spectra shift to lower energies with increasing time delay (0.1 eV shift after 1 ms), the power law decay being faster in alloys containing Ge (Taylor, 1989).

The PL mechanism in hydrogenated a-Si can be studied by frequency-resolved spectroscopy. It was found that (Oheda, 1995) the PL lifetime has two components peaked at about 1 ms and 10 μs, respectively, only the first component being present at low temperatures (12 K). Both components are insensitive to the Urbach energy or the emission energy as long as the excitation energy is low enough to guarantee geminate-pair recombination. This behavior excludes the tunneling recombination mechanism between carriers trapped at tail states after thermalization, and suggests the existence of localized luminescent centers. Moreover, the temperature behavior of the PL lifetime, which changes discontinuously from 1 ms to 10 μs with increasing temperature, together with the fact that the PL intensity remains approximately constant at low temperatures, indicate that there are two kinds of luminescent centers, correlated one with another.

In a-Si:H, the Shuker–Gamon approximation is reasonably accurate, so that both Raman and IR spectra reflect the main features of phonon density of states, although in the amorphous state the Si atoms are arranged in a continuous random network, while in the crystalline state they form a diamond structure. Therefore, it is not possible to infer details of the random network model (number of rings of bonds of various sizes) from Raman or IR measurements (Taylor, 1989). On the contrary, in a-$Si_{1-x}H_x$ alloys, Raman vibrational spectroscopy can help in identifying various structures such as SiH, SiH_2, SiH_3. Several types of modes can exist: bond-stretching modes, which involve changes in the Si-H bond length, bond bending modes, involving changes in the H-Si-H bond angle, or rocking, wagging, or twisting modes describing the rotation of the structures as a rigid unit. Similar considerations apply to H incorporation into a-Ge (Lucovsky and Hayes, 1979).

In Si_xC_{1-x} the optical absorption edge shifts linearly to higher energies in the Si-rich region (Anderson and Spear, 1977). The luminescence spectrum has two bands, the higher-energy one blueshifting with decreasing x, the other remaining constant at the energy of PL maximum in pure Si. The shift of the absorption edge and of one PL peak with x is caused by the coherent overlap of wavefunctions of neighboring Si and C atoms, while the constant position of the other PL peak is caused by Si clusters.

In pure TCAS the unpaired electrons of broken bonds give a strong spin-resonance signal (Solomon, 1979). Localized centers are paramagnetic, irrespective of the position of the Fermi level in the bandgap, whereas centers with a wavefunction spread over several atomic sites are not generally magnetic, with the exception of a small range of energies of the order $k_B T$ around the Fermi level. The spin effects influence the luminescence since the recombining electrons are in strongly correlated spin states. For example, the PL line can be drastically reduced in high magnetic fields due to spin saturation of the recombination centers that enhances nonradiative recombination or, in another model, due to geminate pairs (photocreated electron-hole pairs). The effect is independent of the ratio H/T and thus of the thermal equilibrium polarizations of the spins. Other magnetic-field effects due to spins, observable in PL spectra, are described in Solomon (1979).

a-C and hydrogenated a-C is a material which has recently received much interest, as demonstrated by the experimental determination of the elastic constant in this materials using surface Brillouin scattering data (Ferrari et al., 1999). In a-C:H the PL mechanism is the radiative recombination of electrons and holes trapped in band-tail states, PL quenching occurring due to paramagnetic defects which act as nonradiative centers (Robertson, 1996). The localization degree can be estimated from the PL lifetime for radiative tunneling $\tau_r = \tau_{r0} \exp(2R/a_B)$ where R is the carrier separation and a_B the Bohr radius. The value of this parameter is 10^{-8} s, faster than in a-Si:H where it is about 10^{-3} s due to carrier separation. PL can be also quenched by thermally activated hopping apart of carriers during thermalization; the thermalization depth of carriers in band tails vary with time as $E_d = k_B T \ln(\nu_0 t)$ with ν_0 the hopping attempt frequency. At $t = \tau_r$ carriers above E_d hop to defects and recombine nonradiatively, whereas those below E_d remain trapped and recombine radiatively. The fraction of carriers trapped in band tails, which are described by an exponential density of states of width E_0, is $\eta = \eta_0' \exp(-E_d / E_0) = \eta_0 \exp(-T / T_L)$, in agreement with the observed thermal quenching of PL. For a-$Si_{1-x}C_x$:H, there is a sharp increase in τ_r for $x > 0.5$, for this concentrations the parameter $E_0 / k_B T_L$ decreases strongly, whereas for $x < 0.5$, $E_0 / k_B T_L$ is constant and has a value of about 30. In the absence of hopping the PL intensity depends on the concentration of defects N_d. The electron-hole pair recombines nonradiatively if formed within the tunneling capture radius R_c of a nonradiative recombination center. The probability of hopping when there are no defects within R_c is $\eta = \eta_0 \exp(-4\pi R_c^3 N_d / 3)$. The radiative and nonradiative tunneling rates are equal when $R_c = (a_B / 2) \ln(\nu_0 \tau_r)$, the nonradiative lifetime for tunneling being $1/\tau_{nr} = \nu_0 \exp(-2R/a_B)$. PL quenching in a-C:H occurs if carriers tunnel out of excited clusters via other clusters to defect sites where they recombine. The decay of the tail state from any cluster occurs in the sp^3 matrix (in a-C:H there are sp^2, sp^3 bonds, etc.), which is a fraction $1-z$ of the total volume. PL quenching is then described by $\eta = \eta_0 \exp[-4\pi R_c^3 N_d / 3(1-z)]$. Experimental values for a_B, E_g are 5.7 Å and (3.5–2.5)z, respectively.

Similar to porous Si, a-C also exhibits polarization memory, as a function of the bandgap and excitation energy (Rusli et al., 1996). The polarization memory implies the existence of well-localized states with a low point symmetry, the decay of the polarization memory depending on the difference between the excitation and emission energies. a-C:H and its PL spectrum can be described by a cluster model in which the electron-hole pairs are excited and recombine within clusters of sp^2 sites embedded in sp^3-bonded matrix. The energy dependence of the PL spectrum polarized parallel and perpendicular on the excitation, and that of the polarization degree ρ, are shown in Fig. 7.16

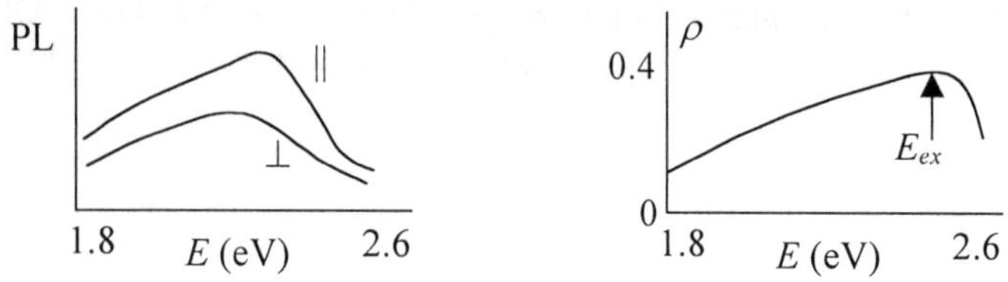

Fig. 7.16. PL spectrum polarized parallel and perpendicular to the excitation, and the energy dependence of the polarization degree ρ in a-C:H

The polarization degree increases with increasing emission energy due to carrier interaction with phonons during thermalization; it is mainly a function of $E_{PL} - E_{ex}$ where E_{ex} is the excitation energy. The polarization memory is lost as carriers hop apart and thermalize down the band tail, since each hop is accompanied by both an energy loss due to phonon emission and a polarization loss q. The loss of the polarization degree is thus a function of the number of emitted phonons multiplied by q, the polarization being described by $P = P_0[1-(E_{ex} - E_{PL})q/\hbar\Omega]$ where P_0 is the initial polarization and $\hbar\Omega$ the phonon energy. The polarization degree is proportional to the density of tail states available for thermalization transitions.

Compounds of tetrahedral materials, such as $Ge_{1-x}Si_x$ alloys have also been studied. Measurements of first-order Raman scattering in such alloys have identified three optical-phonon modes at 300, 400 and 500 cm^{-1}, associated with local Ge-Ge, Ge-Si and Si-Si vibrations, respectively. The frequency of Si-Si (Ge-Ge), phonons show a slight departure from a linear decrease with increasing (decreasing) Ge concentration, and also the Ge-Si frequency does not have a maximum at x = 0.5, since the bonds in alloys are modified due to structural disorder with respect to pure Ge, Si samples (Sui et al., 1993).

7.6.4 Chalcogenide Glasses

Unlike TCAS, which form rigid structures with little possibility for local reorganization of fourfold-coordinated atoms, chalcogenide glasses are characterized by twofold coordination with a greater degree of flexibility and prone to the formation of closed chains of atoms (rings). These materials include amorphous Se (a-Se), Te, S and multicomponent systems containing a large fraction of a chalcogen element. The lower valence and conduction bands are also formed from bonding and antibonding states, but the upper valence band is associated with nonbonding lone-pair states.

The specific optical properties of glasses, which include the metastable and transient optically induced absorption that extends well below the bandgap, the strongly Stokes-shifted PL, and the frequency-independent 'free carrier' absorption, observed only at room temperature in the most highly conducting glasses are determined by the strong electron-phonon coupling. This coupling can outweigh the Coulomb repulsion between electrons at the same site, having as a consequence a strong tendency for electron pairing in bonding configurations. Mathematically, this tendency is described by assuming that the pseudogap also contains two-electron states besides one-electron states. The Hamiltonian that describes the electron-phonon interaction in this case is

$$H = \sum_{i,\sigma} E_i \hat{n}_{i\sigma} + U \sum_i \hat{n}_{i\sigma} \hat{n}_{i-\sigma} + \hbar\omega \sum_i b_i^+ b_i + \alpha \sum_i \hat{n}_i (b_i^+ + b_i), \tag{7.25}$$

where the first two terms represent the electronic Hamiltonian, the third the phonon Hamiltonian and the last one the electron-phonon coupling characterized by the coupling constant α. In the above equation E_i is the energy of the electron at site i, U the electrostatic repulsion energy which appears at a double occupancy of the site i, $\hat{n}_{i\sigma} = c_{i\sigma}^+ c_{i\sigma}$ describes the occupation of an electron at site i with spin σ, and $\hat{n}_i = \hat{n}_{i\sigma} + \hat{n}_{i-\sigma}$, where $c_{i\sigma}^+$, $c_{i\sigma}$ are the creation and annihilation operators for the electron, respectively. The electronic and phonon parts in the above Hamiltonian can be decoupled by the introduction of the operators $d_i = b_i + (\alpha/\hbar\omega)\hat{n}_i$, $d_i^+ = b_i^+ + (\alpha/\hbar\omega)(2 - \hat{n}_i)$ (Liciardello et al., 1981):

$$H = \sum_{i,\sigma} [E_i - (\alpha^2/\hbar\omega)\hat{n}_{i\sigma}]\hat{n}_{i\sigma} + (U - 2\alpha^2/\hbar\omega)\sum_i \hat{n}_{i\sigma}\hat{n}_{i-\sigma} + \hbar\omega \sum_i d_i' d_i. \tag{7.26}$$

The effective correlation energy between two electrons with opposite spins at the same site i becomes $U_{\text{eff}} = U - 2\alpha^2/\hbar\omega$, which can take positive or negative values depending on the strength of the electron-phonon coupling. When U_{eff} is negative, the pairing of two electrons is favored. This can be seen from Fig 7.17 where the energy levels corresponding to the single occupancy E and double occupancy $E + U_{\text{eff}}$ are shown, together with the position of the Fermi level E_F as a function of the concentration N of defect sites. This figure shows that if the effective correlation energy is positive, the lower-energy levels correspond to single occupancy, and electron pairing is not favored. For an electron concentration smaller than N the singly occupied electron levels are filled and the Fermi energy increases slightly with the electron concentration around the lower-energy level until it reaches N. When the electron concentration exceeds N, the Fermi level jumps to the position of the higher-energy levels, across the gap of width U_{eff}, until all levels are occupied for an electron concentration equal to $2N$.

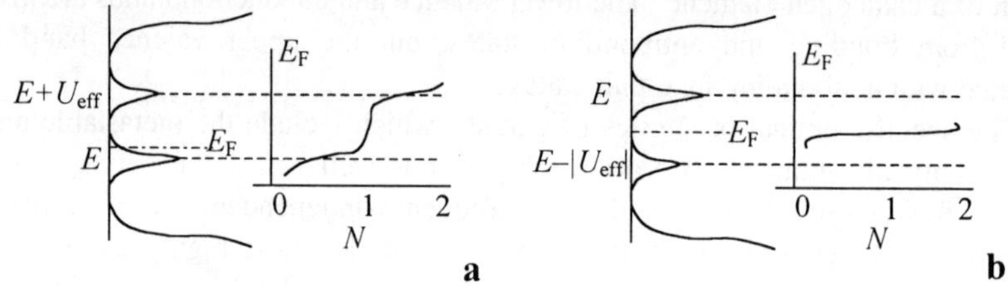

Fig. 7.17. Energy levels for positive (**a**) and negative (**b**) correlation energies

The reverse is true for negative effective correlation energy, where the electrons occupy the defect sites in pairs as the electron concentration increases, the lower energy band being always full, as in the case of intrinsic semiconductors. The Fermi level is always at about half the bandgap, in agreement with experimental data in optical measurements.

In chalcogenide glasses the model described above is applied to the two charged defects D^+ and D^- formed when an electron is transferred from one end to the other in a Se chain, for example, (see Fig. 7.18) (Davis, 1979).

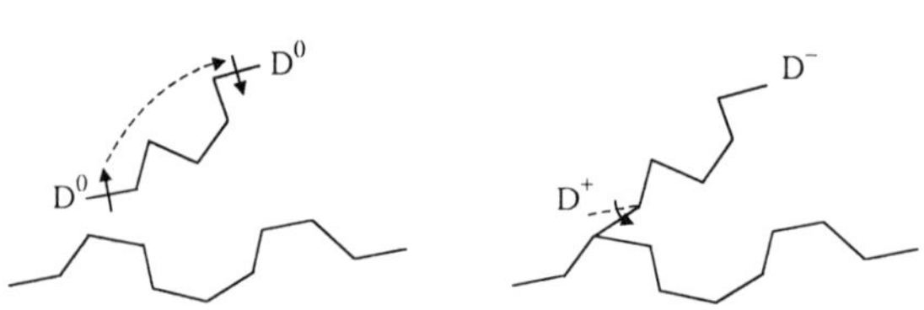

Fig. 7.18. Charged defects D^+ and D^- formed when an electron is transferred from one end of a Se chain to the other

Due to local lattice distortions accompanying the electron transfer, the effective correlation energy becomes negative; the electron lattice interaction leads to pairing of electronic states, even those inside the gap. Lattice distortions occur because at D^+ an extra bond is made with a neighboring chain using the normally nonbonding lone-pair electrons. Therefore the lattice distortion is more significant at D^+ than D^0, and is negligible at D^-. At D^+ the coordination of Se atoms becomes three, whereas at D^- it is one. Because of electron-phonon interaction these defects must contain paired electrons, so the onefold defect is negatively charged, and the threefold defect is positively charged, the defects having a diamagnetic ground state. The D^+ and D^- defects are *frozen-in* after cooling through the glass transition temperature and always have the same concentrations. The diamagnetic ground state can be perturbed thermally or optically to produce excited states that contain paramagnetic forms of defects. As a result of the perturbation, charge is transferred between the two charged defects, and two neutral defects D_0 are created, one of which is singly and the other triply coordinated. The defect formed from the negatively charged defect is called a dangling bond (Taylor, 1989). In compound chalcogenides, like bonds can also be present, but voids are unlikely.

The energy levels associated with D^0, D^+ and D^- are shown in Fig. 7.19 for the case of a thermal (left) or optical (right) excitation. Since at D^- the lattice is not appreciably distorted, the energy level of this defect is close to the energy level of the lone-pair electrons in the normal fully bonded configuration, i.e. close to the valence band; D^- is analogous to a shallow acceptor level. Similarly, D^+ is analogous to a shallow donor level. D^0 is characterized in Fig. 7.19 by two energy levels.

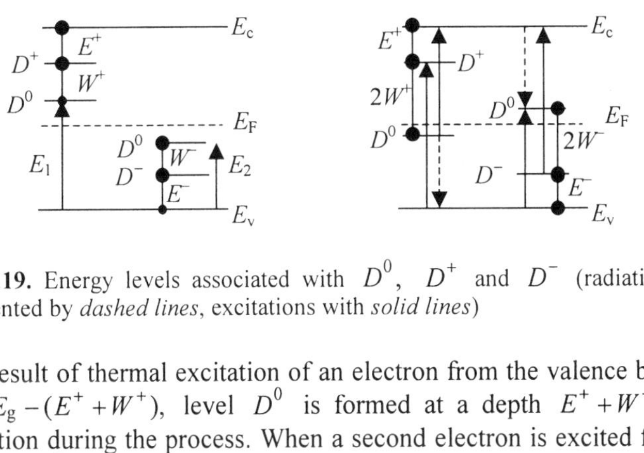

Fig. 7.19. Energy levels associated with D^0, D^+ and D^- (radiative transitions are represented by *dashed lines*, excitations with *solid lines*)

As a result of thermal excitation of an electron from the valence band with energy $E_1 = E_g - (E^+ + W^+)$, level D^0 is formed at a depth $E^+ + W^+$ due to lattice relaxation during the process. When a second electron is excited from the valence band on D^0, level D^- is formed after a relaxation of the lattice with energy W^-. The energy necessary to create D^- by the excitation of a valence electron onto the lower D^0 level is $E_2 = E^- + W^-$. The energy difference between the two states of the D^0 defect is equal to the effective correlation energy between the two electrons on D^0. D^0 can be seen either as a neutral donor or a neutral acceptor; it is thus an amphoteric defect. The Fermi energy level can be defined as the energy necessary to create a hole in the valence band, i.e. as the average of excitation energies E_1 and E_2; it is pinned at about half of the bandgap. When the valence electrons are excited optically, and not thermally, the lattice relaxation cannot occur (according to the Franck–Condon principle) since the optical excitation process occurs in a time that is much too short. Therefore, the energy needed to excite a valence electron to D^+, is $E_1 = E_g - E^+$, the resultant D^0 being created at a depth $E^+ + 2W^+$. The energy necessary for a subsequent excitation of this electron in the conduction band is greater by W^+ than in the case of thermal excitation. Analogously, D^- is created after excitation of a second valence band electron onto D^0; the energy necessary for this process is larger by W^- than in the case of thermal excitation. The position of the two energy levels corresponding to the D^0 level is now reversed: this situation corresponds to the case when the electron-phonon interaction is neglected and the correlation energy is U.

The pair of charged defects D^+ and D^- forms a valence-alternation pair (VAP). The energy necessary to create a VAP, about 1.1 eV, can be reduced if they are created at close proximity, due to the Coulomb energy of attraction. Such bound pairs are called intimate valence-alternation pairs (IVAP). Close to equilibrium (Kastner and Fritzsche, 1978) almost all or almost none of the VAPs from IVAPs. The Fermi energy is pinned at about half the energy gap due to the creation of VAPs; the concentration of VAPs is more sensitive to addition of dopants that form charged centers than IVAPs.

A typical curve for the imaginary part of the dielectric constant of trigonal crystalline (such as As_2S_3) and amorphous chalcogenides is shown in Fig. 7.20 (Connell, 1979).

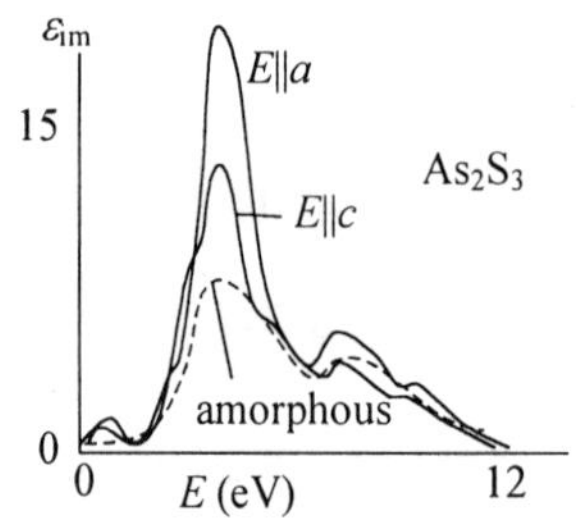

Fig. 7.20. The imaginary part of the dielectric constant of trigonal crystalline (such as As_2S_3) and amorphous chalcogenides

The absorption spectrum loses its sharp features at amorphization, but a two-peaked structure is still visible. This can be explained by the fact that the calculated density of states in the valence bands is very similar in crystalline and amorphous phases and has two components (Fig. 7.21). The two lowest-lying bands correspond to the two p bands (a bonding one and a nonbonding lone-pair band) and the higher, broader band corresponds to s states. The same is true for As_2Se_3. The data of spin-dependent optical response is less studied than in TCAS (for a review see Bishop et al. (1977)). Glasses differ from crystalline solids especially in the far-IR (1–10 meV), where the excess of low-frequency vibration density of states is not described by the Debye approximation, due to the intermediate-range order in glasses.

In the region where the optical absorption is significant, i.e. for interband absorption, its energy dependence has approximately the same form as the imaginary part of the dielectric constant: there are two peaks separated by a broad minimum at about 6–7 eV. From Figs. 7.20 and 7.21 it follows that, unlike in TCAS, the disorder in glasses has only a minor influence on the electronic densities of states and the optical matrix elements. The chalcogenide glasses are very molecular in nature, the interband and vibrational absorption being determined by the 'molecular complexes' and their excited states.

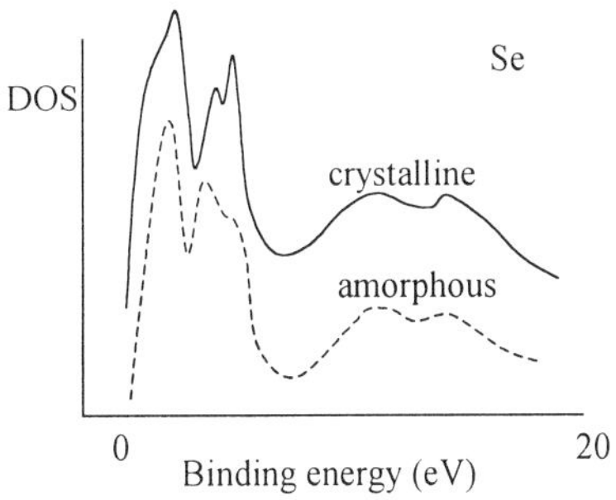

Fig. 7.21. Density of states for crystalline and amorphous Se

Luminescence is quite common in this class of amorphous materials. Contrary to TCAS, the preparation conditions do not influence significantly the luminescence spectra. The PL is due to recombination of electrons with trapped holes with an energy lowered to near half the bandgap by lattice distortions. Several models have been proposed to describe it: charged coordination defects, small polaron model, VAP model and so on (see the references in Taylor (1989)). The PL peak shifts down in energy with increasing delay time after laser excitation, with a power-law decay. At short times after excitation with polarized laser (less than 10^{-4} s) the PL in glassy As_2S_3 remains polarized, but the polarization disappears at longer times. A typical PL spectrum in chalcogenide glasses peaks at an energy close to half the optical bandgap (Fig. 7.22).

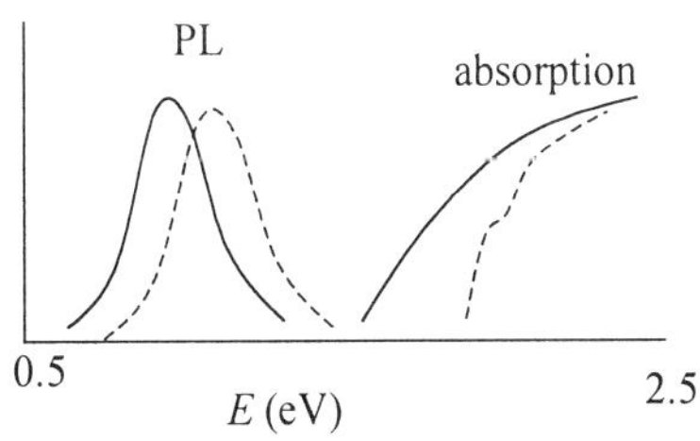

Fig. 7.22. PL and absorption spectra in chalcogenide glasses (*solid line* – amorphous phase; *dashed line* – crystalline phase)

The integrated intensity of PL is greatest when the exciting photons have an energy corresponding to the tail of the optical absorption edge. This behavior can be understood if we consider that luminescence occurs at defects (the excess carriers generally occupy localized states before recombination). For example, absorption of light can take place between valence and conduction band, or between valence band and D^+; thus absorption is significant at or very near to the optical bandgap. On the contrary, the subsequent luminescent transitions are associated to radiative recombinations of the electron and the hole at the defect (between conduction band and D^0); therefore it is peaked at approximately half the bandgap. The Stokes shift between absorption and luminescent energies is caused by the lattice relaxation between the electron excitation and the subsequent luminescent transitions (absorption of a photon changes the configuration of the network).

Se is an exception to the rule that the spectra of amorphous and crystalline phases are similar (Fischer, 1979). In crystalline Se, there is no Stokes shift between luminescence and absorption spectra, since the band-edge excitons are indirect, both absorption and emission implying phonon participation. In a-Se the large Stokes shift is, however, present. The pseudodielectric function of a-Se (equal to the true bulk dielectric function if the surface is perfectly flat and oxidation-free) has been measured between 1.2–5.2 eV at room temperature by ellipsometry (Innami et al., 1999).

Both halfwidth and PL peak energy depend linearly with composition in As-chalcogenide alloys, suggesting a completely random network, with no cluster formation. Luminescent transitions can also occur between the valence band and D^0, corresponding to capture of a hole from the valence band. Also observed are luminescent transitions from VAPs or IVAPs (Kastner and Hudgens, 1978); these transitions confirm the intrinsic nature of luminescence in chalcogenides, because their formation energy is so small that they are present even in crystals. Intrinsic PL can be described by the small polaron model, in which it is assumed that the photoexcited excess carriers form small polarons, which lead to the appearance of polaron bands in the density of states. Luminescence occurs only if the electron and hole polarons are created very close together in real space (which is the case for photons with energies very close to the bandgap), otherwise – for photons with energy larger than the bandgap – radiationless recombination is overwhelming. The polaron and defect models of luminescence can be reconciled since a photoinduced broken bond (by induced transitions from bonding to antibonding states) can be viewed as a limiting case of electron-lattice coupling of the fundamental excitation (Fischer, 1979). The PL efficiency is greatest at low temperatures due to competing nonradiative transitions occurring when the excited electrons escape from the vicinity of the defect, for example by tunneling into the conduction band edge. Qualitatively, the PL intensity follows an exponential law $I(T) = I_0 \exp(-T / T_0)$ (Street et al., 1974). These nonradiative transitions give rise to a decrease of the luminescence intensity during prolonged photoexcitation, phenomenon known as luminescence fatigue, as well as to photoinduced absorption with a threshold at about half of the energy required for the electron excitation into or out of the defect. The temperature dependence of luminescence is probably caused, as in a-Si, by a (non-Coulombic) correlation between excited electrons and holes, which favors the formation of bound excitons. (A binding energy cannot be, however, derived from PL since the temperature dependence is different in the two cases). The electrons must be excited at or very near the defect (or an electron-hole pair must be created near the defect at band-to-band excitation), in order for luminescence to occur. Otherwise, they can easily find a nonradiative recombination channel before reaching the defect, due to the low mobility of carriers in amorphous materials. The important contribution of the defect states is confirmed by the behavior of the PLE as a function of the energy of exciting photons: the excitation spectrum first follows the absorption edge, but with increasing energy it reaches a maximum and drops off, as in Fig. 7.23 (Street et al., 1975).

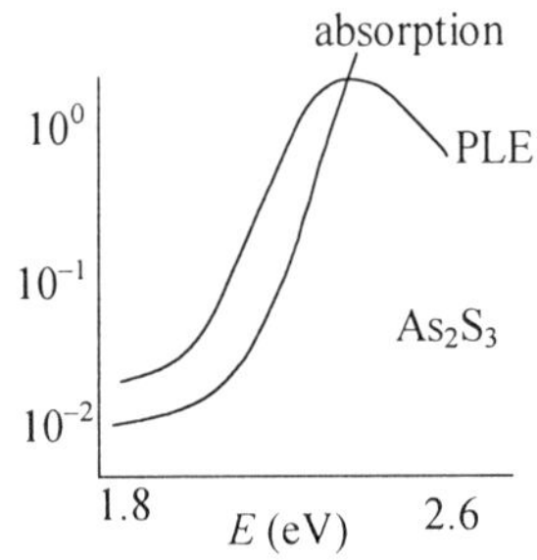

Fig. 7.23. Absorption and PLE in As_2S_3

The fall-off of the PLE at the high-energy side is due to a smaller number of absorbed photons, the absorption being possible to either direct band-to-band transitions or direct absorption at the defects. (Radiative recombinations take place only if the exciton is excited near localized states or defects, energetically close to valence or conduction band edges, which produce the tail in the absorption coefficient. At higher energies the carriers are excited well into the bands and the probability of trapped exciton formation at radiative recombination centers decrease).

The presence of rings or chains of atoms in various proportions in chalcogenide glasses can be inferred from their IR or Raman vibrational spectra since their dominant features are controlled by the local molecular structure. Such studies have been performed in elemental materials, like a-Se and a-Te, as well as for compounds a-$As_2(S,Se,Te)_3$ or a-$As_{1-x}(S,Se)_x$. Also like bonds as S-S and As-As bonds have been evidenced (Lucovski and Hayes, 1979).

In some glasses the low-frequency wing of the fast relaxation spectrum has a power-law behavior with an exponent $\alpha = 0.2 - 0.6$, which depends on both the temperature and the system. This fast relaxation process has a broad quasielastic contribution to scattering, and it can be modeled above the glass-transition temperature T_g by the mode-coupling theory. In the glassy state, the carrier movement is described as thermally activated hopping in asymmetric double-well potentials (ADWP) (Gilroy and Phillips, 1981). In Surovtsev et al. (1998) it is shown that although the structure and physical processes below and above the glass-transition temperature are completely different, the power-law relaxation spectrum persists in the glassy state. Moreover, from light-scattering experiments it is possible to extract the distribution $g(V)$ of barrier heights in the ADWP model. More precisely, the normalized Raman spectrum is $I(\nu)/[\bar{n}(\nu)+1] \approx \nu^{\alpha}$ in the 3–100 GHz region, where $I(\nu)/[\bar{n}(\nu)+1] \approx \int_0^{\infty} 2\pi\nu\tau g(V)\mathrm{d}V/[1+(2\pi\nu\tau)^2]$ with relaxation time $\tau = \tau_0 \exp(V/T)$. If $g(V)$ is sufficiently broad, $I(\nu)/[\bar{n}(\nu)+1] \approx Tg(V)$ and the distribution of barrier heights is $V = T\ln(1/2\pi\nu\tau_0)$.

The model of a two-level tunneling system (Anderson et al., 1972; comparison with experiment Phillips, 1987) predicts that the nonlinear contribution to absorption can be anomalously large at low frequencies for which $\omega\tau_0(T) << 1$, where τ_0 is the minimal relaxation time for two-level systems with an interlevel splitting $\cong k_{\mathrm{B}}T$ (Kirkengen and Galperin, 1997). In dielectric glasses the lowest-order nonlinear contribution is negative, proportional to the wave intensity, and exhibits anomalous frequency and temperature dependences $\Delta\alpha/\alpha_0 \approx [\omega\tau_0(T)]^{-1/2}T^{-2}$. In metallic glasses the nonlinear contribution is also negative, independent of frequency and proportional to the square root of the wave's intensity. Two absorption mechanisms are possible in the two-level system (TLS), which assumes local tunneling states in double-well potentials characterized by a random energy difference Δ between minima, random tunnel splitting 2Λ, and energy spacing $2\varepsilon = 2(\Delta^2 + \Lambda^2)^{1/2}$. One is resonant absorption, i.e. direct absorption of acoustic or microwave quanta accompanied by transitions between levels of TLS and described by $\alpha^{\mathrm{res}} = \alpha(\omega/s)\tanh(\hbar\omega/2k_{\mathrm{B}}T)$ where α is a dimensionless coupling constant and s the sound velocity (the same formula, with another coupling constant is valid for microwave absorption). Another mechanism is relaxational absorption caused by the modulation of populations of TLS levels by the alternating deformation field of the acoustic wave; it dominates when $\omega\tau_0 << 1$. The linear relaxation absorption is $\alpha^{\mathrm{rel}} \approx (\alpha/s)\tau^{-1}$ when $\omega\tau_0 >> 1$ and $(\alpha/s)\omega$ for $\omega\tau_0 << 1$. The nonlinearity is due to the equalization by the sound wave of the populations of the levels, which decreases absorption; it can be observed at low acoustic intensities. At large amplitudes the nonlinear relaxational absorption caused by the strong modulation of interlevel spacing of relevant TLS by the sound wave becomes important. The relevant TLS absorbs energy only during a fraction of the sound period, the total absorption decreasing with the sound amplitude. The nonlinear relaxational absorption has been observed in metallic glasses (PdSiCu, PdSi, PdNiP) (Araki et al., 1980; Park et al., 1981) as a two-stage behavior (successive decrease of resonant and then relaxational absorption as intensity increases).

In chalcogenide glasses the phenomenon of reversible photoinduced anisotropy is common. A possible explanation (Kolobov et al., 1997) is that macroscopically isotropic glasses are composed of anisotropic domains, the light being predominantly absorbed by those domains with electric dipoles parallel with $\boldsymbol{E}$, followed by subsequent recombination and bond rearrangements. The initially random distribution of the domain axes is thus lost, the spatial rearrangement of charged defects or VAPs producing the photoanisotropy. The photoanisotropy is most efficiently induced by subbandgap light absorbed by the defect state, and can be experimentally studied for example by reflection measurements. $\Delta r / r = (r_x - r_y)/[(r_x + r_y)/2]$ for a polarization-inducing beam linearly polarized along x is observed up to much higher energies than the excitation light, and shows positive (negative) values for higher (lower) photon energies. This behavior of $\Delta r / r$ and its reversibility can be explained if we assume that also the main covalent network of glass becomes anisotropic, not only the defects or lone-pair orbitals. The net result would then be a change in topology of covalent bonds. Additionally, two photoexcited lone-pair electrons (mainly parallel to $\boldsymbol{E}$) can form new covalent bonds – the VAPs, causing finally opposite changes in anisotropy at higher and lower energies. This phenomenon appears not only in chalcogenide glasses but also in oxide glasses and polymeric films. It decreases rapidly when the temperature approaches the glass-transition temperature T_g, and is described by $A(t) = A_s[1 - \exp(-t/\tau)^\gamma]$ where A_s is the saturation value and γ a temperature-dependent constant that attains its maximum value $\gamma = 1$ for $T = T_g$. For temperatures near T_g the curve $\ln A_s = f(10^3/T)$ is approximately a straight line from which the activation energy can be calculated (Tikhomirov et al., 1997). The anisotropy of electrical and magnetic dipole moments is essential for the photoinduced anisotropy. IVAP can also participate in this process: photoexcitation produces neutral unstable VAPs that decay into configurations in which the orientation of electrical and magnetic dipoles associated to lone-pair orbitals is changed, the initial isotropic distribution of VAPs being destroyed. The strength of this interaction depends on the relative orientations of the electric vector of the exciting light and lone-pair electrons, the net result being structural changes in the medium-range order.

At temperatures below 100 K metastable absorption can be optically induced with intensities of the order few mW/cm^2, and energies greater than roughly the half of the optical bandgap. A typical optically induced absorption curve looks like that in Fig. 7.24; its form is similar to that for donor level bands (Connell, 1979).

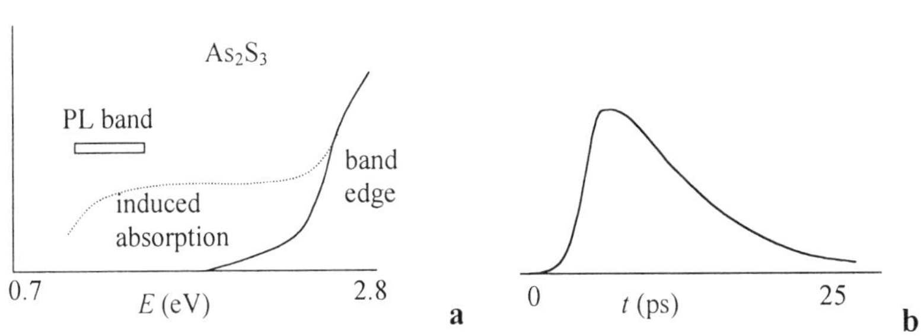

Fig. 7.24. (**a**) Induced absorption and (**b**) transient induced absorption

The optically induced absorption produces metastable D^0 states, which show also electron spin resonance; it can be bleached by thermal cycling at temperatures above 200 K or by illumination with photons with energy between approximately half the optical bandgap and the onset of the fundamental absorption. Not only metastable, but also transient optically induced absorption can be observed in glasses (Fig. 7.23b). The magnitude of the absorption in this case decreases with time after pulsed laser excitation, with a simultaneous shift to higher energies. The rapid decay at low energies (a few ps) is due to geminate recombination of thermalized carriers. For lower temperatures and longer decay times the spectral shape of transient absorption approaches that of metastable optically induced absorption (Taylor, 1989).

Other optically induced changes in the spectra of chalcogenide glasses are the photodarkening and the photostructural processes. The first one manifests as a metastable shift of the optical absorption edge to lower energies after illumination with light of energy greater than or equal to the bandgap; it is caused by rearrangements of some electrons and atoms. The photostructural process is in fact photopolymerization, i.e. a light-induced structural change from predominantly molecular to polymeric.

In dipolar glasses, at low temperatures, the mean relaxation rates deviate from the usual Arrhenius plot $\Gamma(T) = \nu_A \exp(-V_b / k_B T)$, and have an upward curvature attributed to quantum tunneling. A theory that predicts a crossover region between classical activation and quantum tunneling is presented in Lopes dos Santos et al. (2000). The Arrhenius law is expected to hold classically for the escape rate of a system trapped in a metastable potential minimum, at temperatures much lower than the barrier height V_b, i.e. at $k_B T << V_b$. In general, the distribution function of relaxation times $G(\ln \tau, T)$ can be determined from the imaginary part of the dielectric function since

$$\varepsilon_{im}(\omega, T) = \Delta\varepsilon \int_{-\infty}^{\infty} G(\ln \tau, T) \frac{\omega\tau}{1+\omega^2\tau^2} \mathrm{d} \ln \tau, \tag{7.27}$$

where $\Delta\varepsilon = \varepsilon(0) - \varepsilon(\infty)$ and $\int G(\ln \tau, T) \mathrm{d} \ln \tau = 1$. From the symmetric nature of ε_{im}, the general form of G is $G(-y, T) = (1/\sigma(T)) H[(y - y_0(T))/\sigma(T)]$ where $y_0(T)$ is the position of the maximum of the distribution $G(-y, T)$ and $\sigma(T)$ its width. G is thus dependent on a universal function H of one variable, the temperature dependence of all parameters $\Delta\varepsilon$, y_0, σ being determined experimentally, without the need for postulating even the form of H.

Optical measurements are not enough to completely characterize a material. They can often be related, and are complemetary to parameters measured with other means, for example to electrical conductivity. For example, in xRb$_2$O-(1$-x$)GeO$_2$ glasses with low alkali content, ions occupy one type of site (M), while for higher alkali content there are two types of sites (L and H), the frequency of these modes being in the relation $\nu_L < \nu_M < \nu_H$. The ionic conductivity is modeled by $\sigma = [(Ze)^2 n d^2 \nu_0 / 3k_B T] \exp(-E_0 / k_B T)$, where Ze is the charge per mobile ion, n the number of mobile ions per volume, d is the jump distance, ν_0 the ionic oscillation frequencies, $E_0 = (M_c d^2 \nu_0^2)/2$ with M_c the mass of the ion, $d = (V_m / 2xN_A)^{1/3}$, V_m being the molar volume glass, x the mole fraction of Rb_2O and N_A the Avogadro number. Identifying ν_0 with the frequency of each of the three modes we have referred to above, the activation energies E_L, E_M, E_H are found by electrical measurements (Kanutsos et al., 1996).

7.6.5 Other Amorphous Materials

Besides the fourfold and twofold coordinated amorphous materials, there are some elements in which the atomic coordination is three. An example is As and other group V elements. The flexibility of the structure is between that of chalcogenide glasses and TCAS. The threefold coordination favors layered structures in amorphous phases, similar to crystalline structures. For example, As is semimetallic in the crystalline phase and a covalent semiconductor when amorphous. The uppermost valence band of threefold coordinated materials is composed of p-bonding electrons, while the lone-pair s electrons have a lower energy. The lowest-energy structure in a-As corresponds to a normally bonded atom, where three p-like bonds are formed with a certain degree of s-p mixing that varies from site to site. There are several types of defects that can be formed, denoted ${}_lP_n^m$ where m describes the charge: 0 for neutral defects, + for positive charged and − for negatively charged defects, n labels the coordination and l the predominant type of bonding. With these notations the energy levels in the bandgap of a-As are given in Fig. 7.25 (Davis, 1979);

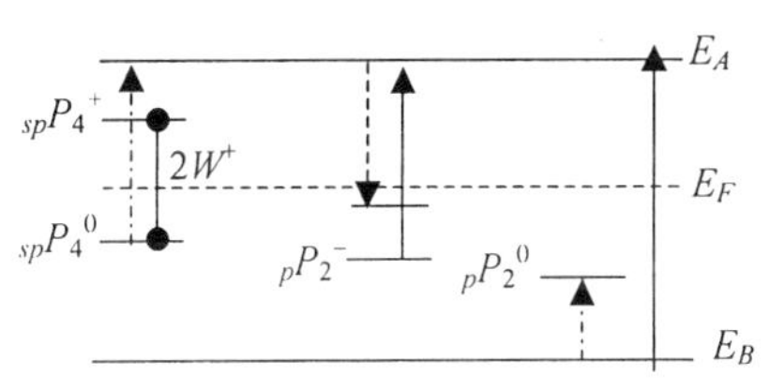

Fig. 7.25. Energy levels in the bandgap of a-As (*dashed lines* - radiative recombinations; *solid lines* - excitation; *dashed-dotted lines* - absorption)

The transitions observed in absorption and luminescent spectra are identified on the figure. The main difference from chalcogenide glasses is that the defect ${}_pP_2^-$, analogous to D^-, lies close to half the optical bandgap instead of being near the top of the valence band. The peaks in PLE correspond to the excitations of the transitions depicted with solid lines, which are analogous to the excitations in chalcogenide materials, but much further apart. Therefore they appear as distinct peaks in the PLE, whereas for chalcogenide glasses these transitions are superimposed in a single peak. As in chalcogenide glasses, the PL band is peaked at about half the optical bandgap, but the PLE spectrum has a double-peaked structure (Fig. 7.26).

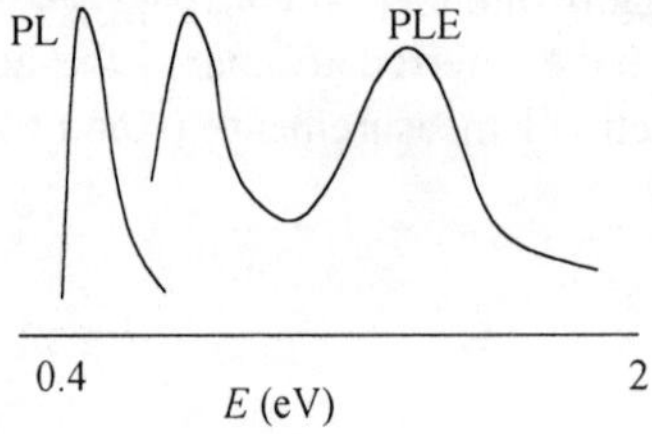

Fig. 7.26. PL and PLE of a-As

The absorption edge cuts the high-energy peak of the PLE spectrum. The low-energy peak of the PLE spectrum corresponds to direct excitations of defects, the Stokes shift being small, about 0.2 eV. IR and Raman vibrational spectroscopy studies have also been made on a-As; they have revealed vibrational modes due to bond coordination defects (Lucovsky and Hayes, 1979).

Other materials with interesting properties are combinations of tetrahedral and twofold coordinated elements, for example Ge_xSe_{1-x} glasses. The low-frequency Raman spectrum of this material has been recently studied by Nakamura et al. (1998). For small x, we have a 1D network structure composed of Se chains, whereas for a stoichiometric composition $GeSe_2$ layered 2D-like fragments are formed. The interesting case is that in-between. In general, low-energy excitations in disordered materials show a plateau in the low-temperature thermal conductivity and a boson peak in the Raman spectra. In the low-frequency region the reduced Raman spectrum $I(\omega,T)/[\overline{n}(\omega,T)+1] = C(\omega)g(\omega)/\omega$ has a power-law dependence ω^{d-1}, where $g(\omega)$ is the vibrational density of states and $C(\omega)$ describes the coupling between lattice and light, which is proportional to ω in the low-frequency region above the boson peak. So, from the reduced Raman spectrum, a noninteger d is found in the glassy states, which implies a fractal structure of the glass. Different values d_b, d_s are found for bending modes, stretching modes, and so on. Raman spectra of other germanium chalcogenides can be found in (Brodsky, 1975).

Many materials are amorphous in thin films under certain growing conditions, and become crystalline for larger layer thicknesses or for higher temperatures. One example is AgI, for which it was found that amorphization influences the crystallization temperature of quenched-deposited thin films. This fact has been demonstrated by measuring the optical density as a function of temperature (Kondo et al., 1998a). The optical density is defined as $OD = \log_{10}[(D-J)/I] - \log_{10}[(D-J_0)/I_0]$ where D is the intensity of the incident light, I_0, J_0 are the transmission and reflected intensity before film deposition, and I, J – after. From the shape of this curve it is easy to determine the crystallization temperature T_c (see Fig. 7.27).

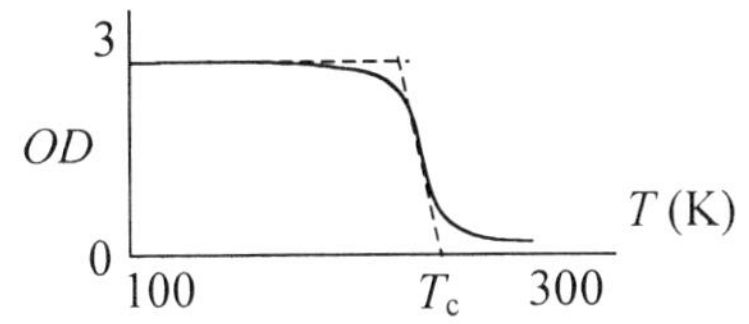

Fig. 7.27. Optical density temperature dependence in quenched-deposited AgI thin films

It is found that the crystallization temperature is lower for thinner films, due to the surface effect in the amorphous film.

In another example, Ce changes its growth mode from amorphous to crystalline at a critical layer thickness, this change being accompanied by the modification of its elastic modulus. It has afterwards a thickness-dependent laminated two-phase structure composed of the two phases of Ce: α and γ phase. For Ce/Fe multilayers with a stacking periodicity Λ, the elastic tensor components can be determined from Brillouin scattering on phonons with a wavelength greater than Λ, using continuum elasticity theory. Both Rayleigh and Sezawa modes are observed in this system (Hassdorf et al., 1997). The surface velocity of Rayleigh waves has a simple relation when the substrate does not contribute to the propagation of surface waves of a multilayer: $v_R = \beta\sqrt{c_{44}/\rho}$ where $\beta = 0.8$–1 is a constant slowly dependent on c_{11}, c_{13}, c_{33}, and $\rho = f_{\alpha\text{-Ce}}\rho_{\alpha\text{-Ce}} + f_{\gamma\text{-Ce}}\rho_{\gamma\text{-Ce}} + f_{\text{Fe}}\rho_{\text{Fe}}$ is the average mass density of the multilayer, with $f_i = t_i/\Lambda$, t_i being the layer thickness. c_{44} is also a weighted sum over the components of the multilayer: $1/c_{44} = f_{\text{Fe}}/c_{44}^{\text{Fe}} + (f_{\text{Ce}}/t_{\text{Ce}})/(t_{\alpha\text{-Ce}}/c_{44}^{\alpha\text{-Ce}} + t_{\gamma\text{-Ce}}/c_{44}^{\gamma\text{-Ce}})$. The α and γ phases of Ce are assumed to be composed of an amorphous and a crystalline part: $1/c_{44}^{\alpha,\gamma\text{-Ce}} = f_{\alpha,\gamma\text{-Ce}}^{\text{am}}/c_{44}^{\alpha,\gamma\text{-Ce,am}} + f_{\alpha,\gamma\text{-}Ce}^{\text{cr}}/c_{44}^{\alpha,\gamma\text{-Ce,cr}}$. From Brillouin measurements it was found that for a multilayer in which the Fe and Ce layers have the same thickness, Ce is completely amorphous if this thickness is less than 40 Å.

Other recently studied materials are $CsPbBr_3$ and $CsPbCl_3$, for which the optical energy gap was found to be largely blueshifted compared to the crystalline phase, and the intensity strongly reduced (Kondo et al., 1998b).

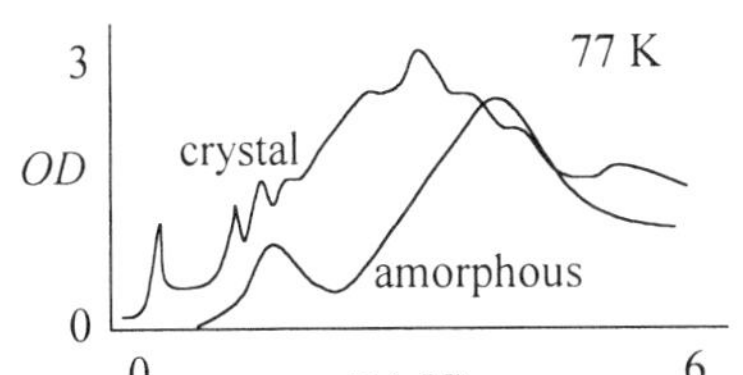

Fig. 7.28. Typical optical density spectra for $CsPbBr_3$ and $CsPbCl_3$

The two peaks of the optical density *OD* (Fig. 7.28) are caused by state localization at amorphization. The two Gaussian bands are due to spin-orbit and dipole-allowed transitions, respectively, between Pb^{2+} 6s and 6p states which are localized in the amorphous phase, being otherwise extended near band edges. The blueshift of the optical gap is due to the disappearance of the $\boldsymbol{k}$ dispersion for these one-electron states. At the crystallization temperature (about 296 K) the transmittance has an abrupt decrease due to nucleation followed by fast grain-growth in the crystalline phase.

References

Adolph, B., K. Tenelsen, V.I. Gavrilenko, and F. Bechstedt (1997) Phys. Rev. B **55**, 1422.

Akulin, V.M., C. Bréchignac, and A. Sarfati (1997) Phys. Rev. B **55**, 3672.

Alsina, F., H.M. Cheong, J.D. Webb, A. Mascarenhas, J.F. Geiz, and M. Olson, (1997) Phys. Rev. B **56**, 13126.

Amer, N.M., A. Skumanich, and W.B. Jackson (1983) Physica **117B+118B**, 897.

Anderson, P. W. (1958) Phys Rev **109**, 1492.

Anderson, P.W., B.I. Halperin, and C.M. Varma (1972) Philos. Mag **25**,1.

Anderson, D.A. and W.E. Spear (1977) Philos Mag **35**, 1.

Anno, E. and M. Tanimoto (1999) Phys. Rev. B **60**, 5009.

Araki, H., G. Park, A. Hikata, C. Elbaum (1980) Phys. Rev.B **21**, 4470.

Barthélémy, M. and D.J. Bergman (1998) Phys. Rev. B **58**, 12770.

Bishop, S.G., U. Strom, and P.C. Taylor (1977) in *Proc. 7th Int. Conf. Amorphous and Liquid Semiconductors*, ed. W.E. Spear, pp.595-606, Univ. Edinburgh.

M. H. Brodsky (1975) in *Light Scattering in Solids*, ed. M. Cardona, p.205, Springer, Berlin, Heidelberg.

Bustarret, E., M.A. Hachicha, and M. Brunel (1988) Appl. Phys. Lett **52**, 165.

Canham, L.T. (1990) Appl. Phys. Lett. **57**, 1046.

Callaway, J. (1991) *Quantum Theory of the Solid State*, Academic Press, New Boston, 2nd edition.

Capozzi, V., G.F. Lorusso, D. Martin, G. Perna, and J.L. Staehli (1996) Phys. Rev.B **54**, 7643.

Cardy, J.L. (1978) J. Phys C **11**, L321.

Chang, I.M., S.C. Pan, and Y.F. Chen (1993) Phys. Rev. B **48**, 8747.

Connell, G.A.N. (1979) in *Amorphous Semiconductors*, ed. M.H. Brodsky, p.73, Springer-Verlag, Berlin Heidelberg.

Connell, G.A.N., R.J. Temkin, and W. Paul (1973) Adv. Phys **22**, 643.

Datta, S. and K.L. Narasimhan (1999) Phys. Rev. B **60,** 8246.

Davis E.A. (1979) in *Amorphous Semiconductors*, ed. M.H. Brodsky, p.41, Springer-Verlag Berlin Heidelberg.

Diener, J., Y.R. Shen, D.I. Kovalev, G. Polliski, and F. Koch (1998) Phys. Rev. B **58** 12629.

Economou, E.N. and M.H. Cohen (1972) Phys. Rev. B **5**, 2931.

Efros, A.L., M. Rosen, B. Averboukh, D. Kovalev, M. Ben-Chorin, and F. Koch (1997) Phys. Rev. B **56,** 3875.

Engemann, D. and R. Fischer (1977) Phys Status Solidi B **79**, 195.

Engemann, D. and R Fisher (1976) in *Structure and Excitations of Amorphous Solids*, eds. G. Lucovsky and F.L. Galeener, p.37, American Inst. of Physics, New York.

Ferrari, A.C., J. Robertson, M.G. Beghi, C.E. Bottani, R. Ferulano, and R. Pastorelli (1999) Appl. Phys. Lett. **75**, 1893.
Fischer, R. (1979) in *Amorphous Semiconductors*, ed. M.H. Brodsky, Springer, Berlin, Heidelberg.
Fishman, G., I. Mihalcescu, and R. Romestain (1993) Phys. Rev. B **48**, 1464.
Freilikher, V., M. Pustilnik, and I. Yurkevich (1997) Phys. Rev. B **56**, 5974.
Gilroy, K.S. and W.A. Phillips (1981) Philos. Mag. B **43**, 735.
Halperin B.I. and M. Lax (1966) Phys. Rev. **148**, 722.
Hassdorf, R., M. Grimsditch, W. Felsch, and O. Schulte (1997) Phys. Rev. B **56**, 5814.
Ingale, A. and K.C. Rustagi (1998) Phys. Rev. B **58**, 7197.
Innami T., T. Miyazaki, and S. Adachi (1999) J. Appl. Phys. **86**, 1382.
Iribarren, A., R. Castro-Rodríguez, V. Sosa, and J.L. Peña (1999) Phys. Rev. B **60**, 4758.
Jäckle, J. (1981) in *Amorphous Solids: Low-Temperature Properties*, ed. W.A. Philips, Springer, New York.
Jellison Jr., G.E and F.A. Modine (1996) Appl. Phys. Lett. **69**, 371.
John, S., C. Soukoulis, M.H. Cohen, and E.N. Economou (1986) Phys. Rev. Lett. **57**, 1777.
John, G.C. and V.A. Singh (1996) Phys. Rev. B **54**, 4416.
Kanemitsu, Y. (1996) Phys. Rev. B **53**, 13515.
Kanutsos, E.I., Y.D. Yiannopoulos, H. Jain, and W.C. Huang (1996) Phys. Rev. B **54**, 9775.
Kastner, M. and H. Fritzsche (1978) Philos. Mag. **37**, 199.
Kastner, M. and S.J. Hudgens (1978) Philos. Mag. B **37**, 665.
Kirkengen, M. and Yu. M. Galperin (1997) Phys. Rev. B **56**, 13 615.
Kolobov, A.V., V. Lyubin, T. Yasuda, and K. Tanaka (1997) Phys. Rev. B **55**, 23.
Kondo, S., T. Itoh, and T. Saito (1998a) Phys. Rev. B **57**, 13235.
Kondo, S., T. Sakai, H. Tanaka, and T. Saito (1998b) Phys. Rev. B **58**, 11401.
Koyama, H., N. Shima, and N. Koshida (1996) Phys. Rev. B **53**, 13291.
Kramer, B. and D. Weaire (1979) in *Amorphous Semiconductors*, ed. M.H. Brodsky, p.9, Springer, Berlin, Heidelberg.
Kuriyama K., K. Ushiyama., K. Ohbora, Y. Miyamoto, and S. Takeda (1998) Phys. Rev. B **58**, 1103.
Laks, D.B., S.H. Wei, and A. Zunger (1992) Phys. Rev. Lett. **69**, 3766.
Liciardello, D.C., D.L. Stein, and F.D. Halane (1981) Philos. Mag. B **43**, 189.
Lucovsky, G. and T.M. Hayes (1979) in *Amorphous Semiconductors*, ed. M.H. Brodsky, p.215, Springer, Berlin, Heidelberg.
Lopes dos Santos, J.M.B., M.L. Santos, M.R. Chaves, A. Almeida, and A. Klöpperpieper (2000) Phys. Rev. B **61**, 3155.
Malý, P., F. Trojánek, J. Kudrna, A. Hospodková, S. Banás, V. Kohlová, J. Valenta, and I. Pelant (1996) Phys. Rev. B **54**, 7929.
Nagels, P. (1979) in *Amorphous Semiconductors*, ed. M.H. Brodsky, p.113, Springer, Berlin, Heidelberg.
Nakamura, M., O. Matsuda, and K. Murase (1998) Phys. Rev. B **57**, 10228.
Noba, K.I. and Y. Kayanuma (1999) Phys. Rev. B **60**, 4418.
O'Donnell, K.A., R. Torre, and C.S. West (1997) Phys. Rev. B **55**, 7985.
O'Donnell, K.A., C.S West, and E.R. Méndez (1998) Phys. Rev. B **57,** 13209.
Oheda, H. (1995) Phys. Rev. B **52**, 16530.
Ossikovski, R. and B. Drévillón (1996) Phys. Rev. B **54,** 10530.
Ovsyuk, N.N. and V.N. Novikov (1998) Phys. Rev. B **57,** 14615.
Park, G., A Hikata, and C. Elbaum (1981) Phys. Rev. B **24**, 7389.
Pellegrini, V., A. Tredicucci, C. Mazzoleni, and L. Pavesi (1995) Phys. Rev. B **52,** 14328.
Penn, D. (1962) Phys. Rev. **128**, 2093.
Périchon, S., V. Lysenko, B. Remaki, D. Barbier, and B. Champagnon (1999) J. Appl. Phys. **86**, 4700.

Phillips, J.C. (1971) Phys Status Solidi B **44**, K1.
Phillips, W.A. (1987) Rep. Prog. Phys **50**, 1657.
Pipino, A.C.R., R.P. Van Duyne, and G.C. Schatz (1996) Phys. Rev. B **53**, 4162.
Pusep, Yu.A., M.T.O. Silva, J.C. Galzerani, N.T. Moshegov, and P. Basmaji (1998) Phys. Rev. B **58**, 10683.
Ramkumar, C., K.P. Jain, and S.C. Abbi (1996a) Phys. Rev. B **53**, 13672.
Ramkumar, C., K.P. Jain, and S.C. Abbi (1996b) Phys. Rev. B **54**, 7921.
Robertson, J. (1996) Phys. Rev. B **53**, 16302.
Rusli, G.A., J. Amaratunga, and J. Robertson (1996) Phys. Rev. B **53**, 16306.
Shalaev, V.M. and A.K. Sarychev (1998) Phys. Rev. B **57**, 13265.
Shuker, R. and R.W. Gamon (1971) in *Proc. 2nd Intern Conf. Light Scattering in Solids*, Paris, p.334, Flammarion Sciences, Paris.
Smolyaninov, I.I., A.V. Zayats, and C.C. Davis (1997) Phys. Rev. B **56**, 9290.
Solin, S.A. (1976) in *Structure and Excitations in Amorphous Solids*, eds. G. Lucovsky and F.L. Galeener, p.205, American Inst of Phys, New York.
Solomon, I. (1979) in *Amorphous Semiconductors*, ed. M.H. Brodsky, p.189, Springer, Berlin, Heidelberg.
Stabler, D.L. and C.R. Wronski (1979) Appl. Phys. Lett. **31**, 292.
Street, R.A., T.M. Searle, and I.G. Austin (1974) in *Amorphous and Liquid Semiconductors*, eds. J. Stuke and W. Brenig, p.953, Taylor and Francis, London.
Street, R.A., T.M. Searle, and I.G. Austin (1975) Philos. Mag. **32**, 431.
Strom, U., J.R. Hendrickson, R.J. Wagner, and P.C. Taylor (1974) Solid State Commun. **15**, 1871.
Stroud, D. and A. Kazaryan (1996) Phys. Rev. B **53**, 7076.
Sui, Z., H.H.. Burke, and I.P. Herman (1993) Phys. Rev. B **48**, 2162.
Surovtsev, N.V., J.A.H. Wiedersich, V.N. Novikov, E. Rössler, and A.P. Sokolov (1998) Phys. Rev. B **58**, 14888.
Tauc, J. (1974) in *Amorphous and Liquid Semiconductors*, ed. J. Tauc, p.159, Plenum, New York.
Taylor, P.C. (1989) in *Laser Spectroscopy of Solids II*, ed. W.M. Yen, Springer, Berlin, Heidelberg, p.257.
Tikhomirov, V.K., G.J. Adriaenssens, and S.R. Elliott (1997) Phys. Rev. B **55**, 660.
Tsang, C. and R.A. Street (1978) Philos. Mag. **37**, 601.
Urbach, F. (1953) Phys. Rev. **92**, 1324.
Wang D.T., A. Göbel, J. Zegenhagen, and M. Cardona (1997) Phys. Rev. B **56**, 13167.
Weaire, D. and M.F. Thorpe (1971) Phys. Rev. B **4**, 2508.
Yamashita K., T. Kita, H. Nakayama, and T. Nishino (1997) Phys. Rev. B **55**, 4411.
Ying, J.Y., J.B. Benziger, and H. Gleiter (1993) Phys. Rev. B **48**, 1830.
Zhang, Y.A., J.A. Strozier, Jr., and A. Ignatiev (1996) Phys. Rev. B **53**, 7434.

8. Optical Properties of High-Temperature Superconductors

The superconducting state is found at low temperatures in materials such as metals and metallic alloys (even disordered alloys), organic compounds and heavy-fermion materials such as UPt_3. These materials are superconducting at temperatures not exceeding 23.2 K; the normal state, stable at higher temperatures, is separated from the superconducting state by a second-order phase transition. Another, distinct category of high-temperature superconductors (HTS) has been discovered in insulating, non-magnetic or antiferromagnetic materials, which show superconducting behavior for temperatures up to 125 K when doped with metal ions and/or oxygen. Recent optical studies concentrate on this last category of superconductors, so we will also focus on them.

Superconductors are characterized by the following specific phenomena:

i) At the critical temperature T_c, due to electron-electron interaction mediated by phonon exchange, electrons with the same wavenumber but opposite wavevector directions and opposite spins couple in Cooper pairs. As a result, all electron pairs condense in a macroscopic quantum state, with the same phase throughout the superconductor. The analogy with a Bose–Einstein condensate is evident. The Fermi surface, characteristic for the metallic state above T_c, disappears at the transition, and an energy gap 2Δ around the Fermi level appears in the energy spectrum of Cooper pairs. This energy gap decreases continuously with temperature and vanishes at T_c (see Fig. 8.1).

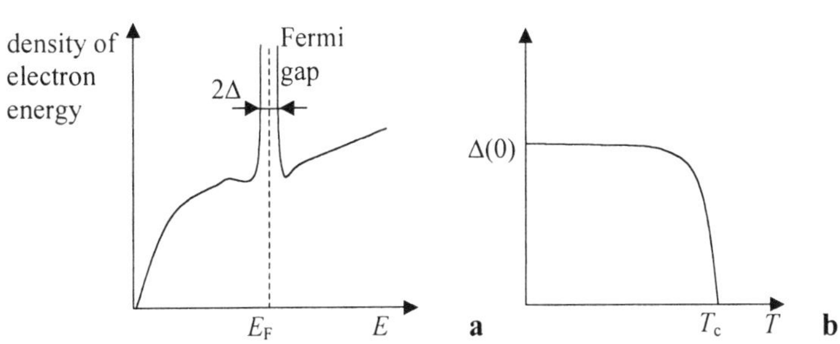

Fig. 8.1. (**a**) The Fermi gap and (**b**) its temperature dependence

ii) The electrical resistivity is zero beyond T_c.
iii) The magnetic flux is expelled at a critical magnetic field H_c, the superconductor becoming diamagnetic. If the magnetic field is further increased the superconducting state is destroyed.
iv) The magnetic flux in superconductors is quantized.

The phonon-exchange-mediated electron-electron interaction, which is the physical origin of superconductor behavior, forms the basis of the BCS (Bardeen, Cooper, Schrieffer) theory. It describes satisfactorily the behavior of common, non-HTS superconductors. Although many HTS characteristics show evident discrepancies with predictions made by the BCS theory, HTS theories are based on the same concepts (electron pairing, energy gap, etc.), with additional assumptions, for example electron correlation. These are non-BCS theories. Still, there is no single theory that explains all characteristics of HTS materials even after fifteen years since their experimental discovery, and there is much controversy about the validity of non-BCS theories caused by the fact that their predictions do not always fit well the experimental results even in the normal state of superconductors. Recent theories have even suggested that in HTS the superconducting transition is not associated to the formation of Cooper pairs but to the onset of long-range phase order, the Cooper pairs existing even above T_c. This theory is more attractive for underdoped cuprate HTS (Emery and Kivelson, 1995). A recent and excellent book that presents the theoretical approaches to HTS is that of Anderson (1997).

We start with a brief description of the BCS theory because some of its predictions, such as electronic Cooper pairs, the existence of the Fermi gap in the excitation spectrum and the isotope effect, are preserved in HTS. Moreover, the need for developing non-BCS theories comes from a complete failure in matching experimental results (including those obtained with optical methods) with the predictions of BCS theory. These predictions must therefore be known.

In the BCS theory it is assumed that the electronic structure of the material is that of a gas of quasiparticles, with no band structure effects. The Hamiltonian that describes the electron-phonon interaction is given by (Callaway, 1991):

$$H = \sum_{k\sigma} E(\boldsymbol{k}) c^{+}_{k\sigma} c_{k\sigma} + \sum_{q} \hbar\omega(\boldsymbol{q}) b^{+}_{q} b_{q} + \sum_{kk'q\sigma\sigma'} [\mathcal{V}(\boldsymbol{q})/2] c^{+}_{k-q\sigma} c^{+}_{k+q\sigma} b_{k'\sigma'} b_{k\sigma}$$
$$+ \sum_{kq\sigma} [D(\boldsymbol{q}) c^{+}_{k+q\sigma} c_{k\sigma} b_{q} + D(-\boldsymbol{q}) c^{+}_{k-q\sigma} c_{k\sigma} b^{+}_{q}], \tag{8.1}$$

where the first two terms represent the system of noninteracting electrons and phonons, respectively, the third describes the electron-electron interaction and the last term – the coupling between electrons and phonons. Considering that electrons couple to only one phonon branch (one-band approximation), and after performing a transformation with a suitably chosen Hermitic operator, our Hamiltonian can be separated into $H = H_0 + H_{\text{int}}$. H_0 denotes the first term in

(8.1) and the interaction Hamiltonian, which involves only electrons with opposite wavevectors and spins ($\boldsymbol{k}\uparrow, -\boldsymbol{k}\downarrow$) that generate the superconducting state, is:

$$H_{\text{int}} = \sum_{kk',\sigma\sigma'} V_{kk'} c^{+}_{k'\sigma'} c^{+}_{-k'\sigma} c_{-k\sigma} c_{k\sigma}, \tag{8.2}$$

where

$$V_{kk'} = 2\,|\,D(\boldsymbol{k}-\boldsymbol{k}')\,|^2\,\hbar\omega(\boldsymbol{k}-\boldsymbol{k}')/\{[E(\boldsymbol{k})-E(\boldsymbol{k}')]^2 - [\hbar\omega(\boldsymbol{k}-\boldsymbol{k}')]^2\} + \mathcal{V}(\boldsymbol{k}-\boldsymbol{k}'), \tag{8.3}$$

with $\mathcal{V}(\boldsymbol{k}-\boldsymbol{k}')$ a matrix element of the Coulomb interaction. To simplify the notations, we incorporate the spin index σ in $\boldsymbol{k}$, by defining $\boldsymbol{k} = \boldsymbol{k}\uparrow$ and $-\boldsymbol{k} = -\boldsymbol{k}\downarrow$. The new Hamiltonian

$$H = \sum_k E(\boldsymbol{k}) c^{+}_k c_k + \sum_{kk'} V_{kk'} c^{+}_k c^{+}_{-k'} c_{-k} c_k, \tag{8.4}$$

can be diagonalized by introducing the operators $\xi_k = \alpha_k c_k - \beta_k c^{+}_{-k}$, $\xi^{+}_k = \alpha_k c^{+}_k - \beta_k c_{-k}$, $\xi_{-k} = \alpha_k c_{-k} + \beta_k c^{+}_k$ and $\xi^{+}_{-k} = \alpha_k c^{+}_{-k} + \beta_k c_k$, which satisfy the fermion commutation rules $\{\xi^{+}_{k'}, \xi_k\} = \delta_{kk'}$, $\{\xi_k, \xi_{k'}\} = 0$. These requirements impose $\alpha_k^2 + \beta_k^2 = 1$.

Expressing the interaction Hamiltonian in terms of the new operators we obtain a diagonal, a second-order nondiagonal, and a forth-order nondiagonal term. Neglecting the latter on the basis that it has a negligible expectation value in the ground state of the BCS theory, the interaction Hamiltonian becomes diagonal if the second-order nondiagonal term vanishes. This happens if

$$2E(k)\alpha_k\beta_k + (\alpha_k^2 - \beta_k^2)\sum_{k'} V_{kk'}\alpha_{k'}\beta_{k'} = 0, \tag{8.5}$$

a relation which can be simplified even further by introducing new variables $\alpha_k = \sin\theta_k$, $\beta_k = \cos\theta_k$ and defining

$$\Delta_k = (-1/2)\sum_{k'} V_{kk'} \sin 2\theta_{k'}. \tag{8.6}$$

Δ_k now satisfies an integral equation $\Delta_k = (-1/2)\sum_{k'} V_{kk'}\Delta_{k'}/[\Delta_{k'}^2 + E^2(\boldsymbol{k}')]^{1/2}$, which has simple solutions if we assume that

$$V_{kk'} = \begin{cases} -V_0, & |\,E(\boldsymbol{k})\,| < \hbar\omega_0,\ |\,E(\boldsymbol{k}')\,| < \hbar\omega_0 \\ 0, & \text{otherwise} \end{cases}, \tag{8.7}$$

i.e. the potential $V_{kk'}$ is negative (attractive) within a small energy range centered around the Fermi level (taken as reference). Δ_k is called the order parameter or the superconducting gap. The solution of (8.5) is zero for $|\,E(\boldsymbol{k})\,| \geq \hbar\omega_0$, around the Fermi level having the value

$$\Delta_0 = \hbar\omega_0 / \sinh(1/V_0G_0) \cong 2\hbar\omega_0 \exp(-1/V_0G_0), \tag{8.8}$$

where $V_0G_0 = (4/9)(E_F/\hbar\omega_D)(m/M)$ if we identify $\hbar\omega$ with the Debye energy $\hbar\omega_D$. With $E_F/\hbar\omega_D \approx 10^2$, m, M the electron and atomic masses, respectively, we obtain $\Delta_0 \approx 10^{-4}$ eV, corresponding to a temperature of 5 K. At finite temperatures, the energy required to create an excitation is dependent on the number of other excitations, i.e. on temperature. Introducing the temperature dependence of the energy gap, (see Callaway (1991) for details), it is found that the superconducting gap is a decreasing function of temperature and vanishes at $T = T_c$ where $k_BT_c \cong 1.13\hbar\omega_0 \exp(-1/V_0G_0)$, i.e. $2\Delta(0)/k_BT_c \cong 3.53$.

The gap dependence on temperature is described by the relation

$$\Delta(T)/\Delta_0 = \exp[\Phi(\Delta/k_BT)], \tag{8.9}$$

where $\Phi(y) = y\int_0^\infty x \sinh^{-1} x \,\mathrm{sech}^2 y(1+x^2)^{1/2}/(1+x^2)^{1/2}\,dx$.

Since $\omega_D \approx 1/M^{1/2}$, V_0G_0 is independent of M, and the theory indicates that $T_c \approx M^{-1/2}$. This dependence of the critical temperature on the atomic mass is called the isotope effect; it was experimentally observed in samples of different isotopes of the same metal element. Writing the isotope effect as $T_c \approx M^{-\alpha}$, metallic superconductors are characterized by $\alpha = 0.5$. The value of this parameter in HTS is $\alpha \cong 0.02$ (for YBCO), in complete disagreement with the BCS theory. In $YBa_{2-x}La_xCu_3O_7$, for example, $\alpha = 0.38$ for $T_c = 38.3$ K, decreases gradually with increasing T_c up to 73 K and then falls sharply to 0.025 for $T_c = 92.3$ K ($x = 0$), its behavior being modeled by $\alpha = (1/2)[1 - T_c(x)/T_{c0}]^{0.59}$ with $T_{c0} = 92.8$ K (Bornemann et al., 1991). The quite schematic and phenomenological BCS theory does not allow the calculation of T_c from first principles; this can be done via the Eliashberg theory (see Callaway, 1991) a much more complicated theory, which relates the self-energy operator with the electron Green's function taking into account the temperature dependence of the electron-phonon contribution to the electron self-energy.

8.1 High-Temperature Superconductors

HTS materials were discovered in 1986 and since then have been extensively studied with regard to their electronic structures and general properties. Optical spectroscopic methods, as well as non-optical measurements, are generally used not only to determine different properties of HTS, but also to test the validity of different theoretical predictions.

There are four categories of HTS (Lynch and Olson, 1999):

i) $Ba_{1-x}K_xBiO_3$, formed from doping the non-magnetic perovskite-structure insulator $BaBiO_3$ with K, has the highest T_c around 30 K for $x = 0.4$.

ii) fullerene (C_{60}) mixtures with alkali metals. The highest T_c (about 33 K) is achieved in Cs_2RbC_{60}.

iii) the intermetallic polycrystalline compound MgB_2, which becomes superconducting at $T_c = 40$ K – the highest critical temperature for a non-copper-oxide bulk superconductor. This superconductor has been discovered very recently, in March 2001 (Nagamatsu et al. 2001), to the amazement of the scientific community, for which it was known that simple metallic compounds show superconductive behavior only at very low temperatures. MgB_2 shows a simple hexagonal structure consisting of alternating honeycomb layers of B and closed-packed Mg (see Fig. 8.2). The unit cell dimensions are $a = 3.086$ Å and $c = 3.524$ Å, the crystal space group being P6/mmm. The preparation of MgB_2 is much simpler compared to copper-oxides superconductors.

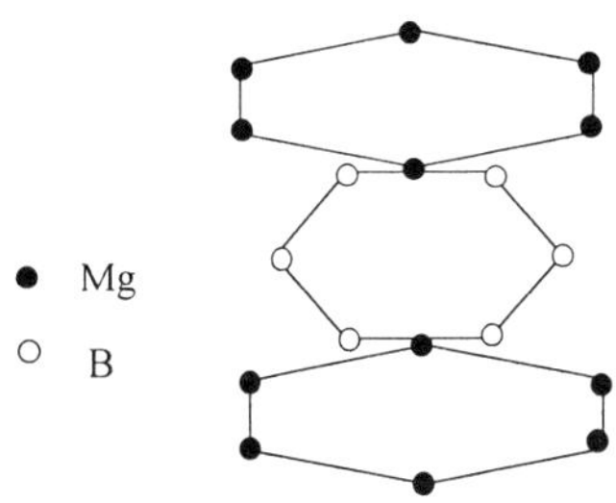

Fig. 8.2. MgB_2 structure

The initial theories and experiments on MgB_2 revealed that this metallic compound behaves like a phonon-mediated superconductor described by the BCS theory in the strong coupling limit. However, the first optical spectroscopic experiments have demonstrated that MgB_2 shows an evident non-BCS behaviour.
(iv) cuprate HTS – the 'true' and the most studied category of HTS – have critical temperatures of more than 100 K. They contain one or more planes of Cu and O atoms/unit cell in the ratio of 1:2 in an almost square lattice (see Fig. 8.2) and all are related to an antiferromagnetic and insulating parent compound. The CuO_2 planar or slightly puckered sheets come in groups of n sheets, each group being separated from the other by isolation planes. In each group, the CuO_2 planes can be adjacent or separated from each other by other sheets. For example, La_2CuO_4 contains only one ($n = 1$) CuO_2 plane, $YBa_2Cu_3O_7$ has $n = 2$ CuO_2 planes separated by a Y plane, while $Tl_2Ba_2Ca_2Cu_3O_{10}$ has $n = 3$ CuO_2 planes and a Ca plane between each CuO_2 plane in the group. The stacking structure of cuprate HTS is represented in Fig. 8.3. The distances between CuO_2 planes in a group are very small (3–4 Å), the groups being separated by larger distances (8–12 Å). The parent compound becomes a HTS by doping with holes; these holes come from oxygen addition or from the substitution of a metallic constituent of the material with another metal with a lower valence than the original atom. With increasing hole concentration the Néel temperature of the antiferromagnetic parent drops until it vanishes, the material becoming a 'strange' metal at higher hole concentrations. On increasing the hole doping further it becomes a HTS. A

schematical phase diagram for HTS cuprates is shown in Fig. 8.4 where AFM denotes an antiferromagnetic insulator, SM – strange metal, SC – superconductor, M – metal. This diagram is, however, not universal, since the normal state can be semiconductor in some doped structures instead of metallic.

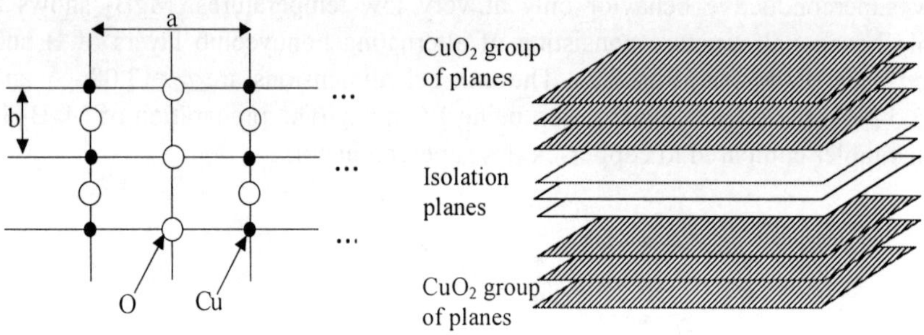

Fig. 8.3. The Cu_2O plane structure

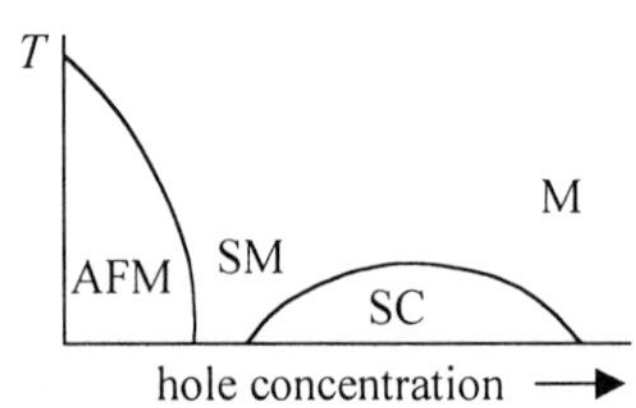

Fig. 8.4. Schematic phase diagram for HTS cuprates

As shown in the phase diagram, the transition temperature T_c between the superconductor and say the metallic phase depends on doping: it is maximum for optimal doping, and decreases for lower and higher hole concentrations than this optimum value, the corresponding samples being called underdoped and overdoped, respectively. The holes can, however, be nonuniformly distributed; they can segregate into stripe domains, with antiferromagnetic domains between them (Emery and Kivelson, 1993). Recent experiments suggest that the hole segregation is dynamic, with fluctuations occurring in this phase, and that the stripe phase can even coexist with superconductive phases (see the references in Lynch and Olson, 1999).

The best known and studied cuprate HTS with $T_c = 92$ K is $YBa_2Cu_3O_7$, also known as Y123 or YBCO. It has an orthorhombic Bravais lattice, whereas $YBa_2Cu_4O_8$ (known as Y124), has an A-orthorhombic Bravais lattice and a critical temperature of 80 K. Y123 has a nearly tetragonal unit cell with lattice parameters $a = 3.82$ Å, $b = 3.88$ Å and $c = 11.65$ Å, the corresponding values of the lattice parameters for Y124 being $a = b = 3.9$ Å and $c = 27.2$ Å. Other examples of HTS materials are $Bi_2Sr_2CaCu_2O_8$ (abbreviated BSCCO or Bi2212) with

T_c = 85 K, $Bi_2Sr_2Ca_2Cu_3O_{10}$ (Bi2223) with T_c = 110 K, $Bi_2Sr_2CuO_6$ (Bi2201) having T_c = 20 K, and $La_{2-x}Sr_xCaCu_2O_6$ with T_c = 60 K.

The knowledge of the electron band structure of HTS materials is extremely important because several properties can be deduced from it, in particular the energy gap behavior. The calculation of HTS electronic structure is, however, an extremely difficult task, almost hopeless, due to the large number of atoms in the unit cell. The electron energy band of the CuO_2 plane, for example, common for all HTS cuprates, can be modeled supposing a square lattice with Cu atoms at the corners and O atoms at the center of each side. The majority of the states of the Cu 3d-O 2p hybrids are filled, the Fermi level being located at an antibonding state composed of hybrids of Cu $3d_{x^2-y^2}$ and O $2p_x$ and $2p_y$ states. The symmetry of these states influences the symmetry of the superconducting energy gap. The tight-binding bands of CuO_2 planes are schematically represented in Fig. 8.5.

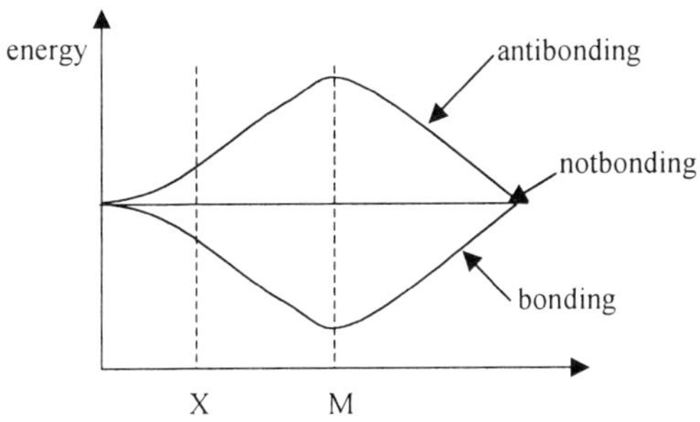

Fig. 8.5. Tight-binding bands of CuO_2 planes

The electronic bands are often calculated using the local-density functional theory, and its simplified version, the local-density approximation (LDA) (Callaway, 1991). In LDA the exchange and correlation potentials are estimated by considering the value of the potential at $\boldsymbol{r}$ equal to the exchange-correlation potential for an interacting electron gas with a uniform density, this uniform density being equal to the charge density of the crystal at $\boldsymbol{r}$. The eigenvalues of the LDA Hamiltonian arc states occupied by electrons below E_F and empty above it. The LDA model, as well as its various extensions, calculates accurately the electronic bands not only in HTS, but also in metals, semiconductors and insulators. It is best suited for metallic systems, self-interaction corrections being needed to treat semiconductors and insulators.

In HTS materials, the LDA model is not applicable when the electrons are well localized, because it treats the electron-electron correlation in an averaged way. For localized electrons, the transfer of an electron from an atom/ion to another atom/ion is made by expending an energy U. The LDA model is not valid when U is larger than the average crystal potential; it is possible that the LDA theory is not suitable to treat the insulating parent compounds, but can be used for the doped metallic superconductors. When the Coulomb repulsion U is included, the band splits into two if there is an odd number of electrons/atoms. The lower band is full while the upper, separated at U, remains empty, the material resembling an insulator treatable with the Hubbard Hamiltonian described below. In HTS U is the energy needed to transfer one d electron from one transition-metal site to another, its value being generally dependent on the multiplet structure of the transition metal. The materials in which U depends on the multiplet structure of the sites between which electrons are transferred are called Mott–Hubard insulators (see Lynch and Olson, 1999 for a brief discussion of these models).

Several improvements of the LDA have been developed. One of them, the LDA+U, incorporates U into the model; others incorporate the electron-electron interactions, using the many-particle picture of the electron gas. The local Coulomb interactions are sometimes better evidenced in cluster or localized models of cuprate HTS. One of the most used localized models is the single-band Hubbard model described by the Hamiltonian (see Callaway, 1991)

$$H = -t\sum_{i,j\sigma} c_{i\sigma}^{+} c_{j\sigma} + U\sum_i n_{i+} n_{i-} , \tag{8.10}$$

where $c_{i\sigma}^{+}$ and $c_{j\sigma}$ are the creation and annihilation operators of an electron of spin σ at site i (3d electrons at Cu sites in cuprate HTS) and $n_{i\sigma}$ is the number operator $c_{i\sigma}^{+} c_{i\sigma}$. The first term in (8.10) represents the kinetic energy with t the transfer or hopping integral between Cu sites, the second term describing the on-site Coulomb repulsion characterized by U; both U and t are positive. Extensions of the single-band Hubbard Hamiltonian for the case of cuprate HTS include the three-band model, which includes the kinetic energies of the O 2p states, the hopping between Cu-O and O-O sites, and Coulomb repulsion terms for O sites (see Lynch and Olson (1999) for references).

For antiferromagnetic materials, if $U >> t$ so that there is only one electron per site, the one-band Hubbard Hamiltonian can be replaced by a version of the Heisenberg Hamiltonian, in which the Coulomb repulsion is replaced by a term containing only spin operators $\boldsymbol{S}$. The Hamiltonian of this t-J model (Callaway, 1991) is

$$H = -t\sum_{i,j,\sigma} c_{i\sigma}^{+} c_{j\sigma} - J\sum_{i,j} \boldsymbol{S}_i \cdot \boldsymbol{S}_j , \tag{8.11}$$

where $J = 4t^2 / U$. In some extensions of the t-J model for cuprate HTS the next-nearest-neighbor hopping t' is taken into account, the effective spin Hamiltonian becoming (MacDonald et al., 1990)

$$\begin{aligned} H_{\text{eff}} &= J\sum_{i,\delta} (\boldsymbol{S}_i \cdot \boldsymbol{S}_{i+\delta} - 1/4) + J'\sum_{i,\delta'} (\boldsymbol{S}_i \cdot \boldsymbol{S}_{i+\delta'} - 1/4) \\ &+ K\sum_{(i,j,k,l)} [(\boldsymbol{S}_i \cdot \boldsymbol{S}_j)(\boldsymbol{S}_k \cdot \boldsymbol{S}_l) + (\boldsymbol{S}_i \cdot \boldsymbol{S}_l)(\boldsymbol{S}_j \cdot \boldsymbol{S}_k) - (\boldsymbol{S}_i \cdot \boldsymbol{S}_k)(\boldsymbol{S}_j \cdot \boldsymbol{S}_l)], \end{aligned} \tag{8.12}$$

where $J = 4t^2 / U - 64t^4 / U^3$, $J' = 4t'^2 / U + 4t^4 / U^3$, $K = 80t^4 / U^3$, the first two sums being performed over the nearest-neighbor sites δ and next-nearest-neighbor sites δ', and the last sum over groups of four spins (i, j, k, l) in the unit square. This Hamiltonian describes for example, the line asymmetry of four-magnon scattering in HTS (Eroles et al., 1999). The discussion of magnetic Raman scattering in cuprates HTS is often done in terms of the London–Fleury extension of the t-J model (Fleury and London, 1968). In this model the charge-transfer exciton produced when the electrons and holes formed by photon scattering are very close in the real space, is taken into account. The value of the Cu-Cu exchange parameter J is calculated in this model from the first moment of the B_{1g} Raman line.

Magnetic impurities in cuprate HTS are treated with the Anderson Hamiltonian

$$H = \sum_{i,\sigma} E_i c_{i\sigma}^{+} c_{i\sigma} + \sum_{\sigma} E_{\sigma} d_{\sigma}^{+} d_{\sigma} + (U/2)\sum_{\sigma} n_{\sigma} n_{-\sigma} + \sum_{i,\sigma} V_k (c_{i\sigma}^{+} d_{\sigma} + d_{\sigma}^{+} c_{i\sigma}), \tag{8.13}$$

where the first two terms describe the conduction electrons, and the electrons localized on an impurity, respectively, the operators d_{σ}^{+}, d_{σ} being the creation and annihilation operators of a 3d electron on a Cu site, considered as an impurity. The third term in (8.15) reflects the Coulomb repulsion between electrons localized on the impurity site and the forth term describes the hybridization between delocalized conduction electrons and the electrons localized on Cu sites.

Which of these Hamiltonians are used to describe the experimental results, depends on the particular HTS. Detailed descriptions of HTS electronic bands have been performed with the models described above, the Fermi level being usually taken as reference. However, there are still cases when these calculations do not fit well the experiments. The structure of energy bands in a HTS is apparently very complicated, but fortunately only those bands that cross the Fermi level are responsible for the superconducting behavior. A more accurate description, especially of the normal state of HTS, would require the consideration of all bands. A schematic representation of the energy bands that cross the Fermi level in Y123 is given in Fig. 8.6. Under the Fermi level, in an energy range of more than 5 eV, there are more than 30 bands due to Cu 3d and O 2p states but only a few cross E_F.

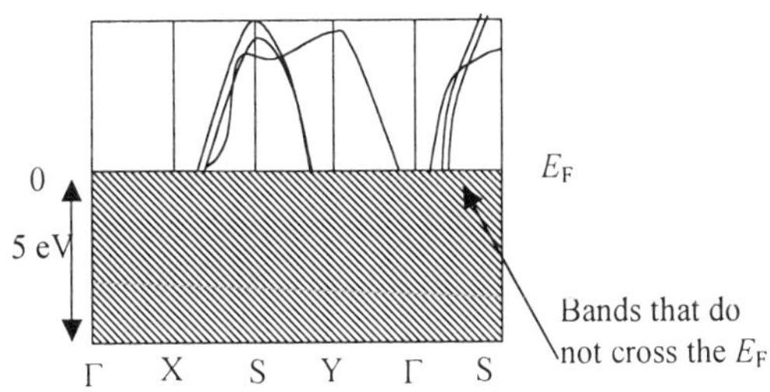

Fig. 8.6. Schematic representation of the energy bands that cross the Fermi level in Y123

The Fermi surface of HTS in either normal or superconducting states can be probed using the de Haas–van Alphen (dH–vA) effect associated with quantization of electron orbits in a magnetic field. Experimental results show that dH–vA oscillations are similar in both superconducting and normal state, the only difference being a higher damping in the superconducting state due to a combined influence of both gapped and gapless regions of the Fermi surface. For example, in YNi_2B_2C, all the 10 dH–vA frequencies found in the normal state, and corresponding to 6 extremal Fermi-surface orbits and several harmonics, survive the transition into the superconducting state induced by changing the magnetic field, and persist over a wide field range deep into the superconducting state (Goll et al., 1996).

The magnetic properties of HTS are similar to those of other superconductors; in particular, the Meissner effect, i.e. the expelling of magnetic lines, is also present in HTS. Magnetic fields penetrate only a small distance into the HTS, being attenuated over a characteristic distance λ called the London penetration depth, defined as

$$\boldsymbol{j} = -(\lambda^2 \mu_0)^{-1} \boldsymbol{A}, \tag{8.14}$$

This definition reflects the fact that in superconductors the current induced by the external field is proportional to the vector potential $\boldsymbol{A}$, rather than to the field itself, as in other solid-state materials.

Maxwell's equations imply then that $\nabla^2 \boldsymbol{H} = \lambda^{-2} \boldsymbol{H}$, an equation that has exponentially decaying solutions $\boldsymbol{H} = \boldsymbol{H}_s \exp(-z/\lambda)$ inside the superconductor, with $\boldsymbol{H}_s$ the magnetic field on the superconductor surface and z the distance along the normal to the surface. Typical values for the penetration depth for different members of the YBCO family are situated in the range 50–250 nm. A microscopic treatment based on the BCS theory (Callaway, 1991) shows that λ^{-2} vanishes for temperatures greater than T_c. Below T_c the penetration depth has a finite value, equal to $\lambda = (m/Ne^2\mu_0)^{1/2} = c/\omega_p$ at $T = 0$.

The superconducting state can be destroyed by applying a magnetic field. The transition to the normal state takes place whenever the energy required for the magnetic flux expulsion from the interior of the sample (proportional to the square of the magnetic field) exceeds the energy gained by condensation into the superconducting state. The critical field at which the transition occurs depends linearly on the energy gap (Callaway, 1991). In a real superconductor material, however, in the presence of a magnetic field the normal and superconducting regions can coexist, enabling the system to reduce the energy needed for the expulsion of the magnetic field and preserving at the same time a part of the condensation energy associated with the superconducting transition. In type I superconductors this intermediate state can exist only when geometric factors tend to increase the energy associated with the magnetic field exclusion, whereas in type II superconductors the mixed state is formed for sufficiently large fields, irrespective of the shape of superconductor. Mathematically, the distinction between type I and type II superconductors is expressed in terms of the ratio λ / ξ where ξ is the coherence length of electrons. A superconductor is of type I if $\lambda / \xi < 1/2^{1/2}$ and of type II when $\lambda / \xi > 1/2^{1/2}$.

8.2 Optical Methods for HTS Investigation

Optical spectroscopy methods are valuable tools for measuring the properties of HTS and for testing the theoretical predictions of different theories. To this end they compete with many other characterization methods such as electrical measurements, microwave absorption, tunneling spectroscopy or photoemission spectroscopy. FIR measurements can be used to test the main features of HTS theories such as the pseudogap, the quantum fluctuations of the superconducting ground states (Ioffe and Millis, 1999) and in general they provide direct evidence of a non-BCS behavior of different HTS. The electronic and vibronic (phonon) spectrum is determined using Raman scattering techniques, and shows strange behaviors even in the normal state. For example, normal metals show peaks in Raman spectra at plasma frequencies corresponding to Raman shifts of a few eV, optimally doped cuprates are characterized by a broad electronic background (from 0 to 1 eV Raman shift) almost independent of temperature and frequency, while underdoped HTS show some temperature dependence at low frequencies explained by quasiparticle lifetime.

In the FIR region, reflectivity or transmission measurements are performed in wide ranges of frequencies and then the real and imaginary parts of the conductivity are determined from the Kramers–Krönig relations. The conductivity can be also obtained without using the Kramers–Krönig relations (which introduce errors due to the extrapolation procedure) by measuring the amplitude and phase of the transmission/reflection by interferometric methods (Pronin et al., 1998). Alternatively, ellipsometric measurements provide both amplitude and phase of the reflected/transmitted light.

8.2.1 Anisotropy of Optical Response

Due to their layered and thus highly anisotropic crystalline structure, it is expected that the optical properties are totally different when measured along the *c* axis or normal to it. Besides the general *a*-*c* anisotropy, some HTS have a small orthorhombic distortion, which induces an additional *a*-*b* anisotropy. For example, in $YBa_2Cu_4O_8$, with the *a*, *b* axes considered with respect to the (001) direction and *c* determined by the (100) direction, the reflectivity curves have the behavior shown in Fig. 8.7 (Kircher et al., 1993), typical for a HTS. From ellipsometric reflectivity measurements in all three directions the dielectric tensor $\hat{\varepsilon}$ can be determined and the plasma edge (related to the transport properties of the material) is calculated. However, the plasma edges have different physical origins depending on the HTS orientation. For example, the plasma edge of $R_{\|a}$ (for the electric field of the incident light parallel to the *a* axis) is due to the contribution of free carriers related to Cu-O layers, while the blueshifted plasma edge in $R_{\|b}$ reflects the additional contribution of free carriers moving along Cu-O chains. The shift towards much lower frequencies and the damping of the plasma edge for $R_{\|c}$ is due to the low value of the electron conductivity normal to the (100) plane. The experimental results are usually fitted with a Drude–Lorentz model in which $R_{\|\alpha} = |[(\tilde{\varepsilon}^{\alpha\alpha})^{1/2} - 1]/[(\tilde{\varepsilon}^{\alpha\alpha})^{1/2} + 1]|^2$ with

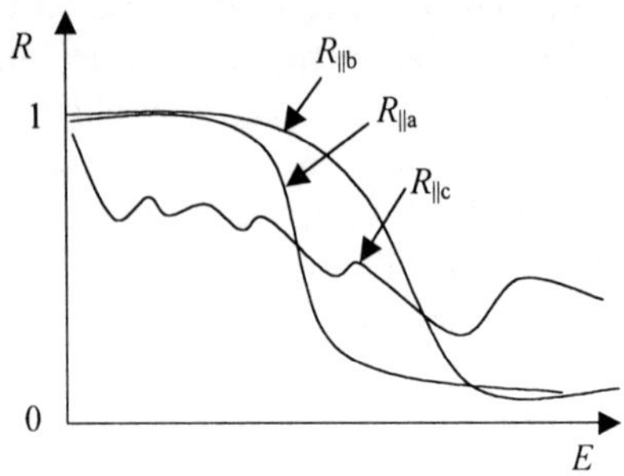

Fig. 8.7. Typical optical anisotropic response in HTS

$$\tilde{\varepsilon}_{\alpha\alpha} = 1 - \Sigma_n (\hbar\omega_{\mathrm{p},n}^{\alpha})^2 / [\hbar\omega(\hbar\omega + \mathrm{i}\hbar\Gamma_n^{\alpha})] + \Sigma_k (F_k^{\alpha})^2 / [(\hbar\omega_k^{\alpha})^2 - (\hbar\omega)^2 - \mathrm{i}\hbar^2\omega\gamma_k], \tag{8.15}$$

where $\omega_{\mathrm{p}}^{\alpha}$ is the unscreened plasma frequency along the α axis. This parameter, as well as the scattering rate Γ, is determined from fitting.

The Lorentzian oscillators introduced in the above formula can have different origins: they can be due to phonons, IR-active vibrations induced by doping, they can model the d-band, mid-IR band or the charge-transfer band. The number and relative importance of the various terms depend on the number of carriers in the Cu-O plane, and so on the stoichiometry of the compound and the amount of doping. For example, in stoichiometric insulating cuprates the mid-IR, IR doping-induced vibrations and the d terms vanish, as does the Drude term, whereas in metallic cuprates the Drude contribution is strong, the phonon and charge-transfer contributions vanishing (Calvani, 1996).

One consequence of the material anisotropy is the presence of two gaps in some superconductors: one for chain-related bands and the other for plane-related bands (Heyen et al., 1991). For example, in $NdBa_2Cu_3O_{7-x}$ Raman scattering with light polarized along CuO_2 planes or along the *c* axis, has revealed two in-plane pair-breaking peaks 2Δ at 330 cm^{-1} for the A_{1g} symmetry and at 580 cm^{-1} for the B_{1g} symmetry. For the *c*-axis spectra, only one pair-braking peak has been found between 400–500 cm^{-1}. These peaks have been determined from the electronic continuum redistribution below T_c. Surprisingly, there is no superconducting gap clearly visible in the in-plane Raman scattering spectrum, which is thresholdless at decreasing temperatures. The broad peak due to Cooper pairs is visible at low temperatures (Misochko et al., 1997).

In HTS with weak interlayer coupling it was shown that σ_c is in fact a probe of in-plane properties, since in this case the interplane coupling is given by convolution of two in-plane Green functions (Anderson, 1991).

8.2.2 Observation of Superconducting Transition

One method to determine the superconducting transition temperature in cuprate HTS is thermal difference reflection (TDR) spectroscopy (Holcomb et al., 1996) which measures the reflectivity variation at a change with the step ΔT of the temperature around T_0

$$\Delta R_T / R \cong 2\Delta T[\partial R(\omega, T_0) / \partial T] / R(\omega, T_0). \tag{8.16}$$

The curve obtained in this way is quite difficult to interpret, with the notable exception of semiconductors, for which it has a characteristic lineshape unique to the critical points in the joint density of states. Another exception applies to metals, for which the TDR curve has a derivative lineshape at the screened plasma frequency $\omega_{sp} = \omega_p / \varepsilon_\infty^{1/2}$. A typical response in cuprate HTS is shown in Fig. 8.8.

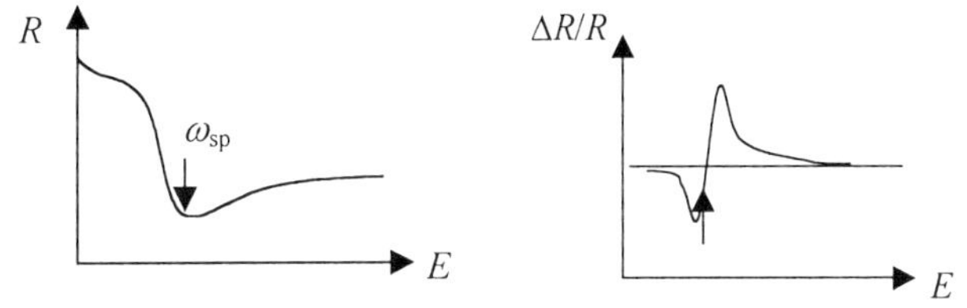

Fig. 8.8. The behavior of TDR for HTS

Since cuprate HTS are not described by a Drude model, but by a modified Drude theory with a temperature- and energy-dependent optical scattering rate $\Gamma(\omega, T)$, the zero-crossing in TDR does not correspond to ω_{sp} except for large Γ values. For smaller Γ values the screened plasma frequency is situated in the negative peak region of the TDR. So, from TDR measurements one can estimate the value of the scattering rate. From the temperature dependence of TDR, due to the temperature dependence of the scattering rate $\Gamma(\omega, T) = A(k_B T)^2 + B\omega$, it is also possible to determine T_c since in the superconducting state the amplitude of TDR decreases with temperature above T_c, whereas in the normal state it increases approximately linearly with the temperature.

The transition to the superconducting state also changes the intensity, frequency and even shape of the vibronic spectrum of HTS. The transition from the normal to the superconducting state produces dramatic effects on phononic spectra. Phonons, strongly coupled to electrons near the Fermi surface are very sensitive to the changes occurring in the vicinity of this surface. The opening of the superconducting gap induces a pronounced redistribution of electronic states in the vicinity of the Fermi surface, and thus a change in the phonon self-energy, i.e. in the contribution of electron-phonon interaction to phonon frequency. It is also accompanied by the appearance of some resonances in the low-energy electronic excitation, for example coupled phonon-electron excitations in the superconducting state. This phonon renormalization is seen in Raman spectroscopy as a change in phonon frequencies, intensities and linewidths. The renormalized phonon frequency, as well as the electron-phonon coupling strength can be obtained by fitting the asymmetric lineshapes of Raman peaks to a Fano profile $I(\omega)=I_c(\varepsilon+q)^2/(1+\varepsilon^2)+I_b(\omega)$, where I_c is a constant, I_b describes the background, and $\varepsilon=(\omega-\omega_r)/\Gamma$. $\omega_r=\omega_{ph}+\delta\omega$ is the renormalized phonon frequency corresponding to the bare phonon frequency ω_{ph} and Γ the linewidth. The Fano parameter q depends on the electron-phonon coupling strength.

The abrupt changes in the frequency and linewidth of some phonons across T_c due to changes of the phonon self-energy induced by the superconducting transition can be expressed in terms of the complex phonon self-energy $\mathcal{E}_\nu$. We have for the νth phonon

$$\Delta\omega_{ph,\nu}/\omega_{ph,\nu}=\lambda_\nu \,\mathrm{Re}(\mathcal{E}_\nu)/2N\,, \quad \Delta\Gamma_\nu/\Gamma_\nu=-\lambda_\nu \,\mathrm{Im}(\mathcal{E}_\nu)/2N\,, \tag{8.17}$$

where N is the normal state density of states/spin at the Fermi level and λ_ν the electron-phonon coupling. The most affected phonons are those with energies close to 2Δ. More specifically, on lowering the temperature from above to below T_c, three types of phenomena are observed (Cardona, 1998):

i) at T_c a downshift occurs if the phonon frequency $\omega_{ph}<2\Delta$, whereas an upshift is observed for $\omega_{ph}>2\Delta$;

ii) the linewidth decreases for $\omega_{ph}\ll 2\Delta$ and increases if $\omega_{ph}\cong 2\Delta$;

iii) the renormalized phonon intensity $I_r=\pi[I(\omega_r)-I_b(\omega_r)]=\pi I_c\Gamma q^2$ changes.

Apart from these modifications, in layered HTS the surface contribution to the Raman lineshape strongly modifies the bulk contribution below 2Δ and introduces additional plasmon modes above 2Δ (Wu and Griffin, 1996). The relative surface contribution depends on the momentum transfer parallel to the layers, the scattering cross-section for out-of-phase modes being, however, not affected.

Fitting the asymmetric Raman lineshapes with the Fano formula, it was found that for $HgBa_2Ca_3Cu_4O_{10+\delta}$ with T_c = 123 K, the parameters of the fit for the 240 cm^{-1} mode vary as in Fig. 8.9. (Hadjiev et al., 1998). The A_{1g} phonons at 240 and 390 cm^{-1}, corresponding to vibrations of plane oxygen atoms with admixture of Ca vibrations, show in an interval 10–15 K below T_c (as $2\Delta(T)$ crosses ω_{ph}) a softening with 6 and 18%, respectively, and a change in linewidth of 10% and 40%, respectively. As also shown in Fig. 8.9, in the normal state the linewidth has a small increase with temperature due to anharmonic decay of phonons with zero wavevector into two phonons with opposite wavevectors and half frequency, described by $\Gamma(T) = \Gamma_0[1 + 2/(\exp(\hbar\omega_{ph}/2k_BT) - 1)]$, from which it is possible to determine the parameter Γ_0. A value of 7.3 cm^{-1} has been found for the 390 cm^{-1} phonon.

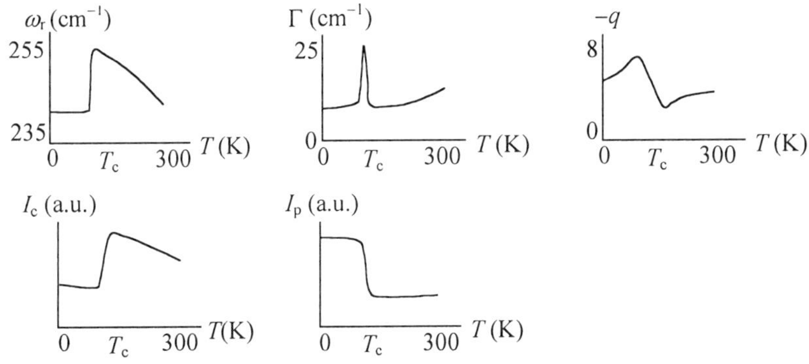

Fig. 8.9. Fano parameters dependence on temperature in $HgBa_2Ca_3Cu_4O_{10+\delta}$

From a microscopic point of view the Fano parameter q is related to the matrix element of the electron-phonon Hamiltonian V as

$$q = [VT_{ph}/T_e + V^2R(\omega)]/[\pi V^2\rho(\omega)], \tag{8.18}$$

where T_{ph}, T_e are the Raman scattering amplitude by phonons and the electronic continuum respectively, $R(\omega) = \mathrm{Re}[\Pi(\omega)]$, and $\rho(\omega) = -(1/\pi)\,\mathrm{Im}[\Pi(\omega)]$, with $\Pi(\omega)$ the retarded electronic polarization at $\boldsymbol{q} = 0$. Considering that the number of electron states in the electron peak around $\omega = \omega_e$ is $\rho(\omega) = N_e\delta(\omega - \omega_e)$, the dimensionless electron-phonon interaction constant for the νth phonon is $\lambda_\nu = (1/\Delta)N_eV_\nu^2/\omega_{ph,\nu}$. This relation allows the estimation of the strength of the electron-phonon interaction from the experimental value of $N_eV^2 = \delta\omega\Delta\omega_{e-ph}$ where $\Delta\omega_{e-ph}$ is the distance between the electron and phonon peaks. Measured values of this parameter are 0.08 for both peaks at 240 and 390 cm^{-1}, which, accounting for the 20 atoms per unit cell that correspond to 60 vibrational modes, gives a final value of 5.4, much larger than in a conventional superconductor described by the BCS theory, for which $\lambda_\nu < 0.5$. This fact confirms again the non-BCS behavior of HTS (Hadjiev et al., 1998)

Similar results have been found in $HgBa_2Ca_3Cu_4O_{8+\delta}$. By fitting the asymmetric peaks with the Fano formula, it was found that the transition from the normal to the superconductor state is accompanied by an overall frequency softening, linewidth narrowing, an increase of $-q$, and an increase of the renormalized phonon intensity $\pi I_c\Gamma q^2$ around T_c (Zhou et al. 1997). For this material in which T_c = 136 K, a softening of 1.5% from 80 K to 130 K of plane oxygen vibrations has been observed in Raman scattering, a three times increase of I from T_c to 10 K, and an increase in the same interval of $-q$ from 2.5 to 6. An anharmonic decay of Raman-active phonons of frequency ω and zero wavevector in two photons with frequencies $\omega/2$ and opposite wavevectors was also observed. Analogous self-energy effects for A_{1g} phonons were evidenced also in $Bi_2Sr_2Cu_2O_{8+\delta}$ at $\delta = 0.13$ and $T_c = 86$ K (Mantin et al., 1997).

The superconducting transition can be triggered not only by varying the temperature, but also by varying the doping concentration. The decrease in T_c by doping changes the position of the superconducting gap relative to phonon energies and can cause the crossing of a certain phonon frequency over the superconducting gap. This produces similar effects on the phonon spectrum, including rapid changes in the frequency and linewidth, as discussed before. For example, in $YBa_2(Cu_{1-x}M_x)_4O_8$ where M = Ni or Zn, the softening of the Ba mode is x dependent due to either strong coupling to the higher gap energy $2\Delta_2$ or the lower gap $2\Delta_1$ with an energy slightly greater than Ba mode. $2\Delta_1$ is the superconducting gap corresponding to chain-related bands and $2\Delta_2$ corresponds to plane-related bands. The frequency and linewidth of the Ba phonon can be fitted again with a Fano profile with a linear background, the value and the symmetry of the superconducting gap being estimated from the temperature dependence of the phonon line (Watanabe and Koshizuko, 1998).

Fano profiles have also been observed in reflectance measurements made on $Nd_{1.96}Ce_{0.04}CuO_{4+y}$ (NCCO), for all four transverse optical E_u phonons in the a-b plane. From the temperature dependence of the renormalized phonon frequencies, obtained from the Fano lineshape, it was inferred that the continuum is not the free-carrier absorption band as in semiconductors, but a polaron band. This polaron band, present in most HTS, softens as the temperature decreases, due to the delocalization of polaronic carriers (Lupi et al., 1998).

The magnetic properties change also at the superconducting transition. Raman experiments in optimal doped $Bi_2Sr_2CaCu_2O_{8-\delta}$ HTS with $T_c = 95$ K, have revealed that the magnon scattering peak B_{1g} increases below T_c, suggesting an increase in the magnetic coherence in the superconducting state, while the A_{1g} peak around 120 meV corresponding to a more incoherent electron contribution, decrease in intensity. The high-energy background around 250 meV suffers a rearrangement of the spectral weight in B_{1g} and $A_{1g} + B_{2g}$ symmetries below T_c. The Raman spectrum in the low-energy region, influenced by the effects at higher energies, shows the opening of a superconducting gap $\Delta = 34$ meV in the B_{1g} symmetry, no gap being observed in the A_{1g} symmetry (Rübhausen et al., 1998)

8.2.3 Symmetry of the Superconducting Gap

An accurate determination of the frequency and wavevector dependence of the superconducting gap is essential for determining the microscopic origin of the pairing process, to each electron pairing mechanisms corresponding a unique $\Delta(\omega, \boldsymbol{k})$. The $\boldsymbol{k}$ symmetry of the superconducting gap can be inferred, in principle, from measurements of optical conductivity. For an isotropic, s-wave gap superconductor, for which $\Delta(\boldsymbol{k})=1$, the optical conductivity should drop rapidly to zero in the superconducting state, approaching zero with decreasing frequency. If the superconducting gap has d-wave symmetry, a finite value of optical conductivity should be found at frequencies approaching zero. However, impurity scattering increases the low-frequency absorption in any superconductor, so that it is not possible to distinguish from c-axis optical conductivity alone if the gap is a strongly anisotropic s-wave, which varies with $\boldsymbol{k}$ as $|k_x^2+k_y^2|$ or a $d_{x^2-y^2}$-wave gap with a dependence $\Delta(\boldsymbol{k})=(k_x a/\pi)^2-(k_y b/\pi)^2$. Neither can Raman scattering distinguish between these two situations since only the modulus of the superconducting gap, not its sign, can be determined from the polarization dependence of the electronic Raman scattering.

The electronic Raman scattering probes the spectrum of electronic excitations near the Fermi level and directly illustrates the condensation in the superconducting state of low-frequency charge-density fluctuations below the gap. In optimally doped HTS the electronic Raman scattering shows at the superconducting transition a redistribution of the continuum spectrum in a broad peak (pair-breaking peak) the intensity and frequency position Ω of which differs for different symmetry components. More specifically, $\Omega(B_{1g})>\Omega(B_{2g})$ $>\Omega(A_{1g})$, where the symmetry of modes are determined from polarization selection rules. For example, if x, y define the CuO_2 plane, with y along the Cu-O bonds, A_{1g} phonons appear in backscattering only for parallel incident and scattered in-plane polarizations, while the B_{1g} phonons appear either in $z(x,x)\bar{z}$ or $z(y,y)\bar{z}$ geometries. The B_{2g} phonons appear in xy polarizations, combinations as $A_{1g}+B_{2g}$ being observed in $x'x'$ polarization, where the axes x', y' are rotated with 45° with respect to x, y.

The cross-section of the Raman signal depends on the superconducting gap through

$$\frac{\mathrm{d}^2\sigma}{\mathrm{d}\omega\mathrm{d}\Omega}\approx[1+n(\omega)]\,\mathrm{Im}\,\chi_{\gamma\gamma}(\boldsymbol{q},\omega), \tag{8.19}$$

where the Raman response function associated, at low frequencies, to intraband transitions in band γ, is given by $\chi_{\gamma\gamma}(\boldsymbol{q},\omega)=\langle\gamma_k^2\lambda_k\rangle-\langle\gamma_k\lambda_k\rangle^2/\langle\lambda_k\rangle$. γ_k, called the Raman vertex, describes the strength of Raman transitions, and the so-called Tsuneto function is defined as

$$\lambda_k(\mathrm{i}\omega)=\frac{\Delta(\boldsymbol{k})^2}{\mathcal{E}(\boldsymbol{k})}\tanh\left(\frac{\mathcal{E}(\boldsymbol{k})}{2k_BT}\right)\left(\frac{1}{2\mathcal{E}(\boldsymbol{k})+\mathrm{i}\omega}+\frac{1}{2\mathcal{E}(\boldsymbol{k})-\mathrm{i}\omega}\right), \tag{8.20}$$

where $\mathcal{E}^2(\boldsymbol{k})=\xi^2(\boldsymbol{k})+\Delta^2(\boldsymbol{k})$ with conduction band energy $\xi(\boldsymbol{k})=E(\boldsymbol{k})-\mu$, and μ the chemical potential. The first term in (8.20) – the bare Raman response – reflects the attractive interaction between Cooper pairs, while the second term describes the screening due to Coulomb repulsion, the average $\langle\ldots\rangle$ over $\boldsymbol{k}$ values being performed over the Fermi surface. The screening term vanishes unless the Raman vertex has the same symmetry as the crystal. Antiscreening, i.e. enhancement of scattering efficiency by screening can occur if the Raman vertex changes sign on the Fermi surface.

For tetragonal symmetry on a square lattice with size a, the scattering symmetries have the following angular dependencies: $\gamma_{A_{1g}} \approx \cos(k_x a) + \cos(k_y a)$, $\gamma_{B_{1g}} \approx \cos(k_x a) - \cos(k_y a)$, $\gamma_{B_{2g}} \approx \sin(k_x a)\sin(k_y a)$. Calculations for the case of a d-wave gap with $\Delta(\boldsymbol{k}) = \Delta_{\max}[\cos(k_x a) - \cos(k_y a)]$ show that the scattering intensity for the B_{1g} symmetry follows the symmetry of the d-wave gap, and the A_{1g} symmetries are the only one screened, with the exception of multisheets Fermi surfaces. In YBCO crystals the B_{1g} scattering symmetry is also screened, the only unscreened symmetry being B_{2g}, whereas in BiSCO the roles of B_{1g} and B_{2g} symmetries change, due to different orientations of crystallographic axes with respect to the Cu-O bonds. The dependencies on the Raman frequency shift ω for a d-wave symmetry in the tetragonal case, for the low-frequency region, are as follows: B_{1g} varies with ω^3, while A_{1g} and B_{2g} have a linear variation with ω. Impurity scattering or orthorhombic distortion modify the frequency dependence of B_{1g} modes, adding a term which varies linearly with frequency. Impurity scattering can, in principle, also decide whether the gap is of $d_{x^2-y^2}$ symmetry, or is an anisotropic s-wave gap. The effect of impurity scattering is a nonvanishing density of states at the Fermi energy for the d-wave pairing case, and a smearing of the gap anisotropy, proportional to impurity concentration, for anisotropic s gaps. Additionally, in the latter case an excitation-free region appears in the electronic Raman spectra below a Raman shift of $2\Delta_{\min}$ (Strohm and Cardona, 1996).

The predictions of this model seem to be in agreement with at least some experimental results, for example those of Gasparov et al. (1998) who performed polarized electronic Raman scattering measurements on $Tl_2Ba_2CuO_{6+\delta}$ optimally, and moderately and strongly doped HTS. The doping dependence of scattering is interesting, because the change in topology of the Fermi surface can be studied in this way, the different scattering components coming from different areas of the Fermi surface. For example, B_{1g} tests phonon dynamics around the Brillouin zone axes, B_{2g} probes the diagonals, and A_{1g} is a sort of average around the Brillouin zone. By observing the peak position of the B_{1g} component at $2\Delta_0$ it was found that the reduced gap $2\Delta_0 / k_B T$ decreases with increasing doping. The low-frequency behavior of B_{1g} and B_{2g} modes indicates a d-symmetry of the superconducting gap. The suppression of the A_{1g} component with respect to B_{1g} in some overdoped Tl-based HTS was attributed to a van Hove singularity. The main results were that $2\Delta_0 / k_B T \cong 7.8$ for optimally doped samples and about 3 for overdoped, indicating a strong non-BCS behavior in the first case. The intensities of the peaks were also strongly dependent on geometry: the A_{1g} peak was strong in optimally doped samples and weak in overdoped ones, the reverse behavior being registered for the B_{1g} peak. In overdoped samples the intensities of B_{1g} and B_{2g} peaks were comparable, whereas B_{1g} was stronger than B_{2g} and

A_{1g} peaks in optimally doped samples. The peak intensities were studied also as a function of temperature, the most sensitive to changes at T_c being the B_{1g} scattering.

The general picture is, however, not so rosy. Although the theoretical calculations can be performed quite easily, the experimental results are not always in agreement with them. Moreover, sometimes the results cannot be easily interpreted. For example, pure electronic Raman spectra bellow T_c for $HgBa_2Ca_3Cu_3O_{8+\delta}$ HTS crystals (Sacuto et al., 1998) show in the A_{1g} symmetry two peaks at $2\Delta \cong 6.4k_BT_c$ and $2\Delta \cong 9.4k_BT_c$, the highest one located at the energy of B_{1g} maximum. These peaks do not soften appreciably as the temperature is raised to T_c, and vanish at T_c. Low-frequency Raman scattering in the B_{1g} symmetry is approximately linear with ω for various excitation lines between 476.5–647.1 nm, indicating that the superconducting gap in this material cannot be of pure $d_{x^2-y^2}$ symmetry, but must be an anisotropic gap with two distinct maxima and nodes outside [110] and $[1\bar{1}0]$ directions in the $\boldsymbol{k}$-space.

Similar conclusions were driven from electronic Raman efficiencies for *a-b* plane polarization geometries in $YBa_2Cu_3O_{7-\delta}$ (Krantz, 1996). The obtained spectra look like those in Fig 8.10.

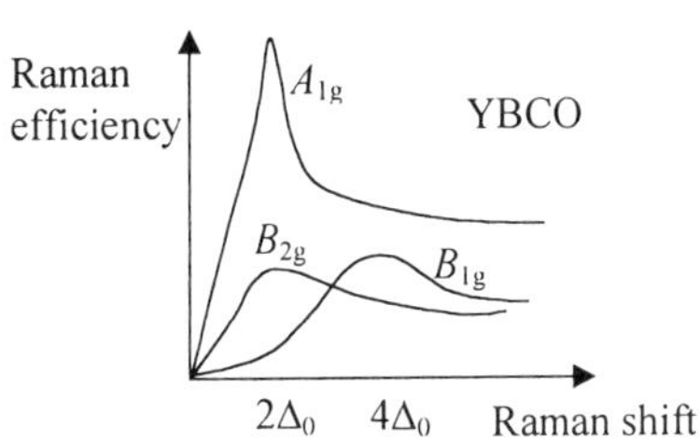

Fig. 8.10. Raman efficiency of A_{1g}, B_{1g} and B_{2g} components in YBCO below T_c at a wavelength of 0.48 μm

For the sample used in the experiments $4\Delta_0 \cong 650\ cm^{-1}$. The *a-b* plane polarized and depolarized spectra show a flat continuum above $1000\ cm^{-1}$ Raman shift due to a collision-limited response of the normal state. Depending on polarization, a gap opens below 200–500 cm^{-1}, the spectral weight being redistributed to pair-breaking peaks at higher frequencies. The pair-breaking peak occurs at 650 cm^{-1} for the B_{1g} polarization and at half of this value for any other symmetry. The absolute electronic Raman efficiencies were compared with calculations performed for both superconducting and normal phases based on a BCS-like theory in the LDA approach, which includes the attractive pairing potential, the repulsive Coulomb potential and assumed a d-wave superconducting gap symmetry independent of the band index (Strohm and Cardona, 1996; Krantz, 1996). The calculations were not able to predict the relative positions of A_{1g} and B_{1g} peaks, which should be situated slightly below $2\Delta_0$ and at $2\Delta_0$, respectively. In the experimental curves the B_{1g} peak is situated at about twice the frequency of

A_{1g} peak. To explain this discrepancy, different parity, d- and s-like gaps were attributed to the bonding and antibonding sheets of the Fermi surface, which did not include the chain component. Reasonable fit with experimental data has been obtained in this way, although the procedure itself is not accurate, since the Fermi surface cannot be broken into bonding and antibonding planes and chain components due to the fact that the sheets are interconnected at certain points of $\boldsymbol{k}$-space (Strohm and Cardona, 1996).

Another approach to reconcile the experiments with theoretical calculations is to replace the point-like impurity potential with extended potentials, which account for electron correlation in both elastic and inelastic scattering potentials. The introduction of this correlation can change the nature of impurity interaction from static to dynamic. Numerical calculations for low temperatures, where the effects of impurity dominate, show not only a remarkable agreement with experiments, but also explain the large discrepancies in the estimation of the magnitude of the scattering rate obtained from different methods: FIR, electronic Raman, and other characterization methods (Devereaux and Kampf, 2000). The root of the problem lies in the fact that IR reflection or conductivity measurements, and Raman spectroscopy gather information from different regions in the Brillouin zone, which have different sensitivities to electron correlation. Calculations for a d-wave superconducting gap symmetry suggest that IR measurements pick up regions governed by small scattering along the Brillouin zone diagonals, and have a weaker dependence on the correlation parameter compared to Raman scattering. IR spectra have a similar behavior as the B_{2g} response; in Raman scattering however, responses along other directions in the Brillouin zones can be gathered in the remaining A_{1g} and B_{1g} channels.

8.2.4 Non-BCS Behavior

When speaking of non-BCS behavior in HTS we must distinguish between cases in which there are contradictory optical response, and cases in which the optical response is in clear contradiction with the BCS theory.

In the first category enter HTS materials for which some characteristics, for example the value of the superconducting gap is in agreement with BCS theory, but other characteristics are not. One such example is $Nd_{1.85}Ce_{0.15}CuO_{4-\delta}$, material which has peculiar properties: (i) the absence of Cu-O chains, (ii) the in-plane normal state dc resistivity depends quadratically on temperature in contrast with other HTS where this dependence is linear, (iii) both electrons and holes participate at the transport in the Cu-O plane. Measurements of FIR transmission in thin films of this material have shown an energy-gap-like peak for the ratio between the transmission in the superconducting and normal states at $E_p \cong 3k_BT_c$, almost like in the BCS theory. This peak also reduces its height on increasing temperature or increasing magnetic fields. However, the agreement stops here, because measurements also show that the energy-gap-like peak does not gradually shift to lower frequency as the temperature is reduced, but shows no shift with either temperature or magnetic field. The peak energy is also not affected by oxygen depletion, which reduces T_c from 19 K to 12 K (Choi et al., 1996).

In this category of materials for which contradictory results are obtained enter also the recently discovered intermetallic polycrystalline compound MgB_2. The FIR reflectivity $R(T)/R(T_c)$ in the range 0–100 cm^{-1} (Gorshunov et al., 2001) increases dramatically for this material (about 100%) below 40 K, thus demonstrating a BCS-like behavior and a similarity with HTS cuprates where the opening of the gap Δ in the density of states at the frequency $2\Delta/\hbar$ implies a decrease of the conductivity at this frequency. As a result the reflectivity increases dramatically below $2\Delta/\hbar$, the obtained experimental values for the bandgap being $2\Delta(0)/hc \cong 70\ cm^{-1}$, $2\Delta(0) \cong 9$ meV and $2\Delta(0)/k_B T_c = 2.6$. Other experiments have found that $2\Delta(0)/k_B T_c = 2.5$–5, no agreement existing up to now even regarding the gap width. Submillimeter investigations in the range 4–30 cm^{-1} on MgB_2 films have found that $2\Delta(0)/k_B T_c = 3.53$ (Pronin et al., 2001). The real part of the optical conductivity dependence on temperature shows a peak below T_c indicating a BCS-like behavior while the imaginary part of the optical conductivity and thus the penetration depth shows a huge discrepancy with BCS theory. Infrared absorption measurements in the range 125–700 cm^{-1} and in the mid-infrared up to 2000 cm^{-1} (Sundar et al., 2001) have identified a peak at 485 cm^{-1} corresponding to the maximum phonon density of states, density of states which has two shoulders at 333 cm^{-1} and 387 cm^{-1}, as predicted by theory. The width of these IR modes, about 40 cm^{-1}, is much smaller than the width of the Raman modes, indicating strong electron-phonon coupling. However, the temperature behavior of these IR modes and their softening below 100 K indicate again a non-BCS character. So, at the moment when these considerations are written (at two months after the discovery of MgB_2 superconducting behavior) the superconducting physical origin of MgB_2 is an open problem despite the structural simplicity of this intermetallic compound.

On the other hand, for $La_{2-x}Sr_xCuO_4$ the c axis reflectivity has a non-BCS behavior, some of its interesting features being not fully understood. For example:

i) In the normal state, for $T > T_c$, the c-axis reflectivity is structureless down to 20 cm^{-1}, as in an ionic insulator.

ii) In the superconducting state the reflection shows a sharp plasma edge at a frequency situated below the superconducting energy gap for which $\mathrm{Re}[\varepsilon(\omega)]|_{\omega=\omega_p} = 0$, a behavior characteristic of classic metals.

iii) The very low c-axis plasma-edge frequency determined from reflectivity measurements corresponds to a carrier mass much larger than the c-axis optical mass computed with LDA.

iv) The plasma-edge frequency in the superconducting state is strongly dependent on temperature and doping, shifting to lower frequencies as T approaches T_c from below.

These properties are not expected from BCS-like theories. The ionic insulator behavior in the normal state (property 1) can be qualitatively understood supposing that the carriers are confined to CuO_2 planes by a strong electronic correlation which decouples the spin and charge degrees of freedom (Anderson, 1997). As the material is cooled, its kinetic energy reduces to the point where it overcomes confinement and drives the superconducting transition associated with condensation of Cooper pairs. The coherent transport in the superconducting state is restored by Josephson tunneling of Cooper pairs, the plasma edge appearing in the superconducting state as a consequence of collective excitations of charge pairs. Another explanation of the *c*-axis reflectivity behavior is based on the assumption that the material is a metal with extremely anisotropic mass and scattering rates for free carriers, the peculiar reflectivity spectrum in the superconducting state being caused by the temperature dependence of both density of superconducting pairs and scattering rates of normal-state carriers (Tamasaku et al., 1992).

To mathematically model this behavior it is essential to introduce a finite-temperature formalism of level-broadening effects in microscopic calculations. The absence of a metallic plasma edge for $T > T_c$ would then be caused by the overdamping of the plasmon in the normal state due to strong disorder scattering, whereas the plasma edge in the superconducting edge, corresponding to a *c*-axis plasma mode, would be preserved since single-particle level broadening does not affect the condensed carriers. In a simple tight-binding model which considers a single band of width $2t_c$ along the *c* axis and a single-particle dispersion relation: $E(k,k_z) = \hbar^2 k^2 / 2m - t_c \cos(k_z d)$ where d is the *c*-axis layer-separation distance, k and m are the wavevector and planar effective mass in the *a-b* plane, respectively, and k_z is the wavevector along the *c* axis, the *c*-axis reflectivity can be calculated assuming a dielectric function expressed as a two-fluid model (Das Sarma and Hwang, 1998):

$$\varepsilon = 1 - \omega_{pn}^2 / (\omega^2 + i\gamma\omega) - \omega_{ps}^2 / \omega^2, \tag{8.21}$$

where ω_{pn} and ω_{ps} are the plasma frequencies for the normal and the superconducting state. No plasma edge appears in the normal state if the level broadening parameter $\gamma \geq \Delta$. Typically, $\omega_{ps} \cong 2t_c - 4t_c$ and $t_c \cong 0.1\Delta - 0.3\Delta$. The hopping amplitude t_c and γ can be determined from normal state measurements of in-plane and *c*-axis dc resistivities as:

$$t_c = (\pi n_{2D} \rho_{ab} / 2d^2 m^2 \rho_c)^{1/2}, \quad \gamma = c^2 n_{2D} d t_c \rho_c / 2. \tag{8.22}$$

Measuring the plasma frequency in the superconducting state we can then determine the penetration depth along the *c* axis and the gap as:

$$\lambda_L = c / \omega_{ps} = \hbar c^2 \rho_c / 4\pi^2 \Delta. \tag{8.23}$$

The behavior of reflectivity calculated from this expression of the dielectric function is in extremely good agreement with the *c*-axis (out-of-plane) polarized reflectivity experiments performed by Uchida et al. (1996) on LSCO

($La_{2-x}Si_xCuO_4$) in both the normal and superconducting states. On increasing the doping it was found that the spectral weight in the out-of-plane spectrum (as in the in-plane spectrum) is transferred from high- to low-energy regions. The transferred weight forms a band at relatively high energies (about 2 eV) whose position is almost unchanged with increasing x. In underdoped compounds ($x < 0.13$), the low-energy c-axis optical conductivity (for $E < 0.3$ eV) is too small to form a Drude peak and is further suppressed when temperature decreases, whereas in overdoped compounds ($x > 0.13$) the c-axis conductivity in this spectral region increases. The Drude peak appears for $x \geq 0.18$. In the superconducting state, a sharp plasma edge is observed; for underdoped compounds, the plasma edge appears within a gap region identified with the Josephson plasma in weakly Josephson-coupled layered superconductors. In overdoped compounds, in the superconducting regime, the Drude-like component is observed even at temperatures below T_c, indicating a damping of the Josephson plasma.

The theoretical description presented above, which assumes the coexistence of paired and unpaired carriers below T_c was also confirmed by c-polarized FIR absorption measurements below the plasma edge at 2 K in $La_{1.87}Sr_{0.13}CuO_4$ (the T_c for this material is 31 K). On fitting the experimental curve with the formula above, it was found that the unscreened plasma frequency in the superconducting state is 230 cm^{-1}; the absorption has been found to decrease below 40 cm^{-1} (the screened Josephson plasma frequency), the decrease being interrupted by an absorption peak near 10 cm^{-1} (see Fig. 8.11).

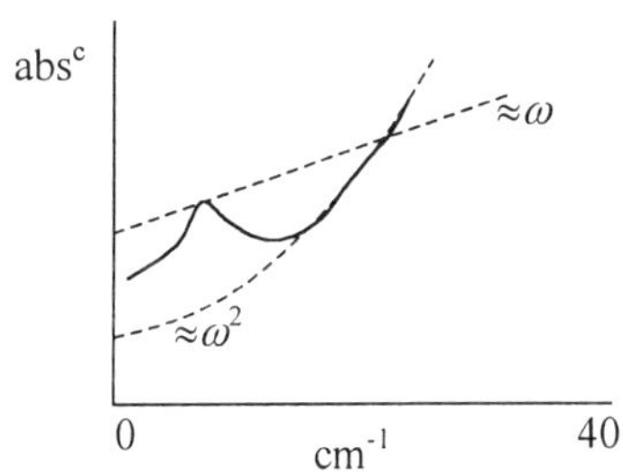

Fig. 8.11. c-polarized FIR absorption in $La_{1.87}Sr_{0.13}CuO_4$

For lower frequencies the absorption coefficient decreases again, but it retains nonzero values down to the lowest measured frequencies. Below 35 cm^{-1} the absorption has an ω^2 behavior, suggesting that a fraction of carriers remains unpaired even below 2 K, this fraction being described as a normal Drude liquid with a finite scattering rate. The existence of this unpaired fraction could be due to pair breaking, which would imply a strongly asymmetric superconducting gap. The peak at 10 cm^{-1} was assigned to an optically active longitudinal plasmon (Birmingham et al., 1999).

8.2.5 BCS-Like Behavior

Several improvements of the BCS theory have been introduced to explain the behavior of some HTS. For example, the finite scattering rate $1/\tau = 2\gamma/\hbar$ of carriers has been incorporated in the theory, the limit $1/\tau >> 2\Delta$ corresponding to the so-called extreme dirty regime, described by the Mattis–Bardeen theory (Mattis and Bardeen, 1958). An example of a HTS material that has a BCS-like behavior is $BaPb_{1-x}Bi_xO_3$, which for $x = 0.25$ shows a BCS-like peak with $2\Delta(0)/k_BT_c = 3.2$ in the FIR transmission in the superconducting state near 21 cm^{-1}. This kind of cubic perovskite HTS has a 3D structure, with a high ratio of T_c to the density of states at the Fermi level, a significant spectral weight in mid-IR and collective states in the insulating parent materials, properties shared also by the 2D HTS. From the experimental data recorded up to 13 Δ a scattering rate of carriers of $1/\tau = 280\ cm^{-1} \cong 14 \times 2\Delta$ has been obtained. This value corresponds to the moderate dirty regime. As the temperature is increased towards T_c the peak reduces intensity and moves to lower frequencies (Jung et al., 1999).

8.2.6 HTS Dynamics

The dynamics of low-frequency phonons in $YBa_2Cu_3O_{7-\delta}$ can be tested using combined time- and frequency-domain spectroscopies (Misochko et al., 2000). Pump-probe optical reflectivity measurements on the fs scale, with pump and probe pulses perpendicularly polarized, and both close to normal incidence, have detected oscillations of low-frequency A_{1g} metal-ion modes in both normal and superconducting states. These oscillations, due to coherent phonons, have an amplitude linearly dependent (above a threshold) on the pump intensity, and are superimposed on a slowly varying background of electronic origin. In the superconducting state the sign of reflectivity change is reversed, the oscillatory part approaching the zero line from above, while in the normal state it approaches this value from below. This change of sign is caused by a different origin of oscillations: in the normal state the amplitude of oscillations is dominated by the high-frequency Cu-mode, whereas in the superconducting state the Fourier transform spectrum of the oscillations is dominated by the low-frequency Ba mode. Moreover, the oscillation decay in the superconducting state is much slower than in the normal state. The time-resolved reflectivity change can be modeled as a sum of two damped and phase shifted oscillators corresponding to the Ba and Cu modes. The ratio between these modes depends on temperature, the intensity of the low-frequency mode starting to grow below the transition temperature. A huge lowering of about 90% of this ratio occurs at the transition from the superconducting to the normal state. No only the amplitude, but also the frequency and initial phase of the coherent vibrations are temperature dependent: the frequency hardens and the dephasing time increases with decreasing temperature. The initial phase shift of the Cu mode is $\pi/4$, the Ba mode being shifted with $\pi/2$. The drastic changes in the time-domain reflectivity at the transition temperature are similar to the changes observed in Raman spectroscopy, but are quantitatively different. One major difference is the polarization dependence of the superconductivity-induced changes in the Raman spectrum, not observed in the pump-probe experiments, where an uncontrolled mixture of polarizations is measured. Another difference is the strong temperature dependence of the linewidth in Raman spectroscopy, caused by anharmonic decay, no such temperature or pump-power dependence being observed in the coherent amplitude decay. The ratio of intensities of the Ba and Cu modes is always smaller for time-domain data, at least in the normal state, all these differences confirming the fact that coherent lattice dynamics is different from the thermal state dynamics probed in Raman spectroscopy.

Ultrafast techniques combined with Raman spectroscopy are able to probe the electronic structure in HTS; such studies have been done for $YBa_2Cu_3O_{7-\delta}$ in insulating (δ = 0.8) and metallic (δ = 0.1) phases (Mertelj et al., 1997). Energy relaxation rates to phonons, obtained by measuring the nonequilibrium phonon relaxation number n_{neq}, can distinguish between relaxation rates via localized states, which are strongly temperature dependent, and temperature-independent relaxation via extended band states. To measure the nonequilibrium phonon relaxation number, pulsed Raman Stokes and anti-Stokes measurements were performed with 1.5 and 70 ps long pulses. Their intensity ratio was used to determine the phonon temperature in nonequilibrium conditions: $T_{\text{p}}(T) = \hbar\omega_{\text{p}} / \{k_{\text{B}} \ln[K I_{\text{S}}(T) / I_{\text{AS}}(T)]\}$, where K is a calibration constant, ω_{ph} the phonon frequency, and I_{S}, I_{AS} are the Stokes and anti-Stokes peak intensities measured at different temperatures in the resonant Raman backscattering geometry. The nonequilibrium phonon occupation number is then $n_{\text{neq}}(T) = n_0 \exp[-E_{\text{a}} / k_{\text{B}} T_{\text{p}}(T)]$, the value of activation energy obtained from fitting being 56–210 meV for the metallic phase and 60–110 meV for the insulator phase. The observed temperature dependence of n_{neq} suggests that at least part of the photoexcited carriers relax via hopping between localized states, mainly coupled to out-of plane vibrations. As the temperature decreases, the energy relaxation rate of the photoexcited carriers decreases also. From studies of the dependence of n_{neq} on the carrier lifetime, obtained from a coupled rate equation for the 1.5 and 70 ps laser pulses, it was inferred that photoexcited carriers relax by emitting low-momentum optical phonons near the center of the Brillouin zone.

8.2.7 Pseudogap

Studies in the normal phase of underdoped cuprate HTS have revealed anomalies in the optical conductivity, which were related to the opening of a pseudogap in the electronic spectrum. In angle-resolved photoemission experiments the pseudogap is evidenced through the anisotropy of the electron spectral density, the maximum value of the pseudogap being observed in the vicinity of the $(\pi,0)$ point in the Brillouin zone, where the Fermi surface is completely destroyed. On the contrary, along the diagonals of the Brillouin zone the pseudogap is absent, the

Fermi surface being conserved. The symmetry of the pseudogap is thus of the same d-wave type as the superconducting gap, the pseudogap being present in the normal state up to a temperature T^* much higher than T_c.

In Raman spectra of $YBa_2Cu_3O_{7-x}$for example, the pseudogap is observed as a decrease in the intensity of a broad feature between 200–700 cm^{-1}, of non-phononic origin. The temperature evolution of the integrated intensity of this electronic background broad peak is shown in Fig. 8.12. The Raman spectrum is recorded with an IR excitation, so that its intensity is determined mainly by the temperature dependence of the electron absorption.

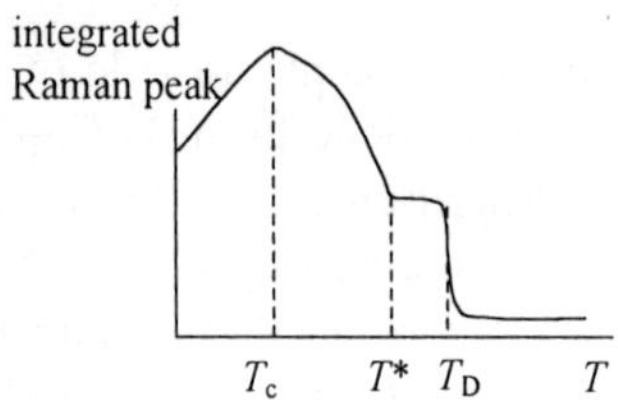

Fig. 8.12. Temperature dependence of the integrated Raman intensity in $YBa_2Cu_3O_7$

At T_c the excitation of the electronic background is coupled to the superconducting order parameter, the integrated intensity of the broad peak being maximum due to electron transitions. The change at T^* is a structural change, the Cu sites being frozen at temperatures smaller than T^*. At T_D a pseudogap opens in the fermionic density of states, the value of T_D decreasing with increasing T_c. The change in Raman spectrum with temperature is due to the change in oscillator strength, described by the Bose–Einstein condensation of bipolarons (Ruani and Ricci, 1997).

In FIR spectroscopy of underdoped $Pb_2Sr_2(Y/Ca)Cu_3O_8$, for example, the pseudogap occurs as a depression in the optical conductivity (Reedyk et al., 1997). For this HTS, with $T_c = 65$ K, the depression in the background c-axis optical conductivity which exists at T_c, maintains its value up to $T^* = 150$ K, where the pseudogap is fully formed, decreasing then with increasing temperature up to 300 K. An anomalously narrow Drude-like peak (a drop in the effective scattering rate) forms in the low-frequency region, accompanied by a rather weak absorption through the pseudogap at higher frequencies. The evolution of the conductivity for the case of $Nd_{1.88}Co_{0.12}CuO_4$ (NCCO) is reproduced schematically in Fig. 8.13; a depression develops in $\sigma(\omega)$ around 1000 cm^{-1} at low temperatures (Lupi et al. 1999). The anomalous Drude peak, associated with the pseudogap in the spectra of metallic cuprates, is described by $\sigma(\omega) = (\omega_D^2/4\pi)/[\Gamma^*(\omega) + (m^*(\omega)/m)^2\omega^2/\Gamma^*(\omega)]$ where ω_D is the constant plasma frequency, the scattering rate Γ^* and the effective mass m^* being functions of ω. This model explains the anomalous behavior in mid-IR $\sigma(\omega) \approx \omega^{-1}$, instead of $\approx \omega^{-2}$ as in conventional metals. Typical high values of the scattering rate in far-IR

$\Gamma^* = 1500\ \mathrm{cm}^{-1}$ are obtained near T^*, lower values of 500–1000 cm^{-1} being observed below T^*.

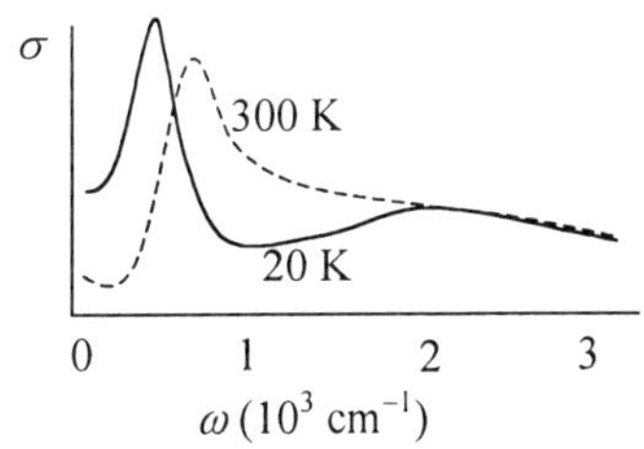

Fig. 8.13. The conductivity spectrum dependence on temperature for $Nd_{1.88}Co_{0.12}CuO_4$

The origin of the pseudogap is unclear. Sadovskii (1999) suggested that it might originate from the fluctuations of antiferromagnetic short-range order. Lupi et al. (2000) suggested that the depression in the optical conductivity is due to the transfer of spectral weight to the FIR peak as this softens and narrows, as requested by ordinary sum rules. The pseudogap would thus not be a real gap, as also suggested by the fact that the value of the pseudogap determined optically is much larger than that from angle-resolved photoemission. The Drude peak would then be caused by the charges bound to the lattice via a polaronic coupling. This last hypothesis was tested by experiments performed on $Bi_2Sr_2CuO_8$ (BSCO) from 400 to 8 K, between 15000 and 10 cm^{-1}. The results suggest that the behavior of $\sigma(\omega)$ cannot be explained by an anomalous Drude behavior alone, and that the pseudogap does not open in the continuum of states of some anomalous free carriers. The suggested explanation implies the coexistence of two types of carriers, the optical pseudogap being created by the temperature-dependent absorption of some weakly bound-charges. This explanation is consistent with the Emery and Kivelson (1993) hypothesis of phase separation. One can imagine that the free carriers are scattered by arrays of bound charges carrying local dipoles. The coexisting of free and weakly bound charges of polaronic nature produces a normal Drude absorption and a broad FIR peak, respectively. The Drude peak disappears below T_c (= 20 K), the anomalous frequency dependence of the optical conductivity being accounted for by the tail of the FIR peak, which has a high-frequency side $\approx \omega^{-1}$ at any temperature above T_c. The pseudogap in mid-IR opens due to the redshift and narrowing of the FIR peak, which produces transfer of spectral weight towards low frequencies. The loss of spectral weight below T_c provides an estimation of the London penetration depth

$$\lambda_L^2 = (c^2/8) / \int_{\omega_{\min}}^{\omega_{\max}} [\sigma_n(\omega, 30\,\mathrm{K}) - \sigma_s(\omega, 8\,\mathrm{K})]\mathrm{d}\omega. \tag{8.24}$$

For $\omega_{\min} = 10\ \mathrm{cm}^{-1}$, $\omega_{\max} = 15\,000\ \mathrm{cm}^{-1}$, $\lambda_L = 300$ nm, in excellent agreement with transport measurements.

Another explanation of the optical conductivity behavior in HTS, supported also by angle-resolved photoemission measurements, suggests a strong anisotropy of quasiparticles as a function of the position around the Fermi surface. This implies a strong $k_{\|}$ dependence of both the scattering rate and c-axis hopping parameter. The strongest scattering would occur at the Fermi surface region closest to the saddle points, the weakest occurring along the zone diagonals. This hypothesis leads to the following frequency dependence of the planar and out-of-plane conductivities

$$\sigma_{xx}(\omega) = \sigma_{xx,0} /(1 - \mathrm{i}\omega\tau)^{1/2}(1 + \Gamma\tau - \mathrm{i}\omega\tau)^{1/2}, \tag{8.25}$$

and

$$\sigma_{zz}(\omega) = \sigma_{zz,0} /\{1 + 2\Omega^2[(\Omega^2 /1 + \Omega^2)^{1/2} - 1]\}, \tag{8.26}$$

respectively, with $1/\tau$ the scattering rate and $\Omega = (1 - \mathrm{i}\omega\tau)^{1/2} /(\Gamma\tau)^{1/2}$, the parameter Γ being obtained from fitting (van der Marel, 1999). Also, it seems that the general formula

$$\sigma_{\mathrm{A}}(\omega) = (\omega_{\mathrm{p}}^2 / 4\pi)\mathrm{i} /[(\omega + \mathrm{i}/\tau)^{1-2\alpha}(\omega + \mathrm{i}\Gamma)^{2\alpha}], \tag{8.27}$$

can be used to analyze the optical conductivity of any non-Fermi liquid, with suitably chosen parameters α and Γ. The low-frequency limit of this formula corresponds to a narrow Drude peak with effective carrier lifetime $\tau^*(0) = (1 - 2\alpha)\tau + 2\alpha / \Gamma$ (van der Marel, 1999).

It is interesting to note that for $\alpha = 0.25$ the formula above corresponds to the predictions of Ioffe and Millis (1998) based on quantum phase fluctuations in the superconducting ground state. The strength of the quantum fluctuations can be experimentally determined from the ratio of the spectral weight in the c-axis superfluid response to the spectral weight lost from the c-axis conductivity as the temperature is decreases from high T to $T = 0$. Quantum fluctuations are important if this ratio is greater than 1. HTS would then be caused by the weak coupling of layers, the pseudogap corresponding to the onset of long-range phase coherence rather than to opening of a gap.

Information about the symmetry of the superconducting gap and of the pseudogap, and about the nonequilibrium quasiparticle recombination can be obtained from ultrafast time domain spectroscopy (Mihailovic and Demsar, 1999). In this method, different overlapping spectral contributions in the low-frequency FIR or Raman electronic excitation spectra can be separated and identified due to their different lifetimes and different behaviors with temperature and doping. The measurements to be carried out are pump-probe measurements of the transient change in the transmission through, or reflection from a HTS, at different temperatures or doping parameters. A typical signal for a near optimally doped $YBa_2Cu_3O_{7-\delta}$ in the superconducting state consists of a slow component modeled by a Heaviside step function, and a fast component described by a single exponential decay with characteristic relaxation time between 0.5 and 3 ps depending on temperature and doping (see Fig. 8.14).

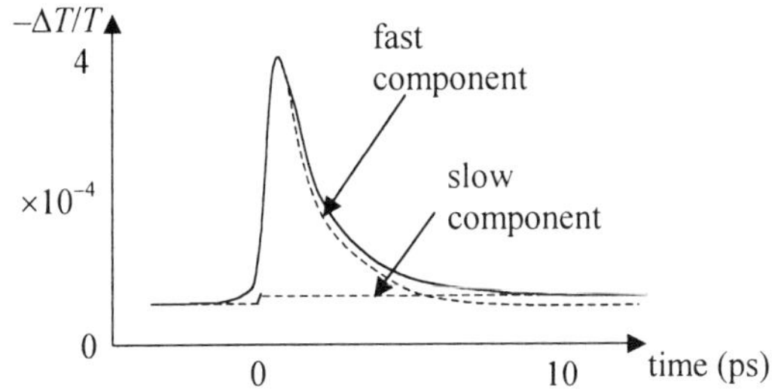

Fig. 8.14. Time variation of pump-probe transmission of near optimally doped $YBa_2Cu_3O_{7-\delta}$ in the superconducting state

Both the amplitude of the signal and the decay lifetime are temperature and doping dependent, the amplitude increasing with decreasing temperature in all cases, while the decay time has different behaviors in near optimally doped ($\delta = 0.08$) and underdoped ($\delta = 0.44$) samples: the relaxation time associated to the exponential decay part has a weak temperature dependence for underdoped samples, whereas in near optimally doped samples it is longer below T_c and has a divergence below T_c. The relaxation time behavior in the two situations is presented in Fig. 8.15. The carrier recombination time is 1–2 orders of magnitude longer than in metals since the opening of the superconducting gap creates a relaxation bottleneck, the carriers accumulating in quasiparticles states above the gap before recombining into pairs. In this case the quasiparticles and high-frequency phonons form a nearly steady-state distribution, the relaxation time being governed by the anharmonic phonon decay. The measured signal is thus an indication of the temporal change in the density of the photoexcited carriers.

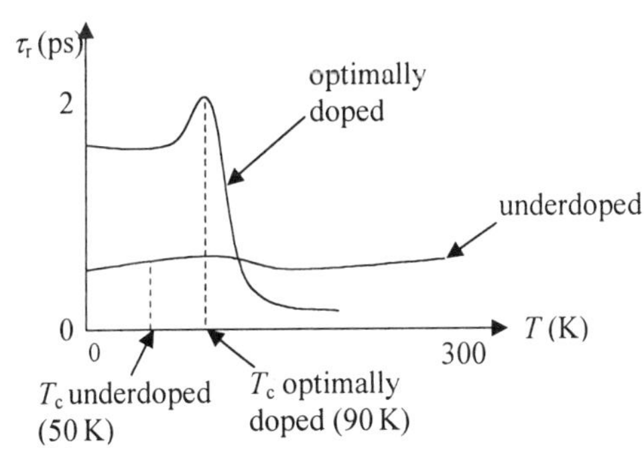

Fig. 8.15. Relaxation time behavior in doped and underdoped YBCO films as a function of temperature

The theoretical models for the temperature dependence of the photoinduced $\Delta T/T$ depend on the temperature dependence of the gap (Kabanov et al., 1999). For a temperature-independent gap

$$|\Delta T/T| \cong (-1/\Delta_0)[1+2\nu\exp(-\Delta_0/k_BT)/N_F\hbar\Omega]^{-1}, \quad (8.28)$$

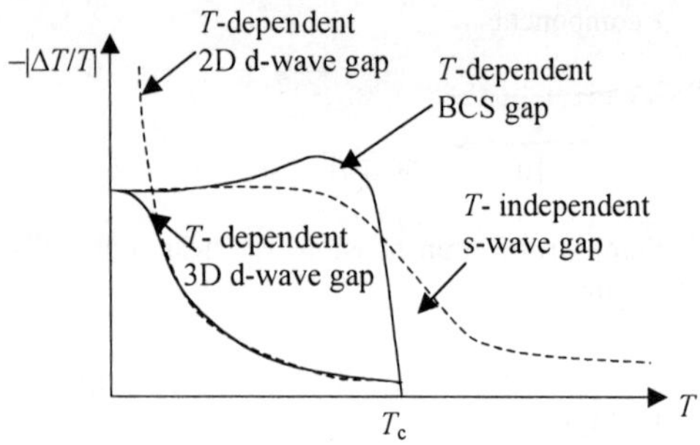

Fig. 8.16. $\Delta T/T$ dependence on temperature for different types of superconducting gaps

where Δ_0 is the energy gap, ν the effective number of phonons/unit cell involved in the recombination processes of quasiparticles, with phonon frequency Ω, and N_F is the density of states at the Fermi level. If the gap is temperature dependent as in the BCS theory, and closes at T_c, the relaxation bottleneck vanishes at the transition temperature and the above formula must be replaced with:

$$|\Delta T/T| \cong (-1/(\Delta(T)+k_BT))[1+2\nu(2k_BT/\pi\Delta(T))^{1/2}\exp(-\Delta(T)/k_BT)/N_F\hbar\Omega]^{-1}, \quad (8.29)$$

The signal $-|\Delta T/T|$ in the two situations is displayed in Fig. 8.16 for different cases. The two situations, of a temperature-independent and temperature-dependent gap can be easily distinguished experimentally. In the first case the transient transmission change persists at much higher temperatures than the transition temperature, while in the second case it vanishes in the normal state.

Comparing the predictions of this model with experimental data for YBCO, it was found that the near optimally doped films follow the $\Delta T/T$ temperature dependence of a temperature dependent BCS-like gap with s-symmetry, while the underdoped films show a behavior characteristic of a non-BCS temperature-independent gap. A pseudogap can be defined in this case as the temperature at which the amplitude of the signal falls to 10% of the maximum value. By modifying the doping of $YBa_2Cu_3O_{7-x}$ samples, it was found that the amplitude of the transmission signal, $|\Delta T/T|$, increases linearly with the hole concentration in the CuO_2 planes $n_h \approx (1-x)$, which suggest that the gap is inversely proportional to the doping $\Delta \approx 1/n_h$. There is no concentration threshold that separates the doping intervals with T-dependent and T-independent gaps, presumably due to inhomogeneities in the hole distribution which allow the coexistence of regions with collective BCS-gaps and local T-independent gaps.

Although the pseudogap is usually associated with underdoped HTS, claims were made for its observation in overdoped samples as well. Since the pseudogap has the same symmetry as the superconducting gap, the depression of spectral weight above T_c was related by Naeini et al. (1998) to precursor pairing. A possible cause is the competition in a nearly antiferromagnetic Fermi liquid (NAFL) between the tendency to order antiferromagnetically and the spin-fluctuation-mediated pairing mechanism for quasiparticles. The antiferromagnetic correlation length ξ is temperature and doping dependent, and so it is possible to define a critical temperature T_{cr} which separates a mean field region $(T > T_{cr})$ from a magnetic pseudoscaling region $(T^* < T \leq T_{cr})$, in addition to the temperature T^* for which a crossover to the pseudogap regime with constant ξ occurs. Both these temperatures decrease with increasing doping, as observed in electronic Raman scattering in overdoped $La_{1.8}Sr_{0.22}CuO_4$. In these measurements the temperature dependence of the low-energy scattering rate for B_{1g} and B_{2g} modes was recorded, and a crossover temperature $T_{cr} \cong 160\,K$ was calculated from experimental data. Since no evidence for a crossover in the pseudogap regime was obtained, it was inferred that $T^* \leq T_c$. The scattering rates $\Gamma(T)$ were determined from the inverse slope of the low-energy Raman response. $\Gamma(T)$ varies almost linearly in the two scattering regimes, but with different slopes, at T_{cr} the correlation length and spin fluctuation energy having a smooth crossover from the values in the pseudoscalling regime to the values in the mean field theory.

The strong influence of antiferromagnetism in HTS has been also confirmed by the hypothesis of spin pseudogap opening above T_c, based on the anomalous temperature dependence of bulk magnetic susceptibility above T_c. For example, in $YBa_2Cu_4O_8$ a crossover temperature of $T_s = 200$ K, was observed, for which there is a change of slope of the linear dependence of the magnetic susceptibility with temperature. Extrapolating for $T = 0$ the straight line obtained for $T_c < T < T_s$ one gets a negative susceptibility value, which can only be accommodated by the introduction of a gap for spin excitations. Such a gap should involve pairing instability of spin-1/2 fermions from the particle-hole continuum, and should be a non-superconducting pairing, in that it should not produce a Meissner effect (Altshuler et al., 1996). So, the mechanism of spin-gap formation would be pairing instability of a Fermi sea of chargeless fermions (spin liquid) in the presence of antiferromagnetic correlations. Experimental evidence for strong temperature-dependent antiferromagnetic correlations in HTS is reviewed in Altshuler et al. (1996). In particular in bilayer HTS, the large inelastic scattering in models of spin liquids leads to strong pair-breaking which can only be overcome by a singular pairing interaction due to coupling between planes, enhanced by in-plane antiferromagnetic correlations. Calculations for two models of spin liquids have shown that T_s scales as a power of the coupling constant between planes, the gap function being sharply peaked at particular regions of the Fermi line. In these models T_s is not a true thermodynamic transition, but only a crossover temperature, the superconducting transition temperature T_c corresponding to a lower temperature at which the charged bosons condense.

The strong interaction between superconductivity and remaining antiferromagnetism was also made responsible for the rearrangement of spectral weight of B_{1g} and A_{1g} Raman peaks below T_c for $Y_{1-x}Pr_xBa_2Cu_3O_7$, at energies about 5–6 times 2Δ (Rübhausen et al., 1997). This redistribution of spectral weight shifts to lower energies with increased doping, a strong renormalization of electronic and phononic structures being observed for optimal doped samples. This phenomenon was attributed to the strong interplay between holes and magnons in the normal and superconducting state, the two-magnon peak specific to the antiferromagnetic parent of the HTS persisting in optimally doped samples, even if it should disappear completely with doping.

References

Anderson, P.W. (1991) Phys. Rev. Lett. **67**, 3844.

Anderson, P.W. (1997) *Theory of Superconductivity in the High-T_c Cuprates*, Princeton Univ. Press, Princeton.

Altshuler, B.L., L.B. Ioffe, and A.J. Millis (1996) Phys. Rev. B **53**, 415.

Birmingham, J.T., S.M. Grannan, P.L. Richards, J. Kircher, M. Cardona, and A. Wittlin (1999) Phys. Rev. B **59**, 647.

Bornemann, H.J. and D.E. Morris (1991) Phys. Rev. B **44**, 5322.

Callaway, J. (1991) *Quantum Theory of the Solid State*, Academic Press, Boston, 2nd edition.

Calvani, P. (1996) in Proc. 1994 GNSM School, *High Temperature Superconductivity. Models and Measurements*, p.233, ed. M. Acquarone, World Scientific, Singapore.

Cardona, M. (1998) in *Ultrafast Dynamics of Quantum Systems; Physical Processes and Spectroscopic Techniques*, NATO ASI Series B 372, ed. B. di Bartolo, Plenum Press, New York.

Choi, E.J., K.P. Steward, S.K. Kaplan, H.D. Drew, S.N. Mao, and T. Venkatessan (1996) Phys. Rev. B **53**, 8859.

Das Sarma, S. and E.W. Hwang (1998) eprint: cond-mat 9802089.

Devereaux, T.P. and A.P. Kampf (2000) Phys. Rev. B **61**, 1490.

Eroles, J., C.D. Batista, S.B. Bacci, and E.R. Gagliano (1999) Phys. Rev. B **59**, 1468.

Emery, V.J. and S.A. Kivelson (1993) Phys. Rev. Lett. **71**, 3701.

Emery, V.J. and S.A. Kivelson (1995) Nature **374**, 434.

Fleury, P.A. and R. London (1968) Phys. Rev. **166**, 514.

Gasparov, L.V., P. Lemmens, N.N. Kolesnikov, and G. Güntherodt (1998) Phys. Rev. B **58,** 11753.

Goll, G., M. Heinecke, A.G.M. Jansen, W. Joss, L. Nguyer, E. Steep, K. Wintzer, and P. Wyder (1996) Phys. Rev. B **53**, 8871.

Gorsunov, B., A.C. Kuntscher, P. Haas, M. Dressel, F.P. Mena, A.B. Kuz'menko, D. van Marel, T. Muranaka, J. Akimitsu (2001) e-print: cond-mat/0103164.

Hadjiev, V.G., X. Zhou, T. Strohm, M. Cardona, Q.M. Lin, and C.W. Chu (1998) Phys. Rev. B **58**, 1043.

Heyen, E.T., M. Cardona, J. Karpinski, E. Kaldis, and S. Rusiecki (1991) Phys. Rev. B **43**, 12958.

Holcomb, M.J., C.L. Perry, J.P. Collman, and W.A. Little (1996) Phys. Rev. B **53**, 6734.
Ioffe, L.B. and A.J. Millis (1998) Phys. Rev. B **58**, 11631.
Ioffe, L.B. and A.J. Millis (1999) Science **285**, 1241.
Jung, U., J.H. Kong, B.H. Park, T.W. Noh, and E.J. Choi (1999) Phys. Rev. B **59**, 8869.
Kabanov, V.V., J. Demsar, B. Podobnik, and D. Mihailovic (1999) Phys. Rev. B **59**, 1497.
Kircher, J., M. Cardona, A. Zibold, H.P. Geserich, E. Kaldis, J. Karpinski, and S. Rusiecki (1993) Phys. Rev. B **48**, 3993.
Krantz, M. (1996) Phys. Rev. B **54**, 1334.
Lupi, S., M. Capizzi, P. Calvani, B. Ruzicka, P. Maselli, P. Dore, and A. Paolone (1998) Phys. Rev. B **57**, 1248.
Lupi, S., P. Maselli, M. Capizzi, P. Calvani, P. Giura, and P. Roy (1999) Phys. Rev. Lett. **83**, 4852.
Lupi, S., P. Calvani, and M. Capizzi (2000) eprint: cond-mat/0001244.
Lynch, D.W. and C.G. Olson (1999) *Photoemission Studies of High-Temperature Superconductors*, Cambridge University Press, Cambridge.
Mac Donald, A.H., S. Girvin, and D. Yoshioka (1990) Phys. Rev. B **41**, 2565.
Mantin, A.A., J.A. Santurjo, K.C. Hewitt, X.Z. Wang, and J.C. Irwin (1997) Phys. Rev. B **56**, 8426.
Mattis, D.C. and J. Bardeen (1958) Phys. Rev. **111**, 412.
Mertelj, T., J. Demsar, B. Podobnik, I. Poberaj, and D. Mihailovic (1997) Phys. Rev. B **55**, 6061.
Mihailovic, D. and J. Demsar (1999) eprint: cond-mat/9902243.
Misochko, O.V., K. Kuroda, and N. Koshizuka (1997) Phys. Rev. B **56**, 9116.
Misochko, O.V., K. Kisoda, K. Sakai, and S. Nakashima (2000) Phys. Rev. B **61**, 4305.
Naeini, J.G., X.K. Chen, K.C. Hewitt, J.C. Irwin, T.P. Devereaux, M. Okuya, T. Kimura, and K. Kishio (1998) Phys. Rev. B **57**, 11077.
Nagamatsu, J., N. Nakagawa, T. Muranaka, Y. Zenitani and Y. Akimitsu (2001) Nature **410**, 63.
Pronin, A.V., M. Dressel, A. Pimeniv, and A. Loidl (1998) Phys. Rev. B **57**, 14416.
Pronin, A.V., A. Pimeniv, A. Lodl and S.I. Krasnosvobodtsev (2001) eprint: cond-mat/0104291.
Reedyk, M., T. Timusk, Y.W. Hsueh, B.W. Statt, J.S. Xue, and J.E. Greedan (1997) Phys. Rev. B **56**, 9129.
Ruani, G. and P. Ricci (1997) Phys. Rev. B **55**, 93.
Rübhausen, M., C.T. Rieck, N. Dieckmann, K.O. Subke, A. Bock, and U. Merkt (1997) Phys. Rev. B **56**, 14797.
Rübhausen, M., P. Guptasarma, D.G. Hinks, and M.V. Klein (1998) Phys. Rev. B **58**, 3462.
Sadovskii, M.V. (1999) eprint: cond-mat/9902192.
Sacuto, A., R. Combescot, N. Bontemps, C.A. Müller, V. Viallet, and D. Colson (1998) Phys. Rev. B **58**, 11721.
Strohm, T. and M. Cardona (1996) eprint: cond-mat 9609143.
Sundar, C.S., A. Bharathi, M. Premila, T.N. Sairam, S. Kolavathi, G.L.N. Reddy, V.S. Sastry, Y. Hariharan, and T.S. Radhakrishnan (2001) eprint: cond-mat/01044354.
Tamasaku, K, Y. Nakamura, and S. Uchida (1992) Phys. Rev. Lett. **69**, 1455.
Uchida, S., K. Tamasaku, and S. Tajima (1996) Phys Rev. B **53**, 14558.
van der Marel, D. (1999) Phys. Rev. B **60**, 765.
Watanabe, N. and N. Koshizuka (1998) Phys. Rev. B **57**, 632.
Wu, T. and A. Griffin (1996) Phys. Rev. B **54**, 6539.
Zhou, X., M. Cardona, D. Colson, and V. Viallet (1997) Phys. Rev.B **55**, 12770.

Index

2D electron gas, 271
band gap renormalization, 320
far-infrared emission, 271
Fermi-edge singularity, 318
in MOS, 283
in type I-type II multiple quantum wells, 291

absorption, 49
band-to-band direct and allowed transitions, 56
band-to-band direct, forbidden transitions, 57
band-to-band indirect and allowed transitions, 57
band-to-band indirect and forbidden transitions, 58
Burstein–Moss shift, 59
excitonic in low-dimensional structures, 266
Franz–Keldysh effect at, 59
in the presence of magnetic fields, 71
indirect excitonic, 63
intraband, 60
intraband in low-dimensional structures, 263
microscopic origin of, 55
of amorphous materials, 375
of donors and acceptors, 170
on free carriers, 63
on phonons, 65
on Wannier–Mott excitons, 62
related to photoluminescence, 77-78
spectra in manganites, 165
two photon and biexcitonic production, 64
Urbach law, 59, 368, 375-377
Alfven waves, 74
amplitude and phase spectroscopy, 105
Anderson transition, 375
anharmonic effects, 190
in bulk materials, 190
in disordered alloys, 355

bandgap, 10
anisotropy, 182
of amorphous materials, 376
of bulk materials, 182-183
of superconductors, 409-410
renormalization, 319-320
BCS, 408
-like behavior, 430
non- behavior, 426
theory, 408
Bloch energy, 260
Bloch equations, 114
applied to FWM, 115
Bloch frequency, 261
Bohr magneton, 23
Bose–Einstein energy gap dependence, 185
Bose–Einstein statistics, 19
Brillouin scattering, 80, 88
geometry configurations, 112

on magnons, 91
on acoustic phonons, 88-91
power and cross-section, 91
Burstein–Moss effect, 59

CARS, 98
charge density waves, 33
Clausius–Mosssoti relation, 358
critical points, 7
in 2D structures, 10
in 3D solids, 9
related to absorption, 56
color centers, 167
Compton scattering, 80
confinement, 273
of carriers, 273
of excitons, 279
of photons, 276
Stark effect, 287
Cooper pairs, 408
Coulomb renormalization, 92
in low-dimensional structures, 270
Coulomb screening, 94
Curie temperature, 23,24

Debye temperature, 148
in glass and crystalline solids, 133
of disordered alloys, 355
of ion-phonon interaction, 148
dephasing time, 100, 115
dielectric constant, 51
anisotropy, 106
complex tensor, 121
in the presence of stress, 195
of alkali halides, 186
of amorphous materials, 374
of calcogenides, 394
of high-temperature superconductors, 418
of ionic insulators, 51
of materials with hoping motion, 183
of metals, 215
of nanoparticles, 338
of polycrystals, 363
of porous Si, 370
of TCAS materials, 382
of thin films, 243
two-fluid model, 429
diluted magnetic semiconductors, 158
heterostructures, 281
quantum wells, 305
superlattices, 339
disordered alloys, 354
Drude–Lorentz theory, 52
in metals, 216
sum rule, 54

effective mass, 12
in low-dimensional structures, 282
of free electrons, 14
of heavy holes, 14
of light holes, 14
magneto-optic determination, 182
nonparapolicity, 282
tensor, 12
Einstein coefficients, 47
energy bands, 2
at interfaces and thin films, 234
general treatment, 3
in amorphous semiconductors, 372
in bulk solids, 181
in calcogenides, 392
in heterostructures, 252
in high-temperature cuprate superconductors, 413
in isolated atoms, 127
in rare-earths, 147
in superconductors, 408
in superlattices, 328
in the presence of magnetic fields, 64
Kane model, 257
numerical methods for calculation, 15
tight-binding approximation, 258
excitons, 26
-biexciton system, 298

-exciton interaction, 294
binding energies, 293
bleaching, 207, 311
biexcitons in low-dimensional structures, 296
bound and complexes, 31, 64
charged states, 292
condensation of, 323
coupling with phonons, 308
dephasing, 294
discrete spectrum, 29
Frenkel excitons, 26
gain, 206
in bulk semiconductors, 186
in microcavities, 309
in quantum wells, 261
in superlattices, 336
indirect absorption, 63
localization in quantum wells, 289
magneto-, 189
molecules, 297
radius of, 29
relaxation, 207, 312
shakeup processes, 294
tunneling of, 322
Wannier–Mott, 27

Fano lineshape, 157
in Mott transitions, 227
in superconducting transitions, 422
Faraday effect, 72
giant rotation, 161
in disordered alloys, 356
Fermi golden rule, 41
Fermi–Dirac statistics, 19
ferro/antiferromagnetic exchange coupling, 247
filled-shell ions, 169
fluorescence line narrowing, 120
lifetime, 155
time-resolved, 140
Fourier transform spectroscopy, 100
four-wave mixing (FWM), 112
quantum and polarization beats, 117
three-beam degenerate, 113
two-beam degenerate, 114
Franck–Condon principle, 2
for defects with large orbits, 137
related to Huang–Rhys factor, 124
Franz–Keldysh effect, 59
in amorphous materials, 376
Fröhlich polarons, 137
Fröhlich resonant coupling, 308

gyromagnetic ratio, 24,196, 297

Hanle effect, 300
heterostructures, 254
quantum dot, 256
quantum well, 254
quantum wire, 255
type I and type II, 258
high-temperature superconductors, 407
cuprates, 411
gap of, 410
pseudogap of, 432
symmetry of gap, 423
Huang–Rhys factor, 123
in nanocrystals, 317
in silicon, 174

impurities in crystals, 129
long-range effects of, 132
short-range effects of, 131
inelastic light scattering, 79
interaction in the presence of magnetic fields, 67
direct interband transitions, 68
in metals, 71
in semimetals, 73
intraband, 68
interfaces, 235
interline spectral transfer, 171
ion-interaction processes, 138
rare-earths, 153
iron-group ion pairs, 158
isotopic effects, 192

Jahn–Teller effect, 138
Judd–Ofelt theory, 142

$\boldsymbol{k} \cdot \boldsymbol{p}$ method, 11
  multiband, 256
Kondo insulator/semiconductor, 224
Kramers-Krönig relation, 50, 93

Langmuir absorption, 241
Larmour precession, 300
Landau levels, 67
Landau transitions, 68
lifetime of states, 115
ligand field, 129
line identification, 122
London–Fleury Hamiltonian, 196
London penetration depth, 416

magnetic-thin films, 246
magneto-optic Kerr effect, 121
  in diluted
    magnetic semiconductors, 164
  nonlinear, 122
  surface-induced transverse, 122
magnons, 24
  Brillouin scattering on, 89
  dispersion relation of, 25
  wavevector, 91
manganites, 165
many-body effects, 250
  in bulk materials, 250
  in low-dimensional structures, 318
metals, 215
  electron-phonon coupling in, 219
  optical properties of, 215
microcavity, 274, 309
  coupled, 311
  of porous silicon, 367
  polariton dispersion in, 309
modulated reflectance, 103, 104, 196

nanoparticles, 347
  gold, 343
  noble metals, 343
  size effects, 346
  superlattices, 333
nanotubes, 324
nonlinear response of solids, 92, 272

oscillator strength, 54
  of confined excitons, 290
  of ions in crystals, 130
  of TO phonons, 51
Overhauser effect, 295

Pauli blocking, 94, 275
penetration depth, 416, 429
phase transitions, 220
phase-space filling, 275
phonon, 16
  -polariton coupling, 202
  bottleneck, 216
  coupling with excitons, 308
  dispersion relation of, 17
  relaxation, 213
  temperature, 436
phonons, 16
  acoustic, 17
  in superconductors, 408
  optical, 17
  statistics, 19
photoluminescence, 76
  anisotropy, 79
  excitation (PLE), 110
  Hanle depolarization of, 300
  hot, 78
  in low-dimensional structures, 267
  measurement, 78
  quenching, 289, 364
  related to absorption, 77
photons, 38
  three- echo, 116
plasmon-polariton, 361
plasmons, 20
  in low-dimensional structures, 271
  magneto-, 272
  two-dimensional intersubband, 306

polaritons, 35, 308
polarizability tensor, 48
  nonlinear, 93
polarization beats, 117
polarons, 31
  in manganites, 165
pump-probe spectroscopy, 104
  REOS, TEOS, 105

quantum beats, 117

quantum wells, 253
  bandgap renormalization in, 320
  Bose–Einstein condensation in, 298
  Coulomb renormalization in, 270
  coupled, 259, 320
  coupled excitations in, 308
  critical points of, 279
  density of states, 254
  excitonic absorption in, 266
  excitons in, 261
  exciton relaxation in, 312
  Fermi-edge singularity in, 319
  Fermi pressure in, 319
  g-factor in, 299
  intraband absorption in, 263
  intersubband absorption in, 269
  intraband transition, 270
  intrasubband plasmons in, 306
  luminescence in, 267
  magnetic properties of, 298
  narrow and wide, 312
  plasmons in, 271
  polaritons in, 309
  spin configuration in, 285
  tunneling in, 312, 321
quantum dots, 257
  carriers relaxation in, 313
  confinement effects in, 277
  diameter of, 290
  density of states in, 256
  excitonic transitions in, 293
  shape determination of, 290
quantum wires, 255
  density of states in, 255
  dephasing of excitons in, 317
  exciton-exciton interaction in, 294
  photonic, 274
  renormalization of, 319
  TPA spectrum in, 293
  superlattices, 338

Rabi coupling, 309
Rabi splitting, 309
Raman scattering, 81
  coherent-anti Stokes (CARS), 98
  cross-section of, 86
  efficiency of, 87
  forward and backward, 82
  hyper-, 97
  in low-dimensional structures, 269
  inverse, 97
  measurement, 111
  nonlinear, 95
  selection rules at, 86
  spectra, 88
  stimulated, 95
  tensor, 84
rare-earths, 141
Rayleigh scattering, 80
reflectivity, 49,102
refractive index, 49
  changes due to stress, 195
  complex, 49
  of Alfven waves, 74
  of metals, 72
  of semimetals, 73
relaxation phenomena, 100
  in bulk semiconductors, 205
  in low-dimensional structures, 311
rough surfaces, 361

Shubnikov–de Hass effect, 75
spectral hole burning, 118
spectral line broadening, 44
  collisional, 95
  Gaussian, 44
  Lorentzian, 46

Voigt, 46
spectroscopy, 100
frequency-resolved, 100
time-resolved, 100
spin-Peierls transition, 228

Stark shift, 134
quantum confined, 287
in doped crystals, 134
in quantum dots, 293
in quantum wells, 269
superlattice, 259
array of, 333
corrugated, 335
disordered, 359
density of states in, 260
excitons in, 336
minibands, 260
mirror-plane type, 332
of diluted magnetic semiconductors, 339
of magnetic materials, 335
optical properties of, 325
type I and type II, 325
susceptibility tensor, 48
linear, 93
second-order, 93
third-order, 94

Tauc law, 376
thin films, 235
Thomson scattering, 80
transition probability, 41
related to absorption, 55
sum rule, 54
time-resolved optical processes, 100
dephasing time in, 100
nonequilibrium carrier distribution in, 101
thermalized distribution in, 101
transmission spectroscopy, 102
tunneling, 312
in coupled quantum wells, 322
in porous silicon, 369
of excitons, 322
of free carriers, 312
tunneling time, 312, 322

Urbach tail, 59
in amorphous materials, 369, 376

Varshni formula, 182
Verwey transition, 222

Wannier–Stark effect, 260
in superlattices, 335

Zeeman splitting, 68
anisotropy, 217
giant, 163
of sublevels of rare-earths, 153

Index of Materials

This index displays the main materials referred to in the book, and their physical properties determined through their interaction with the optical fields. These materials were selected to illustrate the theory of light-solid interaction; any other solid-state material can be characterized using the methods described in the book.

4HSiC
effective mass, 185

$AgGaS_2$
crystal field parameter, 156
$Al_xGa_{1-x}As$ alloys
localization length, 357
Au
relaxation time, 218

$BaPb_{1-x}Bi_x O_3$
scattering rate, 430
BeO
anharmonic phonon-phonon coupling, 190
BiI_3
exciton mean velocity, 242
$Bi_2Sr_2CuO_8$
pseudogap, 433
London penetration depth, 434

C (amorphous)
elastic constants, 393
localization degree, 393
polarization memory, 393
Cd_xMn_xS
subbands, 160
Cd_xMn_xTe
effective mass variation, 159
CdS
valence band, 30
energy gap, 197
time constant of free exciton formation, 206
CdS (dots)
electron g factor, 304
temperature, 304
CdSe (dots)
radius, 286
CdTe (nanocrystals)
exciton gap, 182
crystal field splitting, 182
$CdTe/Cd_{1-x} Mn_xTe$
valence-band offset, 282
g factor, 299
CdSe/ZnSe
exciton recombination life time, 314
exciton formation rate, 314
rate of cooling and localization, 314
Co/Au/Cu(111)
anisotropy constants, 249
Co/Cu(110)
desorption energy, 240

$CoPt_3$
- ferro-paramagnetic phase transition, 230

CuBr (nanocrystals)
- exciton dephasing time, 317
- Huang–Rhys factor, 317

CuCl
- energy gap, 192

CuCl (dots)
- exciton-phonon coupling, 308
- phase-relaxation time, 317

CuCl (nanoparticles)
- mean radius, 342
- exciton and biexciton rise and decay times, 345

$CuGaS_2$
- exciton-phonon coupling constant, 187
- exciton oscillator strength, 187
- shallow localized states, 211

$CuGeO_3$
- elastic properties, 228

EuS
- critical temperature of paramagnetic phase, 198

Fe (dots)
- magnon frequency, 305

FeO_4/MgO
- substrate modes, 244

ferroic crystals,
- incommensurate-commensurate transition, 224

(Ga,In)Sb/GaSb
- biding enrgy of excitons, 290
- oscillator strength of HH exciton, 290

GaAs
- anharmonic effects, 190
- built-in field, 238
- density of vacancies and antisite pairs, 173
- dephasing time, 189
- effective electron density, 185
- electron velocity overshot distribution, 211
- exciton radius, 29
- Fermi-edge singularity, 200
- HH-LH splitting, 207
- Lyman transitions, 172
- paraexcitons, 201
- phonon bottleneck, 214
- phonon decay time, 214
- radiatiative recombination lifetime, 189

GaAs/$Al_xGa_{1-x}As$ (heterostructures)
- acoustic and optic plasmon modes, 298
- alloy fluctuation, 291
- activation energy, 285
- binding energy of biexcitons, 296
- carrier capture efficiency, 285
- carrier density band bending, 242
- charged states of excitons, 294
- diffusive coefficient, 289
- effective mass, 282
- exciton formation time, 292
- g factor, 300
- interface microroughness, 291
- internal electric field, 287
- lateral confinement energy of excitons, 276
- vacancy diffusion coefficient, 289

GaAs/AlAs (superlattices)
- activation energies, 360
- conduction and valence band states, 329
- critical points, 330
- effective mass filtering, 342
- intervalley deformation, 331
- thermal conductivity, 340

$GaAs_{1-x}P_x$
- phonon coupling constant, 356

$GaInP_2$
- ordering, 334
- valence band splitting, 333

GaN
absorption edge, 189
band gap, 182
binding energy of biexcitons, 208
cooling dynamics of excitons, 206
crystal-field, spin-orbit parameters, 184
dielectric constant, 172
effective acoustic deformation potential, 207
effective mass, 172
exciton gain, 206
exciton relaxation, 206
free carrier concentration, 185
free exciton emission, 188
HH exciton oscillator strength, 184
GaN (dots)
internal electric field, 288
GaP/InP (superlattice)
band gap, 329
composition, 329
GaSb
effective mass, 185
GaSe
decay time of cooling temperature, 210
deformation potential, 210
GeO_2
paraelastic to ferroelstic transition, 221

$HgBa_2Ca_3Cu_4O_{7-\delta}$
strength of electron-phonon interaction, 421
$Hg_{1-x-y}Cd_xMn_yTe$,
relaxation time, 159

InAlAs/InP
valence band offset dependence on pressure, 280
InAs
exciton reduced mass, 188
Fermi-edge singularity, 200
generation of LO phonons, 208
InAs/GaAs (dots)
valence band offset dependence on pressure, 280
InAs/GaSb
Fermi pressure, 319
InF_3-GdF_3-GaF3:Cr^{+3}
Fano line shape, 157
$In_xGa_{1-x}As$
lateral confinement energy, 276
InGaAs/GaAs
activation energies, 285, 289
density of states, 303
diffusive coefficient, 289
$In_xGa_{1-x}N$/GaN
critical layer thickness, 245
InP
coupled plasmon-LO phonon, 195
dephasing time, 213
phonon conversion, 214
piezo-optical coefficients, 195
InSb
effective mass, 208

KSCN
coupling constant between two-phonon process and one-phonon state, 221

$La^3F^3:Pr^{3+}$
fluorescence line narrowing, 140
$La_{2-x}Sr_xCuO_4$
Josephson plasma edge, 429
$LiNbO_3$
photon-polariton damping, 203

MgB_2
superconducting gap, 427
$MgO:Cr^{3+}$
circular dicroism, 135

NaV_2O_5
spin Peierls transition temperature, 229

$NdBa_2Cu_3O_{7-x}$
  superconducting gap, 419
NiFe/NiO
  spin relaxation, 247

porous GaP
  depolarization factor, 365
porous Si
  bang gap, 370
  degradation time constant, 367
  distribution of sizes, 370
  polarization memory, 364
  radiative lifetime of
    electron-hole pairs, 365
  porosity, 370
  thermal activation rate, 369
  thermal conductivity, 370

rare-earths
  electron-phonon coupling, 149
  fluorescence line narrowing, 149
  optical properties, 143
  up-conversion, 151
ReS_2
  band gap, 183
  energies and broadening
    parameters, 189
$RTiO_3$ (R=La, Ce, Pr, Nd, Sm, Gd)
  anharmonic effects, 191

Si
  binding energy of bound
    excitons, 173
  Huang–Rhys factors, 174
  self-induced optical anisotropy, 239
Si (amorphous)
  effective charge, 386
  electron-hole distance, 387
  radiative recombination rate, 387
SiC
  optical properties, 362
Si_nGe_n
  band gap, 328

$Si_{1-x}Ge$
  bang gap, 183
Si_3N_4/SiGe
  2D stress map, 239
  LO and TO phonon identification,
    239
Si (nanocristals)-SiO_2 host
  confinement effects, 277
SmB_6
  Kondo temperature transition, 225
$SmBa_2Cu_3O_{7-x}$
  Raman tensor, 111

Tb/Co
  resonant magnetization, 246

V_2O_3
  Mott transition temperature, 227

YAG:Cr^{3+}
  crystal field, 156
  spin-orbit coupling, 156
YB_{12}
  Kondo transition temperature, 224
YBCO
  electronic structure, 431
  phonon temperature, 431
  pseudogap, 432
  relaxation rates, 431
  relaxation times, 436
$YBa_2Cu_3O_{6+\delta}$
  Superexchange constant, 192
$YBa_2Cu_4O_8$
  dielectric tensor, 418
  plasma edge, 418
$YbPO_4$:Yb^{3+}
  elastic constants, 151

$Zn_{1-x}Co_xSe$
  crystal-field level, 161
$Zn_{1-x}Fe_xSe$
  Faraday rotation, 162
$Zn_{1-x}Fe_xTe$
  Exciton splitting, 160

ZnSe
energy gap, 193
deformation potential, 194
Luttinger parameter, 172

ZnSe/Zn(Cd,Mn)Se
nearest-neighbor exchange constant, 305

Printing: Saladruck, Berlin
Binding: H. Stürtz AG, Würzburg